Studies in Fuzziness and Soft Computing

Volume 357

Series editor

Janusz Kacprzyk, Polish Academy of Sciences, Warsaw, Poland
e-mail: kacprzyk@ibspan.waw.pl

About this Series

The series "Studies in Fuzziness and Soft Computing" contains publications on various topics in the area of soft computing, which include fuzzy sets, rough sets, neural networks, evolutionary computation, probabilistic and evidential reasoning, multi-valued logic, and related fields. The publications within "Studies in Fuzziness and Soft Computing" are primarily monographs and edited volumes. They cover significant recent developments in the field, both of a foundational and applicable character. An important feature of the series is its short publication time and world-wide distribution. This permits a rapid and broad dissemination of research results.

More information about this series at http://www.springer.com/series/2941

Mikael Collan · Janusz Kacprzyk
Editors

Soft Computing Applications for Group Decision-making and Consensus Modeling

 Springer

Editors
Mikael Collan
School of Business and Management
Lappeenranta University of Technology
Lappeenranta
Finland

Janusz Kacprzyk
Systems Research Institute
Polish Academy of Sciences
Warsaw
Poland

ISSN 1434-9922   ISSN 1860-0808 (electronic)
Studies in Fuzziness and Soft Computing
ISBN 978-3-319-86805-9   ISBN 978-3-319-60207-3 (eBook)
DOI 10.1007/978-3-319-60207-3

Printed on acid-free paper

This Springer imprint is published by Springer Nature
The registered company is Springer International Publishing AG
The registered company address is: Gewerbestrasse 11, 6330 Cham, Switzerland

This volume is dedicated to Prof. Mario Fedrizzi, Full Professor of the Department of Industrial Engineering at the University of Trento, Italy, as a token of appreciation for his great scientific and scholarly achievements, long time service to many communities, notably those of fuzzy logic, operations research, decision analysis and sciences, and mathematical economics, to name a few. This volume is fuelled by the appreciation for his original thinking and his novel contributions both, concerning the theory and the applications in the above-mentioned fields.

Professor Mario Fedrizzi received the M.Sc. degree in Mathematics in 1973 from the University of Padua, Italy, Ph.D. in Operations Research from the University of Venice (Cà Foscari) in 1976, and then started his academic career at the University of Trento, staying there until his retirement. He served as Head of the Institute of Informatics and as Dean of the Faculty of Economics and Business Administration the University of Trento from 1985 to 1995, and as the Deputy

Rector of the University of Trento from 2004 to 2008.

His research has focused on utility and risk theory, group decision-making, fuzzy decision analysis, consensus modelling, and on decision support systems under uncertain and imprecise (fuzzy) information. He has authored and co-authored numerous books and edited many volumes. His publication record is very rich and contains more than 100 papers, which have appeared in many peer reviewed international journals exemplified by European Journal of Operational Research, Fuzzy Sets and Systems, Decision Support Systems, Information Fusion, Expert Systems with Applications, and Mathematical Social Sciences. He is a member of the editorial boards of International Journal of Uncertainty Fuzziness and Knowledge-Based Systems, Group Decision and Negotiation, International Journal of General Systems, Control and Cybernetics, and Applied Computational Intelligence and Soft Computing.

More specifically, Prof. Fedrizzi has developed novel tools and techniques for the modeling of consensus reaching in a 'soft' environment. In other words, when individual testimonies of experts are expressed as fuzzy preference relations (often expressed by fuzzy linguistic quantifiers) and based on the concept of a degree of consensus (under an imprecisely specified majority). He has also worked on extensions of these models to the dynamic context, with iterative preference changes and on extensions of the same to the multi-criteria context.

Furthermore, an important part of his research, which resulted in a series of original papers, has been the study of fuzzy adjacency relations in social network analysis (SNA), in which the imprecision related to the relationships between the nodes of a social network is modelled by using fuzzy binary adjacency relations and some novel aggregation operators. This makes it possible to continuously characterize, on a scale ranging from non-compensatory to a fully compensatory, the attitudes of the agents (actors) towards their mutual cooperation. The decision-makers, represented by nodes in a social network, are assigned weights via a centrality measure that can describe a wide spectrum of aspects ranging from importance through influence to leadership.

A very important area of research of Prof. Fedrizzi has been related to an important practical problem of fraud detection under uncertainty. He showed how to use the Choquet integral to extend the OWA-based attack-tree approach to fraud detection, assuming that the attack tree is valuated recursively through bottom-up algorithms. This approach has been shown to have a superior numerical complexity. A noticeable example in this line of research is a prototype of a multi-agent system called "Fraud Interactive Decision Expert System" (FIDES). FIDES is focused on the evaluation of behavioral aspects of fraud detection, according to the judgments expressed by two groups of experts, inspectors and auditors respectively. FIDES combines think-maps, attack-trees and fuzzy numbers, within a Delphi-type environment, to provide the users with a "human-consistent" tool for better understanding and managing fraud schemes. This short account of main accomplishments of Prof. Fedrizzi should be completed with the mention of his very important, from theoretical and practical points of view, works on the valuation of giga-investment projects, i.e., projects that require very large amounts of capital and human resources that can bring about big gains, but also create big losses. By employing elements and methods of group decision support, multi-criteria decision-making, and risk analysis in the soft environment qualitatively new results have been obtained.

In addition to prominent positions at the University of Trento, Prof. Fedrizzi has held visiting professorship positions at the Lappeenranta University of Technology in Lappeenranta, Finland, Åbo Akademi University in Turku, Finland, Auckland University of Technology in Auckland, New Zealand, University of Granada in Granada, Spain, Eötvös Lorand University in Budapest, Hungary, and the System Research Institute, Polish Academy of Sciences in Warsaw, Poland. Prof. Fedrizzi has been for years a member of the International Advisory Board of KEDRI at the Auckland University of Technology in New Zealand, and of the Systems Research Institute, Polish Academy of Sciences in Warsaw, Poland.

He has also been involved in consulting activities in the areas of information systems and decision support systems, quality control, project management, and e-learning.

In addition to his academic career, he has also held various prestigious positions in business, notably from 1995 to 2006 he served as the Chairman of Board of Directors of an important regional bank in Trentino Alto Adige and subsequently in the board of one of the largest European banking groups. This experience has clearly had a deep impact on his research work, which had always been a synergistic combination of a high level theory and a feel of what is relevant and necessary in practice.

It is clear that this volume, meant to be a token of appreciation for Prof. Fedrizzi by our entire community, is modest in comparison with his achievements in terms of providing inspiration and important research results, but especially in terms of loyal friendship. We are honored to have had the opportunity to prepare this volume.

Spring 2017   Mikael Collan
                                               Janusz Kacprzyk

Foreword

It gives me great pleasure to write the foreword to this book dedicated to Prof. Mario Fedrizzi. Besides being an esteemed colleague, Mario is a good friend to me.

I met Mario in Trento in 1992, at the Workshop on Current Issues on Fuzzy Technologies (CIFT), which Mario had launched as an intersection point of the Italian and international research communities in fuzzy logic and technologies. At that time I was starting to get close to the fuzzy community, and I was positively surprised by the friendly and collaborative atmosphere, as well as by the perceived enthusiasm of collaborating and contributing to knowledge development. One of the positive consequences of that meeting was the beginning of a scientific collaboration with Mario, which has produced various contributions in the field of group decision-making. Another important outcome was meeting several excellent colleagues, many of whom are contributing to this book.

This volume is unique in many respects. In the first place it witnesses the esteem and friendship for Mario of the international fuzzy community. Mario has been contributing to the field of fuzzy decision-making for many years, and he has collaborated with several reputed scientists worldwide.

Second, it constitutes a collection of important contributions that brings various perspectives in the field of group decision-making and consensus reaching.

The beauty of research is not only in the opportunity it gives to contribute to advancement of knowledge; it is also in the opportunity it gives to grow on a personal side, to be enriched by invaluable human relations.

Spring 2017

Gabriella Pasi
President, European Society for Fuzzy Logic
and Technology (EUSFLAT)
Milan, Italy

Preface

This volume is dedicated to Prof. Mario Fedrizzi, Full Professor at the Department of Industrial Engineering at the University of Trento, Italy, as a token of appreciation for his scientific and scholarly achievements, and his long time service to many communities, notably those of fuzzy logic, operations research, decision analysis and sciences, and mathematical economics. The topic of this volume, chosen after a long deliberation, is related to the areas in which Prof. Fedrizzi has been active for some decades and has obtained valuable results, that is, group decision-making and consensus reaching.

Of course, since these areas are vast and involve aspects that range over an extremely wide spectrum of fields exemplified by psychology, and social and political sciences, decision analysis and decision sciences, data analysis and data mining, computational sciences, artificial and computational intelligence, systems research, just to name a few, the selected contributions cover a range of them. The purpose has been that of providing a bird's view account of some novel directions in the broadly perceived area of group decision-making and consensus modeling. We are glad to note that many well-known researchers and scholars have very positively responded to our initiative to publish this volume and contributed with their latest work.

The volume starts with some philosophical and foundational considerations, then presents some more general economic considerations, before going forward to a the presentation of some promising models for the analysis of data and decision-making processes. The contributions in the volume have collected inspiration from previous research from fields such as social and political sciences, systems modeling, optimization under imprecise (fuzzy) data, artificial and computational intelligence, and naturally consensus reaching modeling and the results are not only conceptual or theoretical, but include also a high real-world implementation potential.

We will now provide a short summary of the contributions to help the interested reader more easily navigate the volume:

Part I contains a selection of more general papers on issues broadly related to social and economic aspects.

Vladik Kreinovich and Thongchai Dumrongpokaphoan consider in their paper "Optimal Group Decision Making Criterion and How it can Help to Decrease Poverty, Inequality, and Discrimination" the traditional approach to group decision-making in economics, in which the goal is to maximize the Gross Domestic Product (GDP), while hoping that the increased wealth will be equally distributed across the society. Though this may, at least theoretically, sometimes happen the reality is usually different in that in spite of an increase in the GDP, much inequality remains, as some (often most) people remain poor, some social groups continue to face economic discrimination, and so forth. The contribution indicates that the maximization of the overall gain may not always be the best criterion in group decision-making connected to social issues. The authors propose a group decision-making criterion, which is in a specific sense optimal and show that by using this optimality criterion one can attain a solution that can indeed help to decrease poverty, inequality, and discrimination.

In his paper "An Overview and Re-Interpretation of Paradoxes of Responsiveness" Hannu Nurmi provides a comprehensive discussion and a desiderata of democratic decision-making, including the "notion" that the political outcomes (policies, elected persons, legislation) should be responsive to popular opinions. In representative forms of governance, this responsiveness is however not expected to pertain to every single outcome, but the very idea of going "to the people" seems to presuppose some degree of responsiveness. In social choice theory, several notions that aim to capture aspects of responsiveness have been introduced and related to other desiderata of social choice, and those which are more relevant are discussed from the point of view of their relevance in democratic decision-making. The author then considers the paradoxes related to non-responsiveness from a novel point of view in that he tries to determine their significance for multiple-criteria decision-making (MCDM). It turns out that the use of some methods of aggregating the performance criteria of policy alternatives can be ruled out, because of their strange and unacceptable behavior in some decision settings.

Jacek Mercik in "Veto in Yes-no and Yes-no-abstain Voting Systems" is concerned, from a theoretical point of view, with a transformation from simple "yes-no" cooperative games to simple cooperative games in which players have more than two actions available to them by introducing abstentions into the yes-no voting system. The results obtained so far in this respect have rather been pessimistic exemplified by Felsenthal and Machover (2013) who even call them "the curious case of the absent abstention". The author discusses the relation between the right of veto, weights of the players and quotas. The results obtained clarify some general properties and make an a priori analysis possible to gain a better understanding of the decision-making mechanism of such decisive bodies. An example of the United Nations Security Council is used to illustrate the discussion.

In "Power Indices for Finance" Cesarino Bertini, Gianfranco Gambarelli, Izabella Stach, and Maurizio Zola consider the use of power indices for the determination of the weight of the share stock of a company to quantify the possibility for each shareholder to get majority positions by coalitions with other shareholders. They indicate the potential of this approach, for building effective and efficient models for forecasting, simulation, and regulation in many financial, political, and economic fields.

Silvia Bortot, Ricardo Alberto Marques Pereira and Anastasia Stamatopoulou ("The Binomial Decomposition of the Single Parameter Family of GB Welfare Functions") consider the binomial decomposition of the generalized Gini welfare functions in terms of the binomial welfare functions and examine the weighting structure of the binomial welfare functions which progressively focus on the poorest part of the population. A parametric family of income distributions is dealt with and the numerical behavior of the single parameter family of GB welfare functions with respect to those income distributions. The binomial decomposition of the GB welfare functions is considered and the dependence of the binomial decomposition coefficients is analyzed in relation to a single parameter which describes the family.

Part II encompasses some papers on formal foundations related to logics and mathematics.

In the paper "The Logic of Information and Processes in System-of-Systems Applications", P. Eklund, M. Johansson and J. Kortelainen propose a new logic and an approach to many-valuedness, which can make it possible to describe underlying logical structures of information as represented in industrial processes, and as a part of their respective markets. The authors emphasize the importance of introducing classification structures in order to devise tools and techniques for the management of information granularity within, and across subsystems in the system-of-systems (SoS) context. The proposal of the logic of information and process is a main contribution of this paper, and the authors' focus on the system-of-systems concept proceeds within the field of energy. In their process view, they look closer into the power market with all its stakeholders, exemplified by those related to renewable energy. Supply, demand, and pricing models are shown to be subjected to logical analyses. The authors show how information and information structures are integrated into processes and their structures. The Business Process Modeling Notation (BPMN) paradigm is adopted for the modeling.

Sarka Hoskova-Mayerova and Antonio Maturo ("Decision-making Process Using Hyperstructures and Fuzzy Structures in Social Sciences") consider algebraic hyper-structures, which is an interesting field of algebra and important, both from the theoretical and the applications point of view. The authors show that a hyper-groupoid structure can be associated with any social relationship. These hyper-groupoids become hyper-groups under some conditions, exemplified by a condition concerning outer individuals. By way of analysis one can establish, in a natural way, when social relationships become optimal. Relations between people are a crucial topic in the social sciences and are usually described by linguistic propositions. Obviously, the binary context is not always sufficient and often a correct and complete modeling of such relations can only be obtained, if a degree of the strength of a relation is used. Assuming such a context, the authors discuss various decision-making problems and their solutions.

Antonio Di Nola, Massimo Squillante, and Gaetano Vitale ("Social Preferences Through Riesz Spaces: A First Approach") propose to employ the Riesz spaces as general framework to be used in the context of pairwise comparison matrices to deal with definable properties, real situations, and the aggregation of preferences. Some significant examples are presented to describe how properties of the Riesz spaces can be used to express preferences. It is shown that the Riesz spaces make it possible to combine advantages of many approaches. The authors also provide a characterization of collective choice-rules, which satisfy some classic criteria assumed in the social choice theory. They also propose an abstract approach to social welfare functions.

In their paper "Coherent Conditional Plausibility: A Tool for Handling Fuzziness and Uncertainty Under Partial Information" Giulianella Coletti and Barbara Vantaggi consider some important issues related to non-additive measures. These may be a powerful tool for the analysis of problems, when only a partial, or indirect, information about the events of interest is available, or when imprecision and ambiguity of agents is considered. The authors focus their attention on one non-additive measure—the plausibility measure. They mainly study inferential processes, where information is expressed in natural language and the uncertainty measure is either partially or imprecisely evaluated.

The cases considered are such, where partial assessments are consistent with a conditional plausibility, and it is assumed that the interpretation of the membership function of a fuzzy set, in terms of coherent conditional plausibility, is regarded as a function of the conditioning event. This kind of interpretation is particularly useful for computing the measure of uncertainty of a fuzzy event, when knowledge about a variable is imprecise and can be managed via a non-additive measure of uncertainty. A simple situation related to a Zadeh's example can be the following: a ball will be drawn from an urn containing balls of different colors and different diameters, but one knows only the distribution of the different colors. The purpose is to compute the uncertainty measure of the fuzzy event "a small ball is drawn" taking into consideration the possible logical constraints among the particular colors and the ball diameter—a new method is proposed.

Krassimir T. Atanassov, Vassia Atanassova, Eulalia Szmidt, and Janusz Kacprzyk ("Intuitionistic Fuzzy Interpretations of Some Formulas for Estimation of Preference Degree") propose a new interpretation of a degree of preference, using some concepts, tools, and techniques of the theory of Atanassov's intuitionistic fuzzy sets. Then, this interpretation is used for the derivation of a degree of consensus in a group of agents following the approach of Fedrizzi et al. and Peneva and Popchev.

Part III is concerned with various aspects of judgments and aggregation, taking into account both human-related and formal aspects.

Jan Stoklasa, Tomàš Talàšek, and Pasi Luukka ("Fuzzified Likert Scales in Group Multiple-criteria Evaluation") discuss Likert scales, which are widely used for the representation of attitudes in many fields of social science. The authors consider their use in multiple-criteria multi-expert evaluations. They propose a methodology that deals with the non-uniformity of the distribution of linguistic labels along the evaluation universe and also with a possible response bias (central tendency and extreme-response tendency). The methodology represents the Likert-type evaluations of an alternative, with respect to various criteria using histograms. The histograms are used in the process of aggregation of information, because the underlying evaluation scale is ordinal. A transformation of the multi-expert multiple-criteria evaluation represented by a histogram into a 3-bin histogram to control for the response bias is performed, and an ideal-evaluation 3-bin histogram is defined. The authors propose a distance-measure to assess the closeness of the overall evaluation to the ideal, and suggest the use of its values interpretation/evaluation fuzzy rules. Some examples showing the effectiveness and efficiency of the proposed approach are presented.

Robert Fullér and Christer Carlsson ("Maximal Entropy and Minimal Variability OWA Operator Weights: A Short Survey of Recent Developments") consider a very important problem of weights of the ordered weighted averaging (OWA) operator. First, a comprehensive and critical analysis of the existing approaches is given. The chapter starts with the early approach by O'Hagan, who proposed to determine a special class of the OWA operators having the maximum entropy of the OWA weights for a given level of the degree of orness, which boiled down to the solution of a constrained optimization problem. Then, Fullér and Majlender proposed the use of the method of Lagrange multipliers that boiled down to the analytical solution of a constrained optimization problem, which gave the optimal weighting vector. Then, the same authors computed the exact minimal variability weighting vector for any level of orness, using the Karush-Kuhn-Tucker second-order sufficiency conditions for optimality. The problem of maximizing an OWA aggregation of a group of variables that are interrelated and constrained by a collection of linear inequalities was first considered by Yager in 1996, where he showed how this problem can be represented by the solution of a mixed integer linear programming problem. Then, Carlsson, Fullér, and Majlender derived an algorithm for solving the constrained OWA aggregation problem under a simple linear constraint that the sum of the variables is less than, or equal, to one. After the above pioneering works, many other approaches have been proposed. The purpose of this paper is to present a survey and analysis of these works essence and properties.

József Mezei and Matteo Brunelli ("A Closer Look at the Relation Between Orness and Entropy of OWA Function") discuss some important problems related to the ordered weighted averaging (OWA) functions that have been extensively used to model the problem of choice and consensus in the presence of multiple experts and decision-makers. Since each OWA operator is associated with a weight vector, a critical problem that has been studied for years and by many authors has focused on the determination of the weight vector. In this paper, the authors consider the degree of orness and entropy, two characterizing measures of the priority vectors, and they study their interplay from a graphical point of view.

Michele Fedrizzi and Silvio Giove ("Rank Reversal in the AHP with Consistent Judgements: A Numerical Study in Single and Group Decision Making") analyze, by using numerical simulations, the influence of some relevant factors on the well-known rank reversal (RR) phenomenon in Saaty's Analytic Hierarchy Process (AHP). The authors consider both the case of a single decision maker and of a group of decision-makers. Their idea is to concentrate on a condition, which preserves RR and on how to relax it step-by-step. First, they study how the estimated probability of RR depends on the distribution of the criteria weights and more precisely on the entropy of this distribution. Entropy is relevant since it is known that RR does not occur in the zero entropy case of weight concentration. The authors show an interesting "increasing behaviour" of the estimated RR probability as a function of the entropy of weights. Additionally, the chapter focuses on the aggregation of local weight vectors and on a more general aggregation rule, based on the weighted power mean, for which the weighted geometric mean and the weighted arithmetic mean are particular cases. Finally, the authors repeat their analysis for the case of a group decision-making problem and observe that the estimated probability of RR decreases by the aggregation of preferences of the decision-makers, suggesting an inverse relationship between consensus and rank reversal, under the assumption that all judgements are totally consistent.

In "Estimating One-Off Operational Risk Events with the Lossless Fuzzy Weighted Average Method" Pasi Luukka, Mikael Collan, Fai Tam, and Yuri Lawryshyn consider the problem of determining an estimate for the size of possible one-off negative operational events. The problem is very important for banks, who are required by the Basel II Accord to report these risks. The typical way to produce these estimates is to use a quantitative value-at-risk methodology that is based on limited data. This is interesting, because also the use of qualitative expert estimate-based methodologies is allowed by the regulations. The final estimations are typically obtained by fusing the input from multiple experts. The authors propose a new approach that is based on the author's original idea of a new lossless fuzzy weighted averaging. They show how to use this method for the problem considered and why it is a good tool for the aggregation of expert estimates in the context of bank risk management. The method proposed is simple to use, intuitive to understand, and does not suffer from the loss of information associated with the use of many other weighted averaging methods.

The paper "Fuzzy Signature Based Methods for Modelling the Structural Condition of Residential Buildings" by Ádám Bukovics, István Á. Harmati and László T. Kóczy deals with some issues related to a very important problem of the conservation, extension, or renovation of residential buildings. The focus is on the proper use of available financial resources and it is shown that an incorrect assessment of the scope and extent of renovation or reconstruction needs may cause a considerable financial loss. The authors present, through a survey of a real stock of buildings, an analysis of what kind of examinations and research should precede the quantitative decision support stage of renovation decisions. The authors introduce three fuzzy signature based methods, which are shown to be suitable for determining the condition of a bigger stock of buildings and for ranking them. These methods are suitable for the aggregation of expert evaluations of different detail and depth. Finally, the authors consider a very relevant issue of the sensitivity analysis of the method and present results of some computational examples.

Ronald R. Yager ("Retrieval from Uncertain Data Bases") investigates tools that can enrich the process of querying databases. He shows how to include soft conditions with the use of fuzzy sets and describes some techniques for aggregating the satisfaction of the individual conditions, based on the inclusion of importance and the use of the OWA operator. The author discusses a method for aggregating the individual satisfactions that can model a lexicographic relation between the individual requirements. Finally, the author looks at querying databases, in which the information can exhibit some probabilistic uncertainty.

Part IV includes chapters on various issues related to decision-making and optimization in various settings and under various kinds of imperfect information.

José Luis García-Lapresta and Raquel González del Pozo ("An Ordinal Multi-criteria Decision-making Procedure in the Context of Uniform Qualitative Scales") propose a multi-criteria decision-making procedure for the purely ordinal context, in which agents evaluate the alternatives regarding several criteria by assigning one, or two, consecutive terms of a uniform ordered qualitative scale to each alternative in each criterion. The weights assigned to criteria are dealt with by replications of the corresponding ratings and alternatives are ranked according to the medians of their ratings, after the replications. The new method and its properties are discussed and illustrated with examples.

In "FRIM—Fuzzy Reference Ideal Method in Multicriteria Decision Making" E. Cables, M.T. Lamata, and J.L. Verdegay consider TOPSIS, one of numerous compensatory multi-criteria decision methods, because of its rationality, intuitive appeal, and easy applicability. This method is based on the concept of scoring alternatives based on their distance to a positive ideal-solution (shorter the better) and simultaneously on their distance to a negative ideal-solution (longer the better). Based on this idea, the Reference Ideal Method (RIM), proposed by the authors, can be considered as an extension of the TOPSIS method, while considering that the ideal solution does not have to be the maximum or the minimum value, but may be a value in-between these. The RIM method yields good solutions, but does not always obtain a solution in the case of using fuzzy numbers. An extension of the RIM is proposed in the paper for making it possible to deal with vagueness and uncertainty, resulting in the Fuzzy Reference Ideal Method (FRIM). Its applicability is shown with practical examples.

In their paper "A new approach for solving CCR data envelopment analysis model under uncertainty" Bindu Bhardwaj, Jagdeep Kaur, and Amit Kumar critically analyze Wang and Chin's approach to the optimistic and pessimistic fuzzy CCR data envelopment analysis (DEA) model, and—after pointing out some if its deficiencies—propose a novel approach that makes it possible to alleviate the observed deficiencies. Moreover, they also propose a new approach to solve the proposed fuzzy CCR DEA models for evaluating the relative geometric efficiency of decision-making units (DMUs).

Then, in the paper "A New Fuzzy CCR Data Envelopment Analysis Model and Its Application to Manufacturing Enterprises" Bindu Bhardwaj, Jagdeep Kaur and Amit Kumar consider the problem of fuzzy data envelopment analysis based on fuzzy arithmetic with an application to the performance assessment of manufacturing enterprises in which the solution is obtained by solving a fuzzy fractional programming problem transformed into a fuzzy linear programming problem as proposed by Wang et al. The authors first show some of the original approach, and propose a new approach to overcome this flaw. Them they present the application a real life planning problem in a manufacturing company.

Part V cover various aspects, approaches, tools and techniques, and algorithms that are relevant for various problems in a large area of multi-person decision-making and consensus reaching.

Francisco Javier Cabrerizo, Francisco Chiclana, Ignacio Javier Pérez, Francisco Mata, Sergio Alonso, and Enrique Herrera-Viedma propose in their paper "A Feedback Mechanism Based on Granular Computing to Improve Consensus in GDM" consider a very important problem of group decision-making, the essence of which is to obtain the best solution to a particular problem according to opinions (testimonies) given by a set of decision-makers, or—more generally—of some agents. Usually, this process proceeds better, when the group is at an agreement, or at consensus. An important issue is the level of consensus achieved among the decision-makers before making a decision. This may be supported by different feedback mechanisms, which can help the decision-makers reach the highest possible degree of consensus, many such mechanisms have been proposed in the previous literature. In this work, the authors present a new feedback mechanism that is based on granular computing and an effective and efficient framework of designing, processing, and interpretation of information granules, to improve (the degree of) consensus in group decision-making problems. The approach proposed provides flexibility required to improve the level of consensus within the group of decision-makers.

In their paper "A Method for the Team Selection Problem Between Two Decision-Makers Using the Ant Colony Optimization" Marilyn Bello, Rafael Bello, Ann Nowé, and María M. García-Lorenzo consider the problem of team selection, which consists of how to perform a personnel selection process to form some collaborating and cooperating teams according to some preferences. This is a clearly important in the field of human resources. This type of selection process usually proceeds by the ranking of candidates based on preferences of decision-makers and by then allowing the decision-makers to select a specific candidate, or candidates. This simple process may be viewed as unfair, because it often leads to an unfair allocation of candidates to different teams, i.e., the quality of the teams formed may not follow the rankings articulated by the decision-makers. A new approach to the team selection problem is proposed, in which two employers form their teams by selecting their members from a set of candidates that is common to both and then each decision maker reveals his or her personal ranking of those candidates. The method proposed is shown to lead to teams of a high quality, according to the valuations of each of the decision-makers, and also to a (more) fair composition of teams. The Ant-Colony Optimization meta-heuristic is employed and its effectiveness and efficiency is shown on some quite large examples.

Mingwei Lin and Zeshui Xu ("Probabilistic Linguistic Distance Measures and Their Applications in Multi-Criteria Group Decision Making") consider a new and interesting topic of probabilistic linguistic term sets, which can express not only the decision-makers' several possible linguistic assessment values, but also the weight of each linguistic assessment value. The authors advocate the use of the proba- bilistic linguistic term sets in various fields, specifically they focus on the distance measures for the probabilistic linguistic term sets and their applications in multi-criteria group decision-making. They first define the distance between two elements of the probabilistic linguistic term set. A variety of distance measures are proposed to calculate the distance between two probabilistic linguistic term sets. Then, these distance measures are further extended to compute the distance between two collections of probabilistic linguistic term sets, by considering the weight information of each criterion. Finally, the concept of the satisfaction degree of an alternative is proposed and used to rank the alternatives in multi-criteria group decision-making. A real-world example is given to show the use of these distance measures and to compare the probabilistic linguistic term sets with hesitant fuzzy linguistic term sets.

Antonio Maturo and Aldo G.S. Ventre ("Fuzzy Numbers and Consensus") consider a frequently encountered case in multi-group decision and consensus reaching processes, when in a group of decision-makers there is a considerable variability in the scores that they assign to given alternatives. The authors propose to represent this variability with fuzzy numbers and then present an algorithm for reaching a consensus in the setting assumed, i.e., based on fuzzy numbers, preorder and order relations in the sets of fuzzy numbers, and on a procedure to decrease the spreads resulting from operations on fuzzy numbers.

Janusz Kacprzyk and Sławomir Zadrożny ("Reaching Consensus in a Group of Agents: Supporting a Moderator Run Process via Linguistic Summaries") present an account and a critical analysis of works on consensus reaching processes driven by a moderator, a "super-agent" who is in charge of running the process in an effective and efficient way. The authors assume the classic approach to the evalu- ation of the degree of consensus of Kacprzyk and Fedrizzi, in which a soft degree of consensus is a degree, in which, e.g., "most of the important individuals agree on with regards to almost all of the relevant options", with the fuzzy "most" majority assumed to be a fuzzy linguistic quantifier, as proposed by Kacprzyk. Typically, this kind of situations are handled via Zadeh's classic calculus of linguistically quantified propositions, or by some other method, notably Yager's OWA (ordered weighted average) operators.

Additional information, which can be very useful to the moderator and the agents, is provided by a novel combination of using a soft degree of consensus alone within a decision support system setting and the linguistic data summaries in Yager's sense, in particular, in its protoform-based version proposed by Kacprzyk and Zadrożny. They are intended to indicate, in a natural language, some interesting relations between individuals and options to help the moderator identify crucial (pairs of) individuals and/options, which pose some threats to the reaching of (a sufficient degree of) consensus. The use of a so-called action rule that in the context of the chapter means "to find best concessions to be offered to the individuals for changing their preferences to increase the degree of consensus" is employed. New results on the use of the concepts of a consensory and dissensory agents are also presented.

In "Consensus in Multiperson Decision Making Using Fuzzy Coalitions" Fabrizio Maturo and Viviana Ventre consider the problem of group decisions, in which the decision-makers have different opinions or interests. The authors propose various metric spaces for the representation of movements of decision-makers for reaching a consensus. They also introduce the concept of a fuzzy coalition for developing an algorithm for building a feasible fuzzy coalition, defined as the union of winning maximum coalitions that solve the issue of consensus among decision-makers.

We would like to express our gratitude to all the authors for their interesting, novel, and inspiring contributions. Peer-reviewers also deserve a deep appreciation, because their insightful and constructive remarks and suggestions have considerably improved many contributions.

And last but not least, we wish to thank Dr. Tom Ditzinger, Dr. Leontina di Cecco, and Mr. Holger Schaepe for their dedication and help to implement and finish this large publication project on time maintaining the highest publication standards.

Lappeenranta, Finland   Mikael Collan
Warsaw, Poland   Janusz Kacprzyk
Spring 2017

Contents

Part I   General Human and Economic Aspects

**Optimal Group Decision Making Criterion and How
It Can Help to Decrease Poverty, Inequality, and Discrimination**   3
Vladik Kreinovich and Thongchai Dumrongpokaphan

**An Overview and Re-interpretation of Paradoxes
of Responsiveness**   21
Hannu Nurmi

Veto in Yes-no and Yes-no-Abstain Voting Systems   33
Jacek Mercik

Power Indices for Finance   45
Cesarino Bertini, Gianfranco Gambarelli, Izabella Stach and Maurizio Zola

**The Binomial Decomposition of the Single Parameter Family
of GB Welfare Functions**   71
Silvia Bortot, Ricardo Alberto Marques Pereira
and Anastasia Stamatopoulou

Part II   General Formal Foundations

**The Logic of Information and Processes in System-of-Systems
Applications**   89
P. Eklund, M. Johansson and J. Kortelainen

**Decision-making Process Using Hyperstructures and Fuzzy Structures
in Social Sciences**   103
Sarka Hoskova-Mayerova and Antonio Maturo

Social Preferences Through Riesz Spaces: A First Approach   113
Antonio Di Nola, Massimo Squillante and Gaetano Vitale

**Coherent Conditional Plausibility: A Tool for Handling Fuzziness
and Uncertainty Under Partial Information** 129
Giulianella Coletti and Barbara Vantaggi

**Intuitionistic Fuzzy Interpretations of Some Formulas
for Estimation of Preference Degree** 153
Krassimir T. Atanassov, Vassia Atanassova, Eulalia Szmidt
and Janusz Kacprzyk

Part III   Judgments and Aggregation

Fuzzified Likert Scales in Group Multiple-Criteria Evaluation 165
Jan Stoklasa, Tomáš Talášek and Pasi Luukka

**Maximal Entropy and Minimal Variability OWA Operator Weights:
A Short Survey of Recent Developments** 187
Christer Carlsson and Robert Fullér

**A Closer Look at the Relation Between Orness and Entropy
of OWA Function** 201
József Mezei and Matteo Brunelli

**Rank Reversal in the AHP with Consistent Judgements:
A Numerical Study in Single and Group Decision Making** 213
Michele Fedrizzi, Silvio Giove and Nicolas Predella

**Estimating One-Off Operational Risk Events
with the Lossless Fuzzy Weighted Average Method** 227
Pasi Luukka, Mikael Collan, Fai Tam and Yuri Lawryshyn

**Fuzzy Signature Based Methods for Modelling the Structural
Condition of Residential Buildings** 237
Ádám Bukovics, István Á. Harmati and László T. Kóczy

Retrieval from Uncertain Data Bases 275
Ronald R. Yager

Part IV   Decision Making and Optimization

**An Ordinal Multi-criteria Decision-Making Procedure
in the Context of Uniform Qualitative Scales** 297
José Luis García-Lapresta and Raquel González del Pozo

**FRIM—Fuzzy Reference Ideal Method in Multicriteria Decision
Making** 305
E. Cables, M.T. Lamata and J.L. Verdegay

**A New Approach for Solving CCR Data Envelopment Analysis
Model Under Uncertainty** 319
Bindu Bhardwaj, Jagdeep Kaur and Amit Kumar

**A New Fuzzy CCR Data Envelopment Analysis Model
and Its Application to Manufacturing Enterprises** 345
Bindu Bhardwaj, Jagdeep Kaur and Amit Kumar

Part V   Multiperson Decision Making and Consensus Reaching

**A Feedback Mechanism Based on Granular Computing
to Improve Consensus in GDM** 371
Francisco Javier Cabrerizo, Francisco Chiclana, Ignacio Javier Pérez,
Francisco Mata, Sergio Alonso and Enrique Herrera-Viedma

**A Method for the Team Selection Problem Between Two
Decision-Makers Using the Ant Colony Optimization** 391
Marilyn Bello, Rafael Bello, Ann Nowé and María M. García-Lorenzo

**Probabilistic Linguistic Distance Measures and Their Applications
in Multi-criteria Group Decision Making** 411
Mingwei Lin and Zeshui Xu

Fuzzy Numbers and Consensus 441
Antonio Maturo and Aldo G.S. Ventre

Consensus in Multiperson Decision Making Using Fuzzy Coalitions 451
Fabrizio Maturo and Viviana Ventre

**Reaching Consensus in a Group of Agents: Supporting
a Moderator Run Process via Linguistic Summaries** 465
Janusz Kacprzyk and Sławomir Zadrożny

Curriculum Vitae 487

Part I
General Human and Economic Aspects

Optimal Group Decision Making Criterion and How It Can Help to Decrease Poverty, Inequality, and Discrimination

Vladik Kreinovich and Thongchai Dumrongpokaphan

Abstract Traditional approach to group decision making in economics is to maximize the GDP, i.e., the overall gain. The hope behind this approach is that the increased wealth will trickle down to everyone. Sometimes, this happens, but often, in spite of an increase in overall GDP, inequality remains: some people remain poor, some groups continue to face economic discrimination, etc. This shows that maximizing the overall gain is probably not always the best criterion in group decision making. In this chapter, we find a group decision making criterion which is optimal (in some reasonable sense), and we show that using this optimal criterion can indeed help to decrease poverty, inequality, and discrimination.

1 Traditional Group Decision Making in Economics And Its Limitations: Formulation of the Problem

Traditional approach to group decision making in economics. Traditional approach to group decision making in economics is to maximize the Gross Domestic Product (GDP), i.e., to maximize the overall gain $\sum\limits_{i=1}^{n} g_i$, where g_i is the gain of the i-th person.

The hope behind this approach. The hope behind the traditional approach is that the rising tide will lift all the boats, that the increased overall wealth will trickle down to everyone.

V. Kreinovich (✉)
Department of Computer Science, University of Texas at El Paso,
500 W. University, El Paso, TX 79968, USA
e-mail: vladik@utep.edu

T. Dumrongpokaphan
Department of Mathematics, Faculty of Science, Chiang Mai University,
Chiang Mai, Thailand
e-mail: tcd43@hotmail.com

© Springer International Publishing AG 2018
M. Collan and J. Kacprzyk (eds.), *Soft Computing Applications for Group Decision-making and Consensus Modeling*, Studies in Fuzziness and Soft Computing 357, DOI 10.1007/978-3-319-60207-3_1

The reality does not not always follow this optimistic vision. Sometimes, the overall increase in wealth indeed increases everyone's income. However, often, in spite of an increase in overall GDP, inequality remains:

- some people remain poor,
- some groups continue to face economic discrimination, etc.

What can we conclude from this fact? The fact that maximizing the overall gain does not always lead to good consequences for everyone indicates that maximizing GDP is probably not always the best criterion in group decision making.

What we do in this paper. To improve the situation, let us try to find a better criterion for group decision making.

To come up with such a better criterion, we formulate the problem of selecting a criterion for group decision making as an optimization problem. We then solve this optimization problem, and we show that the resulting optimal group decision making criterion can indeed help in decreasing poverty, inequality, and discrimination.

2 Individual Decision Making: A Brief Reminder

Need to consider individual decision making. In other to properly describe group decision making, it is necessary to first describe individual decision making.

Main assumptions behind the traditional decision theory. In this section, we will briefly describe the traditional decision theory; see, e.g., [1, 6, 9]. The main assumptions behind the traditional decision theory are as follows.

The first assumption is that for every two alternatives a and b:

- a person either prefers a (we will denote it by $a > b$),
- or prefers b ($b > a$),
- or for this person, alternatives a and b are of equal value (we will denote it by $a \sim b$).

The second assumption is that the person is rational, in the sense that this person's preferences are consistent:

- if $a > b$ and $b > c$, then $a > c$;
- if $a \sim b$ and $b \sim c$, then $a \sim c$;
- if $a > b$ and $b \sim c$, then $a > c$; and
- if $a \sim b$ and $b > c$, then $a > c$.

The main concept of the traditional decision theory: the notion of utility. Under the above assumptions, we can use the following idea to numerically describe the value of each alternative to a person. We select two alternatives:

- we select a very bad alternative a_0, which is much worse than anything that will be really proposed, and

- we select a very good alternative a_1, which is much better than anything that will be really proposed.

Then, for every real number p from the interval $[0, 1]$, we can form a lottery $L(p)$ in which we get a_1 with probability p and a_0 with the remaining probability $1 - p$. The larger p, the better the lottery: if $p < p'$, then $L(p) \prec L(p')$.

For each alternative a, and for every p, we either have $a \prec L(p)$ or $a \succ L(p)$ or $a \sim L(p)$. If $a \prec L(p)$ and $p < p'$, then we will still have $a \prec L(p')$. Similarly, if we have $a \succ L(p)$ and $p > p'$, then we will have $a \succ L(p')$. Thus, we have:

$$\inf\{p : a \prec L(p)\} = \sup\{p : L(p) \prec a\}.$$

This common probability value is called the *utility* of the alternative a. It is usually denoted by $u(a)$.

For the utility, for every $\varepsilon > 0$, we have

$$L(u(a) - \varepsilon) \prec a \prec L(u(a) + \varepsilon).$$

In this sense, a is "almost" equivalent to to the lottery $L(u(a))$; we will denote this equivalence by $a \equiv L(u(a))$.

Utility is defined modulo a linear transformation. The numerical value of the utility depends on our selection of the alternatives a_0 and a_1. If we select a different pair a_0' and a_1', then we, in general, will get different numerical values of the corresponding utility $u'(a)$. What is the relation between the new utility value $u'(a)$ and the original utility value $u(a)$?

To answer this question, let us first consider the case when $a_0 < a_0' < a_1' < a_1$. In this case, each of the new alternatives a_0' and a_1' is equivalent to an appropriate lottery $L(p)$: $a_0' \equiv L(u(a_0'))$ and $a_1' \equiv L(u(a_1'))$ for some values $u(a_0')$ and $u(a_0')$.

By definition of the (a_0', a_1')-based utility $u'(a)$, the alternative a is equivalent to the lottery $L'(u'(a))$ in which we get a_1' with probability $u'(a)$ and a_0' with the remaining probability $1 - u'(a)$.

If we replace a_1' with the equivalent lottery $L(a_1')$ and a_0' with the equivalent lottery $L(a_0')$, then we conclude that the original alternative a is equivalent to the following 2-stage lottery:

- first, we select either 1 (with probability $u'(a)$) or 0 (with probability $1 - u'(a)$);
- then, depending on the value i that we selected in the first stage, we select either a_1 (with probability $u(a_1')$) or a_0 (with probability $u(a_0')$).

As a result of this 2-stage lottery, we get either a_1 or a_0. By using the formula for the complete probability, we can find the probability $u(a)$ for selecting a_1:

$$u(a) = u'(a) \cdot u(a_1') + (1 - u'(a)) \cdot u(a_0').$$

Thus, the original alternative a is equivalent to the lottery in which we select a_1 with this probability $u(a)$ and a_0 with the remaining probability $1 - u(a)$. So, $a \equiv L(u(a))$.

The above formula shows that the utilities $u(a)$ and $u'(a)$ are connected by a linear dependence. This was when we have $a_0 \prec a'_0 \prec a'_1 \prec a_1$.

In the general case, when we want to find a relation between utilities $u(a)$ and $u'(a)$ corresponding to different pairs (a_0, a_1) and (a'_0, a'_1), the way to find this connection is by considering a third pair (a''_0, a''_1) for which $a''_0 \prec a_0$, $a''_0 \prec a'_0$, $a_1 \prec a''_1$, and $a'_1 \prec a''_1$. In this case, the above argument shows that:

- $u''(a)$ is a linear function of $u'(a)$, and
- $u''(a)$ is a linear function of $u(a)$—thus, $u(a)$ is a linear function of $u''(a)$.

Substituting the linear expression of $u''(a)$ in terms of $u'(a)$ into the linear dependence of $u(a)$ on $u''(a)$ and using the fact that a composition of two linear functions is linear, we conclude that $u'(a)$ is indeed a linear function of $u(a)$;

$$u'(a) = k \cdot u(a) + \ell$$

for some $k > 0$ and ℓ.

Now, we have all the preliminaries needed to start discussing group decision making.

3 Group Decision Making: Analysis of the Problem

What is a group decision making criterion? To compare several alternatives, we need to know how valuable is each alternative to each person from the group.

The value of each alternative a to each person i can be described by the i-th utility $u_i(a)$ of this alternative. Thus, to describe the value of the alternative a to the group, we can use the tuple $u(a) = (u_1(a), \ldots, u_n(a))$ consisting of all these utility values.

As we have mentioned earlier, for each person i, his or her utility values are defined modulo arbitrary linear transformation: $u_i(a) \rightarrow k_i \cdot u_i(a) + \ell_i$. We are considering situations in which we start with a certain *status quo* s_0. It is therefore reasonable to re-scale utilities in such a way that the status-quo value of utility corresponds to 0. In other words, instead of the original utility values, we consider the utility gains $u_i(a) - u_i(s_0)$.

In this re-scaled case, for each alternative, we have $u_i(a) > 0$: if $u_i(a) < 0$, i.e., if the i-th person loses with the adoption of the alternative a, this person would then never agree to this alternative. Similarly, if $u_i(a) = 0$, i.e., if the person does not gain anything in the alternative a, why would this person participate in making this decision?

Thus, the only tuples that we have to consider are tuples $u(a)$ in which all the components $u_i(a)$ are positive.

For every two alternatives a and b, based on the corresponding tuples, we need to decide:

- whether a is better than b,

- or b is better than a,
- or that for the group, the alternatives a and b are equivalent.

Such a decision is what we call a group decision making criterion.

Of course, if everyone benefits, i.e., if $u_i < u_i'$ for all i, then clearly, the tuple u' is better than the tuple u. Similarly, if no one loses, i.e., if $u_i \leq u_i'$ for all i, then the tuple u' should be either better or of the same quality as u—but not worse.

Finally, a group decision making criterion should be *continuous* in the following sense: if we have $a^{(k)} \to a$ for some sequence $a^{(k)}$, and if for every k, we have $a^{(k)} \prec b$ or $a^{(k)} \sim b$, then for the limit sequence a, we should also have $a \prec b$ or $a \sim b$.

In other words, if a tuple a is worse than a tuple b, then all the tuples in some neighborhood of a should also be worse than b.

Let us describe all this in precise terms.

Definition 1 By a *pre-ordering* relation on a set A, we mean a pair of binary relations $\prec$ and $\sim$ for which $a \prec b$ for some a and b and for which the following properties hold for all a, b, and c:

- $a \sim a$ and $a \nprec a$;
- if $a \prec b$ and $b \prec c$, then $a \prec c$;
- if $a \sim b$ then $b \sim a$;
- if $a \sim b$ and $b \sim c$, then $a \sim c$;
- if $a \prec b$ and $b \sim c$, then $a \prec c$;
- if $a \sim b$ and $b \prec c$, then $a \prec c$.

Definition 2 A pre-ordering is called *linear* (or *total*) if for every two elements a and b, we have either $a \prec b$, or $b \prec a$, or $a \sim b$.

Definition 3 By a *group decision making criterion*, we mean a linear (total) pre-ordering $(\prec, \sim)$ on the set $\mathbb{R}^n_+$ of all possible n-tuples $u = (u_1, \ldots, u_n)$ of positive real numbers for which:

- if $u_i < u_i'$ for all i, then $u \prec u'$,
- if $u_i \leq u_i'$ for all i, then either $u \prec u'$ or $u \sim u'$, and
- if $a \prec b$, then for all the tuples a' from some neighborhood of a, we should also have $a' \prec b$.

4 We Want to Find the Best Group Decision Making Criterion

Discussion. To select the best group decision making criterion, we need to be able to compare different criteria: some criteria are better, some are worse, some are equivalent. In other words, we need to describe a *linear pre-ordering* $(<, \equiv)$ on the set of all possible criteria.

Then, we should select the *best (optimal)* criterion A, i.e., a criterion for which, for every other criterion B, we should have either $B < A$ or $B \equiv A$.

It is reasonable to require that there should be only one best criterion A. Indeed, if no criterion is the best, then this pre-ordering is of no use. On the other hand, if several criteria are the best, this means that our pre-ordering is not final: we can us this non-uniqueness to optimize something else. For example, if several different group decision making criteria are of equal value based on the economic consequences, it is reasonable to select the only which is the easiest to compute. This means that the original comparison $<$ (based only on economical consequences) is now replaced by a more complex comparison $a <' b$ which holds if:

- either b is better economically,
- or, economically, a and b are of the same quality, but b is easier to compute.

Eventually, we will thus reach the final situation, in which for the resulting linear pre-ordering $(<, \equiv)$, exactly one group decision making criterion is the best.

Another reasonable requirement is that the comparison between different group decision making criteria should not change if we simply re-scale the individual utility values, i.e., replace the original utility values u_i by re-scaled values $u_i' = k_i \cdot u_i$.

Finally, it is reasonable to require that all participants are equal, in the sense that nothing should change if we simply swap two or more participants (and their utilities).

Thus, we arrive at the following definitions.

Definition 4 By a *comparison* between group decision making criteria, we mean a linear (total) pre-ordering $(<, \equiv)$ on the set of all possible group decision making criteria.

Definition 5 We say that a group decision making criterion A is *optimal* with respect to a given comparison $(<, \equiv)$ if for every other group decision making criterion B, we have either $B < A$, or $B \equiv A$.

Definition 6 We say that a comparison $(<, \equiv)$ is *final* if there exists exactly one group decision making criterion which is optimal with respect to this comparison.

Definition 7 Let $k = (k_1, \ldots, k_n)$ be a tuple of positive numbers $k_i > 0$.

- For every tuple $u = (u_1, \ldots, u_n)$, by its *k-rescaling*, we mean a tuple

$$S_k(u) = (k_1 \cdot u_1, \ldots, k_n \cdot u_n).$$

- By a *k-rescaling* $S_k(A) = (\prec_k, \sim_k)$ of a group decision making criterion

$$A = (\prec, \sim),$$

we mean the following criterion:

$$a \prec_k b \Leftrightarrow S_k(a) \prec S_k(b), \text{ and } a \sim_k b \Leftrightarrow S_k(a) \sim S_k(b).$$

Definition 8 We say that a comparison $(<, \equiv)$ is *scale-invariant* if for every tuple k and for every two group decision making criteria A and B, we have:

$$A < B \Leftrightarrow S_k(A) < S_k(B); \text{ and}$$

$$A \equiv B \Leftrightarrow S_k(A) \equiv S_k(B).$$

Definition 9 Let $\pi : \{1, \ldots, n\} \to \{1, \ldots, n\}$ be a permutation.

- For every tuple $u = (u_1, \ldots, u_n)$, by its *permutation*, we mean a tuple

$$\pi(u) = (u_{\pi(1)}, \ldots, u_{\pi(n)}).$$

- By a *permutation* $\pi(A) = (<_\pi, \sim_\pi)$ of a group decision making criterion

$$A = (<, \sim),$$

we mean the following criterion:

$$a <_\pi b \Leftrightarrow \pi(a) < \pi(b), \text{ and } a \sim_\pi b \Leftrightarrow \pi(a) \sim \pi(b).$$

Definition 10 We say that a comparison $(<, \equiv)$ is *permutation-invariant* if for every permutation π and for every two group decision making criteria A and B, we have:

$$A < B \Leftrightarrow \pi(A) < \pi(B); \text{ and}$$

$$A \equiv B \Leftrightarrow \pi(A) \equiv \pi(B).$$

Main Result. For every final scale-invariant permutation-invariant comparison

$$(<, \equiv),$$

the optimal group decision making $(<, \sim)$ has the following form:

$$a = (a_1, \ldots, a_n) < b = (b_1, \ldots, b_n) \Leftrightarrow \prod_{i=1}^{n} a_i < \prod_{i=1}^{n} b_i; \text{ and}$$

$$a = (a_1, \ldots, a_n) \sim b = (b_1, \ldots, b_n) \Leftrightarrow \prod_{i=1}^{n} a_i = \prod_{i=1}^{n} b_i.$$

Comment. For reader's convenience, the proof of this result is given in a special Appendix.

Discussion. According to our result, according to the optimal group decision making criterion, we should select an alternative a for which the product $\prod_{i=1}^{n} u_i(a)$ of all the utilities attains the largest possible value.

This group decision making criterion was first proposed by the Nobelist John Nash [5] (see also [4]) and is thus known as *Nash's bargaining solution*. Thus, what we prove is that Nash's bargaining solution is the optimal group decision making criterion.

5   The Resulting Optimal Criterion for Group Decision Making Is in Good Accordance with Common Sense

The objective of group decision making is to make sure that the first person is happy, *and* that the second person is happy, *and* that the their person is happy, etc. From this viewpoint, selecting the best alternative means selecting the alternative for which the degree to which first person is happy and the second person is happy, etc., is the largest possible.

In line with the usual fuzzy logic techniques (see, e.g., [3, 8, 10]), to describe the degree to which this objective is satisfied, we must:

- first, find the degrees to which each person is happy, and then,
- use an "and"-operation (t-norm) $f_\&(a, b)$ to combine these degrees.

For each alternative a, a reasonable measure of the degree $d_i(a)$ to which the i-th person is happy with this alternative is his/her utility $u_i(a)$. It is therefore reasonable to assume that the degree $d_i(a)$ can be determined if we know the utility $u_i(a)$, i.e., that $d_i(a) = f(u_i(a))$ for an appropriate algorithm $f(x)$.

The simplest possible case is when we simply take $d_i(a) = u_i(a)$, with $f(x) = x$. Among the simplest—and most widely used—t-norms is the algebraic product $f_\&(a, b) = a \cdot b$. If we use this t-norm, then the degree to which all the members of the groups are happy with the given alternative a is equal to the product

$$\prod_{i=1}^{n} d_i(a) = \prod_{i=1}^{n} u_i(a).$$

In this case, selecting the best alternative means selecting the alternative a for which this product attains its largest possible value.

This is exactly Nash's bargaining solution which, as we have shown, is the optimal group decision making criterion. Thus, Nash's bargaining solution is indeed in good accordance with common sense—namely, it is in good accordance with the simplest possible formalization of the above common sense criterion.

6   The Resulting Optimal Criterion for Group Decision Making Can Help In Decreasing Poverty, Inequality, and Discrimination

What happens now: it is possible that maximizing GDP retains poverty, inequality, and discrimination. The problem with the current economic development—which is motivated by the need to maximize the GDP—is that while the GDP is increased and some people get richer, some other people remain at the same poverty level as before.

In other words, while for some people, the gain g_i is positive, for many others, the gain is practically zero: $g_j \approx 0$. Thus, the increase in GDP sometimes increases inequality as well—the rich get richer, but the poor remain equally poor.

Poverty is rarely distributed uniformly: usually, some population groups are poorer. When the income of such groups does not increase, this is an example of an economic discrimination.

How the use of the new criterion for group decision making can help prevent such situations. We would like to avoid the situations in which for the selected alternative a, at least one person j does not get any increase in utility, i.e., $u_j(a) \approx 0$ for some j. Let us show that such alternatives are indeed avoided.

Indeed, if for some alternative a, we have $u_j(a)$ for some person j, then for this alternative a, the product of utilities $\prod_{i=1}^{n} u_i$ is practically equal to 0—i.e., to its smallest possible value. A simple redistribution in which the j-th person (and other people who initially got practically nothing) would get a little more will immediately increase the product.

So, when we select an alternative for which the product of utilities is the largest possible, we will never select an alternative in which one of the persons practically does not get any increase in utility.

In this sense, the new criterion helps decrease inequality. And if there is a whole group of people which is discriminated against—e.g., for which $u_j \approx 0$ for all persons j from this group—then the corresponding alternative will never be selected if we use the new group decision making criterion.

Acknowledgements   This work was supported by Chiang Mai University, Thailand. This work was also supported in part by the National Science Foundation grants HRD-0734825 and HRD-1242122 (Cyber-ShARE Center of Excellence) and DUE-0926721, and by an award "UTEP and Prudential Actuarial Science Academy and Pipeline Initiative" from Prudential Foundation.

Proof of the Main Result

$1°$. In this proof, we will use ideas first described in [2] and in [7].

$2°$. Let us first prove that for every every final scale-invariant permutation-invariant comparison $(<, \equiv)$, the optimal group decision making $A = (\prec, \sim)$ is itself scale-

and permutation-invariant, i.e., $A = S_k(A)$ for every tuple k and $\pi(A) = A$ for every permutation A.

Indeed, let us prove that $A = S_k(A)$ for every tuple $k = (k_1, \ldots, k_n)$. Let us denote $k^{-1} = (k_1^{-1}, \ldots, k_n^{-1})$; then, clearly, $S_k(S_{k^{-1}}(B)) = B$ for all B. Since the group decision making criterion A is optimal, then for every group decision making B, we have $S_{k^{-1}}(B) < A$ or $S_{k^{-1}}(B) \equiv A$. Since the comparison $(<, \equiv)$ is scale-invariant, $S_{k^{-1}}(B) < A$ implies that $S_k(S_{k^{-1}}(B)) < S_k(A)$, i.e., that $B < S_k(A)$. Similarly, $S_{k^{-1}}(B) \equiv A$ implies that $S_k(S_{k^{-1}}(B)) \equiv S_k(A)$, i.e., that $B \equiv S_k(A)$.

Thus, for every group decision making criterion B, we have either $B < S_k(A)$ or $B \equiv S_k(A)$. By definition, this means that the group decision criterion $S_k(A)$ is optimal. However, our comparison $(<, \equiv)$ is final. Thus, there should be only one optimal group decision making criterion. So, we conclude that $S_k(A) = A$.

Similarly, we can prove that $\pi(A) = A$ for any permutation π. Let us denote the inverse permutation by π^{-1}; then, clearly, $\pi(\pi^{-1}(B)) = B$ for all B. Since the group decision making criterion A is optimal, then for every group decision making B, we have $\pi^{-1}(B) < A$ or $\pi^{-1}(B) \equiv A$. Since the comparison $(<, \equiv)$ is permutation-invariant, $\pi^{-1}(B) < A$ implies that $\pi(\pi^{-1}(B)) < \pi(A)$, i.e., that $B < \pi(A)$. Similarly, $\pi^{-1}(B) \equiv A$ implies that $\pi(\pi^{-1}(B)) \equiv \pi(A)$, i.e., that $B \equiv \pi(A)$.

Thus, for every group decision making criterion B, we have either $B < \pi(A)$ or $B \equiv \pi(A)$. By definition, this means that the group decision criterion $\pi(A)$ is optimal. However, our comparison $(<, \equiv)$ is final. Thus, there should be only one optimal group decision making criterion. So, we conclude that $\pi(A) = A$.

3°. Let us now use the scale-invariance of the optimal group decision making criterion $(<, \sim)$. Due to this scale-invariance, for every $y_1, \ldots, y_n, y_1', \ldots, y_n'$, we can take $\lambda_i = \dfrac{1}{y_i}$ and conclude that

$$(y_1', \ldots, y_n') \sim (y_1, \ldots, y_n) \Leftrightarrow \left(\frac{y_1'}{y_1}, \ldots, \frac{y_n'}{y_n} \right) \sim (1, \ldots, 1).$$

Thus, to describe the equivalence relation $\sim$, it is sufficient to describe the set of all the vectors $z = (z_1, \ldots, z_n)$ for which $z \sim (1, \ldots, 1)$. Similarly,

$$(y_1', \ldots, y_n') \succ (y_1, \ldots, y_n) \Leftrightarrow \left(\frac{y_1'}{y_1}, \ldots, \frac{y_n'}{y_n} \right) \succ (1, \ldots, 1).$$

Thus, to describe the ordering relation $\succ$, it is sufficient to describe the set of all the vectors $z = (z_1, \ldots, z_n)$ for which $z \succ (1, \ldots, 1)$.

Alternatively, we can take $\lambda_i = \dfrac{1}{y_i'}$ and conclude that

$$(y_1', \ldots, y_n') \succ (y_1, \ldots, y_n) \Leftrightarrow (1, \ldots, 1) \succ \left(\frac{y_1}{y_1'}, \ldots, \frac{y_n}{y_n'} \right).$$

Thus, it is also sufficient to describe the set of all the vectors $z = (z_1, \ldots, z_n)$ for which $(1, \ldots, 1) \succ z$.

4°. The above equivalence involves division. To simplify the description, we can take into account that in the logarithmic space, division becomes a simple difference: $\ln\left(\dfrac{y'_i}{y_i}\right) = \ln(y'_i) - \ln(y_i)$. To use this simplification, let us consider the logarithms $Y_i \overset{\text{def}}{=} \ln(y_i)$ of different values. In terms of these logarithms, the original values can be reconstructed as $y_i = \exp(Y_i)$. In terms of these logarithms, we thus need to consider:

- the set $S_\sim$ of all the tuples $Z = (Z_1, \ldots, Z_n)$ for which

$$z = (\exp(Z_1), \ldots, \exp(Z_n)) \sim (1, \ldots, 1),$$

  and
- the set $S_\succ$ of all the tuples $Z = (Z_1, \ldots, Z_n)$ for which

$$z = (\exp(Z_1), \ldots, \exp(Z_n)) \succ (1, \ldots, 1).$$

We will also consider the set $S_\prec$ of all the tuples $Z = (Z_1, \ldots, Z_n)$ for which $(1, \ldots, 1) \succ z = (\exp(Z_1), \ldots, \exp(Z_n))$. Since the pre-ordering relation $(\prec, \sim)$ is linear (total), for every tuple z:

- either $z \sim (1, \ldots, 1)$,
- or $z \succ (1, \ldots, 1)$,
- or $(1, \ldots, 1) \succ z$.

In particular, this is true for $z = (\exp(Z_1), \ldots, \exp(Z_n))$. Thus, for every tuple Z:

- either $Z \in S_\sim$,
- or $Z \in S_\succ$,
- or $Z \in S_\prec$.

5°. Let us prove that the set $S_\sim$ is closed under addition, i.e., that if the tuples $Z = (Z_1, \ldots, Z_n)$ and $Z' = (Z'_1, \ldots, Z'_n)$ belong to the set $S_\sim$, then their component-wise sum

$$Z + Z' = (Z_1 + Z'_1, \ldots, Z_n + Z'_n)$$

also belongs to the set $S_\sim$.

Indeed, by definition of the set $S_\sim$, the condition $Z \in S_\sim$ means that

$$(\exp(Z_1), \ldots, \exp(Z_n)) \sim (1, \ldots, 1).$$

Using scale-invariance with $\lambda_i = \exp(Z'_i)$, we conclude that

$$(\exp(Z_1) \cdot \exp(Z'_1), \ldots, \exp(Z_n) \cdot \exp(Z'_n)) \sim (\exp(Z'_1), \ldots, \exp(Z'_n)).$$

On the other hand, the condition $Z' \in S_{\sim}$ means that

$$(\exp(Z'_1), \dots, \exp(Z'_n)) \sim (1, \dots, 1).$$

Thus, due to transitivity of the equivalence relation $\sim$, we conclude that

$$(\exp(Z_1) \cdot \exp(Z'_1), \dots, \exp(Z_n) \cdot \exp(Z'_n)) \sim (1, \dots, 1).$$

Since for every i, we have $\exp(Z_i) \cdot \exp(Z'_i) = \exp(Z_i + Z'_i)$, we thus conclude that

$$(\exp(Z_1 + Z'_1), \dots, \exp(Z_n + Z'_n)) \sim (1, \dots, 1).$$

By definition of the set $S_{\sim}$, this means that the tuple $Z + Z'$ belongs to the set $S_{\sim}$.

6°. Similarly, we can prove that the set $S_{\succ}$ is closed under addition, i.e., that if the tuples $Z = (Z_1, \dots, Z_n)$ and $Z' = (Z'_1, \dots, Z'_n)$ belong to the set $S_{\succ}$, then their component-wise sum

$$Z + Z' = (Z_1 + Z'_1, \dots, Z_n + Z'_n)$$

also belongs to the set $S_{\succ}$.

Indeed, by definition of the set $S_{\succ}$, the condition $Z \in S_{\succ}$ means that

$$(\exp(Z_1), \dots, \exp(Z_n)) \succ (1, \dots, 1).$$

Using scale-invariance with $\lambda_i = \exp(Z'_i)$, we conclude that

$$(\exp(Z_1) \cdot \exp(Z'_1), \dots, \exp(Z_n) \cdot \exp(Z'_n)) \succ (\exp(Z'_1), \dots, \exp(Z'_n)).$$

On the other hand, the condition $Z' \in S_{\succ}$ means that

$$(\exp(Z'_1), \dots, \exp(Z'_n)) \succ (1, \dots, 1).$$

Thus, due to transitivity of the strict preference relation $\succ$, we conclude that

$$(\exp(Z_1) \cdot \exp(Z'_1), \dots, \exp(Z_n) \cdot \exp(Z'_n)) \succ (1, \dots, 1).$$

Since for every i, we have $\exp(Z_i) \cdot \exp(Z'_i) = \exp(Z_i + Z'_i)$, we thus conclude that

$$(\exp(Z_1 + Z'_1), \dots, \exp(Z_n + Z'_n)) \succ (1, \dots, 1).$$

By definition of the set $S_{\succ}$, this means that the tuple $Z + Z'$ belongs to the set $S_{\succ}$.

7°. A similar argument shows that the set $S_{\prec}$ is closed under addition, i.e., that if the tuples $Z = (Z_1, \dots, Z_n)$ and $Z' = (Z'_1, \dots, Z'_n)$ belong to the set $S_{\prec}$, then their component-wise sum

$$Z + Z' = (Z_1 + Z'_1, \dots, Z_n + Z'_n)$$

also belongs to the set $S_<$.

8°. Let us now prove that the set $S_\sim$ is closed under the "unary minus" operation, i.e., that if $Z = (Z_1, \dots, Z_n) \in S_\sim$, then $-Z \stackrel{\text{def}}{=} (-Z_1, \dots, -Z_n)$ also belongs to $S_\sim$.

Indeed, $Z \in S_\sim$ means that

$$(\exp(Z_1), \dots, \exp(Z_n)) \sim (1, \dots, 1).$$

Using scale-invariance with $\lambda_i = \exp(-Z_i) = \dfrac{1}{\exp(Z_i)}$, we conclude that

$$(1, \dots, 1) \sim (\exp(-Z_1), \dots, \exp(-Z_n)),$$

i.e., that $-Z \in S_\sim$.

9°. Let us prove that if $Z = (Z_1, \dots, Z_n) \in S_>$, then $-Z \stackrel{\text{def}}{=} (-Z_1, \dots, -Z_n)$ belongs to $S_<$.

Indeed, $Z \in S_>$ means that

$$(\exp(Z_1), \dots, \exp(Z_n)) > (1, \dots, 1).$$

Using scale-invariance with $\lambda_i = \exp(-Z_i) = \dfrac{1}{\exp(Z_i)}$, we conclude that

$$(1, \dots, 1) > (\exp(-Z_1), \dots, \exp(-Z_n)),$$

i.e., that $-Z \in S_<$.

Similarly, we can show that if $Z \in S_<$, then $-Z \in S_>$.

10°. From Part 5 of this proof, it now follows that if $Z = (Z_1, \dots, Z_n) \in S_\sim$, then $Z + Z \in S_\sim$, that $Z + (Z + Z) \in S_\sim$, etc., i.e., that for every positive integer p, the tuple

$$p \cdot Z = (p \cdot Z_1, \dots, p \cdot Z_n)$$

also belongs to the set $S_\sim$.

By using Part 8, we can also conclude that this is true for negative integers p as well. Finally, by taking into account that the zero tuple $0 \stackrel{\text{def}}{=} (0, \dots, 0)$ can be represented as $Z + (-Z)$, we conclude that $0 \cdot Z = 0$ also belongs to the set $S_\sim$.

Thus, if a tuple Z belongs to the set $S_\sim$, then for every integer p, the tuple $p \cdot Z$ also belongs to the set $S_\sim$.

11°. Similarly, from Parts 6 and 7 of this proof, it follows that

- if $Z = (Z_1, \dots, Z_n) \in S_>$, then for every positive integer p, the tuple $p \cdot Z$ also belongs to the set $S_>$, and

- if $Z = (Z_1, \ldots, Z_n) \in S_<$, then for every positive integer p, the tuple $p \cdot Z$ also belongs to the set $S_<$.

$12°$. Let us prove that for every rational number $r = \dfrac{p}{q}$, where p is an integer and q is a positive integer, if a tuple Z belongs to the set $S_\sim$, then the tuple $r \cdot Z$ also belongs to the set $S_\sim$.

Indeed, according to Part 10, $Z \in S_\sim$ implies that $p \cdot Z \in S_\sim$.

According to Part 4, for the tuple $r \cdot Z$, we have either $r \cdot Z \in S_\sim$, or $r \cdot Z \in S_>$, or $r \cdot Z \in S_<$.

- If $r \cdot Z \in S_>$, then, by Part 11, we would get $p \cdot Z = q \cdot (r \cdot Z) \in S_>$, which contradicts our result that $p \cdot Z \in S_\sim$.
- Similarly, if $r \cdot Z \in S_<$, then, by Part 11, we would get $p \cdot Z = q \cdot (r \cdot Z) \in S_<$, which contradicts our result that $p \cdot Z \in S_\sim$.

Thus, the only remaining option is $r \cdot Z \in S_\sim$. The statement is proven.

$13°$. Let us now use continuity to prove that for every real number x, if a tuple Z belongs to the set $S_\sim$, then the tuple $x \cdot Z$ also belongs to the set $S_\sim$.

Indeed, a real number x can be represented as a limit of rational numbers: $r^{(k)} \to x$. According to Part 12, for every k, we have $r^{(k)} \cdot Z \in S_\sim$, i.e., the tuple $Z^{(k)} \overset{\text{def}}{=} (\exp(r^{(k)} \cdot Z_1), \ldots, \exp(r^{(k)} \cdot Z_n)) \sim (1, \ldots, 1)$. In particular, this means that $Z^{(k)} \geq (1, \ldots, 1)$. In the limit, $Z^{(k)} \to (\exp(x \cdot Z_1), \ldots, \exp(x \cdot Z_n)) \geq (1, \ldots, 1)$. By definition of the sets $S_\sim$ and $S_>$, this means that $x \cdot Z \in S_\sim$ or $x \cdot Z \in S_>$.

Similarly, for $-(x \cdot Z) = (-x) \cdot Z$, we conclude that $-x \cdot Z \in S_\sim$ or $(-x) \cdot Z \in S_>$. If we had $x \cdot Z \in S_>$, then by Part 9 we would get $(-x) \cdot Z \in S_<$, a contradiction. Thus, the case $x \cdot Z \in S_>$ is impossible, and we have $x \cdot Z \in S_\sim$. The statement is proven.

$14°$. According to Parts 5 and 13, the set $S_\sim$ is closed under addition and under multiplication by an arbitrary real number. Thus, if tuples $Z, \ldots, Z'$ belong to the set $S_\sim$, their arbitrary linear combination $x \cdot Z + \cdots + x' \cdot Z'$ also belongs to the set $S_\sim$. So, the set $S_\sim$ is a linear subspace of the n-dimensional space of all the tuples.

$15°$. The subspace $S_\sim$ cannot coincide with the entire n-dimensional space, because then the pre-ordering relation would be trivial. Thus, the dimension of this subspace must be less than or equal to $n - 1$. Let us show that the dimension of this subspace is $n - 1$.

Indeed, let us assume that the dimension is smaller than $n - 1$. Since the pre-ordering is non-trivial, there exist tuples $y = (y_1, \ldots, y_n)$ and $y' = (y'_1, \ldots, y'_n)$ for which $y > y'$ and thus, $Z = (Z_1, \ldots, Z_n) \in S_>$, where $Z_i = \ln\left(\dfrac{y_i}{y'_i}\right)$. From $Z \in S_>$, we conclude that $-Z \in S_<$.

Since the linear space $S_\sim$ is a less than $(n - 1)$-dimensional subspace of an n-dimensional linear space, there is a path connecting $Z \in S_>$ and $-Z \in S_<$ which avoids $S_\sim$. In mathematical terms, this path is a continuous mapping $\gamma : [0, 1] \to R^n$

for which $\gamma(0) = Z$ and $\gamma(1) = -Z$. Since this path avoids $S_\sim$, every point $\gamma(t)$ on this path belongs either to $S_>$ or to $S_<$.

Let $\bar{t}$ denote the supremum (least upper bound) of the set of all the values t for which $\gamma(t) \in S_>$. By definition of the supremum, there exists a sequence $t^{(k)} \to \bar{t}$ for which $\gamma(t^{(k)}) \in S_>$. Similarly to Part 13, we can use continuity to prove that in the limit, $\gamma(\bar{t}) \in S_>$ or $\gamma(\bar{t}) \in S_\sim$. Since the path avoids the set $S_\sim$, we thus get $\gamma(\bar{t}) \in S_>$.

Similarly, since $\gamma(1) \notin S_>$, there exists a sequence $t^{(k)} \downarrow \bar{t}$ for which $\gamma(t^{(k)}) \in S_<$. We can therefore conclude that in the limit, $\gamma(\bar{t}) \in S_>$ or $\gamma(\bar{t}) \in S_\sim$—a contradiction with our previous conclusion that $\gamma(\bar{t}) \in S_>$.

This contradiction shows that the linear space $S_\sim$ cannot have dimension $< n - 1$ and thus, that this space have dimension $n - 1$.

16°. Every $(n - 1)$-dimensional linear subspace of an n-dimensional superspace separates the superspace into two half-spaces. Let us show that one of these half-spaces is $S_>$ and the other is $S_<$.

Indeed, if one of the subspaces contains two tuples Z and Z' for which $Z \in S_>$ and $Z' \in S_<$, then the line segment $\gamma(t) = t \cdot Z + (1 - t) \cdot Z'$ containing these two points also belongs to the same subspace, i.e., avoids the set $S_\sim$. Thus, similarly to Part 15, we would get a contradiction.

So, if one point from a half-space belongs to $S_>$, all other points from this subspace also belong to the set $S_>$. Similarly, if one point from a half-space belongs to $S_<$, all other points from this subspace also belong to the set $S_<$.

17°. Every $(n - 1)$-dimensional linear subspace of an n-dimensional space has the form $\alpha_1 \cdot Z_1 + \cdots + \alpha_n \cdot Z_n = 0$ for some real values α_i, and the corresponding half-spaces have the form $\alpha_1 \cdot Z_1 + \cdots + \alpha_n \cdot Z_n > 0$ and $\alpha_1 \cdot Z_1 + \cdots + \alpha_n \cdot Z_n < 0$.

The set $S_>$ coincides with one of these subspaces. If it coincides with the set of all tuples Z for which $\alpha_1 \cdot Z_1 + \cdots + \alpha_n \cdot Z_n < 0$, then we can rewrite it as

$$(-\alpha_1) \cdot Z_1 + \cdots + (-\alpha_n) \cdot Z_n > 0,$$

i.e., as $\alpha'_1 \cdot Z_1 + \cdots + \alpha'_n \cdot Z_n > 0$ for $\alpha'_i = -\alpha_i$.

Thus, without losing generality, we can conclude that the set $S_>$ coincides with the set of all the tuples Z for which $\alpha_1 \cdot Z_1 + \cdots + \alpha_n \cdot Z_n > 0$. We have mentioned that

$$y' = (y'_1, \ldots, y'_n) > y = (y_1, \ldots, y_n) \Leftrightarrow (Z_1, \ldots, Z_n) \in S_>,$$

where $Z_i = \ln\left(\dfrac{y'_i}{y_i}\right)$. Thus,

$$y' > y \Leftrightarrow \alpha_1 \cdot Z_1 + \cdots + \alpha_n \cdot Z_n = \alpha_1 \cdot \ln\left(\frac{y'_1}{y_1}\right) + \cdots + \alpha_n \cdot \ln\left(\frac{y'_n}{y_n}\right) > 0.$$

Since $\ln\left(\dfrac{y'_i}{y_i}\right) = \ln(y'_i) - \ln(y_i)$, the last inequality is equivalent to

$$\alpha_1 \cdot \ln(y_1') + \cdots + \alpha_n \cdot \ln(y_n') > \alpha_1 \cdot \ln(y_1) + \cdots + \alpha_n \cdot \ln(y_n).$$

Let us take exp of both sides; then, due to the monotonicity of the exponential function, we get an equivalent inequality

$$\exp(\alpha_1 \cdot \ln(y_1') + \cdots + \alpha_n \cdot \ln(y_n')) > \exp(\alpha_1 \cdot \ln(y_1) + \cdots + \alpha_n \cdot \ln(y_n)).$$

Here,

$$\exp(\alpha_1 \cdot \ln(y_1') + \cdots + \alpha_n \cdot \ln(y_n')) = \exp(\alpha_1 \cdot \ln(y_1')) \cdot \cdots \cdot \exp(\alpha_n \cdot \ln(y_n')),$$

where for every i, $e^{\alpha_i \cdot z_i} = (e^{z_i})^{\alpha_i}$, with $z_i \overset{\text{def}}{=} \ln(y_i')$, implies that

$$\exp(\alpha_i \cdot \ln(y_i')) = (\exp(\ln(y_i')))^{\alpha_i} = (y_i')^{\alpha_i},$$

so

$$\exp(\alpha_1 \cdot \ln(y_1') + \cdots + \alpha_n \cdot \ln(y_n')) = (y_1')^{\alpha_1} \cdot \cdots \cdot (y_n')^{\alpha_n}$$

and similarly,

$$\exp(\alpha_1 \cdot \ln(y_1) + \cdots + \alpha_n \cdot \ln(y_n)) = y_1^{\alpha_1} \cdot \cdots \cdot y_n^{\alpha_n}.$$

Thus, the condition $y' > y$ is equivalent to

$$\prod_{i=1}^{n} y_i^{\alpha_i} > \prod_{i=1}^{n} (y_i')^{\alpha_i}.$$

Similarly, we prove that

$$(y_1, \ldots, y_n) \sim y' = (y_1', \ldots, y_n') \Leftrightarrow \prod_{i=1}^{n} y_i^{\alpha_i} = \prod_{i=1}^{n} (y_i')^{\alpha_i}.$$

The condition $\alpha_i > 0$ follows from our assumption that the pre-ordering is monotonic.

18°. Now, from permutation-invariance, we conclude that all the coefficients α_i are equal to each other: $\alpha_i = \alpha$ for all i. Thus, the condition $y' > y$ is equivalent to

$$\prod_{i=1}^{n} y_i^{\alpha} = \left(\prod_{i=1}^{n} y_i \right)^{\alpha} > \prod_{i=1}^{n} (y_i')^{\alpha} = \left(\prod_{i=1}^{n} y_i' \right)^{\alpha},$$

i.e., to

$$\prod_{i=1}^{n} y_i > \prod_{i=1}^{n} y_i'.$$

Similarly, we can prove that the condition $y' \sim y$ is equivalent to

$$\prod_{i=1}^{n} y_i^{\alpha} = \left(\prod_{i=1}^{n} y_i \right)^{\alpha} = \prod_{i=1}^{n} (y_i')^{\alpha} = \left(\prod_{i=1}^{n} y_i' \right)^{\alpha},$$

i.e., to

$$\prod_{i=1}^{n} y_i = \prod_{i=1}^{n} y_i'.$$

This completes the proof of our main result.

References

1. Fishburn PC (1969) Utility theory for decision making. Wiley, New York
2. Jaimes A, Tweedy C, Magoc T, Kreinovich V, Ceberio M (2010) Selecting the best location for a meteorological tower: a case study of multi-objective constraint optimization. J Uncertain Syst 4(4):261–269
3. Klir G, Yuan B (1995) Fuzzy sets and fuzzy logic. Prentice Hall, Upper Saddle River, New Jersey
4. Luce RD, Raiffa R (1989) Games and decisions: introduction and critical survey. Dover, New York
5. Nash J (1953) Two-person cooperative games. Econometrica 21:128–140
6. Nguyen HT, Kosheleva O, Kreinovich V (2009) Decision making beyond Arrow's 'impossibility theorem', with the analysis of effects of collusion and mutual attraction. Int J Intell Syst 24(1):27–47
7. Nguyen HT, Kreinovich V (1997) Applications of continuous mathematics to computer science. Kluwer, Dordrecht
8. Nguyen HT, Walker EA (2006) A first course in fuzzy logic. Chapman and Hall/CRC, Boca Raton, Florida
9. Raiffa H (1970) Decision analysis. Addison-Wesley, Reading, Massachusetts
10. Zadeh LA (1965) Fuzzy sets. Inf Control 8:338–353

An Overview and Re-interpretation of Paradoxes of Responsiveness

Hannu Nurmi

Abstract One of the most obvious desiderata of democratic decision-making is that the political outcomes (policies, elected persons, legislation) be responsive to popular opinions. In representative forms of governance the responsiveness is not expected to pertain to every single outcome, but the very idea of going to the people seems to presuppose some degree of responsiveness. In social choice theory several notions that aim to capture aspects of responsiveness have been introduced and related to other desiderata of social choice. We shall discuss the most common notions and discuss their relevance in democratic decision making. We shall also look at the paradoxes related to non-responsiveness from a novel angle, viz. we try to determine their significance to the multiple criteria decision making (MCDM). It turns out that some methods of aggregating criterion performances of policy alternatives can be ruled out because of their bizarre behavior under some decision settings.

1 Introduction

Responsiveness is one of the most obvious desiderata in democratic rule. At the very least unresponsive rules of governing are certainly not acceptable as the very idea of democracy presupposes that the ruled, the people, can, by expressing by their opinions in legitimate manner, bring about changes in the way public policies are formulated and executed. Elections are the normal institutions to transform the popular views into public policies or other electoral outcomes. The most common of the latter are, of course, those that pertain to composition of parliaments or offices of the president. But what does unresponsiveness, then, mean? A clear example of an unresponsive voting rule is a constant one which results in a fixed outcome, say x, regardless of the distribution of the expressed opinions by the voters. I.e. no

H. Nurmi (✉)
Department of Philosophy, Contemporary History and Political Science,
University of Turku, Turku, Finland
e-mail: hnurmi@utu.fi
URL: http://users.utu.fi/hnurmi/homepage

© Springer International Publishing AG 2018

M. Collan and J. Kacprzyk (eds.), *Soft Computing Applications for Group Decision-making and Consensus Modeling*, Studies in Fuzziness and Soft Computing 357, DOI 10.1007/978-3-319-60207-3_2

matter how the voters vote, x always wins. Clearly, constant rules make the act of voting meaningless in the instrumental sense, that is, as a way of influencing the way public policies are to be pursued. A similarly minimalistic way of defining responsiveness as the exclusion of constant rules is the condition on rules known as citizens' sovereignty. This requires that, given a set of alternatives A of k alternatives and any ranking R of those alternatives, there is a distribution of voter opinions over those alternatives so that R is the outcome resulting from the application of the rule to this distribution. This condition excludes blatant discriminations against some alternatives. This doesn't mean that rules that satisfy citizens' sovereignty are *eo ipso* intuitively responsive to the voter opinions. A case in point is the unanimity rule: the *status quo* alternative, say x, is selected, unless all individuals prefer another alternative, say y, to x. This rule clearly satisfies citizens' sovereignty, but is extremely biased towards the *status quo*.

In what follows we shall investigate some intuitively natural forms of responsiveness of choice rules. The forms will be looked upon as invulnerability to certain kinds of paradoxes. Our primary focus is on variations in the choice sets resulting from rules under changes in individual opinions. Two types of settings are of interest: first, those where the changes in individual opinions happen in a fixed electorate, and second, those where the changes involve enlarging the electorate itself by including new voters in the voter set. The former settings will be called fixed electorate paradoxes and properties, while the latter will be called paradoxes and properties in variable electorates.

2   What Is Responsiveness?

A natural way of approaching the responsiveness problem is to start from comparing the opinions of the electorate to the result of the choice rule. The question then becomes, how well or accurately the latter represents the former. In any given choice situation we could argue that the better the choice result represents the voter opinions, the more responsive the rule. It turns out that nearly all voting rules that transform n-tuples of individual complete and transitive preference relations (rankings) (n being the number of voters) into collective rankings can be seen as the most optimal, i.e. most responsive, rules. What makes them different is their underlying idea of a consensus state, that is, a situation involving no disagreement as to the outcome and the distance metric used in measuring the distance of any preference profile from a consensus state. Such a consensus state can be one where all voters have identical rankings over the alternatives or one where all voters rank the same alternative first or one where a given alternative is the Condorcet winner. Similarly, the distance measure can be the inversion metric counting the number of binary inversions of adjacent alternatives needed to transform one ranking to another or a discrete metric that simply counts those rankings that differ in some respects from one another (see [14, 18]). More precisely, the inversion distance between two rankings R_1 and

R_2 over k alternatives is the smallest number of swaps of two adjacent alternatives required to transform R_1 to R_2.

If the consensus state is one where all voters have an identical ranking over alternatives and if the distance between any two rankings is measured by the inversion metric, then the outcomes ensuing from the application of Kemeny's rule are optimal in the sense of minimizing the distance between the observed profile and the desired consensus state. Similarly, it has been shown by Nitzan that the Borda count outcome represents best the voter opinions if the distance measure is the inversion metric and if the consensus state is one where all voters are unanimous about which alternative should be ranked first [18]. Plurality voting, in turn, can be seen as the optimal representation of the voters' opinions if the distance measure is the discrete metric and the consensus state is the same as in the Borda count. The discrete distance between R_1 and R_2 is defined to be zero if $R_1 = R_2$ and unity, otherwise. Similarly, most voting systems can be defined as optimal distance minimizing rules (see [3, 5, 12, 14]).

Looking at voting rules as distance minimizing devices from a consensus state reveals essential similarities and differences in their underlying motivation. The picture that emerges from this comparison is, however, purely static: the state of consensus—understood in various senses—is being compared with the observed profile of reported preferences of voters. The voting outcome is 'a response' to 'the stimulus' provided by the preference profile. The reasonableness of the response boils down to the plausibility of the consensus states and distance measures associated with various rules. A more nuanced picture of responsiveness of rules emerges when we compare the responses or outcomes of rules under various changes in the stimuli, i.e. preference profiles.

3 Responsiveness in Fixed Electorates: Monotonicity

The very idea of going to the people would seem to imply that the more voters support an alternative, the better chances the latter has for becoming the chosen one. Expressed in this way, the idea allows for several non-equivalent specifications. Firstly, it may mean that if the number of voters supporting an alternative is increased, then the alternative has at least as large a probability of being elected than before the increase. But what do we mean by 'support of an alternative'? At least three different interpretations are possible:

- the number of voters ranking the alternative first is increased,
- the position of the alternative is improved *vis-à-vis* some others, or
- some voters who rank the alternative first join the original electorate.

In fixed electorates, the third interpretation is excluded. In fact, the second interpretation is most common in the theory of voting. It can be further divided into two main concepts: (i) monotonicity, and (ii) Maskin monotonicity. According to the former, the additional support for an alternative means that its position is improved in at least one voter's preference ranking, *ceteris paribus*, i.e. the positions of all other alternatives with respect to each other remain the same. In Maskin monotonicity, in contrast, the additional support means that the position of an alternative is improved *vis-à-vis* some other alternatives, but no restrictions are imposed in the mutual positions of other alternatives.

With these distinctions in mind we can define the best-known responsiveness concept, monotonicity as follows (cf. [10]).

Definition 1 Upward monotonicity. Suppose that in a given profile over a set A of alternatives, $x \in A$ wins when rule D is applied. Suppose now that the profile is modified so that the position of x is improved, *ceteris paribus*, in at least one voter's preference ranking. Now, D is monotonic if and only if x remains the winner in the modified profile.

More recently, Miller has suggested another monotonicity concept well in the spirit of the preceding one. He calls it downward monotonicity, in contradistinction to the above which he calls upward monotonicity [15].

Definition 2 Downward monotonicity. Suppose that in a profile over a set of alternatives A, $x \in A$ wins when D is applied. Suppose moreover that a group of voters change their mind and lower the position of another alternative, y, *ceteris paribus*. Then D is downward monotonic if and only if no such change makes y the winner in the modified profile.

The *ceteris paribus* proviso is essential here. In fact, it is the only thing that distinguishes the upward monotonicity concept from Maskin monotonicity. The latter is defined as follows (cf. [13]).

Definition 3 Suppose that in a given profile over a set A of alternatives, $x \in A$ wins when rule D is applied. Suppose now that the profile is modified so that the position of x with respect to any other alternative y is at least as high in all voters' ranking and perhaps strictly higher for some $z \in A$ and some individuals. Now, D is Maskin monotonic if and only if x remains the winner in the modified profile.

Although *prima facie* the definitions are not very different, their difference is quite dramatic when we apply them to the most common voting procedures. It turns out that none of them is Maskin monotonic, while many are monotonic. In this regard Maskin monotonicity resembles the well-known independence of irrelevant alternatives condition of Arrow's impossibility theorem. For a thorough discussion on the significance of Maskin monotonicity in implementation and choice theory, the reader is referred to [1, 2].

The difference in the two definitions above is illustrated in terms of plurality voting in Table 1. The plurality winner there is x. Suppose that the profile is modified so that y is lifted ahead of z by the voter who ranked z first and y is lifted ahead of both w and z by the voter represented by the right-hand column. These changes do not involve x. Moreover, suppose that the position of x is improved by lifting it ahead of z in the second column from the left and ahead of z and w in the right-most column. So, for the rule to be Maskin monotonic, x would have to remain the winner

Table 1 Plurality voting is not Maskin monotonic

2 voters	1 voter	1 voter	1 voter
x	y	z	w
y	z	y	z
z	x	x	y
w	w	w	x

in the modified profile as well. However, after these modifications the plurality voting elects y since it is ranked first by three voters out of five in the modified profile. So, the plurality voting is not Maskin monotonic. On the other hand, it is obvious that the plurality voting is upward monotonic since lifting the winner ahead of some other alternatives in a profile, *ceteris paribus*, either leaves the number of voters ranking each alternative first unchanged (if the modifications pertain to alternatives ranked lower than first both in the original and modified profile) or increases the number of voters ranking the original winner first by some positive number. So— since no other alternative will be ranked first by more voters than originally—the original winner remains the winner also in the modified profile. For an analysis of some other systems in terms of Maskin monotonicity see [19].

Clearly, if a rule is Maskin monotonic, it is also monotonic, but the converse is not true as the case of plurality voting demonstrates. Overall, the primary significance of the Maskin monotonicity is in implementation theory where it has been shown to be a necessary condition for Nash-implementation (see e.g. [2]). The well-known Muller-Satterthwaite theorem states that all weakly unanimous and Maskin monotonic choice rules are dictatorial [17]. Weak unanimity is also known as Pareto principle: if every voter strictly prefers x to y, then y is not chosen. (For a slightly different formulation of the theorem, see [2]). Obviously, Maskin monotonic and Pareto optimal voting rules—if they exist—are in the dubious company of dictatorial rules. Hence, to avoid confusion, the distinction between monotonic and Maskin monotonic rules should be made explicitly.

It is well-known that the plurality runoff as well as alternative vote are non-monotonic. In fact, they are non-monotonic in a very strong sense: there are profiles where additional support for a winning candidate may turn it into a non-winner *and* diminishing support for a candidate may render it a winner even though it wasn't one in the original profile. Miller calls this a double monotonicity failure and provides the following example (Table 2) [15].

Table 2 Plurality runoff and double monotonicity failure [15]

38 voters	32 voters	30 voters
y	x	z
z	y	x
x	z	y

Here x wins. Suppose now that 9–17 voters from the left-most group lift x ahead of both y and z, *ceteris paribus*. Then, the runoff takes place between x and z, whereupon z wins. This is just another instance of the upward monotonicity failure, but there is another one as well in the same profile. To wit, let three voters of the left-most group drop y to the second place, *ceteris paribus*. Then, x is dropped out of the runoff contest and the winner is y. Hence we have an instance of the double monotonicity failure.

In the definition of upward monotonicity the starting point is a setting where a winner's position is improved, *ceteris paribus*, and one finds out whether the improvement is necessarily accompanied with the winner maintaining its status. One could envisage another, more general, notion of monotonicity whereby an improvement of an alternative's position in some rankings, *ceteris paribus*, is never accompanied with a lower rank for it in the social ranking. This notion would cover not only situations where the focus is on what happens to the winning alternative once its position is improved, but also situations where one looks at the position of non-winners upon an improvement in their ranking in individual preferences, *ceteris paribus*. This more general notion of monotonicity is applicable to social welfare functions, while the above definition applies to social choice functions. Since above definition is by now standard we shall, however, adhere to it.

4 Responsiveness in Variable Electorates: The No-Show Paradox

What the general formulation of monotonicity in the end boils down to is that an improvement of an alternatives position *vis-à-vis* the others, *ceteris paribus*, never lowers the position of the alternative in the collective ranking. This concerns fixed electorates, i.e. those where the modifications occur within the same electorate. In variable electorates, in contrast, the electorate is assumed to expand as a result of new voters joining it. The responsiveness of the voting rule is then typically determined by whether or not it satisfies the property called participation. This property is defined by means of the no-show paradox. The latter occurs whenever a group of identically-minded voters is better off—in terms of the voting outcome—when it abstains than when it votes according to its preferences. Thus, the conclusion that an instance of the no-show paradox occurs is based on comparing two outcomes: one, say O_1, resulting from the application of a given procedure, F, to a set A of alternatives and to a profile R of the set N of voters over A, and the other, say O_2, resulting from applying F to A and to a profile that consists of R augmented by a group of voters each having identical preferences over A. Whenever the added voters prefer O_1 to O_2 we have an instance of the no-show paradox.

This definition includes, as special cases, the two types of no-show paradoxes outlined by Fishburn and Brams [11] (see also [8, 9]). To wit,

Table 3 Plurality runoff and no-show paradox

26 voters	47 voters	2 voters	25 voters
x	y	y	z
z	z	z	x
y	x	x	y

Definition 4 No-show paradox 1: The addition of identical ballots with x ranked last may change the winner from another candidate to x.

And the other type:

Definition 5 No-show paradox 2: One of the candidates elected could have ended a loser if additional people who ranked him in first place had actually voted.

These are clearly non-equivalent definitions with a common feature: the added voters are better off not voting in both cases (see also [16]). In the former definition the fact that they vote according to their preferences brings about the outcome that is their worst, while without their votes, the outcome would have been something more preferable. In the latter definition, the added voters' votes to their most preferred candidate turn their favorite into a non-winner, whereas with their abstaining, it would have won. The non-equivalence of the definitions is demonstrated by the plurality runoff (a.k.a. alternative vote or instant runoff) and the following 100-voter profile (Table 3).

Let us first assume that the group of 47 voters in the second column do not vote at all. Since there are then only 53 voters, none of the three candidates gets elected on the first round. Instead, the second round is arranged between x and z, whereupon z wins with 27 votes against 26. Suppose now that the 47-voter group joins the competition. Its lowest ranked alternative is x. With this group joining the profile is the one depicted in Table 3. Now the second round involves x and y resulting in the victory of x, the worst alternative of the 47-voter group. Thus, the situation is one described Definition 4. On the other hand, the conditions of Definition 5 cannot apply to the plurality runoff system. The reason is the following. Let x be the winner in the original profile and add a group of voters with identical preferences with x ranked first to obtain the augmented profile. Adding the group does not affect the distribution of voters who rank other alternatives first. Thus, whichever alternative was the runoff competitor of x in the original profile, will be its competitor in the augmented profile. Since x defeated its runoff competitor in the original profile, it will defeat it in the augmented one as well since its support has been increased, while that of its competitor hasn't. If x won in the first round in the original profile—i.e. no runoff was required—it clearly does so in the augmented one as well (see also [9]). Thus, Definitions 4 and 5 are non-equivalent.

The case of plurality runoff is instructive in another way as well, viz. it shows that although apparently similar the properties of non-monotonicity and vulnerability to no-show paradox are not identical. Plurality runoff is upward non-monotonic, but not vulnerable to the no-show paradox in the sense of Definition 5. It turns out that—although seemingly closely related—upward non-monotonicity and vulnerability to the no-show paradox in the sense of Definition 5 are largely logically independent: there are systems that are vulnerable and monotonic in this sense and there are vulnerable ones that are non-monotonic. An example of the former combination is Copeland's rule and of the latter, as was just seen in Table 3, the plurality runoff and alternative vote. Of systems that are invulnerable to the no-show paradox in the sense of Definition 5 and are upward monotonic one can mention the Borda count and plurality voting [19]. Finally, Campbell and Kelly provide a constructive proof that there are systems that are upward non-monotonic, but at the same time invulnerable to the no-show paradox in the sense Definition 5 [4].

5 Extreme Forms

The no-show paradoxes can take on various degrees of severity. In particular, it may turn out that the alternative ranked first by a sub-group of unanimous voters will be a winner if they abstain, but a non-winner if they vote according to their preferences, *ceteris paribus*. So, by voting this group may turn their favorite from a winner to a non-winner. This is called the P-TOP paradox by Felsenthal and Tideman [7]. This is obviously the extreme form of Definition 5. It is also known as the strong no-show paradox [20].

The extreme form of Definition 4, on the other hand, occurs when a group of unanimous voters voting according to their preferences where x is ranked lowest, *ceteris paribus*, brings about an outcome where x wins, while, had the group abstained, some other alternative would have won. So, by voting the group changes the outcome from something that is not their worst to something that is. This paradox is called the P-BOT paradox [7]. Table 4 illustrates the strong no-show paradox under Copeland's rule. The example has been originally devised by Fishburn and thereafter utilized by Richelson as well as by Felsenthal and Nurmi [8, 10, 21].

Here v defeats more alternatives in pairwise comparisons (with the majority rule) than any other alternative and is thus the Copeland winner. Suppose we add a voter with the preference ranking $vwxyz$ to the original profile of Table 4. This modification leaves v's Copeland score unchanged, but makes w the Condorcet, and hence

Table 4 Copeland's rule and the P-TOP paradox

2 voters	1 voter	1 voter
w	z	v
v	y	z
x	x	y
y	w	x
z	v	w

Table 5 Copeland's rule and the P-BOT paradox

5 voters	4 voters
y	z
z	v
v	x
x	y

Copeland, winner. Thus, adding a voter ranking the original winner first to the profile promotes another alternative ahead of the original winner.

Table 5 illustrates the vulnerability of Copeland's rule to the other extreme form, the P-BOT paradox.

As y is the strong Condorcet winner, it is *eo ipso* the Copeland winner. If we now add three voters all having the the ranking: $xvyz$, this makes the 12-vote profile cyclic in terms of the majority comparisons. The Copeland winners are now z and v. Thus we have a weak version of the P-BOT paradox whereby the alternative ranked last by the added voters belongs to the choice set in the augmented profile, whereas it was not in the choice set in the original one.

Although we have used just Copeland's rule in the illustration of the paradoxes, it turns out that the P-TOP and P-BOT paradoxes are quite common among voting rules. Indeed, Pérez has shown that nearly all Condorcet extensions are vulnerable to either one or both extreme forms of the no-show paradox, the only exceptions being the Minmax and Young's rules [20]: the former is invulnerable to both P-TOP and P-BOT, while the latter is invulnerable to the P-BOT one [8].

6 The MCDM Context

The vulnerability to monotonicity failures is typically discussed in the context of voting. There these failures confront the voters with contradictory incentives. On the one hand, voting for one's favorite would seem precisely what is needed to increase the probability of his/her getting elected. On the other hand, depending on the procedure used the voting might jeopardize the favorite's chances of being elected or—worse still—might lead to the election of the worst possible candidate. But monotonicity failures can play a role in MCDM as well. To wit, let us assume that we have a choice situation involving n criteria and ordinal measurements of the performance of the k decision alternatives on those criteria. To make the decision one needs a rule which allows the determination of the best alternative or priority ranking over all alternatives to make a decision. The voting rules discussed above can then be used to find a solution to the choice problem. Would this new context affect the significance of the monotonicity failures?

Arguably not. If a non-monotonic aggregation of criterion measurements is being resorted to, this would mean that improving an alternative's position on some criteria *ceteris paribus* might exclude it from the set of chosen alternatives even though it would have been chosen had the improvement not been made. This would apply to upward non-monotonic systems. Similarly, in downward non-monotonic systems the worsening of an alternative's measurement value could lead to its choice. Hence, the use of a (downward or upward) non-monotonic criterion aggregation rule could lead to bizarre outcomes.

The same conclusion holds for monotonicity failures in variable criterion sets. Introducing new criteria—in itself a very common occurrence—may result in strange choices. Table 4 can be viewed from the MCDM angle as an illustration: with four criteria and measurements depicted in the table Copeland's rule results in v. However, if another criterion on which v is on top is added, another alternative—here w—is chosen. To make the case somewhat more concretely, suppose the alternatives are some devices (e.g. fighter jets) with several essential technical qualities (maximum speed, fuel consumption, agility, easiness of service) along which the order of preference can be formed. Suppose that five alternatives can be placed in the order of priority along these technical criteria as in Table 4. Using Copeland's rule v is chosen. Then someone suggests another criterion, overall cost of purchase and maintenance, and it turns out that v is best on this new criterion. If Copeland's rule is used in the new setting of five criteria, w emerges as the winner. This is undoubtedly somewhat counterintuitive.

Equally if not more bizarre is the setting exhibited in Table 5. In the original setting 5 out of 9 criteria place y at the top and by Copeland's rule it is chosen. If three criteria are added as described so that on all of them z is ranked last, the new choice set includes this last ranked alternative.

Although Copeland's rule has been used as an example above, these anomalies characterize many other other rules based on pairwise comparison of alternatives. In particular, it applies to most Condorcet extensions as was pointed out by Pérez. Using these rules in aggregation of criterion measurements is thus questionable.

References

1. Aleskerov FT (1999) Arrovian aggregation models. D. Reidel, Dordrecht
2. Aleskerov FT (2002) Categories of Arrovian voting schemes. In: Arrow KJ, Sen AK, Suzumura K (eds) Handbook of social choice and welfare, vol 1. Elsevier, Amsterdam, pp 95–129
3. Baigent N (1987) Metric rationalisation of social choice functions according to principles of social choice. Math Soc Sci 13:59–65
4. Campbell DE, Kelly JS (2002) Nonmonotonicity does not imply the no-show paradox. Soc Choice Welfare 19:513–515
5. Eckert D, Klamler C (2011) Distance-based aggregation theory. In: Herrera-Viedma E, García-Lapresta JL, Kacprzyk J, Fedrizzi M, Nurmi H, Zadrożny S (eds) Consensual processes. Springer, Berlin, pp 3–22
6. Elkind E, Faliszewski P, Slinko A (2012) Rationalizations of Condorcet-consistent rules via distances of hamming type. Soc Choice Welfare 39:891–905
7. Felsenthal DS, Tideman N (2013) Varieties of failure of monotonicity and participation under five voting methods. Theory Decis 75:59–77
8. Felsenthal DS, Nurmi H (2016) Two types of participation failures under nine voting methods in variable electorates. Pub Choice 168:115–135

9. Felsenthal DS, Nurmi H (2017) Monotonicity failures afflicting voting procedures for electing a single candidate. Springer, Cham
10. Fishburn PC (1977) Condorcet social choice functions. SIAM J Appl Math 33:469–489
11. Fishburn PC, Brams SJ (1983) Paradoxes of preferential voting. Math Mag 56:207–214
12. Klamler C (2005) Borda and Condorcet: some distance results. Theory Decis 59:97–109
13. Maskin E (1985) The theory of implementation in Nash equilibrium. In: Hurwicz L, Schmeidler D, Sonnenschein H (eds) Social goals and social organization: essays in memory of Elisha Pazner. Cambridge University Press, Cambridge, pp 173–203
14. Meskanen T, Nurmi H (2006) Distance from consensus: a theme and variations. In: Simeone B, Pukelsheim F (eds) Mathematics and democracy. Recent advances in voting systems and collective choice. Springer, Berlin, pp 117–132
15. Miller N (2012) Monotonicity failure in IRV elections with three candidates. Presented at the second world congress of the public choice societies, Miami, FL. http://userpages.umbc.edu/~nmiller/MF&IRV.pdf. Accessed 8–11 Mar 2012
16. Moulin H (1988) Condorcets principle implies the no-show paradox. J Econ Theory 45:53–64
17. Muller E, Satterthwaite M (1977) The equivalence of strong positive association and strategy-proofness. J Econ Theory 14:412–418
18. Nitzan S (1981) Some measures of closeness to unanimity and their implications. Theory Decis 13:129–138
19. Nurmi H (2002) Voting procedures under uncertainty. Springer, Berlin
20. Pérez J (2001) The strong no show paradoxes are common flaw in Condorcet voting correspondences. Soc Choice Welfare 18:601–616
21. Richelson JT (1978) A comparative analysis of social choice functions III. Behav Sci 23:169–176

Veto in Yes-no and Yes-no-Abstain Voting Systems

Jacek Mercik

Abstract The paper presents a transformation from simple "yes-no" cooperative games to simple cooperative games where players have more than two actions available to them by introducing abstentions into a yes-no voting system. The results obtained up to now are rather pessimistic (Felsenthal and Machover, Power, voting, and voting power: 30 years after, Part II (2013) [6], even call them "the curious case of the absent abstention"). We discuss in this paper the relation between the right of veto, weights of the players and quotas. Our results clarify some general properties and enable an a priori analysis to gain a better understanding of the decision-making mechanism of such decisive bodies. Examples of the United Nations Security Council and Polish president-parliament cohabitation are used to illustrate our discussion.

1 Introduction

The legislative process in many countries provides for the possibility of the so-called veto, whose task is to strengthen the quality of legislation through the elimination of possible mistakes or imperfections, and on the other hand, in situations of considerable controversy, through strengthening the conditions of its acceptance (rejection of the veto, if permitted, usually requires a larger threshold of votes "for"). For example, the need to break (if permissible within the legal system) the veto results in the law not being passed by a simple majority, but by a qualified majority of legislators. This is the case in Poland, where the passing of a law by the Sejm usually requires a majority of over 50%, while overcoming the veto of the president in the Sejm requires 3/5 of votes "for". In the US, unlike in Poland, overcoming of the veto requires 2/3 of the votes "for" of both chambers.

There appears to be a side effect (so to speak) of granting the right of veto to one instance of the legislative process: its strengthening by increasing the influence of

J. Mercik (✉)
WSB University in Wroclaw, Wroclaw, Poland
e-mail: jacek.mercik@wsb.wroclaw.pl

© Springer International Publishing AG 2018
M. Collan and J. Kacprzyk (eds.), *Soft Computing Applications for Group Decision-making and Consensus Modeling*, Studies in Fuzziness and Soft Computing 357, DOI 10.1007/978-3-319-60207-3_3

this instance on the decision-making process, even when the veto is not actually used. We are considering here the decision-making process in general and not the law-making process in the parliament, albeit veto is universally associated with the so-called presidential veto. However, there are also some analogies in other decision-making processes. Examples include joint stock companies, where the holder of the so-called "golden share" can also stop the whole decision-making process, thus applying the veto in practice. In this case, the veto can not be rejected. It can be thus concluded that there are two types of veto: such that can not be rejected (i.e. veto of the first type) and such that can be rejected (i.e. veto of the second type).

One purpose of this article is to attempt to estimate the "measure" of veto for increasing the impact of the decision maker on the decision-making process. The measurement of the so-called power of veto will allow the estimation of the part of it that is not associated with correcting of legislation (decisions being taken), but is perhaps its unintended "side effect". The so-called power index is a tool to measure the power, including the power of veto. Classic power indices (Shapley-Shubik, Banzhaf and Johnston, to mention the most popular indices) are often modified, in the opinion of the authors, to better capture the feature. For example, one may observe the introduction of many modifications of the classical Shapley-Shubik power index (see, for example, O'Neil [18], Napel and Widgren [17] or Kuziemko and Werker [10]). In our paper, we do not assess which of these modifications are the best. It seems that the study of the effect of veto on the decision-making process needs no such modification.

From a theoretical point of view introduction of veto is a transformation from simple "yes-no" cooperative games to simple cooperative games where players have more than two actions available to them. Felsenthal and Machower [5] introduced the term "tertiary games", introducing abstentions into a yes-no voting system, which for example suits the case of the Security Council. Hence, they generalised the problem, which led to various attempts to better evaluate the role and power of members which are empowered with veto and doesn't empowered with veto in the same decisive body. For example, Tchantchoa et al. [22] analyse satisfaction, Freixas and Zwicker [7] multiple levels of approval and Grabisch and Lange [8] multichoice games; all for yes-no voting with abstention. The results obtained up to now are rather pessimistic: Felsenthal and Machover [6] even call them "the curious case of the absent abstention". Following this stream, we discuss in this paper the relation between the right of veto, weights of the players and quotas. Our results clarify some general properties and enable an a priori analysis to gain a better understanding of the decision-making mechanism of such decisive bodies.

The article is set up as follows. After introduction, the next section outlines the way in which decisions are modelled. This section presents preliminaries connected with the game-theoretical language of modelling and ways of calculating power indices for a simple voting game and how to take vetoes into account in these calculations. The next section presents the calculation of different power indices for games in which voters have equal voting rights, but some have an unconditional veto (voting yes/no). This section describes a procedure to define an equivalent voting game without vetoes in which players have different weights. After that, the next section presents the calculation of the Shapley-Shubik power index in games where voters have equal voting rights and there is yes-no-abstain voting. The last but one section presents the derivation of equivalent games based on a transformation of weights, quotas and the type of voting. Some general evaluations of the power of voting and non-voting players are presented together with examples of the United Nations Security Council and president-parliament system. Finally, there are some conclusions and suggestions for future research.

2 Preliminaries

Let N be a finite set of committee members, q be a quota and w_j be the voting weight of member j, where $j \in N$.

In this paper, we consider a special class of cooperative games called weighted majority games. A weighted majority game G is defined by a quota q and a sequence of nonnegative numbers w_i, $i \in n$, where we may think of w_i as the number of votes, or weight, of player i and q as the threshold, or quota, needed for a coalition to win. We assume that q and w_j are nonnegative integers. A subset of the players is called a coalition.

A game on N is given by a map v: $2^N \rightarrow R$ with $v(\varnothing) = 0$. The space of all games on N is denoted by G. A coalition $T \in 2^N$ is called a carrier of v if $v(S) = v(S \cap T)$ for any $S \in 2^N$. The domain $SG \subset G$ of simple games on N consists of all $v \in G$ such that

(i) $v(S) \in \{0, 1\}$ for all $S \in 2^N$;
(ii) $v(N) = 1$;
(iii) of v is monotonic, i.e. if $S \subset T$ then $v(S) \leq v(T)$.

A coalition S is said to be winning in $v \in SG$ if $v(S) = 1$ and losing otherwise. Therefore, passing a bill, for example, is equivalent to forming a winning coalition consisting of voters. A simple game (N, v) is said to be proper, if and only if the following is satisfied: for all $T \subset N$, if $v(T) = 1$ then $v(N \backslash T) = 0$.

We only analyse simple and proper games where players may vote either yes-no or yes-no-abstain, respectively.

If a given committee member can transform any winning coalition into a non-winning one by using a veto, then that veto is said to be of first degree.

If the veto of a given committee member turns some, but not all, winning coalitions not including that member into non-winning coalitions, then that veto is defined to be of second degree.

3   Measurement of the Power of Decision-Maker

It is generally accepted that measurement of the power of the decision maker, understood as his effect on the final result, is accomplished by means of the so-called power indices. They originate from simple game theory, where they were originally used to determine the division of the so-called payoff, or the value of the game.

We consider the weighted decision-making body of the size n = card{N} in which decisions are taken by vote, with a quorum γ, the sum of weights τ and allocation of weights $\omega = (\omega_1, \ldots, \omega_n)$. We assume that each i-th voter only votes "yes" or "no". Any non-empty subset of players $S \subseteq N$ we call voter setup. For a given allocation ω and quorum γ, we say that $S \subseteq N$ is a winning configuration of voters if $\sum_{i \in S} \omega_i \geq \gamma$ or a losing one if $\sum_{i \in S} \omega_i < \gamma$. let

$$T = \left[(\gamma, \omega) \in R_{n+1} : \sum_{i=1}^{n} \omega_i = \tau, \, \omega_i \geq 0, \, 0 \leq \gamma \leq \tau \right],$$

denote a set of all the coalitions size n with the sum of weights τ and quorum γ.

Most measures of power (referred to as the power indices) are used to measure the so-called a priori power of players constituting a set that is structured only in terms of voting rule. The a priori power index is therefore a vector function $\Pi \colon T \to R_n^+$ mapping the T set of all the n size coalitions in a non-negative set of real numbers R^n. The power index, therefore, describes the expectations of the player associated with decisiveness, in the sense that his voice will affect the final outcome of the vote. There are two possible positions of such a player: pivotal position and the "swing" position.

Let $(i_1, i_2, \ldots, i_n)$ denote a permutation of players from a coalition of n players, and let the player k take the position r in this permutation, i.e. $k = i_r$. We say that if k is a player in a pivotal position with respect to the permutation $(i_1, i_2, \ldots, i_n)$, if

$$\sum_{j-1}^{r} \omega_{i_j} \geq \gamma \quad and \quad \sum_{j-1}^{r} \omega_{i_j} - \omega_{i_r} < \gamma.$$

The most common a priori index, i.e. the Shapley-Shubik index [20, 21], denoted as SS-index, is defined as the following expression:

$$SS(\gamma, \omega) = \frac{p_i}{n!},$$

where:
   pi is the count of cases in which the player i is decisive and,
   n! is the number of possible different powerful orders for the permutation.

For the "swing" position the most commonly used a priori power index is the Penrose- Banzhaf index [2, 19]. Let S be a winning configuration in the coalition $[\gamma, \omega]$, and $\in$ S. Being in the swing position in the configuration S means that the following inequalities are met:

$$\sum_{k \in S} \omega_k \geq \gamma \quad oraz \quad \sum_{k \in S\setminus\{i\}} \omega_k < \gamma.$$

Let s_i denote the number of changes (swings) of the member in total, and in the committee $[\gamma, \omega]$. The Penrose-Banzhaf index for a member of the coalition is defined as follows:

$$PB_i(\gamma, \omega) = \frac{s_i}{\sum_{k \in N} s_k}.$$

Both indices can be used to measure the power of veto, although the Penrose-Banzhaf index, it seems, is better suited for this purpose as, assuming that the coalition is formed and one of its members betrays it (swings), such behaviour is, in a way, a use of the veto. Thus we think that the "swing" should be used in the measurement of the a priori power of veto. The problem is that the Penrose-Banzhaf index is not the only one based on such concept of the position of the decision maker. In addition, we know that the decision-making process in this case is sequential in nature, and only the end result is (or rather may be) vetoed. The use of other indices should also be considered: Coleman index [4] of the coalition maintenance; Coleman index of the power to initiate action; Coleman index of the power of the collectivity to act, Rae index; Zipke index; Brahms-Lake's index; Deegan-Packel index; Holler index; or Johnston index. The latter index is a "swing" type index and according to Brahms [3] or Lorenzo-Freire et al. [11] it best describes the decision-making process, in which a veto can occur.[1]

The concept of Johnston power index [9] is based on the concept of the so-called vulnerable coalition.

The coalition is vulnerable if it includes at least one member in the "swing" position, and whose defection converts the coalition from the winning of losing. For example, in the Polish Sejm, the President, together with 232 members of parliament creates a vulnerable coalition, in which only he has all the power in the sense of Johnson: a defection by the president (the use of the veto) converts the winning coalition into a losing one (and none of the deputies alone is able to do—there is always the remaining majority of 231 deputies). Note that the president and 231 members of parliament also create a vulnerable coalition, however, the president has to share the power with each of the members: each of them is therefore equally powerful.

---

[1]Axiomatic characteristics of strength indices in decision-making bodies of the veto can be found in Mercik [13], while dynamic characteristics can be found in Mercik and Ramsey [16].

Table 1 The Johnston power indices for the game [1–4]

Vulnerable coalitions	Number of vulnerable coalitions	Critical defections			Fractional critical defections		
		3 votes player	2 votes player	1 vote player	3 votes player	2 votes player	1 vote player
(3, 2)	1	1	1	0	½	½	0
(3, 1)	1	1	0	1	½	0	½
(3, 2, 1)	1	1	0	0	1	0	0
Total	3	3	1	1	2	½	½
J(i)					4/6	1/6	1/6

Source Mercik [12]

Let us Consider (Table 1) the following example: the game [4; 3, 2, 1], i.e. voting where there are three voters with 3, 2 and 1 votes each. The majority needed for a decision is 4. The following are vulnerable coalitions in this game: (3, 2), (3, 1) and (3, 2, 1) (vulnerable coalitions must be winning coalitions).

Example 1 The value of the Johnston index for the Polish president-parliament system.

Table 2 shows the results of calculations for the Polish Sejm and the president (the Senate is not involved in the veto process—each majority in the Sejm rejects the veto of the Senate).[2]

The actual presidential veto in Poland is the veto of the second type, i.e. it can be rejected by a majority of 3/5 votes of the Sejm. In theoretical considerations, however, we can also assume the existence of a veto of the first kind, i.e. allow the possibility that the veto of the president can not be rejected.[3] Values for the a priori Johnston index for the situations when the president respectively does not have a right to veto, is equipped with a veto of the first type, and is equipped with a veto of the second type are shown in Table 3.

The applied a priori power index is a standardized index and therefore the results of the calculations show that in the Polish parliamentary system the position of the president can count only in the situation when he is equipped in veto of the first kind, i.e. the kind of veto that can not be rejected. Because such situations are difficult to imagine (and would mean virtual dictatorship of the president) it should suffice to compare the a priori power of the president without the right of veto and with the veto of the second kind. Using the results from Table 3 it can be calculated that the veto itself increases the strength of the president by 1,675 times, which is a significant result, although we observe a considerable imbalance of power between the president and Sejm in favour of the latter (Table 2).

---

[2]Details of the calculations can be found in Mercik [12]. General issues related to the calculation can be found, among other works, in Alonso-Meijide et al. [1].

[3]From the point of view of the winning coalition, this means that the president must be its member.

Table 2 The value of the Johnston power index for the Polish president-parliament system

	Johnston Power index	
	Without the party structure in the Sejm	With the party structure in the Sejm
President	0.9234	0.0067
Sejm	0.0766	0.9933
Senate	0	0

Source Mercik [12]

Table 3 The value of Johnston index for the president equipped with a different kinds of veto

President without the right of veto	President with the veto of the first type	President with the veto of the second type
0.0040	0.1190	0.0067

Source own calculations

Similar calculations can be carried out using other indices of power, although the results seem to be similar. Thus, the right of veto is an important attribute (also with regard to the president in the Polish parliamentary system) and it is possible to "value" it as a priori power of the decision maker.

Example 2 (Mercik and Ramsey [15] A priori power of the UN Security Council and its members (yes-no voting system).

Let us recall that the United Nations Security Council has 5 permanent members with power of veto, 10 non-permanent members without power of veto; in total 15 members ($N = 15$).[4] The quota for a decision to be passed is $q = 9$. All members have equal weight ($w_i = 1$, for $i = 1, ..., 15$). However, the yes-no voting system is not exactly the type of voting in use in the UN Security Council but yes-no voting is a good starting point for a more general analysis.

One may compare the values of a priori indices of veto and non-veto members of the UN Security Council (yes-no version). One may find the results in Table 4.

As we may see from Table 4, the ratio between the power indices of non-permanent and permanent members of the UN Security Council varies from 1 to 10.10 for the Penrose-Banzhaf power index to 1 to 105.24 for the Shapley-Shubik power index. Actually, in the literature on power indices the absolute Penrose-Banzhaf power index is the most commonly accepted one, so we may conclude that a permanent member of the council is approximately 10 times stronger than a non-permanent one if no member uses his or her right to abstain.

Let us now analyse situation when veto is introduced. In fact, abstention by a non-veto player is identical to voting against an issue. This is not the case for veto-players. In a priori analysis, on one hand all veto-players must be included in any winning coalition to prevent a veto being used. On the other hand, when a veto

---

[4]UN Security Council voting System [23] http://www.un.org/en/sc/meetings/voting.shtml (taken on 15.03.2017).

Table 4 A comparison of power indices for permanent and non-permanent members for the yes-no voting version of the game based on the UN Security Council from the perspective of a priori power indices. *Source* own calculations

	Relative Penrose-Banzhaf Power Index	Absolute Penrose-Banzhaf Power Index	Shapley-Shubik Power Index	Johnston Power Index
Permanent member (veto member)	0.166929	0.051758	0.196270	0.177987
Non-permanent (non-veto member)	0.016535	0.005127	0.001865	0.011006
Ratio of non-permanent/permanent members	1:10.10	1:10.10	1:105.24	1:16.17

player abstains, then a given coalition must be enlarged by another non-veto player to substitute the veto player who abstains. Let k_v denote the number of veto-players choosing to abstain, $k_v = 0, 1, 2, \ldots, v$ (for example: each permanent member of the UNSC may abstain. In this case $v = 4$). This means that in the a priori analysis the game (N, q, w) should be replaced by the sequence of games $(N, q + k_v, w)$ for $k_v = 0, \ldots, v$. It is easy to notice that for $k_v = 0$ this game is in fact the game with yes-no voting only.[5]

Example 3 (Mercik and Ramsey [14, 15] A priori power of the UN Security Council and its members (yes-no-abstain voting system).

First, we calculate the Shapley-Shubik power index directly for a given k_v. A non-permanent member is pivotal only if he or she is the ninth player in the coalition, preceded by all non-abstaining $(5 - k_v)$ permanent members and $(3 + k_v)$ non-permanent members. Note that there are $\binom{9}{3 + k_v}$ ways of choosing the players who appear before player i and 8! ways of ordering these players. The number of players that come after player i equals $(6 - k_v)$, i.e. the non-veto players who did not appear before player i. There are $(6 - k_v)!$ orderings of these players. So, for non-veto player i (i > 5) this happens in $\binom{9}{3 + k_v} 8!(6 - k_v)!$ ways. Ignoring the abstaining permanent members, there are $(15 - k_v)!$ sequential coalitions, so the Shapley-Shubik power index for a non-veto player in this game with given k_v equals $\pi_i^{SS}(N, q, w)(k_v) = \frac{1}{(15 - k_v)!} \binom{9}{3 + k_v} 8!(6 - k_v)!$.

Dividing the rest of the power equally among the $5 - k_v$ non-abstaining permanent members, we obtain a Shapley-Shubik power index for them of

---

[5]Note that in the UNSC example we may assume that the abstaining permanent members do not take part in the game, thus we do not need to consider the case $k_v = 5$.

Table 5 Shapley-Shubik power index for veto and non-veto members of the UN Security Council as a function of the number of abstentions among permanent members. *Source* own calculations

Number of veto players abstaining, k_v	Shapley-Shubik power index for non-permanent members	Shapley-Shubik power index for non-abstaining permanent members	Ratio of power between non-permanent and permanent members (in %)
0	0.001864802	0.196270396	0.95
1	0.006993007	0.232517483	3.01
2	0.019580420	0.268065268	7.30
3	0.042424242	0.287878788	14.74
4	0.072727273	0.272727273	26.67

$$\pi_i^{SS}(15,9,1)(k_v) = \frac{1 - \frac{10}{(15-k_v)!}\binom{9}{3+k_v}8!(6-k_v)!}{5-k_v} \qquad i \leq 5.$$

The exact values for different values of k_v are presented in Table 5.

The results shown in Table 5 support our intuition connected with the relaxation of the power of veto, i.e. introduction of abstentions. The relative power of a non-permanent member of the UN Security Council increases from 0.95 to 26.67% of the power of a non-abstaining permanent member of the council as the number of permanent members of the council abstaining increases. One result which seems initially counter-intuitive is that the power index of a lone non-abstaining veto player is actually lower than the power index of two non-abstaining veto players. This is probably due to the fact that a single veto player would need all but two of the non-permanent members to pass a motion. Hence, the non-permanent members are becoming close to veto players.

4 Formal Equivalence of Quota, Weights and Veto

Mercik and Ramsey [14, 15] showed a general formal equivalence in a priori analysis between voting games with first degree vetoes and standard weighted voting games. Consider the n player game where k players have an unconditional veto and the quota is q. We now define a game without veto power, but with weighted votes, that is equivalent. Let w_i be the weight of player i and $\bar{w}$ be the sum of these weights. Assume that the minimum weight of the votes against a motion m necessary to stop it being passed is the same in both games, i.e. $m = n - q + 1$, and non-veto players are given a weight of 1. Hence, we define the quota to be $\bar{w} - n + q$. Note that by giving each of the veto players a weight of m in the new game, then each of them essentially remains a veto player, since the quota cannot be attained if any veto player votes against the motion. Also, from the definition of the

Table 6 The values of quota and weight of veto player for different number of abstains. *Source* own calculations

Number of abstentions	New quota of modified Security Council	Weight of veto player
0	39	7
1	29	6
2	21	5
3	15	4
4	11	3

weights, it is simple to see that if at least $q - k$ non-veto players vote for a motion, in addition to the veto players, then the motion will be passed. It follows that the two games are equivalent, i.e. have the same set of winning coalitions. It should be noted that any higher weights given to the veto players would satisfy this equivalence relationship (as long as the quota is changed appropriately—in fact the veto players could all be given different weights $\geq m$ and the resulting game would still be equivalent).

Example 4 Equivalent simple game for United Nations Security Council.

For the United Nations Security Council where there are 5 permanent members, 10 non-permanent members (in total 15 members), we have n = 15, k = 5 and q = 9. Now, we construct an equivalent simple game without veto players: $m = n - q + 1 = 15 - 9 + 1 = 7$. Hence, the initial representation of the voting game played by the Security Council of the UN is [9; 1*, 1*, 1*, 1*, 1*, 1, 1, 1, 1, 1, 1, 1, 1, 1, 1], where veto players are marked by stars. This is equivalent to the simple game [39; 7, 7, 7, 7, 7, 1, 1, 1, 1, 1, 1, 1, 1, 1, 1].

Given that k_v permanent members abstain, arguing as above we can treat this as a game where $n = 15 - k_v$, $k = 5 - k_v$ and q = 9. The number of votes needed to block a bill is $7 - k_v$. Hence, we ascribe a weight of $7 - k_v$ to each non-abstaining permanent member. It follows that the sum of weights is given by $(7 - k_v)(5 - k_v) + 10$ and the quota is $(7 - k_v)(5 - k_v) + 4 + k_v$. Table 6 presents the appropriate weighted games for different k_v.

5 Summary

Analysis of the impact of the veto attribute on the position of the decision-makers allows to conclude that there are no paradoxical situations and the veto actually increases the decisive power of a certain decision maker. How strong this impact is remains an open question. Measurement of power using power index is an a priori measurement, and therefore relating to situations occurring only at certain frequency. E.g. examining the situation in the UN Security Council, we find that as the most frequent numbers of abstentions equals 0 or 1 (at least in the year 2014). One may say that in the sense of the Shapley-Shubik power index the United Nations

Security Council may be represented by the simple game $(N, q, w) = (15, 39, 7, 7, 7, 7, 7, 1, 1, 1, 1, 1, 1, 1, 1, 1, 1)$ or $(N, q, w) = (14, 29, 6, 6, 6, 6, 1, 1, 1, 1, 1, 1, 1, 1, 1, 1)$ respectively. These weights can be used in a certain sense as a ratio of measures of the power of permanent members compared non-permanent members and the ratio of the force non-permanent members of the Security Council to the permanent ones is at least equal to 1:6, or considering the value of the power index ranges from 1:3.5 to 1:105.25, depending how many permanent members of the council choose to abstain. It is believed that similar proportions will also occur for other decision-making bodies which introduce the veto. That means, without doubt, that the introduction of veto to the decision-making system results in empowering the decision-makers possessing it and should be preceded by an analysis of the balance of power between decision-makers, because the introduction of veto distorts the balance in itself.

Analysing the differences between the yes-no and yes-no-abstain voting systems we find that the introduction of veto in each situation (yes-no and yes-no-abstain) increases the power of the decision-makers possessing it.

References

1. Alonso-Meijide JM, Alvares-Mozos M, Fiestras-Janeiro MG (2009) Values of games with graph restricted communication and a priori unions. Math Soc Sci 58:202–213
2. Banzhaf JF (1965) Weighted voting doesn't work: a mathematical analysis. Rutgers Law Rev 19(2)(winter):317–343
3. Brams SJ (1990) Negotiation games. Routledge, New York, London
4. Coleman JS (1971) Control of collectivities and the power of a collectivity to act. In: Lieberman, B (ed) Social choice. New York, pp 277–287
5. Felsenthal DS, Machover M (1997) Ternary voting games. Int J Game Theory 26(3):335–351
6. Felsenthal DS, Machover M (2013) Models and reality: the curious case of the absent abstention. In: Power, voting, and voting power: 30 years after, Part II. Springer, Heidelberg, pp 73–86. doi:10.1007/978-3-642-35929-3_4
7. Freixas J, Zwicker SW (2009) Anonymous yes–no voting with abstention and multiple levels of approval. Games Econ behav 67(2):428–444
8. Grabisch M, Lange F (2007) Games on lattices, multichoice games and the Shapley value: a new approach. Math Methods Oper Res 65(1):153–167
9. Johnston RJ (1978) On the measurement of power: some reactions to Laver. Environ Plan A 10(8):907–914
10. Kuziemko I, Werker E (2006) How much is a seat on the Security Council worth? Foreign aid and bribery at the United Nations. J Polit Econ 114(5):905–930. doi:10.1086/507155
11. Lorenzo-Freire S, Alonso-Meijide JM, Casas-Méndez B, Fiestras-Janeiro MG (2007) Characterizations of the Deegan–Packel and Johnston power indices. Eur J Oper Res 177(1): 431–444
12. Mercik J (2009) A priori veto power of the president of Poland. Oper Res Decis 19(4):61–75
13. Mercik J (2015) Classification of committees with vetoes and conditions for the stability of power indices. Neurocomputing Part C 149:1143–1148
14. Mercik JW, Ramsey D (2015a) On a simple game theoretical equivalence of voting majority games with vetoes of first and second degrees. In: Nguyen NT, Trawiński B, Kosala R

(eds) Intelligent information and database systems. Springer lectures on computer science (9011). Cham, pp 284–294

15. Mercik J, Ramsey D (2015b) A formal a priori power analysis of the Security Council of the United Nations. In: Kamiński B et al (eds) The 15th international conference on group decision & negotiation letters. Warsaw School of Economics Press, pp 215–223

16. Mercik J, Ramsey D (2016) A dynamic model of a decision-making body where the power of veto can be invoked. In: Petrosyan LA, Mazalov VV (eds) Recent advances in game theory and applications: European meeting on game theory, Saint Petersburg, Russia, 2015, and networking games and management, Petrozavodsk, Russia, 2015. Birkhäuser, cop., Cham pp 131–146

17. Napel S, Widgrén M (2008) Shapley-Shubik vs strategic power: live from the UN Security Council. In: Braham M, Steffen F (eds) Power, freedom, and voting. Springer, Heidelberg, pp 99–117. doi:10.1007/978-3-540-73382-9_6

18. O'Neill B (1996) Power and satisfaction in the United Nations Security Council. J Confl Resol 40(2):219–237. doi:10.1177/0022002796040002001

19. Penrose LS (1946) The elementary statistics of majority voting. J Roy Stat Soc 109:53–57

20. Shapley LS (1953) A value for n-person games. In: Kuhn HW, Tucker AW (eds) Contributions to the theory of games, volume II. Annals of Mathematical Studies, vol 28, pp 307–317

21. Shapley LS, Shubik M (1954) A method of evaluating the distribution of power in a committee system. Am Polit Sci Rev 48(3):787–792. doi:10.2307/1951053

22. Tchantchoa B, Lamboa LD, Pongoub R, Engouloua BM (2008) Voters' power in voting games with abstention: influence relation and ordinal equivalence of power theories. Games Econ Behav 64(1):335–350

23. UN Security Council voting System (2015) http://www.un.org/en/sc/meetings/voting.shtml. Accessed Mar 15 2017

Power Indices for Finance

Cesarino Bertini, Gianfranco Gambarelli, Izabella Stach
and Maurizio Zola

Abstract The weight of the share stock of a company may be described by power indices that quantify the possibility for each shareholder to get majority positions by coalitions with other shareholders. To study such indices allows us to build efficient models for forecasting, simulating, and regulating financial, political, and economic fields. An overview of financial applications of power indices is presented; this was carried out at the University of Bergamo along with partners in Europe and United States. New explanations and examples are added so as to better illustrate the results obtained. Additionally, certain open problems are described.

Keywords Banzhaf index · Finance · Power index · Shapley-Shubik index · Simple games · Takeover · Voting · Weighted majority game

1 Introduction

Let us consider shareholding with only three shareholders (A, B, and C) with the following shares: 30% to each A and B, and 40% to C. If there are no propensities or aversions to any coalition, it is easy to see that all of the players are equivalent to

C. Bertini (✉) · G. Gambarelli (✉) · M. Zola
Department of Management, Economics and Quantitative Methods,
University of Bergamo, Via dei Caniana 2, 24127 Bergamo, Italy
e-mail: cesarino.bertini@unibg.it

G. Gambarelli
e-mail: gambarex@unibg.it
URL: http://dinamico2.unibg.it/dmsia/staff/gambar.html

M. Zola
e-mail: maurizio.zola@gmail.com

I. Stach
AGH University of Science and Technology, Faculty of Management,
Al. Mickiewicza 30, 30-059 Krakow, Poland
e-mail: istach@zarz.agh.edu.pl

© Springer International Publishing AG 2018   45
M. Collan and J. Kacprzyk (eds.), *Soft Computing Applications for Group
Decision-making and Consensus Modeling*, Studies in Fuzziness and Soft
Computing 357, DOI 10.1007/978-3-319-60207-3_4

form possible coalitions for a simple majority. Thus, they have the same "coalition power" (i.e., each has 1/3). The same situation would be if A and B would each have 49% of the shares and C 2%: this last one would possess the same effective power as the others (also if with a nominal power who is very low).

On the contrary, if A possesses 51% of the weight, his power would be 100% (i.e., 1). What could be said if the sharing is 50% to A, 30% to B, and 20% to C? In this case, A does not have the majority; however, each of the other two should join A, because a coalition between B and C is a minority coalition. It is easy to understand that this last one has the same position of power (also if with different quantities of shares). It could be deemed that A has greater power due to his greater percentage of shares; it could be assigned a power division of 2/3, 1/6, or 1/6 after the Shapley-Shubik model (see [61] or [62]). Another possibility is a division of 3/5, 1/5, or 1/5 after the Banzhaf-Coleman-Martin-Penrose model (see [5, 19], or [14, 15]).

2   Some Preliminary Definitions

Let $N = \{1, 2,..., n\}$ be a nonempty finite set. By a game on N, we shall mean real-valued function v whose domain is the set of all subsets of N such that $v(\emptyset) = 0$. We refer to any member of N as a "player" and to any subset of N as a "coalition." Game v is said to be "simple" if function v assumes values only in set $\{0, 1\}$: $v(S) = 0$ or $v(S) = 1$ for all coalitions $S \subseteq N$. In the first case, the coalition is said to be losing; in the second case, winning.

Let us consider an assembly composed of a set $N = \{1,..., n\}$ of members. Each i-th player is given certain "weight" w_i (which can represent votes, seats, shares, and so on). Let t be the sum of weights of all players in N. Given "majority quota" q $(>t/2)$, the elements make up a "weighted majority game," which is usually indicated by symbol $[q; w_1,..., w_n]$ or also by $[q; w]$. In this type of game, each coalition S of members of N is called a "winning coalition" if the sum of the weights of its components is equal to or greater than q; otherwise, it is called a "losing coalition"; that is:

$$v(S) = \begin{cases} 1 & \text{if } \sum_{i \in S} w_i \geq q \\ 0 & \text{otherwise} \end{cases}$$

With reference to the example in the previous section, we have $N = \{A, B, C\}$, $w_A = 50$, $w_B = 30$, $w_C = 20$, $t = 100$, $q = 51$ by using letters instead of numbers; thus, the game is represented as [51; 50, 30, 20]. The winning coalitions are $\{A, B\}$, $\{A, C\}$, and $\{A, B, C\}$, while the losing coalitions are $\emptyset$, $\{A\}$, $\{B\}$, $\{C\}$, and $\{B, C\}$.

In this example, the payoff (the "power") of the total coalition is one: $v(\{A, B, C\}) = 1$. It is clear that, if all of the possible coalitions have the same probability, a reliable power index should give to game [51; 30, 30, 40] the following share of power $(1/3, 1/3, 1/3)$; analogously for game $[51; 49, 49, 2]$. It is also clear that a valuable index should give to game $[51; 51, 39, 10]$ the share of power $(1, 0, 0)$, but it is not so easy to share the total payoff for game $[51; 50, 30, 20]$. How can the problem be solved? A natural solution is based on *cruciality*.

The i-th player is called "crucial" for coalition S if S is a winning coalition, but it becomes a losing coalition without the contribution of this player. Let $C(i, v)$ denote the set of all coalitions (S) for which the i-th player is crucial in game v (i.e., $v(S) = 1$ and $v(S\backslash\{i\}) = 0$).

In the previous example $([51; 50, 30, 20])$, player A is crucial for coalitions $\{A, B\}$, $\{A, C\}$, and $\{A, B, C\}$; player B is crucial only for $\{A, B\}$ and player C only for $\{A, C\}$.

A *value* for a game is function v suitable to share the total payoff $v(N)$ among the n players.

A *power index* is a value for simple games.

A "power index" of a weighted majority game is a function designed to fix a fair division, or to represent a reasonable a priori expectation of the share of a global prize among the players. More specifically, in the case of weighted voting game $(q; w_1, \ldots, w_n)$, power index π assigns voting power $(\pi_1, \ldots, \pi_n)$ to the participants of the voting body.

Let $v: [q; w_1, \ldots, w_n]$ be a weighted majority game. We say that power index π is "locally monotonic" if $w_i > w_j => \pi_i(v) \geq \pi_j(v)$.

Let $v: [q; w_1, \ldots, w_i, \ldots, w_n]$ and $v': [q; w'_1, \ldots, w'_i, \ldots, w'_n]$ be two weighted majority games such that $w_i > w'_i$ for one $i \in N$ and $w_j \leq w'_j$ for all $j \neq i$. We say that power index π is "globally monotonic" if $\pi_i(v) \geq \pi_i(v')$.

In [34], Gambarelli proposed the following definition of strong monotonicity: $\pi(v)$ is "strongly monotonic" if, for all i and all weighted majority games v, v' with $C(i, v) \subset C(i, v')$, $\pi_i(v) < \pi_i(v')$.

3 The Martin-Penrose-Banzhaf-Coleman and Shapley-Shubik Indices

Martin in 1787, Penrose [55], Banzhaf [5], and Coleman [19] introduced indices that, under various aspects, may be considered equivalent (as far as Martin, see [56]). The "normalized Banzhaf-Coleman index" β represents a summary of such indices. This index assigns each player a quota proportional to the number c_i of coalitions for which he is crucial.

For instance, in the previous example $([51; 50, 30, 20])$, A is crucial for three coalitions while B and C are crucial for only one coalition each. As the sum of crucialities is 5, the normalized Banzhaf index assigns $1/5$ of the power to player B, $1/5$ to player C, and $3/5$ to player A; i.e., $\beta_A = 3/5$, $\beta_B = \beta_C = 1/5$.

The "Shapley-Shubik index" Φ (Shapley and Shubik [61]) is an expression of the Shapley value [60] for simple games such as weighted majority games. This index assigns each i-th player an expected payment corresponding to the

Table 1 Computation of the Shapley-Shubik index

	A	B	C	
$A \leftarrow \underline{B} \leftarrow C$		x		
$A \leftarrow \underline{C} \leftarrow B$			x	
$B \leftarrow \underline{A} \leftarrow C$	x			
$B \leftarrow C \leftarrow \underline{A}$	x			
$C \leftarrow A \leftarrow \underline{B}$	x			
$C \leftarrow B \leftarrow \underline{A}$	x			
Totals	4	1	1	= 6

probability of finding himself in a crucial position upon joining an established coalition.

With reference to the previous example ([51; 50, 30, 20]—see Table 1) we start from coalition composted only of singular player A, then player B joins the coalition. The new coalition $\{A, B\}$ becomes a majority coalition; thus, B is crucial for this coalition (see the first row). Analogously (see second row), if C joins the coalition with the lonely A, coalition (A, C) becomes a majority coalition. In the other four cases, A is the crucial player. From the totals, we can derive that the Shapley-Shubik index is 4/6 for A and 1/6 for each B and C; i.e., $\Phi = (2/3, 1/6, 1/6)$.

There is a formula to avoid the computations of Table 1. Indicated as $P(s, n) = (s-1)!(n-s)!/n!$, the Shapley-Shubik index of every i-th player is given by the sum of all $P(s, n)$ extended to all of the coalitions of s members for which the i-th player is crucial:

$$\Phi_i = \sum P(s, n) = \frac{1}{n!} \sum (s-1)! \ (n-s)!$$

In the above example, it is $n = 3$, $0! = 1! = 1$, $2! = 2$, $3! = 6$. As player B is only crucial for a single two-member coalition, his index is $\Phi_B = P(2, 3) = (2-1)!$ $(3-2)!/3! = 1/6$; the same is true for player C. The Shapley-Shubik index of player A is $\Phi_A = P(2, 3) + P(2, 3) + P(3, 3) = 1/6 + 1/6 + 1/3 = 2/3$.

A crucial difference between the normalized Banzhaf and Shapley-Shubik indices lies in the bargaining model: the former does not take into account the order in which a winning coalition is formed, while the latter does (from a mathematics point of view, the first index takes into account the combinations and the second one the permutations).

It is important to note that the first index is not *globally monotonic* in the sense that if a player gains weight from another player, its normalized Banzhaf index *could* decrease. This is a simple example: in a game with 5 players [9; 5, 5, 1, 1, 1], each of the last three players is "dummy" who is it is not crucial for any coalition. To compute the power sharing is easy: (1/2, 1/2, 0, 0, 0). If the first player takes a weight unity from the second one, the game becomes [9; 6, 4, 1, 1, 1] and the Banzhaf-Coleman index is (9/19, 7/19, 1/19, 1/19, 1/19). Thus, the first player gains weight at the second's expense, but his power decreases (after Banzhaf-Coleman) from 1/2 to 9/19.

On the other hand, the Shapley-Shubik index satisfies this monotonicity property. Both indices, however, satisfy local monotonicity: players with higher weights have at least as much voting power as players with lower weights. For some in-depth consideration on this issue, see Turnovec [64] or Holler and Napel [42], for example.

With regard to applications, the normalized Banzhaf index is considered best-suited for inclusion in normative models thanks to its direct proportionality to the number of crucialities. On the other hand, the Shapley-Shubik index is best-suited for forecasting models owing to its properties of monotonicity and stability (belonging to the core in convex games).

Therefore, the Shapley-Shubik index is more suitable to forecast the results of negotiations (for example, the stock exchanges with coalitions that are changing their structure); on the contrary, the Banzhaf-Coleman index is more suitable for normative models (for example, for electoral systems where it is used the pure proportionality). In [57], Rydqvist noticed a strong similarity between the Shapley-Shubik index and the quotation of share equities for equity takeover on the Swedish equity market.

An axiomatic characterization of the Shapley-Shubik index (i.e., a set of requirements that only that index is suitable to fulfill) was given by Dubey in [26]; a characterization of the Banzhaf-Coleman index was given by Owen in [51]. For other lesser-known or lesser-used indices, see [8–10].

4 A First Application

It is easy to understand that the above-described models may be applied to the political field where shares are substituted by seats in a parliament. In this case, it should consider the affinity and aversion of the various members. Guillermo Owen proposed some generalizations of the Shapley-Shubik index [50] and Banzhaf-Coleman index [52] when it is possible to forecast the formation of different coalitions [38].

A political case is reported here due to the interesting computation follow-up.

In this section, using a real-world situation as a starting point, we shall focus our attention on the existence of constant power positions in weighted majority games.

Before examining the numerical results, some observations are required on the reliability of the models in question.

First, some coalitions that are possible in theory are not possible in practice; for instance, those between the extreme right and extreme left parties. For the correct application of an index, it seems necessary to discard all of the coalitions that are unfeasible from the cruciality calculations for each player. However, there are decision situations (e.g., referendum or presidential elections) where numerical strength is more important than political closeness; in these cases, the original model applies.

Table 2 Banzhaf-Coleman and Shapley-Shubik power indices in the Deputies Chamber in Italy

Party	10th legislature			11th legislature		
	Seats	Ba.Co.	Sh.Sh.	Seggi	Ba.Co.	Sh.Sh.
DC	234	35.3	39.3	206	42.7	41.6
PDS	177	21.2	22.1	107	13.3	15.5
PSI	94	21.2	22.1	92	13.0	13.8
Lega Lom.	1	1.3	0.1	55	8.4	7.1
Rif. Com.	0	–	–	35	4.6	4.2
MSI	35	3.9	5.1	34	4.5	4.1
PRI	21	6.5	2.8	27	3.5	3.4
PLI	11	2.0	1.5	17	2.3	2.3
PSDI	17	2.7	2.0	16	2.1	2.1
Verdi	13	2.3	1.7	16	2.1	2.1
Rete	0	–	–	12	1.6	1.5
Pannella	13	2.3	1.7	7	0.9	1.0
SVP	3	0.5	0.3	3	0.5	0.7
Others	8 + 1 + 1 + 1	0.8	1.3	1 + 1 + 1	0.5	0.6
Totals	630	100.0	100.0	630	100.0	100.0

Table 2 shows the powers of the parties after the Italian election taking us from the 10th to the 11th legislature, according to the normalized Banzhaf and Shapley-Shubik indices (the computation was made by the algorithms shown in [37]). It is interesting that the DC party decreased its seat number from 234 to 206 yet increased power after both indices. The different distribution of the seats has favored that party against the others.

Further, a comparison between the PDS and PSI parties shows that, while the former party had almost twice as many seats as the latter, it has the same power (despite the fact that the numerical values differ depending on which index is used). It is easy to explain why if situation (2, 49, 49) is kept in mind, where the third party has the same coalitional power as the other two. This poses some interesting questions: in which cases do the normalized Banzhaf and Shapley-Shubik indices (and possibly other indices as well) behave in a similar way even with different numerical values? How can these properties be used in practical applications? Some answers will be given in the following sections.

5 Financial Applications

Let us now consider that the parties of Table 2 are shareholders of a company. It is worth noting that shareholder PDS owns a lot more shares than shareholder PSI yet has the same decision power. He could then decide to yield some of his shares that are useless to control and to buy other shares of other companies to get a better power position in those companies. In general, the question is whether there is a mathematical model to buy and sell shares so as to give him the maximum expectation of success to control various companies. The problem is very important due to the large quantity of involved money, and it was solved in the '80s by the theory of power indices. Moreover, some common behaviors of the various indices were identified, and some answers to the questions in the previous section were found.

In the following sections, these issues will be examined (starting from the easiest case of two shareholders).

5.1   The Case of Two Players

Using the power indices as a starting point, it is possible to build simulation models of a shareholder's power variations following a modification in share distribution with particular regard to share shifts between one shareholder and another as well as the inclusion of new shareholders who modify share redistribution (the latter case will be examined in a following section). Suppose the shares of a company are initially distributed between only two shareholders in a proportion of 75–25%. In this case, the first party has the majority for all decisions that require a majority of 51%; therefore, the power index is (1, 0), which is to say 1 for the first and 0 for second shareholder. If the first shareholder's shares are 50.5% and the second shareholder's shares are 49.5%, the index is (1/2, 1/2).

In general, the set of points on a Cartesian plane (representing all possible shares distributions) is the oblique segment in Fig. 1, where w_1 and w_2 are the weights of the two shareholders (with conventions $w_1 \geq 0$, $w_2 \geq 0$, and $w_1 + w_2 = 100$). It should be noted that, at all points having abscissa greater than or equal to 51, the power index is (1, 0); at all points having abscissa less than or equal to 49, the power index is (0, 1); at all other points, it is (1/2, 1/2).

Suppose that, starting from position (75, 25), the first shareholder yields shares to the second one (see Fig. 2). As long as the point lies in the lower right segment, the power remains unchanged; therefore, the decrease in power $\Delta\Phi$ of the first shareholder is zero. If the point reaches the central segment, then the decrease of power is $\Delta\Phi = 1/2$, while, if the point passes it, then the decrease of power is $\Delta\Phi = 1$.

5.2   The Case of Three Players

Figure 3 shows a three-player game where vector $w = (w_1, w_2, w_3)$ represents the non-negative weights (shares) of the shareholders under constraints $t = w_1 + w_2 + w_3 = 100$. Owing to the constraints above, vector w lies within the triangle

Fig. 1 All possible
distributions of seats between
two parties

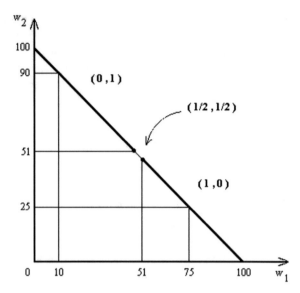

Fig. 2 Power variations
following a modification in
share distribution between
two shareholders

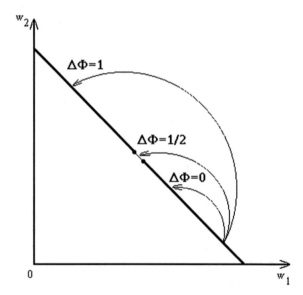

having vertices (100, 0, 0), (0, 100, 0), and (0, 0, 100). For the sake of simplicity,
we will consider a game with simple majority.

Let us consider the parallel plane to the first and third axis, passing through point
(0, 50, 0). At all points to the right of this plane (i.e., where $w_2 > 50$), the second
player has a majority, and therefore his power is 1 in these points. Consider the
smaller triangle having vertices (0, 100, 0), (50, 50, 0), and (0, 50, 50). The power
index is (0, 1, 0) at all points of this triangle (with the exception of the segment

Fig. 3 Power variations
following modification in
share distribution between
three shareholders

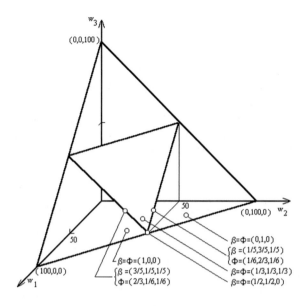

joining the last two vertices). Analogously, the index is $(1, 0, 0)$ at all points on the triangle having vertices $(100, 0, 0)$, $(50, 50, 0)$, and $(50, 0, 50)$. It is also possible to verify that the index is $(1/3, 1/3, 1/3)$ at all points on the central triangle having vertices $(50, 0, 50)$, $(0, 50, 50)$, and $(50, 50, 0)$ (borders excluded). Regarding the borders, the values differ at the internal points of each segment depending on the chosen index. For instance, for the border that joins vertices $(50, 50, 0)$ and $(0, 50, 50)$, the normalized Banzhaf index is $(1/5, 3/5, 1/5)$, while the Shapley-Shubik index is $(1/6, 2/3, 1/6)$; the same (with suitable permutations) applies for the others. Finally, on the vertices of the small central triangle, the index is $(1/2, 1/2, 0)$, $(1/2, 0, 1/2)$, and $(0, 1/2, 1/2)$.

Also note that, when the non-limit case is considered, the diagram would show subdivisions of the large triangle not only as small triangles but also as trapezia. (There are two types of diagrams, depending on whether q is greater or less than $2t/3$.) Each of these polygons has the following property: in all points, the game is constant as regards the coalitions for which each player is crucial. Now, it is possible to give an intuitive answer to one of the questions posed at the end of Sect. 4: do the Shapley-Shubik, normalized Banzhaf, and possibly other indices behave in a similar way even though they have different numerical values?

5.3   *The Case of* n *Players*

Readers can envisage a generalization of what has been examined so far when applied to games with n players. The triangle in Fig. 3 becomes a simplex of Euclidean n-dimensional space, having vertices at all points so that one of the components is the total sum of weights t and all other components are zero (in our case, $t = 100$). This simplex is subdivided into convex polyhedra by hyperplanes parallel to the main axis and where the distance from these is q and $t - q$. In each of these polyhedra, the game is constant; therefore, once a power index has been chosen to represent the real situation being studied, the index remains unchanged at all points on each polyhedron.

For further information and theorems on the matter, see Gambarelli [34]. Other common properties regarding various power indices may be found, for instance, in Freixas and Gambarelli [31].

5.4   Share Shift Between Two Shareholders

Another model was studied to predict changes in power relationships that follow from a shift by a subset of the shares from one shareholder to another (see Gambarelli [34]).

Let us assume that the initial distribution of shares among shareholders A, B, and C in a 3-person weighted majority game is (51, 40, 9) (see Table 3). Given simple majority voting ($q = 51$), a transfer of shares between B and C will not change the situation, as A will remain the majority shareholder. However, let us now analyze what happens if shares are exchanged between A and C. If C receives one share from A, the distribution becomes (50, 40, 10) and the power distribution (according to the Shapley-Shubik index) becomes (2/3, 1/6, 1/6). If C receives two shares from A, the distribution of shares becomes (49, 40, 11) and the power distribution is (1/3, 1/3, 1/3). The division of power remains the same even if C obtains 40 shares from A, as the share distribution becomes (11, 40, 49) in this case, and each player is in the same position as the others. The situation changes only if C receives 41 shares from A: in this case, the distribution becomes (10, 40, 50) and the power of C increases to 2/3. With one more share, C acquires the majority, and his power increases to 100%.

Table 3 shows that the power of C is a monotonic step function of the number of shares acquired by A. The critical stocks that allow C to pass from one position of power to another are 9, 10, 11, 50, and 51. In Gambarelli [34], it was proven that, no matter how a power index is defined (provided that it is strongly monotonic), the sequence of critical stocks corresponding to shares transferred between two players i and j is always the same. The formulae generating these critical stocks d_s in a company with n shareholders are given in this article.

Figure 4 shows the movement of the share vector while shares are being transferred between A and C. As in Table 3, the number of shares owned by B remains unchanged and the relative component is constant; therefore, the point representing the resulting vector of the distribution of shares moves within a plane parallel to the w_A w_C plane.

The resulting segment meets the borders of the small triangles at two points, determining the step function (marked at the top right of the diagram) that shows

Table 3 Exchange of seats between two players (according to the Shapley-Shubik index)

Company	Number of shares C receives from A	Resulting distribution of shares	Resulting distribution of power	Power increase of C
A		51	1	
B		40	0	
C	0	9	0	0
A		50	2/3	
B		40	1/6	
C	1	10	1/6	1/6
A		49	1/3	
B		40	1/3	
C	2	11	1/3	1/3
A		10	1/6	
B		40	1/6	
C	41	50	2/3	2/3
A		9	0	
B		40	0	
C	42	51	1	1

Synthesis

Number of seats (C receives from A)		Resulting increment of power (%)
	1	+16.7
From	2 to 40	+33.3
	41	+66.7
From	42 to 51	+100.0

Fig. 4 Exchange of shares between two shareholders and following resultant distribution of power

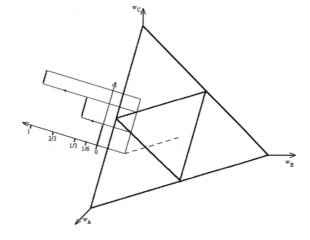

the power increase of C depending on the number of shares received from A. From Fig. 4, we can see how the same number of shares transferred can give different results (in terms of power indices) depending on the shareholder who acquires the shares. If shareholder A yields 40 shares to C instead of to B, A's power will go from 1 to 1/3 (cf. broken segment) instead of from 1 to 0. It is therefore important to know not only the discontinuity points of the step function (i.e., the critical blocks of shares that enable a player to move from one constant power area to another) but also the most "dangerous partner" in a transfer of shares. If the Banzhaf-Coleman index is used instead of the Shapley-Shubik, the power variations change; however, the critical stocks remain the same (see [34]). It should be remembered that the above-mentioned monotonicity holds for the Shapley-Shubik index and is not satisfied by the normalized Banzhaf index.

To determine the critical stocks d_s for companies with n shareholders, the following formulae may be used:

$$d_S = q - \sum_{h=1}^{n} b_h w_h \text{ and } d_S = t - q + 1 - \sum_{h=1}^{n} b_h w_h$$

varying the n-dimensional vectors b whose components take only 0 and 1 values, with the condition $b_i = b_j = 0$. Both summations are, moreover, subjected to the following requirement:

$$0 \le \sum_{h=1}^{n} b_h w_h < H$$

where H is the minimum between q and $(t - q)$.

It is worth noting that formula $d_S = t - q + 1 - \sum_{h=1}^{n} b_h w_h$ is suitable for computing the position of the buying player when he is crucial for all winning coalitions (with integer exchanges of shares).

In the previous example, $t = 100$, $q = 51$, $t - q = 49$, $i = 1$, and $j = 3$. The only binary vectors to be considered are $(0, 0, 0)$ and $(0, 1, 0)$. From these formulae, the following values are obtained: 10, 11, 50, and 51, generating sequences (51, 50, 49, 10, 9) for A and (9–11, 50, 51) for C.

5.5 Trade of Shares Between One Player and an Ocean of Players

Suppose a company has three major shareholders (A, B and C) and an "ocean" of minor shareholders who are not interested in control. Let the initial breakdown of shares between the major shareholders be (20, 15, 4) (see Table 4). What would

Table 4 Trade of shares between one player and the ocean

Player	No. of shares bought by C	Resulting distribution of shares	Resulting majority (A + B+C)/2	Resulting power distribution
A		20		1
B		15		0
C	0	4	19.5	0
A		20		2/3
B		15		1/6
C	1	5	20	1/6
A		20		1/3
B		15		1/3
C	2	6	20.5	1/3
A		20		1/3
B		15		1/3
C	30	34	34.5	1/3
A		20		1/6
B		15		1/6
C	31	35	35	2/3
A		9		0
B		40		0
C	32	36	35.5	1
Synthesis				
	1 share	bought:	+16.7%
From	2 to 30 shares	bought:	+33.3%
With	31 shares	bought:	+66.7%
With	More than 31 shares	bought:	+100.0%

happen if the third shareholder starts to buy shares on the market from minor shareholders (the ocean) to increase his power index in the company?

If C purchased one share from the ocean, the share distribution would become (20, 15, 5), and the power factors (according to Shapley-Shubik) would be (2/3, 1/6, 1/6), as the majority shareholding would go from 19.5 to 20. If C purchased two shares from the ocean, the share distribution would become (20, 15, 6), and the power indices would be (1/3, 1/3, 1/3). This power distribution would remain the same even if C were to purchase 30 shares; the situation would only change if C bought 31 shares. In this case, the share distribution would become (20, 15, 35) and the power factors (1/6, 1/6, 2/3). With the purchase of one more share, C would acquire an absolute majority, and his power factor would be 100%.

Applying the Banzhaf-Coleman index, the critical stocks are the same as those above even though there are different power factors ($0 \rightarrow 1/5 \rightarrow 1/3 \rightarrow 3/5 \rightarrow 1$).

In [34], it was proven, however, that the power index is also defined in these cases (as long as it is monotonic, as in the case of the Shapley-Shubik index), and the power of the raider (i-th player) to form coalitions is a monotonic step function of the number of shares purchased from minor shareholders. The critical stocks d_S are generated using the following formula (where q, t, and w_h have the conditions indicated in the previous section):

$$d_S = -\frac{M}{tb_i - q} + b_i$$

where

$$M = \sum_{\substack{h=1 \\ h \neq i}}^{n} (tb_h - q)w_h$$

With the condition $M \geq 0$ for $b_i = 0$; $M \leq 0$ for $b_i = 1$.

In this example, the third player is involved ($i = 3$). Initially $t = 39$, $q = 19.5$, $w = (4, 15, 20)$, and these change as w_3 increases. The formula (with the necessary roundings) generates critical points 5, 6, 35, and 36.

Note that the model proposed here differs from classic oceanic games (see, for example, [46, 48]), as it supposes that all of the power is held by major shareholders. It is, therefore, more suitable for incomplete information imperfect markets, where the minor shareholders are obviously excluded from the board of directors and where the means and the information the raider has renders the power of the ocean (which is not able to form a coalition) completely ineffective. (This model also describes this type of situation, because the i-th player could be a syndicate of shareholders.)

5.6 Remarks on the Prices

The takeover can be done with the agreement of the present control group (who is interested in getting a new shareholder due to company politics, development outlook, etc.) or against the control group. In this last case, the raider should expect an increase of the share offer price by increasing the requested quantity. Such an increase is artificial with respect to the real value of the shares that are only a part of the considered company, because it is only an added value that the raider wants to pay to gain control and the subsequent benefits. This could ensue an increase in the value of the company (for example, by better managing politics) or damage it (for example, by choosing worse suppliers, managers, and company politics connected in other way to the raider, or by using confidential information to pursue different goals). The new controller will then indirectly affect the value of the shares; however, such an influence is more or less far from a temporary increase in the quotation connected to the takeover.

Another advantage for the raider is to sell the whole share pack to the present control group (of course, at a major price): this fact will decrease the quotation of the share, of which the small shareholders will pay the consequences. For further remarks, see [18, 20].

During the acquisition phase, the first shares are normally bought on the market of small shareholders with silent operations so to avoid alarming the control group. After eventual agreements with some of the big shareholders, a takeover bid is presented with a fixed price and a pledge to buy only if a pre-determined quantity is reached.

An "a priori" evaluation of the price to pay for this operation is very important for the raider. It is based on objective information (available share quantity on the market, closeness to the majority quote, economic power of the present control group, eventual undercutting of the share, and so on) and subjective considerations (power and cohesion of the present control group, possible collateral benefits to favor destabilizing agreements, and so on).

It has to be noted that the trend of the price versus the demand on the perfect market should coincide with the trend of the power position defined by a suitable index. On the contrary, in the most usual cases where small shareholders do not have any possibility of control, the two curves do not coincide: the raider is playing on the precise evaluation of this difference. The model presented in the previous section can also be used to describe the effects of the formation of a syndicate of small shareholders who want to defend their position.

5.7 Steadiness of Control

Another particular problem concerns the steadiness of the control position reached. It is not sufficient to receive the minimum number of shares to enter into a power position in order to be able to exercise the power. The present controllers can, in turn, buy some more shares at an increased price so to reject the new shareholder from the power position (from a geometrical point of view outside the polyhedrum at constant power). It is then necessary to buy a further "security amount" Δs in relation to each discontinuity point s. How to determine such a quantity? It is clear that a purchase to get the absolute majority quote could be enough to defend against counter-actions, but it is also clear that the cost of such an operation could nullify the advantages.

In [36], a method is shown based on the following considerations. When the present controllers try to buy shares on the market to regain their lost position, they would have trouble finding them, and they could pay a higher and higher price for those shares. The raider himself could offer to sell them the lacking shares at such a price to cover the overcharge he paid to buy the shares. With a reliable forecasting model of the quotation $p(s)$ versus the number of exchanged shares starting from the initial price p_o, the investor could calculate the unknown quantity Δs by equating the sum of the paid overcharged prices for each share (from 0 to $s + \Delta s$) to the sum

Fig. 5 Determination of security amount of shares Δs

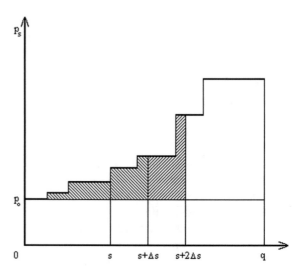

of the following overcharged prices to ask for each share (from $s + \Delta s$ to $s + 2\Delta s$). In a model in the continuum space, the two shaded areas in Fig. 5 should be equated by obtaining the unknown Δs from the equation:

$$\int_0^{s+\Delta s} p(s)ds - p_o(s + \Delta s) = \int_{s+\Delta s}^{s+2\Delta s} p(s)ds - p_o\Delta s$$

Being $P(s)$ the integral function of $p(s)$ in the considered interval, the problem is to find the minimum Δs that is the solution of the following equation:

$$P(s + 2\Delta s) - 2P(s + \Delta s) = -p_o s - P(0)$$

5.8 Indirect Control

A particularly interesting problem concerns those cases in which an investor has a shareholding in a certain company that, in turn, holds shares in another company (and so on). In situations of this nature, it may be useful to calculate the power in the whole system.

Let a shareholder hold 20% of the shares of a company whose remaining shares are divided equally (40 and 40%) between two other shareholders. Let this company own 51% of the shares of another company that owns a quarter of the shares of a third company, whose remaining shares are equally divided among three other shareholders. What is the power of the first shareholder in this last company?

It could be answered that the shareholder holds a third of the power in the first company, which has total control of the second one; thus, he has a third of the power in the second one. This last one has a quarter of the power in the last one; thus, the shareholder has $(1/3) \cdot (1) \cdot (1/4) = 1/12$ of the power in this last one. It seems logical to assign indirect control power equal to the product of the power indices to each shareholder. There are counter-examples that show how this way of proceeding in calculation can bring a total of shares in a company that is different than 100%. Then, another way should be found.

This problem was tackled in [39] by transforming the set of inter-connected games into just one game, using the *multi-linear extensions* introduced by Owen [49] (see also [53]). The advantage of this method is that the power index considered to be most suitable in describing the situation in question can then be applied to the unified game.

Gambarelli and Owen [39] and Denti and Prati [23, 24] focused on determining the winning coalitions in a control structure. An algorithm for the automatic computation of such situations was elaborated by Denti and Prati in [23, 24]. Kołodziej and Stach [45] proposed a computer program based on the approach of Denti and Prati that enabled simulations. On the other hand, the works of Hu and Shapley [43], Crama and Leruth [21, 22], Karos and Peters [44], as well as Mercik and Lobos [47] are dedicated to modeling indirect control relationships in corporate structures and using power indices to evaluate the power of players.

Karos and Peters [44] developed a theory to compute power indices for indirect control in general cases, giving a unique solution when dealing with invariant mutual control structures. In a mutual control structure, agents exercise control over each other, and a mutual control structure is invariant if it incorporates all indirect control relationships.

Mercik and Lobos [47] proposed a measure of reciprocal ownership as a modification of the Johnston power index. This measure, called the implicit power index, takes into account not only the power of the individual entities constituting the companies but also the impact of the companies themselves on implicit relationships.

For further studies about the subject, see [17, 22, 58] or [13]. Bertini et al. [13] critically examined the models of Gambarelli and Owen [39], Denti and Prati [23, 24], Crama and Leruth [21, 22], Karos and Peters [44], and Mercik and Lobos [47].

5.9 A Global Index of de-Stability

Let us consider a set of companies that could be subjected to takeover. Is it possible to state which one is more vulnerable or to give a numerical index stating the stability of each company? The answer was given in [36]; let us see how to proceed. Let n be the number of big investors holding shares of at least one such company, whereas all of the other shares belong to the ocean of small shareholders.

Let A be the matrix in which generic element a_{hk} represents the share quantity of the h-th shareholder $(1 \leq h \leq n)$ or of the h-th company $(n + 1 \leq h \leq n + m)$ in the k-th company. Let B be the matrix of which generic element b_{hk} represents the Shapley-Shubik index of the h-th shareholder $(1 \leq h$ $n)$ in the k-th company (being the power distributed only among the big share-holders, excluding the other companies).

Let C be the matrix in which generic element c_{hk} represents the effective power (Shapley-Shubik index) of the representatives of the h-th shareholder in the board of directors of the k-th company. Generic element d_{hk} of matrix $D = C - B$ represents the difference between the theoretical and effective power; higher values represent greater dissatisfaction of the h-th shareholder for the situation in the k-th company. To calculate the above-defined indices, the presence of special friendships among the big shareholders should be taken into account (the generalization given in [50] should then be used). Let d_k represent the maximum value of the k-th column of matrix D. Such a value represents maximum dissatisfaction in the considered company (the k-th one) and contributes to the formation of the *de-stability index* proposed in [36]. Other data necessary to define such an index (with reference to each company) is as follows (for sake of simplicity, index k is omitted):

w_r number of shares owned by the "raider"
w_c number of shares owned by the control group $(0 \leq w_r < w_c)$
q the majority quote
p_z a former reference quotation
p_o a present quotation
s the power (politic and economic power) of the present control group; this parameter gives indications on the relevant reaction capacity $(0 \leq s \leq 1)$

The above-cited values contribute to the formation of the following preliminary indices (each taking values from 0 [= the maximum stability] to 1 [= the minimum stability] of the company):

$c = w_r/w_c$ ratio between the numbers of shares of the raider and the control group
$m = (t - w_r - w_c)/t$ availability of residual shares on the market
$v = (q - w_c)/q$ the vicinity of the absolute majority quota by the controlling shareholders
$f = \max(0, (p_z - p_o)/p_z)$ is the drop of present quotation p_o with respect to reference quotation p_z

Thus, global index i is given by:

$$i = d^{a_1} \cdot s^{a_2} \cdot c^{a_3} \cdot m^{a_4} \cdot v^{a_5} \cdot f^{a_6}$$

where $a_1, \ldots, a_6$ are positive exogenous parameters that can be estimated using statistical methods on historical series of past takeovers.

It has to be noted that the resulting index is still limited to between 0 and 1 (0 for minimal and 1 for maximum de-stability).

5.10 Portfolio Theory

Certain developments of the results above even involve the Theory of Portfolio Selection. It is known that traditional portfolio models imply the diversification of investments to minimize risk: the classic models of Portfolio Selection advise the saver to diversify his share portfolio in such a way as to efficiently reduce risk (the problem is solved by multi-task optimization by maximizing the expected return and minimizing risk (see [63]). This, however, is in conflict with the relevant amount of a single stock that needs to be acquired to carry out hostile take-over bids (TOB). The connection between takeover and portfolio theories was initially approached by Amihud and Barnea [2] and Batteau [7], who found a hindrance in determining the control function; this function was determined at the beginning of the '80s in [33, 34]. A method of linking these two theories has been proposed by means of a control propensity index that can be linked to the risk aversion index (see [33, 40]).

To summarize, the optimal composition of a portfolio is determined by taking into account not only the expected return and variance of the classical investments but also of the investments with ordinary shares to be used for control.

One of the difficulties for this generalization is that the price is aleatory in the new model, whereas a fixed price for buying the shares was assumed in classic models.

The method is as follows:

- To identify the "index of inclination to the control" of the investor that can be connected to his risk aversion as it is used in classic models;
- To share the capital in two classes of investments by using the new index (those classic and those for control);
- To identify the company to takeover (or companies, if small—with respect to their available capital) and to identify the more-suitable power quotes in each company;
- To eliminate from the group of companies, used for classic investments, those already chosen for takeover and those with a strong correlation with the final one;
- To undertake in a silent way the purchase of shares for the takeover;
- To finalize the operation.

To apply the model, the algorithms described below are used. For further application of the games theory to the portfolio, see [6, 40].

To conclude this section, attention should be paid to the recent work by Crama and Leruth [22] in which they show how techniques such as power indices are more-suitable than cut-off methods for describing power-sharing among shareholders.

5.11   Algorithms

Many open problems are to be found in the search for properties common to various power indices, especially with reference to alternative share shifts. On the basis of considerations regarding the geometrical properties of the Shapley value, an algorithm was used for calculating this value in super-additive games (see [32]), which is in games where the payoff of each coalition is not lower than the sum of payoffs of any partition of it. The algorithm was generalized in Gambarelli [35] for subadditive games. The algorithm is linear in the number of significant coalitions and uses a theorem of early stop, based on reaching the desired degree of precision. In majority games having a low total sum of weights, the Shapley value (which assumes the role of the Shapley-Shubik index in these cases) can be better-calculated using the algorithm proposed by Mann and Shapley [46]. This algorithm was suggested to Shapley by an idea by Cantor (see also [16]).

The generation of "power" function relative to share exchanges between parties necessitates the repeated use of Mann and Shapley's algorithm in each of the constant power regions. A subsequent algorithm by Arcaini and Gambarelli [3] enables further savings in calculation, as it directly generates the increase in the index starting from each point of discontinuity, taking into account the information that was used to calculate the preceding value.

A similar technique was applied in Gambarelli [37] to generate the power function in the case of the normalized Banzhaf index. This algorithm also provides a direct method of calculating this index. This method turned out to be faster than the one used previously (i.e., the one suggested by Banzhaf [5]). Certain improvements in the calculation of the normalized Banzhaf index have been subsequently proposed (see, for instance, [1]); however, the possibility of further savings in time remains open, especially in the case of seat shifts.

In the end, a program is now available to generate the power functions in the cases of both of Shapley-Shubik and Banzhaf-Coleman indices for the exchange of shares between two shareholders or between a shareholder and the ocean. Such a program needs a small amount of memory (as it manages only a few vectors) and can be coupled with other algorithms for the portfolio choice.

6   Some Open Problems

Various problems are still open both in the theory and application of power indices: in the following section, some of these issues are listed.

6.1   Open Problems in Theory

As far as the theory of power indices, very interesting issues are as follows:

- To apply the results of [35] on the Shapley value as a center of gravity to other values with the development of more-effective algorithms for relevant computation;
- To complete the studies of [8, 9] on the comparison among various indices in literature to identify a more-suitable index for the specific application.

6.2   Open Problems in Finance

The possibility for a group of shareholders to take over a company with a share majority can lead to considerable economic advantages. For this reason, we sometimes see share transactions that are not linked to traditional objectives of expected returns and risk but rather aimed at acquiring control of a firm. Until a few decades ago, traditional financial theories did not deal with these problems, mainly because the theory of power indices was little-known in the field. The first models have been developed more recently.

An interesting application in the financial field could be to find out a better description of how to form control coalitions by applying models applied only to political applications up until now; for example [54].

Here in the following, some specific problems are identified; for further issues regarding the application of power indices to the financial field, see [11].

A complete unexplored area of research is the use of the fuzzy sets theory in order to model the uncertainty in the evaluation of power indices [28–30].

6.2.1   Moving Shares to Gain Control

Formulae have been devised to determine changes in an investor's power in a company following an exchange of shares with others (see [31, 34]), and successive work by Freixas). Moreover, algorithms have been drawn up to calculate the Shapley value [35] and the variations of both Shapley-Shubik and Banzhaf-Coleman indices following exchanges of shares (see [37]). These tools may be useful not only to the bidder but also to the current controller, because they enable him to assess the stability of his position in relation to potential takeovers (see [36]).

Some financial institutions have begun using these techniques, though obviously without divulging their related results. Therefore, a comparison between theoretical models and their application remains an open problem at the official level.

Moreover, the above-mentioned formulae concern the exchange of shares between two shareholders or among one shareholder and an ocean of small shareholders who cannot control the firm (we note that some studies regarding small shareholders who can control the firm have been developed, starting with Milnor and Shapley [48] onwards. It would be useful to widen such research to include other types of buying and selling.

In the ocean games, the limit behavior of the power indices is studied when the weight of the greater of the small shareholders goes to zero. Some results of those types of games shown in [4, 25, 27, 48, 59] could be compared with the results of the present paper, and the main differences could be discussed.

Some practical work is moreover necessary to calibrate the parameters for the stability index shown in Sect. 5.9 by the use of historical series of past takeover situations; also, some more work is needed for models dealing with portfolio selection. The confidentiality of real data makes such studies difficult.

6.2.2   Indirect Control

It could be very useful to develop algorithms for the computation of indirect control with reference to the studies of [21, 24, 39, 44, 47].

A particularly interesting problem concerns those cases where an investor has a shareholding in a certain company that, in turn, holds shares in another company (and so on). In situations of this kind, it may be useful to calculate the power in the whole system. This problem was tackled in Gambarelli and Owen [39] by transforming the set of inter-connected games into just one game, using the multi-linear extensions introduced by Owen in [49]. The power index considered to be most-suitable in describing the situation in question can then be applied to the unified game.

Karos and Peters [44] axiomatically developed a large class of power indices that satisfy some axioms and can measure the power of players in a shareholding network. They also indicated several interesting possibilities for further theoretical research in this area. One open problem is what becomes feasible in terms of power indices if the axioms are changed. The next further development refers to a mutual control structure that can be modeled as a hypergraph, and a value for transferable utility games combined with such a hypergraph can take the imposed control relationships into consideration.

An open problem is the cumbersome computation of the algorithm that should be improved for practical applications. An algorithm for the automatic computation of such situations was elaborated by Denti and Prati [24] and Crama and Leruth [21], but it is hoped that even-more-efficient techniques will be devised.

6.2.3 Portfolio Theory

Possible developments of the Theory of Portfolio Selection remain open. Furthermore, analogously to what has been mentioned above, a comparison between the model and the applicative phase at an official level is still open.

Acknowledgements This work is sponsored by research grants from the University of Bergamo by the Group GNAMPA of INDAM and the statutory funds (no. 11/11.200.322) of the AGH University of Science and Technology. Some parts are taken from previous papers of the same authors; for example [11, 12, 41]. The authors thank the editors for the relevant authorizations.

A special thanks is due to Manfred Holler for having given permission to take some figures from [41].

References

1. Algaba E, Bilbao JM, Fernández JR, López JJ (2003) Computing power indices in weighted multiple majority games. Math Soc Sci 46(1):63–80
2. Amihud Y, Barnea A (1974) Portfolio selection for managerial control. Omega 2:775–783
3. Arcaini G, Gambarelli G (1986) Algorithm for automatic computation of the power variations in share tradings. Calcolo 23(1):13–19
4. Aumann RJ, Shapley LS (1974) Values of non-atomic games. Princeton University Press, Princeton
5. Banzhaf JF (1965) Weighted voting doesn't work: a mathematical analysis. Rutgers law review 19:317–343
6. Bassetti A, Torricelli C (1992) Optimal portfolio selection as a solution to an axiomatic bargaining game. In: Feichtinger G (ed) Dynamic economic models and optimal control. Elsevier Publisher, Amsterdam, pp 373–384
7. Batteau P (1980) Approches formelles du problème du control des firmes et sociétés par actions. Revue de l'Association Française de Finance 1:1–26
8. Bertini C, Freixas J, Gambarelli G, Stach I (2013) Comparing power indices. In: Fragnelli V, Gambarelli G (eds) Open problems in the theory of cooperative games, special issue of international game theory review, vol 15, no 2, pp 1340004-1–1340004-19
9. Bertini C, Freixas J, Gambarelli G, Stach I (2013) Some open problems in simple games. In: Fragnelli V, Gambarelli G (eds) Open problems in the theory of cooperative games. Special issue of international game theory review, vol 15, no 2, pp 1340005-1–1340005-18
10. Bertini C, Gambarelli G, Stach I (2008) A public help index. In: Braham M, Steffen F (eds) Power, freedom, and voting. Springer, Berlin-Haidelberg, pp 83–98
11. Bertini C, Gambarelli G, Stach I (2015) Some open problems in the application of power indices to politics and finance. Homo Oeconomicus 32(1):147–156
12. Bertini C, Gambarelli G, Stach I (2016) Indici di potere in politica e in finanza, Bollettino dei docenti di matematica, 72:9–34. Repubblica e Cantone Ticino Ed, Bellinzona - Svizzera. Power indices in politics and finance, (in Italian)
13. Bertini C, Mercik J, Stach I (2016) Indirect control and power. Oper Res Decis 26(2):7–30
14. Bertini C, Stach I (2011) Banzhaf voting power measure. In: Dowding K (ed) Encyclopedia of power. SAGE Publications, Los Angeles, pp 54–55
15. Bertini C, Stach I (2011) Coleman index. In: Dowding K (ed) Encyclopedia of power. SAGE Publications, Los Angeles, pp 117–119
16. Brams SJ, Affuso PJ (1976) Power and size: a new paradox. Theory Decis 7:29–56

17. Brioschi F, Buzzacchi L, Colombo MG (1990) Gruppi di imprese e mercato finanziario. La Nuova Italia Scientifica, Roma

18. Buzzacchi L, Mosconi R (1993) Azioni di risparmio e valore di controllo: alcune evidenze empiriche e spunti di ricerca. Centro di Economia Monetaria e Finanziaria Paolo Baffi, Milano 73

19. Coleman JS (1971) Control of collectivities and the power of collectivity to act. In: Liberman B (ed) Social choice. Gordon and Breach, New York, pp 269–300

20. Corielli F, Nicodano G, Rindi B (1993) The value of non-voting shares, the structure of corporate control and market liquidity. Centro di Economia Monetaria e Finanziaria Paolo Baffi, Milano

21. Crama Y, Leruth L (2007) Control and voting power in corporate networks: concepts and computational aspects. Eur J Oper Res 178:879–893

22. Crama Y, Leruth L (2013) Power indices and the measurement of control in corporate structures. In: Fragnelli V, Gambarelli G (eds) Open problems in the applications of cooperative games. Special issue of international game theory review, vol 15, no 3, pp 1340017-1–1340017-15

23. Denti E, Prati N (2001) An algorithm for winning coalitions in indirect control of corporations. Decis Econ Finan 24(2):153–158

24. Denti E, Prati N (2004) Relevance of winning coalitions in indirect control of corporations. In: Gambarelli G (ed) Essays on cooperative games—in honor of guillermo owen. Special issue of theory and decision 36:183–192

25. Dragan I, Gambarelli G (1990) The compensatory bargaining set of a big boss game. Libertas Math 10:53–61

26. Dubey P (1975) On the uniqueness of the Shapley value. Int J Game Theory 4:131–139

27. Dubey P (1975) Some results on values of finite and infinite games. Technical Report, Center of Applied Mathematics, Cornell University, Ithaca, 14853

28. Fedrizzi M, Kacprzyk J (1990) Vague notions in the theory of voting. In Fedrizzi M, Kacprzyk J (eds) Multiperson decision making models using fuzzy sets and possibility theory. Theory and decision library. Series B, Mathematical and statistical methods. Kluwer academic, Dordrecht, Boston, New York, N.Y., pp 43–52

29. Fedrizzi M, Kacprzyk J (1995) Brief introduction to fuzzy sets. In Kacprzyk J, Onisawa T (eds) Reliability and safety analyses under fuzziness. Studies in Fuzziness. Physica-Verlag, Heidelberg, New York, N.Y., pp 31–39

30. Fedrizzi M, Kacprzyk J (1999) A brief introduction to fuzzy sets and fuzzy systems. In Cardoso J, Camargo H (eds) Fuzziness in Petri nets. Studies in Fuzziness and Soft Computing. Physica-Verlag, Heidelberg, New York, N.Y., pp 25–51

31. Freixas J, Gambarelli G (1997) Common properties among power indices. Control Cybern 26(4):591–603

32. Gambarelli G (1980) Algorithm for the numerical computation of the Shapley value of a game. Rivista di Statistica Apllicata 13(1):35–41

33. Gambarelli G (1982) Portfolio selection and firms' control. Finance 3(1):69–83

34. Gambarelli G (1983) Common behaviour of power indices. Int J Game Theory 12(4):237–244

35. Gambarelli G (1990) A new approach for evaluating the Shapley value. Optimization 21 (3):445–452

36. Gambarelli G (1993) An index of de-stability for controlling shareholders. In: Flavell R (ed) Modelling reality and personal modelling. Physica-Verlag, Heidelberg, pp 116–127

37. Gambarelli G (1996) Takeover algorithms. In: Bertocchi M et al (eds) Modelling techniques for financial markets and bank management, proceedings of the 16-th and 17-th Euro working group of financial modelling meetings. Physica Verlag, Heidelberg, pp 212–22

38. Gambarelli G, Hołubiec J (1990) Power indices and democratic apportionments. In: Fedrizzi M, Kacprzyk J (eds) Proceedings of the 8-th Italian-Polish symposium on systems analysis and decision support in economics and technology. Onnitech Press, Varsavia, pp 240–255

39. Gambarelli G, Owen G (1994) Indirect control of corporations. Int J Game Theory 23 (4):287–302
40. Gambarelli G, Pesce S (2004) Takeover prices and portfolio theory. In Gambarelli G (ed) Essays on cooperative games—in honor of guillermo owen. Special issue of theory and decision, vol 36. Kluwer Academic Publishers, Dordrecht, pp 193–203
41. Gambarelli G, Stach I (2009) Power indices in politics: some results and open problems. Essays in Honor of Hannu Nurmi, Homo Oeconomicus 26(3/4):417–441
42. Holler MJ, Napel S (2004) Monotonicity of power and power measures. Theory Decis 56(2_2):93–111
43. Hu X, Shapley LS (2003) On authority distributions in organizations: controls. Game Econ Behav 45(1):153
44. Karos D, Peters H (2015) Indirect control and power in mutual control structures. Game Econ Behav 92:150–165
45. Kołodziej M, Stach I (2016) Control sharing analysis and simulation. In: Sawik T (ed) Conference proceedings: ICIL 2016: 13th international conference on industial logistics, 28 September–1 October, Zakopane, Poland, pp 101–108
46. Mann I, Shapley LS (1962) Values of large games, VI: evaluating the electoral college exactly. Rand Corporation, RM 3158, Santa Monica
47. Mercik J, Łobos K (2016) Index of implicit power as a measure of reciprocal ownership. Springer lecture notes in computer science, vol 9760, pp 132–145
48. Milnor JW, Shapley LS (1961) Values of large games II: oceanic games. Rand Corporation, RM 2646, Santa Monica
49. Owen G (1972) Multilinear extensions of games. Manage Sci 18:64–79
50. Owen G (1977) Values of games with a priori unions. Lect Notes Econ Math 141:76–88
51. Owen G (1978) Characterization of the Banzhaf-Coleman index. SIAM J Appl Math 35:315–327
52. Owen G (1981) Modification of the Banzhaf-Coleman index for games with a priori unions. In: Holler MJ (ed) Power. Voting and Voting Power, Physica, Würzburg, pp 232–238
53. Owen G (1995) (III ed.) Game theory. Academic Press, San Diego
54. Owen G, Shapley LS (1989) Optimal location of candidates in ideological space. Int J Game Theory 18(3):339–356
55. Penrose LS (1946) The elementary statistics of majority voting. J Roy Stat Soc 109:53–57
56. Riker WH (1986) The first power index. Soc Choice Welfare 3:293–295
57. Rydqvist K (1985) The pricing of shares with different voting power and the theory of oceanic games. Stockolm School of Economics 4:1–70
58. Salvemini MT, Simeone B, Succi R (1995) Analisi del possesso integrato nei gruppi di imprese mediante grafi. L'industria 16(4):641–662
59. Shapiro NZ, Shapley LS (1960) Values of large games, I: a limit theorem. Rand Corporation, RM 2646, Santa Monica
60. Shapley LS (1953) A value for n-person games. In: Tucker AW, Kuhn HW (eds) Contributions to the theory of games II. Princeton University Press, Princeton, pp 307–317
61. Shapley LS, Shubik M (1954) A method for evaluating the distributions of power in a committee system. Am Polit Sci Rev 48:787–792
62. Stach I (2011) Shapley-Shubik index. In: Dowding K (ed) Encyclopedia of power. SAGE Publications, Los Angeles, pp 603–606
63. Szegö G (1980) Portfolio theory. Academic Press, New York
64. Turnovec F (1997) Monotonicity of power indices. East European series 41. Institute for Advanced Studies

The Binomial Decomposition of the Single Parameter Family of GB Welfare Functions

Silvia Bortot, Ricardo Alberto Marques Pereira
and Anastasia Stamatopoulou

Abstract We consider the binomial decomposition of generalized Gini welfare functions in terms of the binomial welfare functions $C_j, j = 1, \ldots, n$ and we examine the weighting structure of the binomial welfare functions $C_j, j = 1, \ldots, n$ which progressively focus on the poorest part of the population. We introduce a parametric family of income distributions and we illustrate the numerical behavior of the single parameter family of GB welfare functions with respect to those income distributions. Moreover, we investigate the binomial decomposition of the GB welfare functions and we illustrate the dependence of the binomial decomposition coefficients in relation with the single parameter which describes the family.

Keywords Generalized Gini welfare functions · Binomial decomposition of the single parameter family of GB welfare functions

1 Introduction

The generalized Gini welfare functions introduced by Weymark [61], and the associated inequality indices in Atkinson-Kolm-Sen's (AKS) framework, see Atkinson [5], Kolm [48, 49], and Sen [56], are related by Blackorby and Donaldson's correspondence formula [13, 14], $A(x) = \bar{x} - G(x)$, where $A(x)$ denotes a generalized Gini welfare function, $G(x)$ is the associated absolute inequality index, and $\bar{x}$ is the plain mean of the income distribution $x = (x_1, \ldots, x_n) \in \mathbb{D}^n$ of a population of $n \geq 2$ individuals, with income domain $\mathbb{D} = [0, \infty)$.

The generalized Gini welfare functions [61] have the form $A(x) = \sum_{i=1}^{n} w_i x_{(i)}$ where $x_{(1)} \leq x_{(2)} \leq \cdots \leq x_{(n)}$ and, as required by the principle of inequality aversion, $w_1 \geq w_2 \geq \cdots \geq w_n \geq 0$ with $\sum_{i=1}^{n} w_i = 1$. These welfare functions correspond to a particular class of the ordered weighted averaging (OWA) functions introduced by

S. Bortot (✉) · R.A. Marques Pereira · A. Stamatopoulou
Department of Economics and Management, University of Trento,
Via Inama, 5, 38122 Trento, Italy
e-mail: silvia.bortot@unitn.it

© Springer International Publishing AG 2018
M. Collan and J. Kacprzyk (eds.), *Soft Computing Applications for Group Decision-making and Consensus Modeling*, Studies in Fuzziness and Soft Computing 357, DOI 10.1007/978-3-319-60207-3_5

Yager [64], which in turn correspond [35] to the Choquet integrals associated with symmetric capacities.

In this paper we recall the binomial decomposition of generalized Gini welfare functions due to Calvo and De Baets [22], see also Bortot and Marques Pereira [20]. The binomial decomposition is formulated in terms of the functional basis formed by the binomial welfare functions.

The binomial welfare functions, denoted C_j with $j = 1, \ldots, n$, have null weights associated with the $j - 1$ richest individuals in the population and therefore they are progressively focused on the poorest part of the population.

The paper is organized as follows. In Sect. 2 we review the notions of generalized Gini welfare function and the associated generalized Gini inequality index for populations of $n \geq 2$ individuals.

In Sect. 3 we consider the binomial decomposition of generalized Gini welfare functions in terms of the binomial welfare functions $C_j, j = 1, \ldots, n$. We examine the weights of the binomial welfare functions $C_j, j = 1, \ldots, n$ which progressively focus on the poorest part of the population.

In Sect. 4 we investigate the single parameter family of GB welfare functions, particularly in the context of the binomial decomposition. In Sect. 4.1, we illustrate the weighting structure and the numerical behavior of the GB welfare functions in relation with a parametric family of income distributions. Moreover, in Sect. 4.2, we study the binomial decomposition of the GB welfare functions in the cases $n = 2, 4, 6, 8$. Finally, Sect. 5 contains some conclusive remarks.

2   Generalized Gini Welfare Functions and Inequality Indices

In this section we consider populations of $n \geq 2$ individuals and we briefly review the notions of generalized Gini welfare function and generalized Gini inequality index over the income domain $\mathbb{D} = [0, \infty)$. The income distributions in this framework are represented by points $x, y \in \mathbb{D}^n$.

We begin by presenting notation and basic definitions regarding averaging functions on the domain $\mathbb{D}^n$, with $n \geq 2$ throughout the text. Comprehensive reviews of averaging functions can be found in Chisini [27], Fodor and Roubens [34], Calvo et al. [23], Beliakov et al. [9], Grabisch et al. [46], and Beliakov et al. [10].

Notation. Points in $\mathbb{D}^n$ are denoted $x = (x_1, \ldots, x_n)$, with $\mathbf{1} = (1, \ldots, 1)$, $\mathbf{0} = (0, \ldots, 0)$. Accordingly, for every $x \in \mathbb{D}$, we have $x \cdot \mathbf{1} = (x, \ldots, x)$. Given $x, y \in \mathbb{D}^n$, by $x \geq y$ we mean $x_i \geq y_i$ for every $i = 1, \ldots, n$, and by $x > y$ we mean $x \geq y$ and $x \neq y$. Given $x \in \mathbb{D}^n$, the increasing and decreasing reorderings of the coordinates of x are indicated as $x_{(1)} \leq \cdots \leq x_{(n)}$ and $x_{[1]} \geq \cdots \geq x_{[n]}$, respectively. In particular, $x_{(1)} = \min\{x_1, \ldots, x_n\} = x_{[n]}$ and $x_{(n)} = \max\{x_1, \ldots, x_n\} = x_{[1]}$. In general, given a permutation σ on $\{1, \ldots, n\}$, we denote $x_\sigma = (x_{\sigma(1)}, \ldots, x_{\sigma(n)})$. Finally, the arithmetic mean is denoted $\bar{x} = (x_1 + \cdots + x_n)/n$.

Definition 1 Let $A : \mathbb{D}^n \longrightarrow \mathbb{D}$ be a function. We say that

1. A is *monotonic* if $x \geq y \Rightarrow A(x) \geq A(y)$, for all $x, y \in \mathbb{D}^n$. Moreover, A is *strictly monotonic* if $x > y \Rightarrow A(x) > A(y)$, for all $x, y \in \mathbb{D}^n$.
2. A is *idempotent* if $A(x \cdot 1) = x$, for all $x \in \mathbb{D}$. On the other hand, A is *nilpotent* if $A(x \cdot 1) = 0$, for all $x \in \mathbb{D}$.
3. A is *symmetric* if $A(x_\sigma) = A(x)$, for any permutation σ on $\{1, \dots, n\}$ and all $x \in \mathbb{D}^n$.
4. A is *invariant for translations* if $A(x + t \cdot 1) = A(x)$, for all $t \in \mathbb{D}$ and $x \in \mathbb{D}^n$. On the other hand, A is *stable for translations* if $A(x + t \cdot 1) = A(x) + t$, for all $t \in \mathbb{D}$ and $x \in \mathbb{D}^n$.
5. A is *invariant for dilations* if $A(t \cdot x) = A(x)$, for all $t \in \mathbb{D}$ and $x \in \mathbb{D}^n$. On the other hand, A is *stable for dilations* if $A(t \cdot x) = t A(x)$, for all $t \in \mathbb{D}$ and $x \in \mathbb{D}^n$.

The terms positive (negative), increasing (decreasing), and monotonic are used in the weak sense. Otherwise these properties are said to be strict.

Definition 2 A function $A : \mathbb{D}^n \longrightarrow \mathbb{D}$ is an *averaging function* if it is monotonic and idempotent. An averaging function is said to be *strict* if it is strictly monotonic. Note that monotonicity and idempotency implies that $\min(x) \leq A(x) \leq \max(x)$, for all $x \in \mathbb{D}^n$.

Particular cases of averaging functions are weighted averaging (WA) functions, ordered weighted averaging (OWA) functions, and Choquet integrals, which contain the former as special cases.

Definition 3 Given a weighting vector $w = (w_1, \dots, w_n) \in [0, 1]^n$, with $\sum_{i=1}^n w_i = 1$, the *Weighted Averaging (WA) function* associated with w is the averaging function $A : \mathbb{D}^n \longrightarrow \mathbb{D}$ defined as

$$A(x) = \sum_{i=1}^n w_i x_i. \tag{1}$$

Definition 4 Given a weighting vector $w = (w_1, \dots, w_n) \in [0, 1]^n$, with $\sum_{i=1}^n w_i = 1$, the *Ordered Weighted Averaging (OWA) function* associated with w is the averaging function $A : \mathbb{D}^n \longrightarrow \mathbb{D}$ defined as

$$A(w) = \sum_{i=1}^n w_i x_{(i)}. \tag{2}$$

The traditional form of OWA functions as introduced by Yager [64] is as follows, $A(x) = \sum_{i=1}^n \tilde{w}_i x_{[i]}$ where $\tilde{w}_i = w_{n-i+1}$. In [65, 66] the theory and applications of OWA functions are discussed in detail.

The following is a classical result particulary relevant in our framework. This result regards a form of dominance relation between OWA functions and the associated weighting structures, see for instance Bortot and Marques Pereira [20] and references therein.

Proposition 1 *Consider two OWA functions* $A, B : \mathbb{D}^n \longrightarrow \mathbb{D}$ *associated with weighting vectors* $\boldsymbol{u} = (u_1, \ldots, u_n) \in [0, 1]^n$ *and* $\boldsymbol{v} = (v_1, \ldots, v_n) \in [0, 1]^n$, *respectively. It holds that* $A(\boldsymbol{x}) \leq B(\boldsymbol{x})$ *for all* $\boldsymbol{x} \in \mathbb{D}^n$ *if and only if*

$$\sum_{i=1}^{k} u_i \geq \sum_{i=1}^{k} v_i \quad \text{for} \quad k = 1, \ldots, n \tag{3}$$

where the case $k = n$ *is an equality due to weight normalization.*

A class of welfare functions which plays a central role in this paper is that of the generalized Gini welfare functions introduced by Weymark [61], see also Mehran [51], Donaldson and Weymark [30, 31], Yaari [62, 63], Ebert [33], Quiggin [55], Ben-Porath and Gilboa [12].

Definition 5 Given a weighting vector $\boldsymbol{w} = (w_1, \ldots, w_n) \in [0, 1]^n$, with $w_1 \geq \cdots \geq w_n \geq 0$ and $\sum_{i=1}^{n} w_i = 1$, the *generalized Gini welfare function* associated with $\boldsymbol{w}$ is the function $A : \mathbb{D}^n \longrightarrow \mathbb{D}$ defined as

$$A(\boldsymbol{x}) = \sum_{i=1}^{n} w_i x_{(i)} \tag{4}$$

and, in the AKS framework, the associated *generalized Gini inequality index* is defined as

$$G(\boldsymbol{x}) = \bar{x} - A(\boldsymbol{x}) = -\sum_{i=1}^{n} \left(w_i - \frac{1}{n} \right) x_{(i)}. \tag{5}$$

The generalized Gini welfare functions are *strict* if and only if $w_1 > \cdots > w_n > 0$. The generalized Gini welfare functions are stable for translations and the associated generalized Gini inequality indices are invariant for translations. Both are stable for dilations.

The classical Gini [37–39], Bonferroni [18, 19], and De Vergottini [28, 29] welfare functions and associated inequality indices are important instances of the AKS generalized Gini framework. The substantial subsequent research on the three classical cases of generalized Gini welfare functions can be found, for instance, in Kolm [47], Atkinson [5], Sen [56, 57] Piesch [53, 54], Mehran [51], Blackorby and Donaldson [13–16], Lorenzen [50], Donaldson and Weymark [30, 31], Nygård and Sandström [52], Blackorby et al. [17], Weymark [61], Yitzhaki [67], Giorgi [40, 41], Benedetti [11], Ebert [32], Shorrocks and Foster [58], Yaari [63], Silber [59], Bossert [21], Tarsitano [60], Ben Porath and Gilboa [12], Zoli [69], Gajdos [36], Aaberge [1–3], Giorgi and Crescenzi [42, 43], Chakravarty and Muliere [26], Chakravarty [24, 25], Bárcena and Imedio [6], Giorgi and Nadarajah [44], Bárcena and Silber [7, 8], Aristondo et al. [4], and Zenga [68].

The central instance of the AKS generalized Gini framework is the classical Gini welfare function $A_G(\boldsymbol{x})$ and the associated classical Gini inequality index $G(\boldsymbol{x}) = \bar{x} - A_G(\boldsymbol{x})$,

$$A_G(x) = \sum_{i=1}^{n} w_i^G x_{(i)} \qquad w_i^G = \frac{2(n-i)+1}{n^2} \qquad (6)$$

where the weights of $A_G(x)$ are positive and strictly decreasing with unit sum, $\sum_{i=1}^{n} w_i^G = 1$, and

$$G(x) = \sum_{i=1}^{n} \left(\frac{1}{n} - w_i^G \right) x_{(i)} = -\sum_{i=1}^{n} \frac{n-2i+1}{n^2} x_{(i)} \qquad (7)$$

where the coefficients of $G(x)$ have zero sum.

The classical absolute Gini inequality index G is traditionally defined as

$$G(x) = \frac{1}{2n^2} \sum_{i,j=1}^{n} |x_i - x_j| = -\frac{1}{n^2} \sum_{i=1}^{n-1} \sum_{j=i+1}^{n} \left(x_{(i)} - x_{(j)} \right) \qquad (8)$$

where the double summation expression for $-n^2 G(x)$ in (8) can be written as

$$(-(n-1))x_{(1)} + (1-(n-2))x_{(2)} + \cdots + ((n-2)-1)x_{(n-1)} + (n-1)x_{(n)} \qquad (9)$$

which corresponds to (7).

Another important instance of the AKS generalized Gini framework is the classical Bonferroni welfare function $A_B(x)$ and the associated classical Bonferroni inequality index $B(x) = \bar{x} - A_B(x)$,

$$A_B(x) = \sum_{i=1}^{n} w_i^B x_{(i)} \qquad w_i^B = \sum_{j=i}^{n} \frac{1}{jn} \qquad (10)$$

where the weights of $A_B(x)$ are positive and strictly decreasing with unit sum, $\sum_{i=1}^{n} w_i^B = 1$, and

$$B(x) = \sum_{i=1}^{n} \left(\frac{1}{n} - w_i^B \right) x_{(i)} \qquad (11)$$

where the coefficients of $B(x)$ have zero sum.

The classical absolute Bonferroni inequality index B is traditionally defined as

$$B(x) = \bar{x} - \frac{1}{n} \sum_{i=1}^{n} m_i(x) \qquad (12)$$

where the mean income of the i poorest individuals in the population is given by

$$m_i(x) = \frac{1}{i} \sum_{j=1}^{i} x_{(j)} \qquad \text{for} \quad i = 1, \dots, n. \tag{13}$$

Therefore we have

$$A_B(x) = \frac{1}{n} \sum_{i=1}^{n} m_i(x) \tag{14}$$

$$= \frac{1}{n} \left[\left(x_{(1)} \right) + \frac{1}{2} \left(x_{(1)} + x_{(2)} \right) + \dots + \frac{1}{n} \left(x_{(1)} + \dots + x_{(n)} \right) \right] \tag{15}$$

$$= \frac{1}{n} \left[\sum_{j=1}^{n} \frac{1}{j} x_{(1)} + \sum_{j=2}^{n} \frac{1}{j} x_{(2)} + \dots \sum_{j=n}^{n} \frac{1}{j} x_{(n)} \right] \tag{16}$$

which corresponds to (10).

3 The Binomial Decomposition

In this section we review the binomial decomposition of generalized Gini welfare functions due to Calvo and De Baets [22] and Bortot and Marques Pereira [20]. We examine the weighting structures of the binomial welfare functions C_j, with $j = 1, \dots, n$.

Definition 6 The *binomial welfare functions* $C_j : \mathbb{D}^n \longrightarrow \mathbb{D}$, with $j = 1, \dots, n$, are defined as

$$C_j(x) = \sum_{i=1}^{n} w_{ji} x_{(i)} = \sum_{i=1}^{n} \frac{\binom{n-i}{j-1}}{\binom{n}{j}} x_{(i)} \qquad j = 1, \dots, n \tag{17}$$

where the binomial weights w_{ji}, $i, j = 1, \dots, n$ are null when $i + j > n + 1$, according to the usual convention that $\binom{p}{q} = 0$ when $p < q$, with $p, q = 0, 1, \dots$ Given that the binomial weights are decreasing, $w_{j1} \geq w_{j2} \geq \dots \geq w_{jn}$ for $j = 1, \dots, n$, the binomial welfare functions are generalized Gini welfare functions.

Except for $C_1(x) = \bar{x}$, the binomial welfare functions C_j, $j = 2, \dots, n$ have an increasing number of null weights, in correspondence with $x_{(n-j+2)}, \dots, x_{(n)}$. The weight normalization of the binomial welfare functions, $\sum_{i=1}^{n} w_{ji} = 1$ for $j = 1, \dots, n$, is due to the column-sum property of binomial coefficients,

$$\sum_{i=1}^{n} \binom{n-i}{j-1} = \sum_{i=0}^{n-1} \binom{i}{j-1} = \binom{n}{j} \qquad j = 1, \dots, n. \tag{18}$$

The binomial welfare functions $C_j, j = 1, \ldots, n$ are continuous, idempotent, and stable for translations, where the latter two properties follow immediately from $\sum_{i=1}^{n} w_{ji} = 1$ for $j = 1, \ldots, n$.

The following interesting result concerning the cumulative properties of binomial weights is due to Calvo and De Baets [22], see also Bortot and Marques Pereira [20].

Proposition 2 *The binomial weights $w_{ji} \in [0, 1]$, with $i, j = 1, \ldots, n$, have the following cumulative property,*

$$\sum_{i=1}^{k} w_{j-1,i} \leq \sum_{i=1}^{k} w_{ji} \qquad k = 1, \ldots, n \tag{19}$$

for each $j = 2, \ldots, n$.

Given that the binomial weights have the cumulative property (19), Proposition 1 implies that the binomial welfare functions $C_j, j = 1, \ldots, n$ satisfy the relations $\bar{x} = C_1(x) \geq C_2(x) \geq \cdots \geq C_n(x) \geq 0$, for any $x \in \mathbb{D}^n$.

Proposition 3 *Generalized Gini welfare functions $A : \mathbb{D}^n \longrightarrow \mathbb{D}$ can be written uniquely as*

$$A(x) = \alpha_1 C_1(x) + \alpha_2 C_2(x) + \cdots + \alpha_n C_n(x) \tag{20}$$

where the coefficients $\alpha_j, j = 1, \ldots, n$ are subject to the following conditions,

$$\alpha_1 = 1 - \sum_{j=2}^{n} \alpha_j \geq 0 \tag{21}$$

$$\sum_{j=2}^{n} \left[1 - n \frac{\binom{i-1}{j-1}}{\binom{n}{j}} \right] \alpha_j \leq 1 \qquad i = 2, \ldots, n \tag{22}$$

$$\sum_{j=2}^{n} \frac{\binom{n-i}{j-2}}{\binom{n}{j}} \alpha_j \geq 0 \qquad i = 2, \ldots, n. \tag{23}$$

The binomial welfare functions constitute therefore a functional basis for the generalized Gini welfare functions, which can be uniquely expressed as $A(x) = \sum_{j=1}^{n} \alpha_j C_j(x)$ where the coefficients $\alpha_j, j = 1, \ldots, n$ satisfy the constraints (21)–(22)–(23), one of which is $\sum_{j=1}^{n} \alpha_j = 1$. However, the binomial decomposition does not express a simple convex combination of the binomial welfare functions, as the condition $\alpha_1 + \cdots + \alpha_n = 1$ might suggest. In fact, condition (21) ensures $\alpha_1 \geq 0$ but conditions (22)–(23) allow for negative $\alpha_2, \ldots, \alpha_n$ values.

Notice that the only strict binomial welfare function is $C_1(x) = \bar{x}$ for all $x \in \mathbb{D}^n$. On the other hand, $C_2(x)$ has $n - 1$ positive linearly decreasing weights and one null last weight. In terms of the classical Gini welfare function we have that $A^c(x) = \frac{1}{n} C_1(x) + \frac{n-1}{n} C_2(x)$. The remaining $C_j(x), j = 3, \ldots, n$, have $n - j + 1$ positive nonlinear decreasing weights and $j - 1$ null last weights.

In dimensions $n = 2, 3, 4, 5, 6, 7, 8$ the weights $w_{ij} \in [0, 1]$, $i, j = 1, \ldots, n$ of the binomial welfare functions $C_j, j = 1, \ldots, n$ are as follows,

$n = 2$

$C_1 : (\frac{1}{2}, \frac{1}{2})$
$C_2 : (1, 0)$

$n = 3$

$C_1 : (\frac{1}{3}, \frac{1}{3}, \frac{1}{3})$
$C_2 : (\frac{2}{3}, \frac{1}{3}, 0)$
$C_3 : (1, 0, 0)$

$n = 4$

$C_1 : (\frac{1}{4}, \frac{1}{4}, \frac{1}{4}, \frac{1}{4})$
$C_2 : (\frac{3}{6}, \frac{2}{6}, \frac{1}{6}, 0)$
$C_3 : (\frac{3}{4}, \frac{1}{4}, 0, 0)$
$C_4 : (1, 0, 0, 0)$

$n = 5$

$C_1 : (\frac{1}{5}, \frac{1}{5}, \frac{1}{5}, \frac{1}{5}, \frac{1}{5})$
$C_2 : (\frac{4}{10}, \frac{3}{10}, \frac{2}{10}, \frac{1}{10}, 0)$
$C_3 : (\frac{6}{10}, \frac{3}{10}, \frac{1}{10}, 0, 0)$
$C_4 : (\frac{4}{5}, \frac{1}{5}, 0, 0, 0)$
$C_5 : (1, 0, 0, 0, 0)$

$n = 6$

$C_1 : (\frac{1}{6}, \frac{1}{6}, \frac{1}{6}, \frac{1}{6}, \frac{1}{6}, \frac{1}{6})$
$C_2 : (\frac{5}{15}, \frac{4}{15}, \frac{3}{15}, \frac{2}{15}, \frac{1}{15}, 0)$
$C_3 : (\frac{10}{20}, \frac{6}{20}, \frac{3}{20}, \frac{1}{20}, 0, 0)$
$C_4 : (\frac{10}{15}, \frac{4}{15}, \frac{1}{15}, 0, 0, 0)$
$C_5 : (\frac{5}{6}, \frac{1}{6}, 0, 0, 0, 0)$
$C_6 : (1, 0, 0, 0, 0, 0)$

$n = 7$

$C_1 : (\frac{1}{7}, \frac{1}{7}, \frac{1}{7}, \frac{1}{7}, \frac{1}{7}, \frac{1}{7}, \frac{1}{7})$
$C_2 : (\frac{6}{21}, \frac{5}{21}, \frac{4}{21}, \frac{3}{21}, \frac{2}{21}, \frac{1}{21}, 0)$
$C_3 : (\frac{15}{35}, \frac{10}{35}, \frac{6}{35}, \frac{3}{35}, \frac{1}{35}, 0, 0)$
$C_4 : (\frac{20}{35}, \frac{10}{35}, \frac{4}{35}, \frac{1}{35}, 0, 0, 0)$
$C_5 : (\frac{15}{21}, \frac{5}{21}, \frac{1}{21}, 0, 0, 0, 0)$
$C_6 : (\frac{6}{7}, \frac{1}{7}, 0, 0, 0, 0, 0)$
$C_7 : (1, 0, 0, 0, 0, 0, 0)$

$n = 8$

$C_1 : (\frac{1}{8}, \frac{1}{8}, \frac{1}{8}, \frac{1}{8}, \frac{1}{8}, \frac{1}{8}, \frac{1}{8}, \frac{1}{8})$
$C_2 : (\frac{7}{28}, \frac{6}{28}, \frac{5}{28}, \frac{4}{28}, \frac{3}{28}, \frac{2}{28}, \frac{1}{28}, 0)$
$C_3 : (\frac{21}{56}, \frac{15}{56}, \frac{10}{56}, \frac{6}{56}, \frac{3}{56}, \frac{1}{56}, 0, 0)$
$C_4 : (\frac{35}{70}, \frac{20}{70}, \frac{10}{70}, \frac{4}{70}, \frac{1}{70}, 0, 0, 0)$
$C_5 : (\frac{35}{56}, \frac{15}{56}, \frac{5}{56}, \frac{1}{56}, 0, 0, 0, 0)$
$C_6 : (\frac{21}{28}, \frac{6}{28}, \frac{1}{28}, 0, 0, 0, 0, 0)$
$C_7 : (\frac{7}{8}, \frac{1}{8}, 0, 0, 0, 0, 0, 0)$
$C_8 : (1, 0, 0, 0, 0, 0, 0, 0)$

The binomial welfare functions $C_j, j = 1, \ldots, n$ have null weights associated with the $j - 1$ richest individuals in the population and therefore, as j increases from 1 to n, they behave in analogy with poverty measures which progressively focus on the poorest part of the population.

4 The Single Parameter GB Welfare Functions

The single parameter family of GB welfare functions which interpolates between the classical Gini and Bonferroni cases has been proposed and discussed by Don-

aldson and Weymark [30], Yitzhaki [67], Bossert [21], Aaberge [1], and Bárcena
and Silber [8].

The welfare functions of this family are of the form

$$A_{GB}(x) = \sum_{i=1}^{n} w_i^{GB} x_{(i)} \tag{24}$$

with

$$w_i^{GB} = (1/n^2)\left[n - i(n/i)^\gamma + \sum_{j=i}^{n} (n/j)^\gamma\right] \qquad \gamma \in [0, 1] \tag{25}$$

where the classical Gini and Bonferroni welfare functions are special cases with
$\gamma = 0, 1$. Note that when $\gamma = 0$ we obtain the "equally distributed equivalent level
of income" corresponding to the Gini welfare function, while when $\gamma = 1$ we obtain
the "equally distributed equivalent level of income" corresponding to the Bonferroni
welfare function.

Given that the weights of the GB welfare functions are strictly decreasing, $w_1^{GB} >
w_2^{GB} > \ldots > w_n^{GB} = 1/n^2$, the GB welfare functions are generalized Gini welfare
functions. The weighting structure of the GB welfare functions is illustrated in Fig. 1
in the cases $n = 2, 4, 6, 8$.

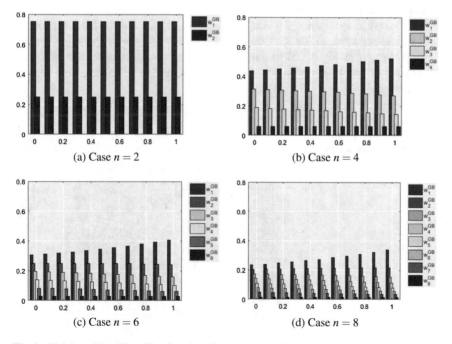

Fig. 1 Weights of the GB welfare functions for parameter values $\gamma = 0, 0.1, \ldots, 1$

4.1 A Parametric Family of Income Distributions

We now examine the GB welfare functions in relation with a parametric family of income distributions in the cases $n = 2, 4, 6, 8$. This family of income distributions, each with unit average income, is defined on the basis of the parametric Lorenz curve associated with the generating function

$$f_\beta(r) = re^{-\beta(1-r)} \qquad r \in [0, 1] \tag{26}$$

where the parameter $\beta \geq 0$ is related with inequality. Figure 2 provides a graphical illustration of the parametric Lorenz curve for parameter values $\beta = 0, 1, \ldots, 8$.

Consider a population with n individuals. The family of income distributions $x = (x_1, x_2, \ldots, x_n)$ with unit average income $\bar{x} = 1$ associated with the parametric Lorenz curve above is given by

$$x_{(i)} = n\left[f_\beta\left(\frac{i}{n}\right) - f_\beta\left(\frac{i-1}{n}\right)\right] \qquad i = 1, \ldots, n. \tag{27}$$

The values of the GB welfare functions in relation to the family of income distributions (27), in the cases $n = 2, 4, 6, 8$, is illustrated in Fig. 3.

The pattern of the numerical data suggests that the values of the GB welfare functions are decreasing with respect to the parameter $\gamma \in [0, 1]$, in the context of the parametric income distribution (27).

We observe in fact that $\gamma \leq \gamma'$ implies $A_{GB}(x) \leq A'_{GB}(x)$ for the income distributions considered here, with $x \in \mathbb{D}^n$ and $\bar{x} = 1$. This is consistent with the fact that the single parameter family of GB welfare functions interpolates between the Gini case, with higher orness $1/3 - 1/6n$, and the Bonferroni case, with lower orness $1/4$, see Aristondo et al. [4].

Fig. 2 Parametric Lorenz curve for parameter values $\beta = 0, 1, \ldots, 8$

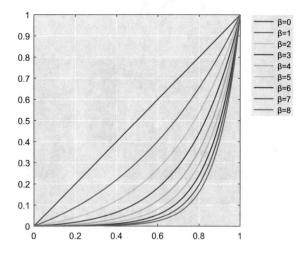

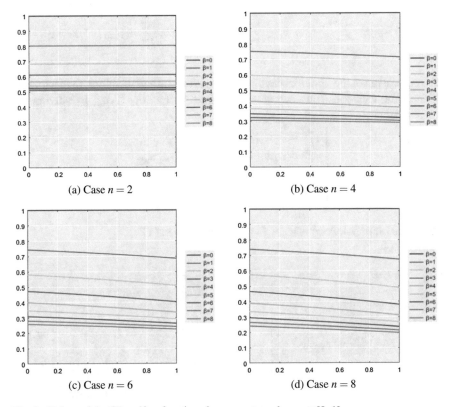

Fig. 3 Values of the GB welfare functions for parameter values $\gamma \in [0, 1]$

Moreover, considering the parametric Lorenz curves depicted in Fig. 2, the values taken by the GB welfare functions with $n = 2, 4, 6, 8$ for $\beta = 0, 1, \ldots, 8$ illustrate clearly the effect of the parameter $\beta \geq 0$ in relation with increasing inequality.

4.2 The Binomial Decomposition of GB Welfare Functions

In the framework of the binomial decomposition (20), each GB welfare function $A_{GB}(x)$ can be uniquely expressed in terms of the binomial Gini welfare functions $C_1, C_2, \ldots C_n$ as follows,

$$A_{GB}(x) = \alpha_1 C_1(x) + \alpha_2 C_2(x) + \cdots + \alpha_n C_n(x) \qquad \gamma \in [0, 1] \qquad (28)$$

which can be written as

$$\sum_{i=1}^{n} w_i^{GB} x_{(i)} = \alpha_1 \sum_{i=1}^{n} w_{1i} x_{(i)} + \alpha_2 \sum_{i=1}^{n} w_{2i} x_{(i)} + \cdots + \alpha_n \sum_{i=1}^{n} w_{ni} x_{(i)} \qquad \gamma \in [0, 1].$$

(29)

The expression of the binomial decomposition is unique and therefore, for each value of the parameter $\gamma \in [0, 1]$, we obtain a unique vector $(\alpha_1, \ldots, \alpha_n)$ by solving the linear system

$$\begin{cases} w_1^{GB} = \alpha_1 w_{11} + \alpha_2 w_{21} + \cdots + \alpha_n w_{n1} \\ w_2^{GB} = \alpha_1 w_{12} + \alpha_2 w_{22} + \cdots + \alpha_n w_{n2} \\ \quad \cdots \\ w_n^{GB} = \alpha_1 w_{1n} + \alpha_2 w_{2n} + \cdots + \alpha_n w_{nn} \end{cases}$$

(30)

where the binomial weights w_{ji}, $i, j = 1, \ldots, n$ are as in (17).

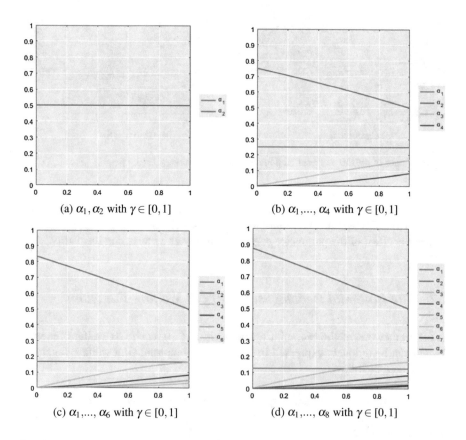

(a) α_1, α_2 with $\gamma \in [0, 1]$

(b) $\alpha_1, \ldots, \alpha_4$ with $\gamma \in [0, 1]$

(c) $\alpha_1, \ldots, \alpha_6$ with $\gamma \in [0, 1]$

(d) $\alpha_1, \ldots, \alpha_8$ with $\gamma \in [0, 1]$

Fig. 4 Coefficients of the binomial decomposition for $n = 2, 4, 6, 8$

In Fig. 4 we depict the vector $(\alpha_1, \ldots, \alpha_n)$ as a function of the parameter $\gamma \in [0, 1]$ in the cases $n = 2, 4, 6, 8$.

We observe, as expected, that $\alpha_1 = 1/n$ is independent of the parameter $\gamma \in [0, 1]$ since, in the last equation of the linear system (30), we have $w_n^{GB} = 1/n^2$ and $w_{1n} = 1/n$ and $w_{2n} = \cdots = w_{nn} = 0$.

On the other hand, we observe that only α_2 is decreasing, whereas $\alpha_3, \ldots \alpha_n$ are increasing with respect to $\gamma \in [0, 1]$.

It is well known that the classical Gini welfare function is 2-additive, see for instance Grabisch [45] and Bortot and Marques Pereira [20] and references therein. On the other hand, the classical Bonferroni welfare function is n-additive. In fact in Fig. 4 we observe that only $\alpha_1, \alpha_2 \neq 0$ in the classical Gini case $\gamma = 0$, and $\alpha_1, \ldots, \alpha_n \neq 0$ in the classical Bonferroni case $\gamma = 1$.

5 Conclusions

We have examined the binomial decomposition of the single parameter family of GB welfare functions. An interesting fact regarding the coefficients $\alpha_1, \ldots, \alpha_n$ of the binomial decomposition, which has been observed in Fig. 4, is that the coefficient α_1 is constant in the parameter $\gamma \in [0, 1]$ whereas the decreasingness of the coefficient α_2 compensates exactly the increasingness of the higher order coefficients, given that $\alpha_1 + \alpha_2 + \cdots + \alpha_n = 1$. This fact, which naturally relates with the weighting structure of GB welfare functions, is presently under investigation.

References

1. Aaberge R (2000) Characterization of Lorenz curves and income distributions. Soc Choice Welf 17:639–653
2. Aaberge R (2001) Axiomatic characterization of the Gini coefficient and Lorenz curve orderings. J Econ Theory 101:115–132
3. Aaberge R (2007) Ginis nuclear family. J Econ Inequal 5:305–322
4. Aristondo O, García Lapresta JL, Lasso de la Vega C, Marques Pereira RA (2013) Classical inequality indices, welfare and illfare functions, and the dual decomposition. Fuzzy Sets Syst 228:114–136
5. Atkinson AB (1970) On the measurement of inequality. J Econ Theory 2:244–263
6. Bárcena Martín E, Imedio LJ (2008) The Bonferroni, Gini and De Vergottini indices. Inequality, welfare and deprivation in the European Union in 2000. Res Econ Inequal 16:231–257
7. Bárcena Martín E, Silber J (2011) On the concepts of Bonferroni segregation index and curve. Rivista Italiana di Economia Demografia e Statistica LXV(2):57–76
8. Bárcena Martín E, Silber J (2013) On the generalization and decomposition of the Bonferroni index. Soc Choice Welf 41(4):763–787
9. Beliakov G, Pradera A, Calvo T: Aggregation functions: a guide for practitioners. Studies in fuzziness and soft computing, vol 221. Springer, Heidelberg (2007)
10. Beliakov G, Bustince HS, Calvo T (2016) A practical guide to averaging functions. Springer, Berlin

11. Benedetti C (1986) Sulla interpretazione benesseriale di noti indici di concentrazione e di altri. METRON Int J Stat XLIV:421–429
12. Porath Ben, Gilboa I (1994) Linear measures, the Gini index, and the income-equality trade-off. J Econ Theory 2(64):443–467
13. Blackorby C, Donaldson D (1978) Measures of relative equality and their meaning in terms of social welfare. J Econ Theory 18:59–80
14. Blackorby C, Donaldson D (1980) A theoretical treatment of indices of absolute inequality. Int Econ Rev 21(1):107–136
15. Blackorby C, Donaldson D (1980) Ethical indices for the measurement of poverty. Econometrica 48(4):1053–1060
16. Blackorby C, Donaldson D (1984) Ethical social index numbers and the measurement of effective tax/benefit progressivity. Can J Econ 17(4):683–694
17. Blackorby C, Donaldson D, Auersperg M (1981) A new procedure for the measurement of inequality within and among population subgroups. Can J Econ 14(4):665–685
18. Bonferroni CE (1930) Elementi di Statistica Generale. Seeber, Firenze
19. Bonferroni CE (1933) Elementi di Statistica Generale. Ristampa con aggiunte, anno accademico 1932–1933. Litografia Felice Gili, Torino
20. Bortot S, Marques Pereira RA (2014) The binomial Gini inequality indices and the binomial decomposition of welfare functions. Fuzzy Sets Syst 255:92–114
21. Bossert W (1990) An axiomatization of the single-series Ginis. J Econ Theory 50(1):82–92
22. Calvo T, De Baets B (1998) Aggregation operators defined by k-order additive/maxitive fuzzy measures. Int J Uncertain Fuzzyness Knowl-Based Syst 6(6):533–550
23. Calvo T, Kolesárova A, Komorníková M, Mesiar R (2002) Aggregation operators: properties, classes and construction methods. In: Calvo T, Mayor G, Mesiar R (eds) Aggregation operators: new trends and applications. Physica-Verlag, Heidelberg, pp 3–104
24. Chakravarty SR (2005) The Bonferroni indices of inequality. In: International conference in memory of C. Gini and M. O. Lorenz. Università degli Studi di Siena
25. Chakravarty SR (2007) A deprivation-based axiomatic characterization of the absolute Bonferroni index of inequality. J Econ Inequal 5(3):339–351
26. Chakravarty SR, Muliere P (2003) Welfare indicators: a review and new perspectives. Measurement of inequality. METRON Int J Stat 61(3):457–497
27. Chisini O (1929) Sul concetto di media. Periodico di Matematiche 4:106–116
28. De Vergottini M (1940) Sul significato di alcuni indici di concentrazione. Giornale degli economisti e annali di economia 2(5/6):317–347
29. De Vergottini M (1950) Sugli indici di concentrazione Statistica 10(4):445–454
30. Donaldson D, Weymark JA (1980) A single-parameter generalization of the Gini indices of inequality. J Econ Theory 22(2):67–86
31. Donaldson D, Weymark JA (1983) Ethically flexible Gini indices for income distributions in the continuum. J Econ Theory 29(2):353–358
32. Ebert U (1987) Size and distribution of incomes as determinants of social welfare. J Econ Theory 41(1):23–33
33. Ebert U (1988) Measurement of inequality: an attempt at unification and generalization. Soc Choice Welf 5(2):147–169
34. Fodor J, Roubens M (1994) Fuzzy preference modelling and multicriteria decision support. Kluwer Academic Publishers, Dordrecht
35. Fodor J, Marichal JL, Roubens M (1995) Characterization of the ordered weighted averaging operators. IEEE Trans Fuzzy Syst 3(2):236–240
36. Gajdos T (2002) Measuring inequalities without linearity in envy: Choquet integrals for symmetric capacities. J Econ Theory 106(1):190–200
37. Gini C (1912) Variabilità e Mutabilità. Contributo allo Studio delle Distribuzioni e delle Relazioni Statistiche. Paolo Cuppini, Bologna
38. Gini C (1914) Sulla misura della concentrazione e della variabilità dei caratteri. Atti del Reale Istituto Veneto di Scienze, Lettere ed Arti LXXIII:1203–1248 (1914) (English translation given in METRON Int J Stat LXIII:1–38 (2005))

39. Gini C (1921) Measurement of inequality of incomes. Econ J 31(121):124–126
40. Giorgi GM (1984) A methodological survey of recent studies for the measurement of inequality of economic welfare carried out by some Italian statisticians. Econ Notes 13(1):145–157
41. Giorgi GM (1998) Concentration index, Bonferroni. In: Kotz S, Read CB, Banks DL (eds) Encyclopedia of statistical sciences. Wiley, New York, pp 141–146
42. Giorgi GM, Crescenzi M (2001) A look at the Bonferroni inequality measure in a reliability framework. Statistica 61(4):571–583
43. Giorgi GM, Crescenzi M (2001) A proposal of poverty measures based on the Bonferroni inequality index. METRON Int J Stat 59(4):3–16
44. Giorgi GM, Nadarajah S (2010) Bonferroni and Gini indices for various parametric families of distributions. METRON Int J Stat 68(1):23–46
45. Grabisch M (1997) k-order additive discrete fuzzy measures and their representation. Fuzzy Sets Syst 92(2):167–189
46. Grabisch M, Marichal JL, Mesiar R, Pap E (2009) Aggregation functions, encyclopedia of mathematics and its applications, vol 127. Cambridge University Press
47. Kolm SC (1969) The optimal production of social justice. In: Margolis J, Guitton H (eds) Public economics. Macmillan, London, pp 145–200
48. Kolm SC (1976) Unequal inequalities I. J Econ Theory 12:416–442
49. Kolm SC (1976) Unequal inequalities II. J Econ Theory 13:82–111
50. Lorenzen G (1979) A generalized Gini-coefficient. Universität Hamburg, Institut für Statistik und Ökometrie
51. Mehran F (1976) Linear measures of income inequality. Econometrica 44(4):805–809
52. Nygård F, Sandström A (1981) Measuring income inequality. Almqvist & Wiksell International, Stockholm
53. Piesch W (1975) Statistische Konzentrationsmaße: Formale Eigenschaften und verteilungstheoretische Zusammenhänge. Mohr
54. Piesch W (2005) Bonferroni-index und De Vergottini-index. Diskussionspapiere aus dem Institut fr Volkswirtschaftslehre der Universitt Hohenheim, vol 259. Department of Economics, University of Hohenheim, Germany
55. Quiggin J (1993) Generalized expected utility theory: the rank-dependent model. Kluwer Academic Publisher, Dordrecht
56. Sen A (1973) On economic inequality. Clarendon Press, Oxford
57. Sen A (1978) Ethical measurement of inequality: some difficulties. In: Krelle W, Shorrocks AF (eds) Personal income distribution. North-Holland, Amsterdam
58. Shorrocks AF, Foster JE (1987) Transfer sensitive inequality measures. Rev Econ Stud 54(3):485–497
59. Silber J (1989) Factor components, population subgroups and the computation of the Gini index of inequality. Rev Econ Stat 71(1):107–115
60. Tarsitano A (1990) The Bonferroni index of income inequality. In: Dargum C, Zenga M (eds) Income and wealth distribution, inequality and poverty. Springer, Heidelberg, pp 228–242
61. Weymark JA (1981) Generalized Gini inequality indices. Math Soc Sci 1(4):409–430
62. Yaari M (1987) The dual theory of choice under risk. Econometrica 55(1):95–115
63. Yaari M (1988) A controversial proposal concerning inequality measurement. J Econ Theory 44(2):381–397
64. Yager RR (1988) On ordered weighted averaging aggregation operators in multicriteria decision making. IEEE Trans Syst Man Cybern 18(1):183–190
65. Yager RR, Kacprzyk J (1997) The ordered weighted averaging operators. Theory and applications. Kluwer Academic Publisher, Dordrecht
66. Yager RR, Kacprzyk J, Beliakov G (2011) Recent developments in the ordered weighted averaging operators: theory and practice. Studies in fuzziness and soft computing, vol 265. Springer, Heidelberg
67. Yitzhaki S (1983) On an extension of the Gini inequality index. Int Econ Rev 24(3):617–628
68. Zenga MM (2013) Decomposition by sources of the Gini, Bonferroni and Zenga inequality indexes. Statistica Applicazioni 11(2):133–161
69. Zoli C (1999) Intersecting generalized Lorenz curves and the Gini index. Soc Choice Welf 16(2):183–196

Part II
General Formal Foundations

The Logic of Information and Processes in System-of-Systems Applications

P. Eklund, M. Johansson and J. Kortelainen

Abstract Logic and many-valuedness as proposed in this paper enables to describe underlying logical structures of information as represented within industrial processes, and as part of their respective markets. We underlines the importance of introducing classification structures in order to enable management of information granularity within and across subsystems in aa system-of-systems. The *logic of information and process* is a main contribution of this paper, and our illumination of a system-of-systems is drawn within the field of energy. In our process view we look closer into the power market with all its stakeholders, and e.g. as related to renewable energy. Supply, demand and pricing models are shown to become subjected to logical considerations. In our approach we show how information and their structures are integrated into processes and their structures. Information structures build upon our many-valued logic modelling, and for process modelling we adopt the BPMN (Business Process Modeling Notation) paradigm.

Keywords Business process modeling notation · Energy · Lative logic

1 Introduction

In system-of-systems (SoS) applications, engineers typically see *information* as residing in products and subsystems taxonomies, and even specifically as emerging from measuring devices. However, in the case of energy, it is part of a market with a variety of stakeholders. Public opinion and governance, as well as related rules and regulations, affect this market in one way or another. Energy can roughly be seen as produced, transmitted, distributed and consumed, where the sources of

P. Eklund (✉) · M. Johansson
Department of Computing Science, Umeå University, Umeå, Sweden
e-mail: peklund@cs.umu.se

J. Kortelainen
Department of Information Technology, South-Eastern Finland University
of Applied Sciences, Mikkeli, Finland

© Springer International Publishing AG 2018
M. Collan and J. Kacprzyk (eds.), *Soft Computing Applications for Group Decision-making and Consensus Modeling*, Studies in Fuzziness and Soft Computing 357, DOI 10.1007/978-3-319-60207-3_6

energy are renewable or non-renewable. The system as a whole enables to identify risk and opportunity, and fine-granular information structures enable accuracy and completeness of information models.

Uncertainty is often viewed statistically as a phenomenon of variance. Therefore, uncertainty is more seldom modelled by logical many-valuedness, which adds complexity to this information representation. Furthermore, uncertainty as part of many-valued logic in systems-of-systems applications, is required when observing conditions as a basis for providing efficient and effective decision-making e.g. as related to service and maintenance.

In this paper we will show how uncertainty resides in various elements in logic, thus constituting an overall many-valued logic for systems-of-systems.

2 The Logic of Processes

Categorial frameworks provide suitable formalism for logical structures, with term monads [11] and sentence functors [12] playing fundamental roles in these respects.

When departing from bivalence to many-valuedness, and in particular for the set of valuations, *quantales* [21, 22], as algebraic structures, have been shown to provide suitable structures in particular given their capability to embrace non-commutativity.

2.1 The Role of Classifications for Information and Processes

In [14, 16] we showed how quantales providing valuation in disorder (ICD) and functioning (ICF) is arranged tensorially in a setting for disorder and functioning within health care. For modelling many-valued and quantale based valuation of faults (*EnFa*) and functioning (*EnFu*) in the energy SoS and its design structures, we have a similar tensor

$$EnFu = EnFa \otimes EnFa,$$

where *EnFa* typically is a three-valued non-commutative quantale, and the tensored *EnFu* becomes a six-point non-commutative quantale. For detail, see [14]. This tensor clearly reflects the situation that a valuation of a multiple fault system-of-systems fault-fault interaction of *EnFa* encoding corresponds to the way valuation of *EnFu* based functioning is done with respect to *EnFu* encoding. More generally, encoding in this manner can potentially be used for integrating modelling standards like UML, SysML and BPMN.

This paper builds upon and further extends a *logic of BPMN* approach adopted in [15] for modelling process generally in crisis management, and emergency care in particular. Engineering, procurement and construction (EPC) as related to plant project management for fossil fuel engine based power plants was developed in [20, 23].

Within the overall energy SoS, there are several important subsystem to be identified with respect to intensity of logic. One is the smart grid and smart transmission where we build upon smart grids as introduced in [1]. The descriptions in [1] are quite general and informal, but do cover the entire spectrum of the electrical system, from transmission to distribution and delivery. Our process model framework enables to embrace a BPMN subview also of smart grids.

2.2 Lative Logic

The notion of logic as a structure embraces signatures and constructed terms and sentences *latively* constructed [7, 13] as based on these terms. Similarly, sentence and conglomerates of sentences are fundamental for entailments, models and satisfactions, in turn latively to become part of axioms, theories and proof calculi. This lativity is always produced and maintained by functors and monads, and as acting over underlying categories in form of monoidal categories. Category theory is thus a suitable metalanguage for logic, in particular when applications and typing of information must be considered.

Uncertainty may reside in generalized powerset functors, and may be internalized in underlying categories. In both cases, suitable algebras must motor this uncertainty representation, and quantales are very suitable in this context [14].

Substitutions as morphisms in Kleisli categories of term monads, carry data and information within and across subsystems, where each subsystem is seen as a logical theory. Thereby we have the distinction between expression and statement within the SoS. Expression is a term produced by a term functor over a signature, and over an underlying category. We have separate and specific signatures within all subsystems. A statement is a sentence produced by a sentence functor [12].

Quantales are well suited for describing multivalence in many-valued logic, when valuation of uncertain information is subjected to various algebraic operations. This provides a unique situation where proper logical and mathematical foundation will meet the requirement of richness needed in real-world applications. Noncommutativity in these operations is a typically important consideration from application point of view. It represents a causality which intuitively resides between commutative conjunction and non-commutative logical implication.

In the following we briefly introduce notation and constructions needed in our descriptions related to our energy SoS signatures and terms. The many-sorted term monad $\mathbf{T}_\Sigma$ over Set_S, the many-sorted category of sets and functions, where $\Sigma = (S, \Omega)$ is a signature, can briefly be described as follows. For a sort (i.e. type) $s \in S$, we have sort specific functors $\mathsf{T}_{\Sigma,s} : \mathrm{Set}_S \to \mathrm{Set}$, so that

$$\mathsf{T}_\Sigma(X_s)_{s \in S} = (\mathsf{T}_{\Sigma,s}(X_s)_{s \in S})_{s \in S}.$$

The important recursive step in the term construction is

$$T'_{\Sigma,s}(X_s)_{s \in S} =$$

$$\coprod_{s_1,\ldots,s_m} (\Omega^{s_1 \times \cdots \times s_m \to s})_{\mathrm{Set}_S} \times \mathrm{arg}^{s_1 \times \cdots \times s_m} \circ \bigcup_{\kappa < \iota} T^{\kappa}_{\Sigma}(X_s)_{s \in S}$$

and then with

$$T'_{\Sigma}(X_s)_{s \in S} = (T'_{\Sigma,s}(X_s)_{s \in S})_{s \in S},$$

we finally arrive at the term functor

$$T_{\Sigma} = \bigcup_{\iota < \bar{k}} T'_{\Sigma}.$$

The purely categorical construction of the corresponding term monad can be seen in [11].

2.3 Process Modelling

For process modelling, BPMN diagrams build syntactically upon four basic categories of elements, namely Flow Objects, Connecting Objects, Artifacts and Lanes. Flow Objects, represented by Events, Activities and Gateways, define the behaviour of processes. Start and End are typical Event elements. Task and Sub-Process are the most common Activities. There are three Connecting Objects, namely Sequence Flow, Message Flow and Association. Gateways, as Event elements, handle branching, forking, merging, and joining of paths.

A Data Object is an Artifact, and having no effect on Sequence Flow or Message Flow. Data Objects are indeed seen to "represent" data, even if BPMN does not at all specify these representation formats or rules for such representations. However, Data Objects are expected to *provide information about what activities require to be performed and/or what they produce* [2]. Information produced is in our sense the result of a reduction or inference, with related substitutions.

Notion like '*service* provision' or 'failure *report*' in terms of their content and data formats is often well understood but this is not the case when considering provision and reports as structured *documents*, as a whole. To better understand the documents as a whole we must consider in detail the notions of documents, document structures, and document templates. In the categorical framework outlined above we can indeed identify *a document over a signature* Σ with the notion of a ground term over Σ, i.e., a term containing no variables. In this interpretation a *document template over a signature* Σ then is a non-ground Σ-term over some set of variables X. That is, $T_{\Sigma}(X) \backslash T_{\Sigma}(\emptyset)$ is the set of all document templates. We may underline here that the report structure really *is* the signature Σ?

Our suggestion for information semantics [10] is then that BPMN's Data Object coincides with document and, by extension, is a ground term. For documentations and document refinements, this means in reality that we call a document draft a 'document template' all the way until it has been matured to become the "finished" document, where all variables have been instantiated with relevant information (ground terms). Similarly, 'token' coincides with variable substitution. From this view of BPMN information semantics, we are able to extract at any point in the data flow a valid variable substitution that precisely represents an information snapshot of the process at that particular point. An activity in the BPMN sense can then be viewed as a composition of variable substitutions with the initial token or variable substitution being the Kleisli category identity morphism $\eta : X \to \mathsf{T}_\Sigma X$.

In order to provide examples, let us briefly outline how the underlying signatures as 'owned' and recognized by respective disciplines might look like. In our example we may start with the signatures,

$$\Sigma_{transmission} = (S_{transmission}, \Omega_{transmission})$$
$$\Sigma_{distribution} = (S_{distribution}, \Omega_{distribution}),$$

respectively, for the subsystems of energy transmission and distribution stakeholders.

We may aim at providing failure reports as terms t being of sort $\mathsf{s}_{FailureReport} \in S_{distribution}$. This may also be denoted by $t :: \mathsf{s}_{FailureReport}$. This term t is then seen as produced by a number of operators, manipulating and attaching terms in form of various 'subdocuments', like a specific cable failure report to be integrated with the overall failure report. Assume then we have the cable report, as a term being $u :: \mathsf{s}_{CableReport}$, with $\mathsf{s}_{CableReport} \in S_{transmission}$. The term u is then typically delivered by operations as a response to a referral, $v :: \mathsf{s}_{ReferralToTransmissionMaintenanceReport}$, with $\mathsf{s}_{ReferralToTransmissionMaintenanceReport} \in S_{transmission}$, from a first responder onsite where the failure has been identified.

The refinement of failure and maintenance reports are assumed to include detailed energy engineering knowledge as typically appearing in diagnostic and maintenance guidelines of various kind and with different level of detail.

Report production, enhancement and enrichment is then enabled, e.g., by operators like $\mathtt{RefineReport}_\mathsf{s} : \mathsf{s} \to \mathsf{s}$, where in the case of $\mathsf{s} \in S_{transmission}$ it is report production authored by maintenance and service.

Suppose now we are given classifications for failures, ClEnFa, and functioning, ClEnFu. Failure taxonomies typically as fault trees usually exist, whereas functioning classification are rare.

We would arrange these within an over BPMN *data object signature*, so that we have ClEnFa, ClEnFu $\in S_{DataObjectSorts}$, e.g., with constant c $: \to$ ClEnFa representing a particular specific fault recognized in the fault tree.

As an example, consider a failure report. We may have as a template of such a report a non-ground term

$$\mathtt{DiagnosisAndReassessment}(\dots, x_{\mathsf{s}_{CableReport}}, \dots)$$

so that a particular activity performed by the first responder (having a unique personal ID, say 4321) will then give rise to a many-sorted substitution σ such that

$$\sigma_s(x) = \begin{cases} t & \text{if } s = \texttt{CableReport} \\ x & \text{otherwise} \end{cases}$$

where

$$t = \texttt{CableReport(frId(4321),}$$
$$\texttt{Cables...}$$
$$\texttt{Conductors...}$$
$$\texttt{DeadEndTower...}$$
$$\texttt{...).}$$

The term resulting from this substitution—which may still be a template, but nevertheless is closer to a document—will be

$$\texttt{FailureReport(...,CableReport(...), ...)}$$

and we can view it in the following, alternative, form

```
|----- Failure Report | |  ... transmission line location ... | |
|----- Cable Report |    | frID = 4321 |   | Cables ... |      |
Conductors ... |    | Dead-end tower ... |     |----- | ... |-----
```

Our view of BPMN information semantics in this paper differ from [15] where Data Objects were taken to be variable substitutions.

Uncertainty can be modelled using composition of many-valued power monads with the term monad, i.e., $\mathbf{Q} \cdot \mathbf{T}$, where $\mathbf{T}$ is the term monad and $\mathbf{Q}$ as the many-valued powerset monad based on an underlying quantale Q. The structure of substitutions,

$$(\text{Hom}(X, QTX), +, \cdot, {}^*, 0, 1),$$

i.e., the set of morphisms in the Kleisli category $\text{Set}_{\mathbf{Q} \cdot \mathbf{T}}$, is a Kleene algebra. See [15] for detail.

For substitutions $\sigma_1, \sigma_2 \in \text{Hom}(X, QTX)$, we have

$$\sigma_1 + \sigma_2 = \sigma_1 \vee \sigma_2,$$

and

$$\sigma_1 \cdot \sigma_2 = \sigma_1 \circ \sigma_2$$

where $\sigma_1 \circ \sigma_2 = \mu_X^{\mathbf{Q} \cdot \mathbf{T}} \circ QT\sigma_2 \circ \sigma_1$ is the composition of morphisms in the corresponding Kleisli category of $\mathbf{Q} \cdot \mathbf{T}$.

A "partial algebra of documents" can now be provided as follows. Let $t_{CableReport}$ $:: s_{CableReport}$ be a template, or "document in progress", as part of an overall term

$$(t_s^{ScopeOfReport})_{s \in S_{DataObjectSorts}}$$

in

$$T_{\Sigma_{DataObjectSignature}}(X_s)_{s \in S_{DataObjectSorts}},$$

with substitutions $\sigma_i, i = 1, 2$. Then

$$\mu \circ T(\sigma_1 + \sigma_2)((t_s^{ScopeOfReport})_s)$$

is a concatenation or composition of information along a path of maintenance tasks, of $\mu \circ T(\sigma_1)((t_s^{ScopeOfReport})_s)$ and $\mu \circ T(\sigma_2)((t_s^{ScopeOfReport})_s)$, whereas

$$\mu \circ T(\sigma_1 \cdot \sigma_2)((t_s^{ScopeOfReport})_s)$$

is a corresponding 'sharpening the uncertainties', or enhancing truth values residing in that report.

3 Energy

Energy sources in our nature are often subdivided into renewable and non-renewable sources. Opportunistic and coarsest-granular politics may say only "we need more focus on renewable". However true it may be, it hides detail and granularity of the underlying information structure of pros and cons. Awareness raising among the public enables finer-granular opinion making, which in turn enables the consumer to understand how to affect supply and demand.

Renewable sources include geothermal, hydro, water and wind. Non-renewable source. include fossil and nuclear fuel.

Concerning renewable energy, who owns sunlight, flow of water, underground heat or windy air? We all do. We can all exploit it, under certain rules and regulations, given opinions and policies. And we all do, sometimes even so that rules and regulations are updated, and more opinion and policy is created.

3.1 The Logic and Fungibility of the Energy Market

In the following we will show some snapshots from an overall BPMN view of *The Lative Logic and Fungibility of ENERGY*, using the following stakeholders in respective BPMN Lanes:

- Power SOURCE
- Central or Federal GOVERNMENT
- National and Regional AUTHORITY
- Local GOVERNMENT
- Power Model
- Power MARKET
- Power Plant EPC
- Power Plant MAINTENANCE
- Power Plant EPC
- Power GENERATION
- Power TRANSMISSION
- Power DISTRIBUTION
- Industrial CONSUMER
- Public CONSUMER
- Household CONSUMER

Opinion, policy making and governance, in dialogue and interaction over time, are important parts of the energy market SoS (Fig. 1). Those dialogues and interactions obviously involve political and policy-making type of consensus [17, 18], consensus reaching [3, 5, 9] among a variety of stakeholders, and negotiation in various form, as representatives in BPMN swimlanes try to meet their respective objectives.

3.2 Energy to Current and Back to Energy

Force and energy residing in flow of water and wind create rotation so that attached generators can convert mechanical energy to electrical energy. Electric charge is a fundamental conserved property as appearing in electromagnetic interaction.

This is the ideal electrophysical situation. In practice, current is lost in transmission, and for a wide variety of reasons. Improved efficiency and reliability in transmission becomes important. In ideal situations, the power P, in watts, is equal to the current I, in amps, times the voltage V, in volts, i.e., $P = I \times V$. Currency loss is modelled using a power factor pf, so that in a realistic situation we have

$$P = pf \times I \times V.$$

In practice, the calculation and estimation of pf is non-trivial. and the value itself can seldom point at where and why currency is lost. A number of techniques help to improve the power factor. Measuring the power factor, however, is usually an average condition, which means it does not reveal if a power factor decrease is a widespread or a localized problem. Furthermore, electric energy distribution using overhead and underground power lines makes the power factor differently transparent. Overhead power lines are more economical, but also more susceptible to damage. Repairing physical damage is very expensive. From logic point of view, pf is still a numerical

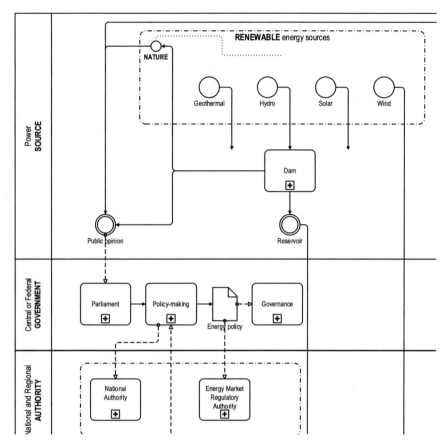

Fig. 1 Policy making and public opinion in dialogue

value, but not a constant. It is a many-valued logical term (expression) $\mathrm{pf}(t_1, \dots, t_n)$, where its value in turn depends on subterms $t_1, \dots, t_n$, so that outcome power is logically computed as

$$P = \mathrm{pf}(t_1, \dots, t_n) \otimes (I \times V).$$

Obviously, this is still an informal expression as it mixes logical and numerical computation, but it clearly enriches the value to become an valuated expression, where the reason e.g. for power factor loss can be contributed to observations integrated into the terms $t_1, \dots, t_n$, some of which may be bivalent and instant, while others are dynamic and multivalent. e Tangible and intangible impacts of current loss altogether makes it difficult for the electric utility industry e.g. to justify the placing of overhead power lines underground. Power factor considerations differ and need to be considered also for underground cables [19]. This is a simple yet very concrete example where investments in transmission efficiency and safety has a direct effect on needs to adjust energy transmission pricing.

Assumptions are required to arrive at what would be a guess at best. Predicting the performance of an underground line is difficult, yet the maintenance costs associated with an underground line are significant and one of the major impediments to the more extensive use of underground construction. Major factors that impact the maintenance costs for underground transmission lines include: deterioration because of the loading cycles the lines undergo during their lifetimes. As time passes, the cables' insulation weakens, which increases the potential for a line fault. If the cables are installed properly, this debilitating process can take years and might be avoided. If and when a fault occurs, however, the cost of finding its location, trenching, cable splicing, and re-embedment is sometimes 5–10 times more expensive than repairing a fault in an overhead line where the conductors are visible, readily accessible and easier to repair. repairs to the underground cables. experience of repair personnel. The typical repair duration of cross-linked polyethelene (XLPE), a solid dielectric type of underground cable, ranges from five to nine days. Outages are longer for lines that use other nonsolid dielectric underground cables such as high-pressure, gas-filled (HPGF) pipe-type cable, high-pressure, fluid-filled (HPFF) pipe-type cable, and self-contained, fluid-filled (SCFF)-type cable. In comparison, a fault or break in an overhead conductor usually can be located almost immediately and repaired within hours or a day or two at most. redundant feeders, but the duration of such outages is still longer than those associated with overhead lines, and they have additional costs associated with them. have been installed. Such modifications to underground power lines are more expensive because of the inability to readily access lines or relocate sections of lines. designed, constructed and made available for connection to the new home in a relatively short time. Service drops to new residences can be installed within a day or two after the service request is submitted to the utility.

3.3 Failure and Recovery

Failure described only by name, without structured information about the failure, has the consequence that time, to and between failures [4], is the only value pertinent to a failure. In reliability engineering, failure rate is seen simply as the frequency with which an (sub)system or component fails. This means counting *how many* failures per unit of time, rather than in addition explaining *how* it occurs and *what* precisely it pertains. Using only time as the only pertinent characteristic of a failure leads to risk analysis being mostly based on probability of occurrence. Intrinsic reasons of failure are hidden behind observation of failure frequency.

Our approach leans on failure and how it is valuated as described at a level of granularity, which is sufficient for providing required solutions e.g. in service and maintenance. Granularity of that description resides in the granularity of the underlying signature for the logic of the system. A signature with one sort for 'failure', and constant operators simply as names for failures, indeed means time is the only value pertinent related to failure.

Many-valuedness adds further granularity to valuation of failure and recovery. In some cases, a failure can be instantaneous, but in general, a failure is a progression from normal to failed, passing through a number of stages which are either reversible or from which a process can recover before reaching a final failure state. Recovery is then a *dual* progression, similarly passing from failure through a number of stages to full recovery, and from there back to normal operation.

A failure may also be instantaneous and total, but local and residing in a certain subsystem, so that the failure status on the system-of-systems level is less critical. Recovery as described on local level is therefore also not to be identified with recovery taking place on more global level.

Further, a production process in transition to failure may sometimes lose only level of quality. but may maintain level of quantity, so that recovery makes quality return to normal level, whereas quantity levels remain unchanged.

Failure is a special for of *crisis*, where the description of a crisis is much more complicated. Roughly speaking, a failure can be more easily valuated, whereas a crisis is more of a process. A valuation like "mean time to or between crisis" obviously makes less sense, if any sense at all. *Recovery* from failure and *mitigation* from crisis is therefore not to be confused.

Which values and value structures then are most pertinent to failure and recovery in a specific system? A system build upon its underlying signature, so value expressions build upon sorts and operators in that signature. The *total effect* of a failure is then also a more complicated matter where *generalized integrals* need to be developed. Syntactic derivatives based on underlying signatures was developed in [8], and can potentially be used to develop corresponding generalized integrals for the purpose of valuation and total effect of failure. This is, however, outside the scope of this paper.

Maintainability and availability modelling, beyond pertinence just involving time, will obviously involve suitable and adapted maintainability and availability assessment frameworks, where the underlying logic of it is expected to resemblance assessment frameworks e.g. as appearing in health care [6].

A program that has complex control flow will require more tests to achieve good code coverage and will be less maintainable (functor is an identity). Therefore be used to calculate time wasted, but it is not usable to calculate quantity and/or quality of production (as) value wasted.

3.4 Time and Location

Considerations related to maintenance involve time and location. When did a failure occur, where is it located, and what is involved. *Time and geographic location* of a fault on a higher level in the SoS relates also to Points-of-Interest (PoI) approaches in geographic information systems. *Location and description* of a fault is even more challenging.

Further, locations are often uncertain, and so is prediction of time in preventive actions. An important many-valued extension is therefore the generalization of points not just to sets of points, but indeed to many-valued sets of points of interest (MvSPoI).

3.5   Who Profits and Who Pays?

On harvesting the wind, scale is commercial typically when over 100 kW, where electricity is sold rather than used on-site. In smaller scale, ownership related to that harvesting can manifest in form of lease of land, community ownership, or ownership of (a small) turbine. Note how land ownership is wind flow ownership, but for flowing water is different.

Intangibility of impact for consumers has become tangible as consumers are reimbursed for loss of current in the energy supply chain. Consumer opinion must therefore be part of the pricing models. We no longer have a simplified numerical

$$\texttt{EnergyPrice}(supply,\ demand)$$

but also supply and demand depending on several factors, and logically explained as terms:

$$\texttt{supply}(s_1, \dots, s_m)$$

$$\texttt{demand}(d_1, \dots, d_l)$$

Several factors and phenomena affect this overall situation.

Construction and maintenance time and cable cost are typical. Excavation costs are also considerable, and in all, and the question about who bears all these costs remains unanswered. The cost of new distribution services are carried by developers, but may quickly be passed on to municipalities and in the end to consumers. Allocation of cost associated with placing cables underground is also unclear. Costs are absorbed in various points of the market chain, and if allowed by regulatory agencies, consumer will eventually pay.

The power price is still seen as a balance between supply and demand. However, such pricing models overlook public opinion and policy making (Fig. 2). Renewable resources of energy, that nobody owns as such, are transformed into economic resources, that are subject to the power market.

Regulatory agencies play an important role, and this combines with governmental policy making.

Restrictions enforced on the energy SoS by the regulatory agencies has as its objective to protect the interests of consumers, so that consumers are not to carry too much of the burden arising from SoS improvements, in particular if

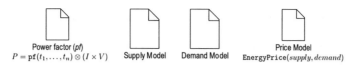

Power factor (*pf*)   Supply Model   Demand Model   Price Model
$P = \mathtt{pf}(t_1, \ldots, t_n) \otimes (I \times V)$ $\mathtt{EnergyPrice}(supply, demand)$

Fig. 2 Multifactorial energy price modelling

improvement benefit consumers unequally. Regulatory agencies might also still base their delivery models using out-dated constructions, in particular in the case where there is a regional shift from overhead to underground transmission.

4 Conclusion

We have described a many-valued logical framework for information ontology as part of business process structures. A main contribution is to show where uncertain but well-structured information resides within a process, and how information is canonically integrated rather than amalgamated in ad hoc approaches. Developments include a rigorous, yet flexible, modelling approach with energy supply as part of the energy market as whole involving all stakeholders.

Acknowledgements This work is partly supported by the *WindCoE* (Nordic Wind Energy Centre), a competence centre within wind power. WindCoe is funded within Interreg IVA (European Cross-Border cooperation) Botnia-Atlantica, as part of European Territorial Cooperation (ETC). ETC provides a framework for the implementation of joint actions and policy exchanges between national, regional and local actors from different European Member States.
This work is partly supported by the Logic in Manufacturing (LiM) project, funded by the Swedish Innovation Agency (VINNOVA) PRODUKTION2030 programme.

References

1. Amin S, Wollenberg BF (2005) Toward a smart grid: power delivery for the 21st century. IEEE Power Energy Mag 3:3441
2. Business Process Modeling Notation (BPMN) Version 1.2 (2009) Object Management Group
3. Carlsson C, Ehrenberg D, Eklund P, Fedrizzi M, Gustafsson P, Merkuryeva G, Riissanen T, Ventre AGS (1992) Consensus in distributed soft environments. European J Oper Res 61:165–185
4. Cavallaro J, Walker I (1994) A survey of NASA and military standards on fault tolerance and reliability applied to robotics. In: Proceedings of AIAA/NASA conference on intelligent robots in field, factory, service, and space, 282286
5. Ehrenberg D, Eklund P, Fedrizzi M, Ventre AGS (1989) Consensus in distributed soft environments. Rapporter från Åbo Akademi, inf.beh. & matem., Ser. A, nr 88
6. Eklund P (2010) Signatures for assessment, diagnosis and decision-making in ageing. In: üllermeier EH, Kruse R, Hoffmann F (eds) IPMU 2010, Part II, CCIS 81. Springer, Heidelberg, pp 271–279

7. Eklund P Lative logic accomodating the WHO family of international classifications, encyclo-pedia of E-Health and telemedicine. In: Maria Manuela Cruz-Cunha, Isabel Miranda (eds) IGI Global, 2016, in print. http://www.igi-global.com/book/encyclopedia-health-telemedicine/141916

8. Eklund H-T, Eklund P (2014) Process drama based information management for assessment and classification in learning, social networks: a framework of computational intelligence. In: Pedrycz W, Chen S-M (eds) Studies in computational intelligence, Springer, 526, pp 333–352

9. Eklund P, Gustafsson P, Lindholm P, Tony Riissanen (1991) An architecture for topological concensus reaching. In: Proceedings of AIRO91, Riva del Garda, 18–20 September 1991, pp 110-113

10. Eklund P, Helgesson R (2012) Information ontology and process structure in social and health care, unpublished, 2012. Appears partly as Chapter 6 in R. Helgesson, *Generalized General Logics*, PhD thesis, Umeå University, 2013

11. Eklund P, Galán MA, Helgesson R, Kortelainen J (2014) Fuzzy terms. Fuzzy Set Syst 256:211–235

12. Eklund P, Galán MA, Helgesson R, Kortelainen J, Moreno G, Vasquez C (2013) Towards cat-egorical fuzzy logic programming, WILF 2013—10th international workshop on fuzzy logic and applications. In: Masulli F, Pasi G, Yager R (eds) Lecture Notes in Computer Science, vol 8256, pp 109–121

13. Eklund P, Höhle U, Kortelainen J (2014) The fundamentals of lative logic, Abstract in LINZ2014, 35th Linz Seminar on Fuzzy Set Theory, Linz, Austria, February 18–22

14. Eklund P, Höhle U, Kortelainen J (2016) Non-commutative quantales for many-valuedness in applications. Carvalho JP et al. (eds) IPMU 2016, Part I, CCIS 610, Springer, Heidelberg, pp 437–449

15. Eklund P, Johansson M, Karlsson J, Åström R (2009) BPMN and its semantics for information management in emergency care. In: Fourth 2009 international conference on convergence and hybrid information technology (ICCIT 2009), IEEE Computer Society, pp 273–278

16. Eklund P, Löfstrand M (2016) Many-logic in manufacturing. In: Position papers of the feder-ated conference on computer science and information systems (FedCSIS), ACSIS 9, p 1117. doi:10.15439/2016F73

17. Eklund P, Rusinowska A, de Swart HC (2007) Consensus reaching in committees. European J Op Res 178:185–193

18. Eklund P, Rusinowska A, de Swart H (2008) A consensus model of political decision-making. Annal Op Res 158:5–20

19. Eteruddin H, bin Mohd Zin AA (2012) Reduced dielectric losses for underground cable dis-tribution systems. Int J Appl Power Eng (IJAPE) 1 37–46

20. Holm A, Karlsson J, Eklund P (1994) Information technology for project management: progress management and communication support. In: Proceedings of multiple paradigms for artificial intelligence, Turku (Åbo), Finland, 29–31 August 1994, pp 243–248

21. Mulvey CJ, Pelletier JW (1992) On the quantisation of the calculus of relations, CMS Proceed-ings 13. Amer Math Soc Providence RI 345–360

22. Pelletier JW, Rosický J (1997) Simple involutive quantales. J Algebra 195:367–386

23. von Schoultz F, Eklund P (1994) Information technology for project management: intelligent systems and decision support. In: Proceedings of multiple paradigms for artificial intelligence, Turku (Åbo), Finland, August 29–31, pp 238–242

Decision-making Process Using Hyperstructures and Fuzzy Structures in Social Sciences

Sarka Hoskova-Mayerova and Antonio Maturo

Abstract Algebraic hyperstructures represent an interesting field of algebra, important both from the theoretical point of view and also for their applications. A hypergroupoid structure can be associated with any social relationship. These hypergroupoids become hypergroups in some particular conditions, among them a condition concerning outer individuals. By analyzing it we can establish in a natural way when social relationships become optimal. The relations among persons are one of the bases of Social Sciences that usually are described by linguistic propositions. If U is a set of individuals (the universe set to consider), usually it is not possible to affirm that for an ordered pair (x, y) of persons belonging to U a relation R given in a linguistic form holds or not, as happens in a binary context. A correct and overall complete modeling of each of these relations is obtained only if we assign to every pair $(x, y) \in U \times U$ a real number $R(x, y) = xRy$ belonging to interval [0, 1] that is the measure to which the decision maker believes the relation holds.

Keywords Hyperstructures · Fuzzy relations · Mathematical models in social sciences

S. Hoskova-Mayerova (✉)
Faculty of Military Technology, Department of Mathematics and Physics,
University of Defence, Brno, Czech Republic
e-mail: Sarka.mayerova@unob.cz

A. Maturo
Department of Architecture, University of Chieti-Pescara, Viale Pindaro 42,
65127 Pescara, Italy
e-mail: antomato75@gmail.com

© Springer International Publishing AG 2018
M. Collan and J. Kacprzyk (eds.), *Soft Computing Applications for Group Decision-making and Consensus Modeling*, Studies in Fuzziness and Soft Computing 357, DOI 10.1007/978-3-319-60207-3_7

103

1 Introduction

Social science is an academic set of disciplines concerned with society and human behavior. "Social science" is commonly used as an umbrella term to refer to anthropology, archaeology, criminology, economics, education, history, linguistics, communication studies, political science and international relations, sociology, geography, law, and psychology. In modern academic practice, researchers are often eclectic, using multiple methodologies (for instance, by combining the quantitative and qualitative techniques). The term social research has also acquired a degree of autonomy as practitioners from various disciplines share in its aims and methods [17, 25].

Mathematics, particularly analyses, probability, statistics or graph theory, have been used in order to construct social and psychological theories. By constructing mathematical models, some assumptions are made about social psychology and they are expressed in formal mathematics, providing so an empirical interpretation. With statistical procedures and fuzzy logic, properties of models are deduced and they are compared with empirical data [26, 30–32].

Using mathematical sociology and psychology, it can be understood how predictable local interactions are able to give an idea about global models of social structure. It has often been useful in science to apply developed mathematical theory to real-world processes, and the first step in this is to show how the real-world process can be translated into the concepts of the theory [20].

Hypergroups are generalizations of groups (set with a binary operation on it meeting a number of conditions). If this binary operation is taken to be multivalued, then we get a hypergroup. The motivation for generalization of the notion of group resulted naturally from various problems in non-commutative algebra, another motivation for such an investigation came from geometry. The theory of hypergroups, created in 1934 in Marty's paper at the VIII Congress of Scandinavian Mathematicians in Stockholm [21] was subsequently developed around the 40s with the contribution of various authors e.g. [7, 9, 10], Wall (who introduced a generalization of hypergroups, where the hyperproduct is a multiset, i.e. a set in which every element has a certain multiplicity).

In the 50s and 60s they worked on hyperstructures, in Romania Benado, in Czech Republic Drbohlav, in France Koskas, Sureau, In Greece Mittas, Stratigopoulos, in Italy Orsatti, Boccioni, in USA Prenowitz, Graetzer, Pickett, McAlister, in Japan Nakano, in Yugoslavia Dacic.

But it is above all since 70' that a more luxuriant flourishing of hyperstructures has been and is seen in Europe, Asia, America, Australia, e.g. [3, 22].

In this congress Marty defined the hypergroups, giving some of their applications to non-commutative groups, algebraic function sand rational fractions. Over the following decades, new and interesting results again appeared, but it is, above all, a more luxuriant flourishing of hyperstructures that has been seen in the last 25 years.

Algebraic hyperstructures represent an interesting field of algebra, important both from the theoretical point of view and also for their applications. Hypergroups have been used in algebra, geometry, convexity, automata theory, combinatorial problems of coloring, lattice theory, Boolean algebras, logic, probability, fuzzy logic, etc. [1, 2, 4–6, 11, 14, 27, 32]. The most complete bibliography up to 1991 can be found in Corsini's monograph: Prolegomena of Hypergroup Theory [3], the main applications were described in the book [4]. Using hyperstructures, mathematical models of a social group can be constructed; some assumptions can be made about sociology and psychology [17, 29].

A hypergroupoid structure can be associated with any social relationship. These hypergroupoids become hypergroups in some particular conditions, among them a condition concerning outer individuals. By analyzing it we can establish in a natural way when social relationships become optimal [20].

2 Basic Definitions

In order to make this topic precise, we need some preliminary concepts. Used terms and definitions are e.g. [13, 17, 33, 34].

On a nonempty set H, we consider a hyperoperation • that is a map that associates a nonempty subset $a • b$ of H with any two elements a and b of H. If A and B are nonempty subsets of H, then by A • B we intend the union of all $a • b$, when a belongs to A and b belongs to B. The pair (H, •) is called a hypergroupoid.

A hypergroupoid (H, •) is called a *hypergroup* if the hyperoperation • is:

1. associative, (which means that for all a, b, c of H, $(a • b) • c = a • (b • c)$);
2. and reproductive. (which means that for all a of H, $H • a = a • H = H$).

A hypergroupoid (H, •) is called *quasi-hypergroup* if $\forall\ (a, b) \in H \times H$, $\exists$ $(x, y) \in H \times H$ such that $a \in b • x$ and $a \in y • b$.

3 Hypergroups Inspired by Social Relationships

In this paragraph we present several hypergroups inspired by religion and friendship, hierachical organization of a group [28, 17, 20]. In group social behaviour study, sociologists very often have to deal with questions like these:

Which student would you like to do the project with?

Which student do you want to invite for a party?

With whom do you wish to cooperate on the task?

Etc.

These, at the first sight, simple and trivial questions say a lot about the relations in the social group (e.g. in the class, department, ...) to the sociologist. Such type of relations can be described using hyperstructures. Some properties of hyperstructures have a particular social meaning. For instance:

- Quasi hypergroups mean that no person is alone in the group!
- Transitive relation can be used as a mediator between two people.

But there are open questions, too; e.g.

Does the semihypergroup have any meaning in sociology?

Is it possible to model a specific situation or fact in sociology with the help of semihypergroup?

The following examples are overtaken from [17] as they can answer and explain some of the above-mentioned questions and problems.

Example 1 Let S be the set of persons. Let R be a "*friendship relation*" between the pairs of persons. By the notation a R b we mean the relation of $a \in$ S towards person $b \in$ S. To indicate that person $a \in$ S has the relation R (e.g. "*likes*", "*loves*",...) towards person $b \in$ S number 1 is used. If a is not in relation with b, 0 is used.

Example 2 For every relation R on S let us define the *symmetric relation* associated to R as the relation $\bar{R} = R \bigcup R^{-1}$. We assume aRa, for all a $\in$ S. We define:

1. the *active R-neighborhood* of $a \in$ S as the set $A^R(a) = \{x \in$ S: a R $x\}$;
2. the *active R-chain of length* n > 1 *from* $a \in$ S as the set

$A^R(a) = \{x \in$ S: exist $x_1, x_2, ... x_{n-1} \in$ S: a R x_1, for all i $<$ n $- 2$, x_i R x_{i+1}, x_{n-1} R $x\}$;

3. the *passive R-neighborhood* of $a \in$ S as the set $P^R(a) = \{x \in$ S: x R $a\}$;
4. the *passive R-chain of length* n > 1 *from* $a \in$ S as the set

$P_n^R(a) = \{x \in$ S: exist $x_1, x_2, ... x_{n-1} \in$ S: x R x_1, for all i $<$ n $- 2$, x_i R x_{i+1}, x_{n-1} R $a\}$.

The active neighborhood of a is considered as an active chain of length 1 from a and the passive neighborhood as a passive A chain of length 1. For every n $\in$ N, X $\subseteq$ S, we put

$$A_n^R(X) = \cup_{a \in X} A_n^R(a), P_n^R(X) = \cup_{a \in X} P_n^R(a).$$

From the social point of view an active neighborhood of a can be seen as the set of all people that a chooses according to the rules given by the relation R, while an active chain of length 2 from a is the set of choices x or by a or by an intermediary y selected by a. Moreover, a passive neighborhood of b is the set of people that choose b.

Example 3 Let S be a set of persons. We define a hyperoperation • on S, as follows:

$\forall$ a, b $\in$ S, $a \bullet b = \{z \in$ S, such that the person a suggests z for cooperation and a person b accept the invitation of the person $z\}$.

It is clear that the commutative property does not hold in this case.

The constructed hyperstructures and also the associated fuzzy structures can be used in decision-making process. More examples can be found e.g. in [12, 18, 23].

Now we present the necessary mathematical context. Let R be a binary relation with full domain on a nonempty set H, which means that R is a subset of the Cartesian product H $\times$ H, such that for all a in H, there exists b in H, such (a, b) belongs to R. The *range* of R is the set of all elements b of H, such that there exists an element a in H, for which (a, b) belongs to R. We often denote a R b instead of (a, b) belongs to R.

Let R be an *equivalence relation* R, i.e. a *reflexive* (which means that for all a in H, a R a), *symmetric* (which means that a R b implies that b R a) and *transitive* (which means that a R b and b R c imply that a R c) relation.

We define *hypergroupoid* (H, $\bullet_R$) *associated* to R, as follows:

(1) $\forall a \in H$, $a \bullet_R a = \{y \mid (a, y)$ belongs to R$\}$;
(2) $\forall a, b$ of H, $a \bullet_R b$ is the union *of a* $\bullet_R$ *a* and $b \bullet_R b$.

An element c of H is called an *outer element* of R if there exists an element b in H such that (b, c) does not belong to R^2, where $R^2 = \{(a, c) \mid$ there exists b in H, such that (a, b) and (b, c) are in R$\}$.

The following definition and example are overtaken from [20].

Definition 1 A social relationship R is *optimal* if the hypergroupoid (H, $\bullet_R$) is a hypergroup.

Hence hypergroup structures can be used as a mathematical tool to decide if a social relationship is optimal or not and to establish when it is so.

Example 4 We consider again H a community and the relation R defined as follows:

a R b if and only if a and b are relatives.

We can consider that each element of H is a relative of himself and clearly, the relation R is also symmetric.

But, is it transitive? In other words, *the relatives of the relatives of an individual are his relatives?*

According to dictionary, a relative means a person who is connected with another or others by blood or marriage. We can imagine enough complicated situations, for instance the children of a women by a previous relationship are not related to the children of his husband, by a previous marriage, too. At least, they are not related for ever, because in case the woman and the man divorce, nothing directly connects them anymore. Even if we consider relatives only by blood, the relation is not transitive. Indeed, for instance if a is a cousin of b from b mother part and c is a cousin of b from b father part, then a and c are not relatives by blood.

So, we can consider that the community H contains outer individuals, with respect to the relation R defined above. This means that there are individuals in H that are not relatives to some individuals of H. The third condition claims that for any outer individual, the relatives of his relatives must be his relatives. For more details see [20].

4 Modelling Social Relations with Fuzzy Sets

4.1 Basic Definition on Fuzzy Sets and Fuzzy Relations

Fuzzy Sets were introduced in the 60s by an Iranian scientist Zadeh, who lives in USA [40]. He and others, in the following decades, found surprising applications to almost every field of science and knowledge: from engineering to sociology, from agronomy to linguistic, from biology to computer science, from medicine to economy, from psychology to statistics and so on. They are now cultivated in the entire world. Let's remember what a fuzzy set is.

Let U be a nonempty set (in Social Science it can be, e.g., a set of individuals or a set of media). A fuzzy set μ in U is a function $\mu: U \rightarrow [0, 1]$. The meaning of such a function is that it is considered a *linguistic property* in the set U, (e.g., the property of individuals to be *clever, old, educated*, etc.), and for every individual $x \in U$, $\mu(x)$ is the degree to which the property holds. If $\mu(x) = 1$, then the individual x has the property in the maximum possible degree. If $\mu(x) = 0$ then the individual does not meet the property at all. Let us denote by F(U) the set of all fuzzy sets on U.

Zadeh in many papers [40] managed to impose, showing significant applications and consistent, the theory of fuzzy sets. Some important books on fuzzy sets and fuzzy logic are in [8, 19, 36]. Some interesting applications of fuzzy sets to Social Sciences are considered by Ragin in [26, 31, 35].

A fuzzy relation R on U is a fuzzy set in $U \times U$, i.e., a function $R: U \times U \rightarrow [0, 1]$. For every $(x, y) \in U$ we write $x R y$ to denote $R(x, y)$. The meaning of a fuzzy relation is the mathematical representation of a linguistic relation in the set U, (e.g., the property that individual x to be *friend, trustful, relative*, to the individual y, or *dominated* by y etc.).

If $x R y = 1$, then the individual x has the relation with y in the maximum possible degree. On the contrary, if $x R y = 0$ then the individual x does not have the relation with y at all. We denote by R(U) the set of all fuzzy relations on U.

4.2 Fuzzy Relations in Social Sciences

One of the bases of Social Sciences are the relations among persons that usually are described by linguistic propositions. If U is a set of individuals (the universe set to consider), usually it is not possible to affirm that for an ordered pair (x, y) of persons belonging to U a relation R given in a linguistic form holds or not, as happens in a binary context.

For instance, let us consider one of the linguistic relations "friendship", "cooperation between colleagues", "affection for a person", "confidence in a person for a job" and other interpersonal relations. A correct and overall complete modeling of each of these relations is obtained only if we assign to every pair $(x, y) \in U \times U$ a real number $R(x, y) = x \, R \, y$ belonging to interval $[0, 1]$ that is the measure to which the decision maker believes the relation holds.

This leads us to model the relation, as a fuzzy set with domain $U \times U$, that is a function

$$R: (x, y) \in U \times U \to R(x, y) \in [0, 1]. \tag{4.1}$$

Another important aspects consists in a fuzzy relation between a set U of persons and a set Ω of objects. It is formalized by a fuzzy set with domain $U \times \Omega$, i.e. a function:

$$\rho: (x, \omega) \in U \times \Omega \to \rho(x, \omega) \in [0, 1]. \tag{4.2}$$

For instance, let U be a set of children and Ω a set of media. Some example of media that transmit information to children and then are in relation with children are *books, television, internet, cinemas, parents, teachers*, and so on. Some possible fuzzy relations are: "confidence in the media", "dependence on the media", "concentration in the use of the media", and so on.

If ρ is one of these fuzzy relations, $\rho(x, \omega)$ means the degree to which x is in the relation ρ with ω. For more details see e.g. [12, 15, 16, 18, 24, 37–39].

5 Conclusion

Algebraic hyperstructures represent an interesting field of algebra, important both from the theoretical point of view and also for their applications. A hyperstructure can be associated with any social relationship. These hyperstructures, called hypergroupoids, become hypergroups in some particular conditions. By analyzing them, we can establish in a natural way when social relationships become optimal.

In this paper, we have extended the area of possible applications of hypergroups theory. The possible application of hyperstructure theory in social sciences was shown. Several examples of hyperstructures constructed in a social sphere and on human relations were presented, e.g. dealing with relations between persons and their properties. Moreover, the properties of defined hyperoperations were studied. Such hyperoperations and resulted sets can be very helpful in everyday life of sociologists, especially in their decision-making process when studying the relations of a particular social group.

Acknowledgements The work presented in this paper was supported within the project "Development of basic and applied research in the long term developed by the departments of theoretical and applied foundation FMT" (Project code VÝZKUMFVT) supported by the Ministry of Defence the Czech Republic.

References

1. Ameri R, Hoskova-Mayerova S, Amiri-Bideshki M, Borumand Saeid A (2016) Prime filters of hyperlattices. An Stiint Univ "Ovidius" Constanta Ser Mat 24(2):15–26
2. Ameri R, Kordi A, Hoskova-Mayerova, S (2017) Multiplicative hyperring of fractions and coprime hyperideals. An Stiint Univ "Ovidius" Constanta Ser Mat 25(1):5–23
3. Corsini P (1993) Prolegomena of hypergroup theory. Aviani Editore, USA
4. Corsini P, Leoreanu V (2003) Applications of hyperstructure theory. Kluwer Academic Publishers, Dordrecht, Hardbound
5. Cristea I (2009) Hyperstructures and fuzzy sets endowed with two membership functions. Fuzzy Sets Syst 160:1114–1124
6. Cristea I, Hoskova S (2009) Fuzzy topological hypergroupoids. Iran J Fuzzy Syst 6(4):13–21
7. Dresher M, Ore O (1938) Theory of multigroups. Am J Math 60:705–733
8. Dubois D, Prade H (1988) Fuzzy numbers: an overview. In: Bedzek JC (ed) Analysis of fuzzy information, vol 2. CRC-Press, Boca Raton, FL, pp 3–39
9. Eaton JE (1940) Remarks on multigroups. Am J Math 62:67–71
10. Griffiths LW (1938) On hypergroups, multigroups and product systems. Am J Math 60: 345–354
11. Hoskova S, Chvalina J (2008) Discrete transformation hypergroups and transformation hypergroups with phase tolerance space. Discret Math 308(18):4133–4143
12. Hofmann A, Hoskova-Mayerova S, Talhofer V (2013) Usage of fuzzy spatial theory for modelling of terrain passability. In: Advances in fuzzy systems. Hindawi Publishing Corporation, Article ID 506406, p 7. doi:10.1155/2013/506406
13. Hoskova S (2005) Discrete transformation hypergroups. In: Proceeding of international conference aplimat, FX spol. s r.o., Bratislava, pp 275–279
14. Hoskova S, Chvalina J (2009) A survey of investigations of the Brno reseach group in the hyperstructure theory since the last AHA Congress. In: Proceeding of AHA 2008, University of Defence, Brno, pp 71–83
15. Hoskova-Mayerova S (2012) Topological hypergroupoids. Comput Math Appl 64(9): 2845–2849
16. Hoskova-Mayerova S, Maturo A (2014) An analysis of social relations and social group behaviors with fuzzy sets and hyperstructures. Int J Algebraic Hyperstruct Appl
17. Hoskova-Mayerova S, Maturo A (2013) Hyperstructures in social sciences. AWER Procedia Inf Technol Comput Sci 3(2013):547–552
18. Hoskova-Mayerova S, Talhofer V, Hofmann A (2013) Decision-making process with respect to the reliability of geo-database. In: Ventre AG, Maturo A, Hoskova-Mayerova Š, Kacprzyk J (eds) Multicriteria and multiagent decision making with applications to economics and social sciences. Studies in fuzziness and soft computing. Springer, Berlin pp 179–195
19. Klir G, Yuan B (1995) Fuzzy sets and fuzzy logic: theory and applications. Prentice Hall, Upper Saddle River, NJ
20. Leoreanu-Fotea VPC (2010) Hypergroups determined by social relationship. Ratio Sociol 3(1):89–94
21. Marty F (1935) Sur une generalisation de la notion de groupe. In: 8th Scandinavian Congress of Mathematicians. H. Ohlssons boktryckeri, Lund, pp 45–49
22. Massouros ChG, Mittas J (2009) On the theory of generalized M-polysymmetric hypergroups. Proceeding of AHA 2008. University of Defence, Brno, pp 217–228

23. Maturo A, Ventre A (2009) Multipersonal decision making, concensus and associated hyperstructures. In: Proceeding of AHA 2008. University of Defence, Brno, pp 241–250
24. Maturo A (2009) Coherent conditional previsions and geometric hypergroupoids. Fuzzy sets, rough sets and multivalued operations and applications 1(1):51–62
25. Maturo A, Maturo F (2014) Finite geometric spaces, steiner systems and cooperative games. Analele Universitatii Ovidius Constanta. Seria Matematica 22(1):189–205 ISSN: Online 1844-0835. doi:10.2478/auom-2014-0015
26. Maturo A, Maturo F (2013) Research in social sciences: fuzzy regression and causal complexity. Springer, Berlin, pp 237–249. doi:10.1007/978-3-642-35635-3_18
27. Maturo A, Maturo F (2017) Fuzzy events, fuzzy probability and applications in economic and social sciences. Springer International Publishing, Cham, pp 223–233. doi:10.1007/978-3-319-40585-8_20
28. Maturo A, Sciarra E, Tofan I (2008) A formalization of some aspects of the Social Organization by means of the fuzzy set theory. Ratio Sociologica 1(2008):5–20
29. Maturo F (2016) Dealing with randomness and vagueness in business and management sciences: the fuzzy probabilistic approach as a tool for the study of statistical relationships between imprecise variables. Ratio Mathematica 30:45–58
30. Maturo F (2016) La regressione fuzzy. Fuzziness: teorie E applicazioni. Aracne Editrice, Roma, Italy, pp 99–110
31. Maturo F, Fortuna F (2016). Bell-shaped fuzzy numbers associated with the normal curve. Springer International Publishing, Cham, pp 131–144. doi:10.1007/978-3-319-44093-4_13
32. Maturo F, Hoskova-Mayerova S (2017). Fuzzy regression models and alternative operations for economic and social sciences. Springer International Publishing, Cham. pp 235–247. doi:10.1007/978-3-319-40585-8_21
33. Novák M (2014) n-ary hyperstructures constructed from binary quasi-ordered semigroups. An Stiint Univ "Ovidius" Constanta Ser Mat 22:147–168
34. Novák M (2015) On EL-semihypergroups. Eur J Comb 44(B):274–286
35. Ragin CC (2000) Fuzzy-set social science. University Chicago Press, Chicago, USA
36. Ross TJ (1995) Fuzzy logic with engineering applications, McGraw-Hill, New York
37. Rybansky M, Hofmann A, Hubacek M, Kovarik V, Talhofer V (2015) Modelling of cross-country transport in raster format. Environ Earth Sci 74(10):7049–7058.
38. Talhofer V, Hofmann A, Hoskova-Mayerova S (2015) Application of fuzzy membership function in mathematical models for estimation of vehicle trafficability in terrain. In B. L. Szarkova D. (ed.) 14th Conference on applied mathematics APLIMAT 2015, Slovak University of Technology in Bratislava, Bratislava, pp 711–719
39. Talhofer V, Hoskova S, Hofmann A, Kratochvil V (2009) The system of the evaluation of integrated digital spatial data realibility. In: 6th conference on mathematics and physics at technical universities, University of Defence, Brno, pp 281–288
40. Zadeh LA (1965) Fuzzy sets. Inf Control 8:338–358

Social Preferences Through Riesz Spaces: A First Approach

Antonio Di Nola, Massimo Squillante and Gaetano Vitale

Abstract In this paper we propose Riesz spaces as general framework in the context of pairwise comparison matrices, to deal with definable properties, real situations and aggregation of preferences. Some significant examples are presented to describe how properties of Riesz spaces can be used to express preferences. Riesz spaces allow us to combine the advantages of many approaches. We also provide a characterization of collective choice rules which satisfy some classical criteria in social choice theory and an abstract approach to social welfare functions.

1 Introduction

Our choices are strictly related to our ability to compare alternatives according to different criteria, e.g. price, utility, feelings, life goals, social conventions, personal values, etc. This means that in each situation we have different *best alternatives* with respect to many criteria; usually, the context gives us the *most suitable* criteria, but no one says that there is a unique criterion. Even when we want to make a decision according to the opinions of the experts in a field we may not have a unique advice. To sum up, we have to be able to define our *balance* between different criteria and opinions, to give to each comparison a weight which describes the importance, credibility or goodness and then to include all these information in a mixed criteria. As

A. Di Nola · G. Vitale (✉)
Department of Mathematics, University of Salerno, Via Giovanni Paolo II, 132, 84084 Fisciano, SA, Italy
e-mail: gvitale@unisa.it

A. Di Nola
e-mail: adinola@unisa.it

M. Squillante
Department of Law, Economics, Management and Quantitative Methods, University of Sannio, Piazza Guerrazzi, 1, 82100 Benevento, BN, Italy
e-mail: squillan@unisannio.it

© Springer International Publishing AG 2018
M. Collan and J. Kacprzyk (eds.), *Soft Computing Applications for Group Decision-making and Consensus Modeling*, Studies in Fuzziness and Soft Computing 357, DOI 10.1007/978-3-319-60207-3_8

usual, we need a formalization which gives us tools to solve these problems; properties of this formalization are well summarized by Saaty in [25], according to whom

> [it] must include enough relevant detail to: represent the problem as thoroughly as possible, but not so thoroughly as to lose sensitivity to change in the elements; consider the environment surrounding the problem; identify the issues or attributes that contribute to the solution; identify the participants associated with the problem.

Riesz spaces, with their double nature of both weighted and ordered spaces, seem to be the natural framework to deal with multi-criteria methods; in fact, in real problems we want to obtain an order starting from weights and to compute weights having an order.

We remark that:

- Riesz spaces are already studied and widely applied in economics, mainly supported by works of Aliprantis (see [1–3]);
- contrary to the main lines of research, which prefer to propose ad-hoc models for each problems, this paper is devoted to analyze and propose a general framework to work with and to be able, in the future, to provide a universal translator of various approaches.

We introduce basic definitions and properties of Riesz spaces with a possible interpretation of them in the context of pairwise comparison matrices, focusing on aggregation procedures. As main results of the paper we have:

1. a characterization of collective choice rules satisfying Arrow's axioms (Theorem 1);
2. established an antitone Galois correspondence between total preorders and cones of a Riesz space (Theorem 2);
3. a categorical duality between categories of preorders and of particular cones of a Riesz space (Theorem 3).

This paper is organized in 6 sections. Sections 1 and 6 are, respectively, a brief introduction and a concluding section with some remarks and perspectives. In Sect. 2 we recall some basic definitions of Riesz space and of pairwise comparison matrix. Section 3 is devoted to explain, also with meaningful examples, the main ideas that led us to propose Riesz spaces as suitable framework in the context of decision making; in particular it will explained how properties of Riesz spaces can be appropriate to model, and to deal with, real problems. In Sects. 4 and 5 we focus on a particular method of decision making theory, i.e. PCMs; we pay special attention to:

- collective choice rules;
- classical social axioms (Arrow's axioms);
- total preorder spaces;
- duality between total preorders and geometric objects.

2   Preliminaries

We will use $\mathbb{N}$, $\mathbb{Z}$ and $\mathbb{R}$ to indicate, respectively, the set of natural, integer and real numbers. We will indicate with $<$ and $\leq$ the usual (strict and non-strict) orders and $\preceq$ will be the order of the considering example and it will be defined in each context.

2.1   Riesz Spaces

Definition 1   A structure $\mathscr{R} = (R, +, \cdot, \bar{0}, \preceq)$ is a Riesz space (or a vector lattice) if and only if:

- $\mathscr{R} = (R, +, \cdot, \bar{0})$ is a vector space over the field $\mathbb{R}$;
- $(R, \preceq)$ is a lattice;
- $\forall a, b, c \in R$ if $a \preceq b$ then $a + c \preceq b + c$;
- $\forall \lambda \in \mathbb{R}^+$ if $a \preceq b$ then $\lambda \cdot a \preceq \lambda \cdot b$.

A Riesz space $(R, +, \cdot, \bar{0}, \preceq)$ is said to be *archimedean* iff for every $x, y \in R$ with $n \cdot x \preceq y$ for every $n \in \mathbb{N}$ we have $x \preceq \bar{0}$. A Riesz space $(R, +, \cdot, \bar{0}, \preceq)$ is said to be linearly ordered iff $(R, \preceq)$ is totally ordered. We will denote by R^+ the subset of positive elements of R Riesz space (the *positive cone*), i.e. $R^+ = \{a \in R \mid \bar{0} \preceq a\}$. We say that u is a *strong unit* of R iff for every $a \in R$ there is a positive integer n with $|a| \preceq n \cdot u$, where $|a| = (a) \vee (-a)$.

Examples:

1. An example of non-linearly ordered Riesz space is the vector space $\mathbb{R}^n$ equipped with the order $\preceq$ such that $(a_1, \ldots, a_n) \preceq (b_1, \ldots, b_n)$ if and only if $a_i \leq b_i$ for all $i = 1, \ldots, n$; it is also possible to consider $(1, \ldots, 1)$ as strong unit.
2. A non-archimedean example is $\mathbb{R} \times_{LEX} \mathbb{R}$ with the lexicographical order, i.e. $(a_1, a_2) \preceq (b_1, b_2)$ if and only if $a_1 < b_1$ or $(a_1 = b_1$ and $a_2 \leq b_2)$; in this case $(1, 0)$ is a strong unit.
3. $(\mathbb{R}, +, \cdot, 0, \leq)$, which is the only (up to isomorphism) archimedean linearly ordered Riesz space, as showed in [20]; obviously 1 can be seen as the standard strong unit.
4. $(\mathbb{R}^C, +, \cdot, \mathbf{0}, \preceq)$ the space of (not necessarily continuous) functions from C compact subset of $\mathbb{R}$, e.g. the closed interval $[0, 1]$, to $\mathbb{R}$, such that for every $f, g \in \mathbb{R}^C$ and $\alpha \in \mathbb{R}$ we have $(f + g)(x) = f(x) + g(x)$, $(\alpha \cdot f)(x) = \alpha f(x)$, $f \preceq g \Leftrightarrow f(x) \leq g(x)$ $\forall x \in C$ and $\mathbf{0}$ is the zero-constant function; if we consider continuous functions the one-constant function $\mathbf{1}$ is a strong unit.
5. $(M_n(R), +, \cdot, 0_{n \times n}, \preceq)$ the space of $n \times n$ matrices over R Riesz space with component-wise operations and order as in example (1).

Definition 2 A cone in R^n is a subset K of R^n which is invariant under multiplication by positive scalars. A polyhedral cone is convex if it is obtained by finite intersections of half-spaces.

Cones play a crucial role in Riesz spaces theory, as showed in [4] with also some applications (e.g. to linear programming [4, Corollary 3.43]). Another remarkable example of this fruitful tool is the well-known Baker-Beynon duality (see [7]), which shows that the category of finitely presented Riesz spaces is dually equivalent to the category of (polyhedral) cones in some Euclidean space. Analogously to Euclidean spaces, in R^n (with R generic Riesz space) we can consider *orthants*, i.e. a subset of R^n defined by constraining each Cartesian coordinate to be $x_i \leq \bar{0}$ or $x_i \geq \bar{0}$. Here we introduce the definition of *TP-cones*, which will be useful in the sequel.

Definition 3 Let us consider L cone. We say that L is a *TP-cone* if it is the empty-set, or an orthant or an intersection of them.

2.2 Pairwise Comparison Matrices

Let $N = \{1, 2, ..., n\}$ be a set of alternatives. Pairwise comparison matrices (PCMs) are one of the way in which we can express preferences. A PCM has the form:

$$X = \begin{pmatrix} x_{11} & x_{12} & \cdots & x_{1n} \\ x_{21} & x_{22} & & x_{2n} \\ \vdots & & \ddots & \vdots \\ x_{n1} & \cdots & \cdots & x_{nn} \end{pmatrix}. \tag{1}$$

The generic element x_{ij} express a vis-à-vis comparison, the intensity of the preference of the element i compared with j. The request is that from these matrices we can deduce a vector which represents preferences; more in general we want to provide an order $\lesssim_X$. In literature there are many formalizations and definitions of PCMs, e.g. preference ratios, additive and fuzzy approaches. In [8] authors introduce PCMs over abelian linearly ordered group, showing that all these approaches use the same algebraic structure. A forthcoming paper provides a more general framework, archimedean linearly ordered Riesz spaces to deal with *aggregation* of PCMs. In this paper we want to go beyond the archimedean property and the linear order. Using different Riesz spaces with various characteristics it is possible to describe and solve a plethora of concrete issues.

PCMs are used in the Analytic Hierarchy Process (AHP) introduced by Saaty in [24]; it is successfully applied to many Multi-Criteria Decision Making (MCDM) problems, such as facility location planning, marketing, energetic and environmental requalification and many others (see [5, 19, 23, 28]).

As interpretation in the context of PCMs we will say that alternative i is preferred to j if and only if $\bar{0} \leq x_{ij}$.

3 Preferences via Riesz Spaces

Why should we use an element of a Riesz space to express the intensity of a preference? As showed in [8, 10], Riesz spaces provide a general framework to present at-once all approaches and to describe properties in the context of PCMs. Preferences via Riesz spaces are *universal*, in the sense that *(I)* they can express a ratio or a difference or a fuzzy relation, *(II)* the obtained results are true in every formalization and *(III)* Riesz spaces are a common language which can be used as a bridge between different points of view.

What does it mean non-linear intensity? In multi-criteria methods decision makers deals with many (maybe conflicting) objectives and intensity of preferences is expressed by a (real) number in each criteria. In AHP we have different PCMs, which describe different criteria; if we consider $\mathbb{R}^n$ (see example (2) above) we are just writing all these matrices as a unique matrix with vectors as elements. Actually, we can consider each component of a vector as the standard way to represent the intensity preference and the vector itself as the natural representation of multidimensional (i.e. multi-criteria) comparison. This construction has its highest expression in the subfield of MCDM called Multi-Attribute Decision Making, which has several models and applications in military system efficiency, facility location, investment decision making and many others (e.g. see [6, 26, 29, 30])

Does it make sense to consider non-archimedean Riesz space in this context? Let us consider the following example. A worker with economic problems has to buy a car. We can consider the following hierarchy:

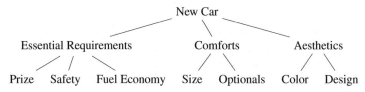

It is clear that Essential Requirements (ER), Comforts (C) and Aesthetics (A) cannot be just weighted and combined as usual. In fact, we may have the following two cases:

- we put probability different to zero on (C) and (A) and in the process can happen that the selected car is not the most economically convenient or even too expensive for him (remember that the worker has a low budget and he has to buy a car), and this is an undesired result.

- conversely, to skip the case above, we can just consider (ER) as unique criterion and neglect (C) and (A). Also in this case we have a non-realistic model, indeed our hierarchy does not take into account that if two cars have the same rank in (ER) then the worker will choose the car with more optionals or with a comfortable size for his purposes.

In a such situation it seems to be natural to consider a lexicographic order (see example (2) above) such as $(\mathbb{R} \times_{LEX} \mathbb{R}) \times_{LEX} \mathbb{R}$, where each component of a vector $(x, y, z) \in (\mathbb{R} \times_{LEX} \mathbb{R}) \times_{LEX} \mathbb{R}$ is a preference intensity in (RE), (C) and (A) respectively (we may shortly indicate the hierarchy with $(RE) \times_{LEX} (C) \times_{LEX} (A)$). We remark that *lexicographic preferences* cannot be represented by any continuous utility function (see [12]).

Which kinds of intensity can we express with functions? This approach is one of the most popular and widely studied one, under the definition of *utility functions*. These functions provide a cardinal presentation of preferences, which allows to work with choices using a plethora of different tools, related to the model (e.g. see [17, 18, 21]). We want to stress that in example (4) we consider functions from a compact to $\mathbb{R}$, without giving a meaning of the domain, which can be seen as a time interval, i.e. in this framework it is also possible to deal with Discounted Utility Model and intertemporal choices (e.g. see [16]). Manipulation of a particular class of these functions (i.e. piecewise-linear functions defined over $[0, 1]^n$) in the context of Riesz MV-algebras is presented in [14]. Furthermore, it is possible to consider more complex examples, for instance we can consider the space $\mathbb{R}^F$ of functionals, where F is a general archimedean Riesz space with strong unit (e.g. see [11]).

4   On Collective Choice Rules for PCMs and Arrow's Axioms

In this section we want to formalize and characterize Collective Choice Rules f in the context of *generalized PCMs*, i.e. PCMs with elements in a Riesz space, which satisfy classical conditions in social choice theory.

Let R be a Riesz space. Let us consider m experts/decision makers and n alternatives. A collective choice rule f is a function

$$f : GM_n^m \rightarrow GM_n$$

such that

$$f(X^{(1)}, \dots, X^{(m)}) = X$$

where X is a *social* matrix, GM_n is the set of all matrices (PCMs) over R with n alternatives such that for every $i \in \{1, \dots, n\}$ $x_{ii} = \bar{0}$. f can be seen also as follows:

$$f = (\tilde{f}_{ij})_{1 \leq i,j \leq n},$$

where

$$\tilde{f}_{ij} : GM_n^m \to R.$$

Note that GM_n is a subspace of $M_n(R)$ (see example (5)), i.e. it is a Riesz space. Let us introduce properties related with axioms of democratic legitimacy and informational efficiency required in Arrow's theorem.

$$\forall i,j \; (\exists f_{ij} : R^m \to R \; : \; \tilde{f}_{ij}(X^{(1)}, \dots, X^{(m)}) = f_{ij}(x_{ij}^{(1)}, \dots, x_{ij}^{(m)})) \qquad \text{(Property } I^*)$$

$$\forall i,j \; (f_{ij}((R^m)^+) \subseteq R^+) \qquad \text{(Property } P^*)$$

$$\nexists i \in \{1, \dots, m\} \; : \; \forall X^{(j)}, \text{with } j \neq i \; (f(X^{(1)}, \dots, X^{(i)}, \dots, X^{(m)}) = X) \; \text{(Property } D^*)$$

Theorem 1 *Let R be a Riesz space and let f be a function $f : (R^{n^2})^m \to R^{n^2}$. f is a collective choice rule satisfying Axioms of Arrow's theorem if and only if f has properties I^*, P^* and D^*.*

Proof Unrestricted Domain (Axiom U). The first axiom asserts that f has to be defined on all the space GM_n^m, i.e. decision makers (DMs) can provide every possible matrix as input. This is equivalent to say that f is defined on $(R^{n^2})^m$.

Independence from irrelevant alternatives (Axiom I). The second axiom says that the relation between two alternatives is influenced only by these alternatives and not by other ones, i.e. it is necessary and sufficient to know how DMs compare just these two alternative. This is equivalent to property I^*.

Pareto principle (Axiom P). The third axiom states that f has to compute a preference if it is expressed unanimously by DMs. This is equivalent to property P^*.

Non-dictatorship (Axiom D). The last axiom requires democracy, that is no one has the right to impose his preferences to the entire society. This is equivalent to property D^*.

In Theorem 1 it is presented a characterization of collective social rules which respect Arrow's axioms; however it does not guarantee that the social matrix produce a *consistent* preference, in fact not all PCMs provide an order on the set of alternatives. We will study this feature in Sect. 5.

5   On Social Welfare Function Features

Social welfare functions (SWFs) are all the collective choice rules which provide a total preorder on the set of alternatives. We can decompose a SWF g as follows:

$$g = \omega \circ f,$$

where f is a collective choice rule having properties I^*, P^* and D^*, and ω is a function such that

$$\omega : \ GM_n \rightarrow \textbf{TP},$$

where **TP** is the set of total preorders on the set of alternatives. Let us consider a social matrix $X = f(X^{(1)}, \dots, X^{(m)})$. We want to characterize property of ω such that g is a social welfare function.

Let us recall the definition of transitive PCM.

Definition 4 [9, Definition 3.1] A pairwise comparison matrix X is transitive if and only if $(\bar{0} \leq x_{ij} \ and \ \bar{0} \leq x_{jk}) \Rightarrow \bar{0} \leq x_{ik}$

It is trivial to check that if X is *transitive*, then it is possible to directly compute an order which expresses the preferences over alternatives. In fact, let X be a GM_n, it has two properties:

$$\begin{array}{ll} (\rho) \ x_{ii} = \bar{0}, & \text{(Reflexivity)} \\ (\gamma) \ \forall i, j \in \{1, \dots, n\} x_{ij} \in R. & \text{(Completeness)} \end{array}$$

If we have also that

$$(\tau) \ (\bar{0} \leq x_{ij} \ and \ \bar{0} \leq x_{jk}) \Rightarrow \bar{0} \leq x_{ik} \qquad \text{(Transitivity)}$$

We say that an order $\lesssim_X$ is *compatible with X* if and only if we have that:

$$\bar{0} \leq x_{ij} \quad \Leftrightarrow \quad j \lesssim_X i.$$

An analogous definition is proposed in [27] in the context of utility functions.

Proposition 1 *Let X be a transitive GM_n (TGM_n) then there exists a unique total preorder $\lesssim_X$ compatible with X. Or equivalently, the correspondence*

$$\theta : \ TGM_n \ \rightarrow \ \textbf{TP}$$

which associates to each $X \in TGM_n$ a preorder $\lesssim_X$ compatible with X itself is a surjective function. Moreover $\lesssim_X \equiv \lesssim_{\alpha \cdot X}$ for every $\alpha \in \mathbb{R}^+$, and $\lesssim_X \equiv \gtrsim_{\alpha \cdot X}$ for every $\alpha \in \mathbb{R}^-$.

Let $\mathscr{C}(R) = \{A \subseteq R \mid A \text{ is a cone}\}$ be the set of all closed cones of R Riesz space. By Proposition 1 we can consider the function Φ

$$\Phi : \mathbf{TP} \rightarrow \mathscr{C}(TGM_n)$$

such that

$$\Phi(\lesssim) = \{X \in TGM_n \mid \lesssim \text{ is compatible with } X\}$$

Proposition 2 *The function Φ is injective.*

We can define an order relation $\ll$ over **TP** as follows:

$$\lesssim_1 \ll \lesssim_2 \quad \Leftrightarrow \quad i \lesssim_2 j \rightarrow i \lesssim_1 j .$$

It is also possible to denote with $\lesssim = \lesssim_1 \vee \lesssim_2$ as the total preorder such that

$$i \lesssim j \quad \Leftrightarrow \quad i \lesssim_1 j \text{ and } i \lesssim_2 j.$$

Remark 1 By easy considerations, we have that $\Phi(\lesssim_1) \cap \Phi(\lesssim_2) = \Phi(\lesssim_1 \vee \lesssim_2)$. Moreover, note that **TP** is closed with respect to $\vee$, i.e. $(\mathbf{TP}, \vee)$ is a *join-semilattice*.

Examples
Let us consider n alternatives. The spaces of total preorder with $n = 2$ and $n = 3$ have the following configurations:

$$a_1 = a_2 = 0$$

$$a_1 = a_2$$

$$a_1 \leq a_2 \qquad a_1 \geq a_2$$

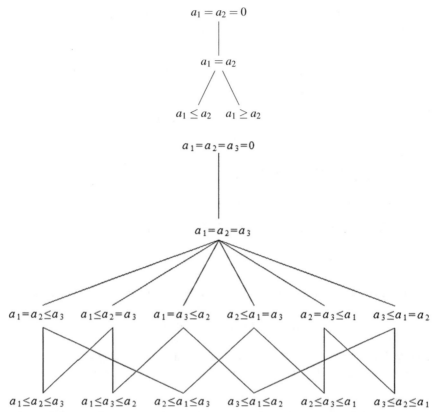

Note that in each space we have exactly one atom which expresses indifference. We call *basic total preorder* an element which is minimal in $(\mathbf{TP}, \ll)$.

Remark 2 In order to deal with aggregation of many TGM_n we added a root $(\top)$, which can be interpreted as impossibility to make a social decision (related to *Condorcet's paradox* and *Arrow's impossibility theorem* in the context of PCMs). We put

$$\Phi(\top) = \emptyset.$$

Proposition 3 *Every $\lesssim$ total preorder different from $\top$ can be written as $\bigvee_i \lesssim_i$, where $\lesssim_i$ are basic total preorders.*

Proof If $\lesssim$ has no identities then it is a basic total preorders. For each identity $a_i = a_j$ in $\lesssim$ we can consider $\lesssim_h \vee \lesssim_k$, with $\lesssim_h$ and $\lesssim_k$ basic total preorders such that $a_i \lesssim_h a_j$, $a_j \lesssim_k a_i$ and preserve all the other relations of $\lesssim$.

Proposition 4 *Let $\lesssim$ be a basic total preorder over n elements. We have that $\Phi(\lesssim)$ is an orthant in TGM_n.*

Proof By the fact that $\lesssim$ is a basic total preorder we have that $a_i \lesssim a_j$ or $a_j \lesssim a_i$ for each alternatives a_i and a_j, i.e. $x_{ij} \geq \bar{0}$ or $x_{ij} \leq \bar{0}$.

Analogously to θ we can define Θ in this way:

$$\Theta : \mathscr{C}(TGM_n) \rightarrow \mathbf{TP}$$

where $\Theta(\emptyset) = \top$ and

$$\Theta(K) = \Phi^{-1}\left(\bigcap_{\substack{C \in \Phi(\mathbf{TP}) \\ C \cap K \neq \emptyset}} C\right).$$

By Remark 1 we have that the function is well-defined.

Definition 5 Let $(A, \leq_A)$ and $(B, \leq_B)$ be two partially ordered sets. An antitone Galois correspondence consists of two monotone functions: $F : A \rightarrow B$ and $G : B \rightarrow A$, such that for all a in A and b in B, we have $F(a) \leq_B b \quad \Leftrightarrow \quad a \geq_A G(b)$.

Now we can state the following result.

Theorem 2 *The couple* (Θ, Φ) *is an antitone Galois correspondence between* $(\mathscr{C}(TGM_n), \subseteq)$ *and* $(\mathbf{TP}, \ll)$.

Proof Let K be an element of $\mathscr{C}(TGM_n)$ and $\lesssim$ an element of $\mathbf{TP}$. Let $\lesssim_K$ be $\Theta(K)$. The proof follows by this chain of equivalence:

$$\Theta(K) \ll \lesssim \quad \Leftrightarrow \quad (i \lesssim j \rightarrow i \lesssim_k j) \quad \Leftrightarrow \quad (X \in \Phi(\lesssim) \rightarrow X \in K) \quad \Leftrightarrow \quad K \supseteq \Phi(\lesssim).$$

We denote by K_n the subset of $\mathscr{C}(TGM_n)$ of all the cones L such that $L \in \Phi(\mathbf{TP})$.

Proposition 5 *Let L be a cone of TGM_n. We have that*

$$L \in \Phi(\mathbf{TP}) \quad \Leftrightarrow \quad L \text{ is a } TP - cone.$$

Proof ($\Rightarrow$) Let L be in $\Phi(\mathbf{TP})$, this means that $L = \emptyset$ or $L = \Phi(\lesssim)$ for some $\lesssim$ total preorder. Using Proposition 3 and Remark 1 we have:

$$L = \Phi(\lesssim) = \Phi(\bigvee_i \lesssim_i) = \bigcap_i \Phi(\lesssim_i),$$

where $\lesssim_i$ are basic total preorders. By Proposition 4 and Definition 3 we have that L is a TP-cone.

($\Leftarrow$) Let L be a TP-cone. We have that:

- if $L = \emptyset$ then $L \in \Phi(\mathbf{TP})$;
- if L is an orthant then for each i and j $x_{ij} \geq \bar{0}$ or $x_{ij} \leq \bar{0}$, which is equivalent to say that there exists $\lesssim$ (basic) total preorder such that $a_i \lesssim a_j$ or $a_j \lesssim a_i$, i.e. $L \in \Phi(\mathbf{TP})$;

- if L is an intersection of O_i orthants then

$$L = \bigcap_i O_i = \bigcap_i \Phi(\lesssim_i) = \Phi\left(\bigvee_i \lesssim_i\right),$$

for some $\lesssim_i$ basic total preorders, i.e. $L \in \Phi(\mathbf{TP})$.

5.1   Categorical Duality

In this subsection we provide a categorical duality between the categories of total preorders and of TP-cones (for basic definition on categories see [22]).

Let us define the categories $\mathbb{TP}_n$ (of total preorders) and $\mathbb{K}_n$ (of TP-cones in TGM_n). In $\mathbb{TP}_n$ the objects are total preorder on n elements and arrows are defined by order $\ll$, i.e.

$$\lesssim_1 \to \lesssim_2 \quad \Leftrightarrow \quad \lesssim_1 \ll \lesssim_2 .$$

In a similar way we define $\mathbb{K}_n$ whose objects are TP-cones in the space TGM_n and arrows are defined by inclusion.

Theorem 3 *Categories of preorders and of TP-cones are dually isomorphic.*

Proof of Theorem 3 descends from lemmas below.

Lemma 1 *The maps* $\Theta : \mathbb{K}_n \to \mathbb{TP}_n$ *and* $\Phi : \mathbb{TP}_n \to \mathbb{K}_n$ *defined as follows*

- $\Theta(C) = \Theta(C)$
- $\Theta(\to) = \leftarrow$
- $\Phi(\lesssim) = \Phi(\lesssim)$
- $\Phi(\to) = \leftarrow$

are contravariant functors.

Proof Let us consider C and D TP-cones, such that $C \to D$. We have that:

$$C \to D \quad \Leftrightarrow \quad C \subseteq D \quad \Leftrightarrow \quad \Theta(C) \gg \Theta(C) \quad \Leftrightarrow \quad \Theta(C) \leftarrow \Theta(D).$$

Analogously, if we consider $\lesssim_1$ and $\lesssim_2$ total preorders over n elements, such that $\lesssim_1 \to \lesssim_2$, then:

$$\lesssim_1 \to \lesssim_2 \quad \Leftrightarrow \quad \lesssim_1 \ll \lesssim_2 \quad \Leftrightarrow \quad \Phi(\lesssim_1) \supseteq \Phi(\lesssim_2) \quad \Leftrightarrow \quad \Phi(\lesssim_1) \leftarrow \Phi(\lesssim_2).$$

Lemma 2 *The composed functors* $\Phi\Theta : \mathbb{K}_n \to \mathbb{K}_n$ *and* $\Theta\Phi : \mathbb{TP}_n \to \mathbb{TP}_n$ *are the identity functors of the categories* $\mathbb{K}_n$ *and* $\mathbb{TP}_n$ *respectively.*

Proof Let us consider K TP-cone, we have that

$$\Phi\Theta(K) = \Phi(\Theta(K)) = \Phi\left(\Phi^{-1}\left(\bigcap_{\substack{C \in \Phi(\mathbf{TP}) \\ C \cap K \neq \emptyset}} C\right)\right) = \bigcap_{\substack{C \in \Phi(\mathbf{TP}) \\ C \cap K \neq \emptyset}} C,$$

but K is a TP-cone, i.e. $K \in \Phi(\mathbf{TP})$, hence

$$\bigcap_{\substack{C \in \Phi(\mathbf{TP}) \\ C \cap K \neq \emptyset}} C = K.$$

Vice versa, let $\precsim$ be a total preorder, then

$$\Theta\Phi(\precsim) = \Theta(\Phi(\precsim)) = \Theta(\{X \in TGM_n \mid \precsim \text{ is compatible with } X\}).$$

Let us denote by $K_{\precsim} = \{X \in TGM_n \mid \precsim \text{ is compatible with } X\}$, therefore we have:

$$\Theta(K_{\precsim}) = \Phi^{-1}\left(\bigcap_{\substack{C \in \Phi(\mathbf{TP}) \\ C \cap K \neq \emptyset}} C\right) = \Phi^{-1}(K_{\precsim}) = \precsim .$$

In both cases arrows are preserved by Lemma 1.

6 Conclusions

This paper proposes a first attempt to use Riesz spaces as general framework for decision making methods, in particular we focused on pairwise comparison matrices and AHP. We provide a characterization of collective choice rules satisfying Arrow's axioms (Theorem 1); we established an antitone Galois correspondence between total preorders and cones of a Riesz space (Theorem 2) and a categorical duality between categories of preorders and of particular cones of a Riesz space (Theorem 3). After some motivations and examples (presented in Sect. 3) it is proposed a systematical investigation on connections between PCMs, total preorders and Riesz spaces; we deal with collective choice rules and classical social axioms (Arrow's axioms), supported by a duality between total preorders and geometric objects.

We stress also that Riesz spaces are categorically equivalent to Riesz MV-algebras (see [15]); this implies that generalized PCMs provide a semantic for (an extension of) Łukasiewicz logic, which is already applied also in artificial neural network field (e.g. see [13]).

References

1. Abramovich Y, Aliprantis C, Zame W (1995) A representation theorem for Riesz spaces and its applications to economics. Econ Theory 5(3):527–535
2. Aliprantis CD, Brown DJ (1983) Equilibria in markets with a Riesz space of commodities. J Math Econ 11(2):189–207
3. Aliprantis CD, Burkinshaw O (2003) Locally solid Riesz spaces with applications to economics. Am Math Soc 105
4. Aliprantis CD, Tourky R (2007) Cones and duality. Am Math Soc 84
5. Badri MA (1999) Combining the analytic hierarchy process and goal programming for global facility location-allocation problem. Int J Prod Econ 62(3):237–248
6. Belton V (1986) A comparison of the analytic hierarchy process and a simple multi-attribute value function. Eur J Oper Res 26(1):7–21
7. Beynon WM (1975) Duality theorems for finitely generated vector lattices. Proc Lond Math Soc 31(3):114–128. Citeseer
8. Cavallo B, D'Apuzzo L (2009) A general unified framework for pairwise comparison matrices in multicriterial methods. Int J Intell Syst 24(4):377–398
9. Cavallo B, D'Apuzzo L (2015) Reciprocal transitive matrices over abelian linearly ordered groups: characterizations and application to multi-criteria decision problems. Fuzzy Sets Syst 266:33–46
10. Cavallo B, Vitale G, D'Apuzzo L (2016) Aggregation of pairwise comparison matrices and Arrow's conditions. Submitt Fuzzy Sets Syst
11. Cerreia-Vioglio S, Maccheroni F, Marinacci M, Montrucchio L (2015) Choquet integration on Riesz spaces and dual comonotonicity. Trans Am Math Soc 367(12):8521–8542
12. Debreu G (1954) Representation of a preference ordering by a numerical function. Decis Process 3:159–165
13. Di Nola A, Lenzi G, Vitale G (2016) Łukasiewicz equivalent neural networks. In: Advances in neural networks. Springer, pp 161–168
14. Di Nola A, Lenzi G, Vitale G (2016) Riesz–McNaughton functions and Riesz MV-algebras of nonlinear functions. Fuzzy Sets Syst
15. Di Nola A, Leuştean I (2014) Łukasiewicz logic and Riesz spaces. Soft Comput 18(12):2349–2363
16. Frederick S, Loewenstein G, O'donoghue T (2002) Time discounting and time preference: a critical review. J Econ Lit 40(2):351–401
17. Harsanyi JC (1953) Cardinal utility in welfare economics and in the theory of risk-taking. J Polit Econ 61(5):434–435
18. Houthakker HS (1950) Revealed preference and the utility function. Economica 17(66):159–174
19. Hua Lu M, Madu CN, Kuei Ch, Winokur D (1994) Integrating QFD, AHP and benchmarking in strategic marketing. J Bus Ind Mark 9(1):41–50
20. Labuschagne C, Van Alten C (2007) On the variety of Riesz spaces. Indagationes Mathematicae 18(1):61–68
21. Levy H, Markowitz HM (1979) Approximating expected utility by a function of mean and variance. Am Econ Rev 69(3):308–317
22. Mac Lane S (1978) Categories for the working mathematician, vol 5. Springer Science & Business Media
23. Racioppi V, Marcarelli G, Squillante M (2015) Modelling a sustainable requalification problem by analytic hierarchy process. Qual Quant 49(4):1661–1677
24. Saaty TL (1977) A scaling method for priorities in hierarchical structures. J Math Psychol 15(3):234–281
25. Saaty TL (1990) How to make a decision: the analytic hierarchy process. Eur J Oper Res 48(1):9–26

26. Torrance GW, Feeny DH, Furlong WJ, Barr RD, Zhang Y, Wang Q (1996) Multiattribute utility function for a comprehensive health status classification system: health utilities index mark 2. Med Care 34(7):702–722
27. Trockel W (1998) Group actions on spaces of preferences and economic applications. Util Funct Ordered Spaces 5:159–175
28. Vaidya OS, Kumar S (2006) Analytic hierarchy process: an overview of applications. Eur J Oper Res 169(1):1–29
29. Xu Z (2015) Uncertain multi-attribute decision making: methods and applications. Springer
30. Zanakis SH, Solomon A, Wishart N, Dublish S (1998) Multi-attribute decision making: a simulation comparison of select methods. Eur J Oper Res 107(3):507–529

Coherent Conditional Plausibility: A Tool for Handling Fuzziness and Uncertainty Under Partial Information

Giulianella Coletti and Barbara Vantaggi

Abstract Non-additive measures, such as plausibility, are meaningful when only a partial or indirect information on the events of interest is available, or when imprecision and ambiguity of agents are considered. Our main aim is to study inferential processes, like the Bayesian one, when the information is expressed in natural language and the uncertainty measure is either partially or imprecisely assessed. We deal with partial assessments consistent with a conditional plausibility, and adopt the interpretation of the membership of a fuzzy set in terms of coherent conditional plausibility, regarded as a function of the conditioning events. This kind of interpretation, inspired to that given in terms of coherent conditional probability, is particularly useful for computing the measure of the uncertainty of fuzzy events, when the knowledge on the variable is imprecise and can be managed with a non-additive measure of uncertainty. A simple situation related to a Zadeh's example can be the following: a ball will be drawn from an urn containing balls of different colours and different diameters, but one knows only the distribution of the different colours. The aim is to compute the uncertainty measure of the fuzzy event "a small ball is drawn", taking in considerations possible logical constrains among the colours and the diameters.

1 Introduction

Combining uncertainty due to heterogeneous sources of knowledge, or to ambiguity of agents, is usually performed under different frameworks (e.g. probability theory, theory of evidence, fuzzy set theory). This is necessary in many fields such as

G. Coletti (✉)
Dipartimento di Matematica e Informatica, Università di Perugia,
via Vanvitelli, 1, 06123 Perugia, Italy
e-mail: giulianella.coletti@unipg.it

B. Vantaggi
Dipartimento S.B.A.I, "La Sapienza" Università di Roma,
via Scarpa, 00185 Rome, Italy
e-mail: barbara.vantaggi@sbai.uniroma1.it

© Springer International Publishing AG 2018   129
M. Collan and J. Kacprzyk (eds.), *Soft Computing Applications for Group Decision-making and Consensus Modeling*, Studies in Fuzziness and Soft Computing 357, DOI 10.1007/978-3-319-60207-3_9

decision making, operational research, artificial intelligence, management science, expert systems (as already discussed in seminal works such as [3, 32, 35, 36, 54, 57]) and more recently in semantic web and in bioinformatics (see for instance [42, 44, 48, 56]).

Fuzzy set theory, introduced by Zadeh [53], for handling vagueness and imprecision due to natural language, has become very popular and it provides an advanced formalization of some concepts expressed by means of natural language. Different interpretations of fuzzy sets have been given [31, 39, 43, 47] in terms of (conditional) probabilities, we refer to that given in [11–13], where the membership function of a fuzzy subset is interpreted in terms of a coherent conditional probability assessment. This interpretation, as shown in [6, 19, 20], is particularly meaningful when fuzzy and statistical information is simultaneously available [55].

Nevertheless sometimes the statistical information is imprecise, so it is unavoidable to act under ambiguity since uncertainty must be evaluated by means of classes of probabilities and their envelopes, instead of a single probability distribution. For example, such measures can be obtained through an extension process: when the probabilistic information is related to events different from those of interest and where the fuzzy information is available, the probabilistic assessment needs to be extended, a la de Finetti [23, 51], obtaining a family of probabilities whose upper envelope could be a plausibility [5, 15, 26, 38, 50] or also a possibility [14, 25, 33].

In this paper we consider the above problems by focusing mainly on plausibility measures, for which many proposals of conditioning are present. We follow mainly the approach based on Dempster rule [26], generalized in [45?]. This rule assures a "weak disintegration rule" and admits as particular case conditional probabilities a la de Finetti [23] and T-conditional possibility, with T the t-norm of product [2, 17, 18].

It is well known that conditional plausibilities, defined through Dempster rule, cannot be obtained as the upper envelope of a class of conditional probabilities, even if, for any conditioning event, they can be obtained as the upper envelope of the extensions of a suitable probability [7].

In the first part of the paper, in order to consider a generalized Bayesian inferential procedure, by using the concept of coherence (that is the consistency of a partial assessment with a conditional possibility or plausibility), we study the properties of likelihood functions, both as point and set functions, in the different frameworks. Then, we extend some results provided in [8, 9, 21] and we study the coherence of a likelihood and a plausibility (or possibility) "a prior" measure.

In Sect. 4 we give an interpretation of the membership of fuzzy sets as a plausibilistic likelihood function and we study which properties of fuzzy set theory are maintained. The semantic of the interpretation is very similar to those made in terms of either probabilistic or possiblistic likelihood: if φ is a property, related to a variable X, the meaning associated to the membership $\mu_\varphi(\cdot)$ on x consists into the plausibility that You claim that X is φ under the hypothesis that X assumes the value x.

We show, from a syntactical point of view, the differences and common features under the various frameworks. Among those is relevant to emphasize that the degree of fuzziness, expressed through the membership, of the union $\mu_{\varphi\vee\psi}$ of two fuzzy sets, with memberships μ_φ and μ_ψ, is not strictly linked to the membership of intersections $\mu_{\varphi \wedge \psi}$ by the Frank equation [37], as in the probabilistic interpretation. On the other hands $\mu_{\varphi \vee \psi}$ is not univocally determined as the $\max\{\mu_\varphi, \mu_\psi\}$, independently of $\mu_{\varphi \wedge \psi}$, as in possibilistic interpretation. While in the case of plausibilistic framework it is not univocally determined, but $\mu_{\varphi \vee \psi}(x)$ must be between the bounds $\max\{\mu_\varphi(x), \mu_\psi(x)\}$ and $\min\{\mu_\varphi(x) + \mu_\psi(x)\} - \mu_{\varphi \wedge \psi}(x), 1\}$.

In all these interpretations the fuzzy membership μ_φ coincides with a likelihood and the fuzzy event E_φ is the Boolean event "You claim that X is φ"; moreover for the measure of uncertainty of E_φ when the prior on X is subjected to imprecision and so it could be given by means of a plausibility, we get an upper bound, while when the prior is a possibility we give an analytic formula depending on the chosen t-norm.

2 The Uncertainty Framework of Reference

Our framework of reference is that of conditional plausibility. Among the definitions present in the literature we choose the axiomatic one. It is directly defined as a function on sets of conditional events $E|H$, that is ordered pairs (E, H) of events which can be both true or false with $H \neq \emptyset$, but that play a different role (the conditioning one is assumed as a hypothesis). Any event E can be seen as the conditional event $E|\Omega$ (where Ω is the sure event). In the definition of conditional plausibility (such as for the conditional probability) the set of conditional events is required to have a proper algebraic structure and the function must satisfy a set of rules.

In order to combine uncertainty information with vagueness, we need to manage assessments in arbitrary sets of conditional events and so we need to refer to the notion of coherence, which guaranties an effective tool for controlling global consistency with a conditional plausibility and ruling the inferential procedures.

We recall that a general inferential problem can be simply seen in fact as an extension of an assessment to other events, maintaining consistency with the framework of reference.

2.1 Conditional Measures

Usually in literature conditional measures are presented as a derived notion of unconditional ones, by introducing a law (equation) involving the joint measure and its marginal. Nevertheless, this could be restrictive, since for some pair of events the solution of the equation (the conditional measure) can either not exist or not be unique. So, in analogy with conditional probability [24], it is preferable to define conditional measures in an axiomatic way, directly as a function defined on a suitable set of conditional events, satisfying a set of rules (axioms). We recall here the notion of conditional plausibility axiomatically defined as follows (see [15]):

Definition 1 Let $\mathcal{B}$ be a Boolean algebra and $\mathcal{H} \subseteq \mathcal{B} \setminus \{\emptyset\}$ an additive set. A function Pl defined on $\mathcal{C} = \mathcal{B} \times \mathcal{H}$ is a conditional plausibility if it satisfies the following conditions

(i) $Pl(E|H) = Pl(E \wedge H|H)$;
(ii) $Pl(\cdot|H)$ is a plausibility function $\forall H \in \mathcal{H}$;
(iii) For every $E \in \mathcal{B}$ and $H, K \in \mathcal{H}$

$$Pl(E \wedge H|K) = Pl(E|H \wedge K) \cdot Pl(H|K).$$

Moreover, given a conditional plausibility, a conditional belief function $Bel(\cdot|\cdot)$ is defined by duality as follows: for every event $E|H \in \mathcal{C}$

$$Bel(E|H) = 1 - Pl(E^c|H). \tag{1}$$

Conditions *(i)* and *(ii)* require that $Pl(\Omega|H) = Pl(H|H) = 1$ and $Pl(\emptyset|H) = 0$ and moreover, for any n, $Pl(\cdot|H)$ is n-alternating [26, 27], that is, for any $A_1, \dots, A_n \in A$ and $A = \bigwedge_{i=1}^{n} A_i$:

$$Pl(A|H) \leq \sum_{\emptyset \neq I \subseteq \{1,\dots,n\}} (-1)^{|I|+1} Pl(\bigvee_{i \in I} A_i|H) \tag{2}$$

Then, $Bel(\cdot|H)$ is n-monotone, for any n.
The above definition extends the Dempster's rule [26], i.e.

$$Bel(F|H) = 1 - \frac{Pl(F^c \wedge H)}{Pl(H)},$$

for all conditioning events H such that $Pl(H) > 0$.

We notice that, when all the conditioning events have positive plausibility, i.e.for any $H \in \mathcal{H}$ with $Pl(H|H^0) > 0$ (where $H^0 = \bigvee_{H \in \mathcal{H}} H$), the above notions of conditional plausibility and belief coincide with those given in [28]. In fact, if $Pl(H) > 0$ it follows

$$Bel(F|H) = \frac{Bel(F \vee H^c) - Bel(H^c)}{Pl(H)}. \tag{3}$$

An easy consequence of Definition 1 is a weak form of disintegration formula [4] for the plausibility of any conditional event $E|H$ with respect to a partition $H_1, \dots, H_N$ of H

$$Pl(E|H) \leq \sum_{k=1}^{N} Pl(H_k|H) Pl(E|H_k) \tag{4}$$

Other different definitions of conditioning are present in the literature: the most interesting and famous is that due to Jaffray and Walley (see [34, 40, 50, 51]) defined by the following equation:

$$Pl(F|H) = \frac{Pl(F \wedge H)}{Pl(F \wedge H) + Bel(F^c \wedge H)} \tag{5}$$

and obtained as upper envelope of particular classes of conditional probabilities.

We notice that conditional plausibility defined by Eq. (5) does not satisfy axiom *(iii)* of Definition 1 and then, in particular, it does not satisfy (4).

Finally we point out that the class of conditional plausibilities (defined in Definition 1) contains in particular important classes: conditional probabilities, as introduced in [24, 29, 41], and T-conditional possibilities [2, 18], with the t-norm T equal to the usual product (in symbols P-conditional possibility).

The next result (proved in [4]) shows that every conditional plausibility on $\mathscr{B} \times \mathscr{H}$ can be extended (not uniquely) to a full conditional plausibility on $\mathscr{B}$ (i.e., a conditional plausibility on $\mathscr{B} \times \mathscr{B}^0$) and to a full conditional plausibility on $\mathscr{B}'$ for any finite super-algebra $\mathscr{B}' \supset \mathscr{B}$.

Theorem 1 *Let $\mathscr{B}$ be a finite algebra. If Pl on $\mathscr{B} \times \mathscr{H} \to [0, 1]$ is a conditional plausibility, then there exists a conditional plausibility Pl' : $\mathscr{B} \times \mathscr{B}^0 \to [0, 1]$ such that $Pl'_{|\mathscr{B} \times \mathscr{H}} = Pl$. Furthermore, for any finite superalgebra $\mathscr{B}' \supseteq \mathscr{B}$, there exists a full conditional plausibility Pl' : $\mathscr{B}' \times \mathscr{B}'^0 \to [0, 1]$ such that $Pl'_{|\mathscr{B} \times \mathscr{H}} = Pl$.*

Note that the full conditional plausibility Pl' on $\mathscr{B}'$ extending the given conditional plausibility Pl is not unique.

2.2 Coherent Conditional Plausibility

In probability theory the concept of coherence [24] has been introduced for handling conditional probability assessed on arbitrary set of conditional events $\mathscr{C}$. Coherence assures the consistency of the assessment with a conditional probability defined on a superset with the logical requisites required in Definition 1 and rules its coherent extensions on any superset $\mathscr{C}'$ of $\mathscr{C}$. It is possible to introduce a similar notion also for plausibility functions as already made for T-conditional possibilities in [2, 18].

Definition 2 A function (or assessment) γ : $\mathscr{C} \to [0, 1]$, on a set of conditional events $\mathscr{C}$, is a coherent conditional plausibility iff there exists a full conditional plausibility Pl on an algebra $\mathscr{B}$ such that $\mathscr{C} \subseteq \mathscr{B} \times \mathscr{B}^0$ and the restriction of Pl on $\mathscr{C}$ coincides with γ.

Both for conditional probabilities and for P-conditional possibilities a characterization of coherent assessment has been given in terms of suitable classes of unconditional probabilities or possibilities, respectively, or in terms of solvability of a sequence of systems (see [11, 18]).

The following result proved in [4] extends these results to (coherent) conditional plausibility functions.

Theorem 2 *Let* $\mathscr{F} = \{E_1|F_1, E_2|F_2, \ldots, E_m|F_m\}$ *and denote by* $\mathscr{B}$ *the algebra generated by* $\{E_1, \ldots, E_m, F_1, \ldots, F_m\}$, $H_0^0 = \vee_{j=1}^m F_j$.

For $Pl : \mathscr{F} \rightarrow [0, 1]$ *the following statements are equivalent:*

(a) *Pl is a coherent conditional plausibility;*

(b) *there exists a class* $\mathscr{P} = \{Pl_\alpha\}$ *of plausibility functions such that* $Pl_\alpha(H_0^\alpha) = 1$
and $H_0^\alpha \subset H_0^\beta$ *for all* $\beta < \alpha$, *where* H_0^α *is the greatest element of* $\mathscr{K}$ *for which* $Pl_{(\alpha-1)}(H_0^\alpha) = 0$.
Moreover, for every $E_i|F_i$, *there exists a unique index* α *such that* $Pl_\beta(F_i) = 0$
for all $\alpha > \beta$, $Pl_\alpha(F_i) > 0$ *and*

$$Pl(E_i|F_i) = \frac{Pl_\alpha(E_i \wedge F_i)}{Pl_\alpha(F_i)}, \tag{6}$$

(c) *all the following systems* (S_Pl^α), *with* $\alpha = 0, 1, 2, \ldots, k \leq n$, *admit a solution*
$\mathbf{X}^\alpha = (\mathbf{x}_1^\alpha, \ldots, \mathbf{x}_{j_\alpha}^\alpha)$ *with* $\mathbf{x}_j^\alpha = m_\alpha(H_j)$ $(j = 1, \ldots, j_\alpha)$:

$$(S_{Pl}^\alpha) = \begin{cases} \displaystyle\sum_{H_k \wedge F_i \neq \emptyset} x_k^\alpha \cdot Pl(E_i|F_i) = \sum_{H_k \wedge E_i \wedge F_i \neq \emptyset} x_k^\alpha, \forall F_i \subseteq H_0^\alpha \\ \displaystyle\sum_{H_k \in H_0^\alpha} x_k^\alpha = 1 \\ x_k^\alpha \geq 0, \hspace{3cm} \forall H_k \subseteq H_0^\alpha \end{cases}$$

where H_0^α *is the greatest element of* $\mathscr{K}$ *such that* $\displaystyle\sum_{H_i \wedge H_0^\alpha \neq \emptyset} m_{(\alpha-1)}(H_i) = 0$.

In particular, conditions *(b)* and *(c)* put in evidence that this conditional measure can be written in terms of a suitable class of basic assignments, instead of just one as in the classical case, where all the conditioning events have positive plausibility. Whenever there are events in $\mathscr{K}$ with zero plausibility the class of unconditional plausibilities contains more than one element and we can say that Pl_1 gives a refinement of those events judged with zero plausibility under Pl_0.

Every class $\mathscr{P}$ (condition *(b)* of Theorem 2) is said to be *agreeing* with conditional plausibility *Pl*.

3   Likelihood Functions

In this section we recall some results related to a comparative analysis of likelihood functions under different frameworks: probability, possibility. Moreover, we investigate the relation with likelihoods as coherent conditional plausibility.

Let us consider a finite partition $\mathscr{L} = \{H_i\}_{i \in I}$ of Ω together with two Boolean algebras $\mathscr{A}_\mathscr{L}$ and $\mathscr{A}$ where $\mathscr{A}_\mathscr{L} = \langle \mathscr{L} \rangle$ is the algebra generated by $\mathscr{L}$ and $\mathscr{A}$ a super-algebra such that $\mathscr{A}_\mathscr{L} \subseteq \mathscr{A}$.

A *plausibilistic strategy* is a map $\sigma : \mathscr{A} \times \mathscr{L} \to [0, 1]$ satisfying the following conditions for every $H_i \in \mathscr{L}$:

(S1) $\sigma(E|H_i) = 0$ if $E \wedge H_i = \emptyset$ and $\sigma(E|H_i) = 1$ if $E \wedge H_i = H_i$, for every $E \in \mathscr{A}$;
(S2) $\sigma(\cdot|H_i)$ is a plausibility on $\mathscr{A}$.

Plausibilistic strategies differ from probabilistic (see [29]) and possibilistic strategies (see [8]) on the requirement *(S2)*: this condition is replaced by the requirement that $\sigma(\cdot|H_i)$ on $\mathscr{A}$ is a finitely additive probability or a finitely maxitive possibility, respectively.

When the assignment is not given on a structure as $\mathscr{A} \times \mathscr{L}$, then we need to require a coherence condition. In the particular case that the assignment is related to an evidence, expressed by an event E, coherence is assured in all the frameworks, as showed in the following Theorem 3. In the following we refer to the following notion of likelihood:

Definition 3 Given an event E and a finite partition $\mathscr{L}$, a likelihood function is an assessment on $\{E|H_i : H_i \in \mathscr{L}\}$ (that is a function $f : \{E\} \times \mathscr{L} \to [0, 1]$) satisfying only the following trivial condition:

(L1) $f(E|H_i) = 0$ if $E \wedge H_i = \emptyset$ and $f(E|H_i) = 1$ if $H_i \subseteq E$

Theorem 3 *Let* $\mathscr{L} = \{H_1, \ldots, H_n\}$ *be a finite partition of* Ω *and* E *an event. For every function* $f : \{E\} \times \mathscr{L} \to [0, 1]$ *satisfying the condition (L1) the following statements hold:*

(a) f is a coherent conditional probability;
(b) f is a coherent T-conditional possibility (for every continuous t-norm T);
(c) f is a coherent conditional plausibility.

Proof Condition *(a)* and *(b)* have been proved in [13] and [8], respectively.
Condition *(c)* derives from the fact that any coherent conditional probability (or equivalently any coherent T-conditional possibility, with T the usual product) is a coherent conditional plausibility.   □

Corollary 1 *Let* $\mathscr{L} = \{H_1, \ldots, H_n\}$ *be a finite partition of* Ω *and* E *an event. If the only coherent conditional plausibility (probability, possibility) f takes values in* $\{0, 1\}$, *then it is* $H_i \wedge E = \emptyset$ *for every* H_i *such that* $f(E|H_i) = 0$ *and it is* $H_i \subseteq E$ *for every* H_i *such that* $f(E|H_i) = 1$.

Proof It follows directly from Theorem 2 and the characterization theorem for coherent conditional probabilities [11] and T-conditional possibilities [18].   □

The previous result points out that "syntactically" a probabilistic likelihood function is indistinguishable from a possibilistic likelihood function or a plausibilistic likelihood function, i.e., any function f satisfying the minimal requirement of consistency *(L1)* can be extended either as a probabilistic strategy or as a possibilistic strategy or as a plausibility strategy.

For this, we drop the adjective probabilistic, possibilistic or plausibilistic and we call such functions simply likelihood.

Obviously, the extensions are syntactically different [10, 46], as aforementioned, so a criterion for choosing the framework is need to be determined. This criterion could be guided from semantic motivations or related to syntactically reasons. In the last case the choice could be guided from the "prior" information. This situation is analysed in the following session.

3.1 Likelihood and Prior

In some situations where imprecision or ambiguity of agents is presented there is no a unique prior distribution on a given partition $\mathscr{L}$ of interest. In these situations a class of prior distribution are available on $\mathscr{A}_{\mathscr{L}}$. These prior distribution could arise from the extension process of a given probability (see, for example, [14, 25, 33, 51]). In this paper we deal with the situation where the upper envelope of this class of prior distribution is a plausibility.

The aim is now to make inference according to a Bayesian-like procedure, so we need to deal with an initial assessment consisting of a "prior" φ on an algebra $\mathscr{A}_{\mathscr{L}}$ and a "likelihood function" f related to the set of conditional events $E|H_i$'s, with E any event and $H_i \in \mathscr{L}$. This topic has been deeply discussed in [50, 51] by considering several interesting examples.

First of all we need to test the consistency of the global assessment

$$\{f, v\} = \{f(E|H_i), v(H_i) \, : \, H_i \in \mathscr{L}\}$$

with respect to the framework of reference, pointing out by an uncertainty measure (plausibility or more specific measures such as probability or P-possibility).

The choice of the framework of reference could be essentially decided by the prior, since as shown in Theorem 3, a likelihood can be re-read in any framework.

Actually in this session we present essentially some results given in [21].

Theorem 4 *Let $\mathscr{L}$ be a partition of Ω, consider a likelihood f related to an event E on $\mathscr{L}$ and consider a probability P, a plausibility Pl and a possibility Π, respectively, on the algebra $\mathscr{A}_{\mathscr{L}}$ generated by $\mathscr{L}$. Then the following statements hold:*

(a) the global assessment $\{f, P\}$ is a coherent conditional probability;
(b) the global assessment $\{f, Pl\}$ is a coherent conditional plausibility;
(c) the global assessment $\{f, \Pi\}$ is a coherent T-conditional possibility (for every continuous t-norm T);

Proof Condition *(a)* is well known and condition *(c)* has been proved in [1].

Concerning condition *(b)* note that Pl on $\mathscr{A}_{\mathscr{L}}$ defines a unique basic assignment function m_0 on $\mathscr{A}_{\mathscr{L}}$ that is the unique solution of S_{Pl}^0 concerning the coherence of Pl.

Then, the assessment $\{f, Pl\}$ can be seen as an assessment and we need to establish whether it is a coherent conditional plausibility. Therefore, we need to check whether the relevant system $S^0_{Pl,f}$ has solution and so whether there is a class of basic assignment $\{m'_\alpha\}$ on $\langle E, \mathscr{L} \rangle$.

Note that if the system $S^0_{Pl,f}$ has solution, then coherence with respect to conditional plausibility follows: in fact if there is some $H_i \in \mathscr{L}$ such that $Pl(H_i) = 0$, we need to build the system $S^1_{Pl,f}$ by considering equations related to $f(E|H_i)$ (with $Pl(H_i) = 0$) and coherence follows since the likelihood is a coherent conditional plausibility.

Actually, the atoms in $\langle E, \mathscr{L} \rangle$ are all the events $E \wedge H_i, E^c \wedge H_i$ with $H_i \in \mathscr{L}$.

From [22] any plausibility on $\mathscr{A}_{\mathscr{L}}$ induces a unique function, called basic plausibility assignment, ν (possibly taking also negative values) on $\mathscr{A}_{\mathscr{L}}$ such that $\sum_{A \in \mathscr{A}_{\mathscr{L}}} \nu(A) = 1$ and $\sum_{A \in \mathscr{A}_{\mathscr{L}}: A \subseteq B} \nu(A) = Pl(B)$.

Let μ be on $\mathscr{A}_{\mathscr{L}}$ be the plausibility assignment induced by Pl, consider μ' defined on $\langle E, L \rangle$ as follows

$$
\begin{aligned}
&\mu'(H_i) = 0, \\
&\mu'(E \wedge H_i) = f(E|H_i)Pl(H_i) \\
&\mu'(E^c \wedge H_i) = \mu(H_i) - \mu'(E \wedge H_i), \\
&\mu(A) = \mu'(A), \qquad\qquad \text{for any } A \in \mathscr{A}_{\mathscr{L}} \setminus \mathscr{L}
\end{aligned}
$$

By construction $\sum_{A \in \langle E, \mathscr{L} \rangle} \mu'(A) = 1$. For any B in $\langle E, \mathscr{L} \rangle$ but not in $(\mathscr{A}_{\mathscr{L}} \cup \{E \wedge H_i, E^c \wedge H_i : H_i \in \mathscr{L}\})$ one has $\mu'(B) = 0$. Then, the function f on $\langle E, L \rangle$ defined as $\sum_{A \in \langle E, \mathscr{L} \rangle: A \subseteq B} \mu'(A) = f(B)$ is such that by construction, for any $B \in \mathscr{A}_{\mathscr{L}}$,

$$
f(B) = \sum_{A \in \langle E, \mathscr{L} \rangle: A \subseteq B} \mu'(A) =
$$

$$
\sum_{A \in \mathscr{A}_{\mathscr{L}}: A \subseteq B} \mu'(E \wedge A) + \mu'(E^c \wedge A) + \mu'(A) = \sum_{A \in \mathscr{A}_{\mathscr{L}}: A \subseteq B} \mu(A) = Pl(B)
$$

then f extends Pl.

We need to prove that f is a plausibility: the proof can be made by induction, we prove here that is 2-alternating, the property of n-alternance follows by induction.

For any event $A \in \langle E, \mathscr{L} \rangle$ there is an event $\bar{A} \in \mathscr{A}_{\mathscr{L}}$ such that $\bar{A} \subseteq A$ and no event $B \in \mathscr{A}_{\mathscr{L}}$ such that $\bar{A} \subset B \subseteq A$, that is the maximal event of $\mathscr{A}_{\mathscr{L}}$ contained in A. Then, given any pair of events $A, B \in \langle E, \mathscr{L} \rangle$ let $\bar{A}, \bar{B} \in \mathscr{A}_{\mathscr{L}}$ be the two maximal events contained, respectively in A and B. Thus,

$$
f(A \vee B) = \sum_{C \in \langle E, \mathscr{L} \rangle: C \subseteq A \vee B} \mu'(C) = \sum_{E \wedge H_i \subseteq A \vee B} \mu'(E \wedge H_i) + \sum_{E^c \wedge H_i \subseteq A \vee B} \mu'(E^c \wedge H_i) + \sum_{C \in \langle \mathscr{L} \rangle \setminus \mathscr{L}, C \subseteq A \vee B} \mu'(C) =
$$

$$
= \sum_{H_i \subseteq A \vee B} \mu(H_i) + \sum_{E \wedge H_i \subseteq A \vee B, E^c \wedge H_i \not\subseteq A \vee B} \mu'(E \wedge H_i) + \sum_{E^c \wedge H_i \subseteq A \vee B, E \wedge H_i \not\subseteq A \vee B} \mu'(E^c \wedge H_i) + \sum_{C \in \langle \mathscr{L} \rangle \setminus \mathscr{L}, C \subseteq A \vee B} \mu(C) =
$$

$$= Pl(\overline{A \vee B}) + \sum_{E \wedge H_i \subseteq A \vee B, E^c \wedge H_i \not\subseteq A \vee B} \mu'(E \wedge H_i) + \sum_{E^c \wedge H_i \subseteq A \vee B, E \wedge H_i \not\subseteq A \vee B} \mu'(E^c \wedge H_i) =$$

$$= Pl(\bar{A} \vee \bar{B}) + \sum_{H_i \subseteq \overline{A \vee B}, H_i \not\subseteq \bar{A} \vee \bar{B}} \mu(H_i) + \sum_{E \wedge H_i \subseteq A \vee B, E^c \wedge H_i \not\subseteq A \vee B} \mu'(E \wedge H_i) + \sum_{E^c \wedge H_i \subseteq A \vee B, E \wedge H_i \not\subseteq A \vee B} \mu'(E^c \wedge H_i).$$

Note that $A = \bar{A} \vee \bigvee_{H_i \in \mathscr{L} : H_i \not\subseteq A}((E \wedge H_i \wedge A) \vee (E^c \wedge H_i \wedge A))$ and analogously for B. Obviously, $\bar{A} \vee \bar{B} \subseteq A \vee B$ and $\bar{A} \wedge \bar{B}$ coincides with $\overline{A \wedge B}$.

Moreover, $\bar{A} \vee \bar{B}$ is included into $\overline{A \vee B}$ but does not coincide with it, in fact $H_i \in \mathscr{L}$ could be included in $A \vee B$, but H_i is not included neither in A nor in B (e.g. $E \wedge H_i \subseteq A$ and $E^c \wedge H_i \subseteq B$).

Hence,

$$f(A \vee B) \leq Pl(\bar{A}) + Pl(\bar{B}) - Pl(\bar{A} \wedge \bar{B}) + \sum_{H_i \subseteq A \vee B, H_i \not\subseteq \bar{A} \vee \bar{B}} \mu(H_i) + \sum_{E \wedge H_i \subseteq A \vee B, E^c \wedge H_i \not\subseteq A \vee B} \mu'(E \wedge H_i)$$

$$+ \sum_{E^c \wedge H_i \subseteq A \vee B, E \wedge H_i \not\subseteq A \vee B} \mu'(E^c \wedge H_i) \leq$$

$$\leq Pl(\bar{A}) + Pl(\bar{B}) - Pl(\bar{A} \wedge \bar{B}) + \sum_{H_i \subseteq \overline{A \vee B}, H_i \not\subseteq \bar{A} \vee \bar{B}} (\mu'(E \wedge H_i) + \mu'(E^c \wedge H_i)) + \sum_{E \wedge H_i \subseteq A \vee B, E^c \wedge H_i \not\subseteq A \vee B} \mu'(E \wedge H_i) +$$

$$+ \sum_{E^c \wedge H_i \subseteq A \vee B, E \wedge H_i \not\subseteq A \vee B} \mu'(E^c \wedge H_i) =$$

$$= f(A) + f(B) - Pl(\bar{A} \wedge \bar{B}) - \sum_{E \wedge H_i \subseteq A \wedge B, E^c \wedge H_i \not\subseteq A \vee B} \mu'(E \wedge H_i) - \sum_{E^c \wedge H_i \subseteq A \wedge B, E \wedge H_i \not\subseteq A \vee B} \mu'(E^c \wedge H_i)$$

$$= f(A) + f(B) - f(A \wedge B)$$

Finally, f induces a conditional plausibility, that we continue to denote with f, on $\langle E, \mathscr{L} \rangle \times \mathscr{H}$ where H is the additive set generated by $H_i \in \mathscr{L}$ such that $f(H_i) > 0$. For any $H_i \in \mathscr{L}$ one has

$$f(E|H_i) = \frac{f(E \wedge H_i)}{f(H_i)} = \frac{\mu'(E \wedge H_i)}{Pl(H_i)} = f(E|H_i).$$

This implies that the system $S^0_{Pl,f}$ admits a solution and so for the above consideration the assessment $\{Pl, f\}$ is a coherent conditional plausibility. ☐

3.2 Aggregated Likelihoods

The interest come from inferential problems in which the available information consists of a plausibilistic (or probabilistic or possibilistic) "prior" on a partition $\{K_j\}$

and a likelihood related to the events of another partition refining the previous one. So first of all we need to aggregate the likelihood function preserving coherence with the framework of reference.

In what follows $g : \{E\} \times \mathcal{H} \to [0, 1]$ denotes a function such that the restriction $g_{|\{E\} \times \mathcal{L}}$ of g to $\{E\} \times \mathcal{L}$ coincides with f.

We recall a common feature of probabilistic and possibility framework: any aggregated likelihood g, seen as a coherent conditional probability or a coherent T-conditional possibility, satisfies the following condition for every $K \in \mathcal{H}$:

$$\min_{H_i \subseteq K} f(E|H_i) \leq g(E|K) \leq \max_{H_i \subseteq K} f(E|H_i). \tag{7}$$

Now the question is to investigate whether an aggregated likelihood seen as a coherent conditional plausibility must satisfy the same constraints.

In the following example we show that the quantity $\max_{H_i \subseteq K} f(E|H_i)$ is not an upper bound.

Example 1 Let $\mathcal{L} = \{H_1, H_2\}$ be a partition and E an event logically independent of the events $H_i \in \mathcal{L}$. Consider the following likelihood on $\mathcal{L}$

$$f(E|H_1) = \frac{1}{4}; \ f(E|H_2) = \frac{1}{2}$$

and let g be a function extending f on $\{E\} \times \mathcal{H}$ such that $g(E|H_1 \vee H_2) = \frac{3}{4} = f(E|H_1) + f(E|H_2)$.

From Eq. (7) it follows that g is not a coherent T-conditional possibility or conditional probability; we prove that it is indeed a coherent conditional plausibility. For that let us consider the following system with unknowns $m_0(C)$, where $C \in \langle E, \mathcal{L} \rangle$

$$(S^0) = \begin{cases} 1/4 \cdot \sum\limits_{H_1 \wedge C \neq \emptyset} m_0(C) = \sum\limits_{H_1 \wedge E \wedge C \neq \emptyset} m_0(C), \\ 1/2 \cdot \sum\limits_{H_2 \wedge C \neq \emptyset} m_0(C) = \sum\limits_{H_2 \wedge E \wedge C \neq \emptyset} m_0(C), \\ 3/4 \cdot \sum\limits_{(H_1 \vee H_2) \wedge C \neq \emptyset} m_0(C) = \sum\limits_{(H_1 \vee H_2) \wedge E \wedge C \neq \emptyset} m_0(C), \\ \sum\limits_{C \subseteq H_1 \vee H_2} m_0(C) = 1 \\ m_0(C) \geq 0, \qquad\qquad\qquad\qquad \forall C \in \langle E, \mathcal{L} \rangle \end{cases}$$

It is easy to see that the basic assignment:

$$m_0((E \wedge H_1) \vee (E^c \wedge H_2)) = m_0(H_1 \vee (E^c \wedge H_2)) = \frac{1}{8},$$

$$m_0((E^c \wedge H_1) \vee (E \wedge H_2)) = m_0((E^c \wedge H_1) \vee H_2) =$$

$$m_0(E^c \wedge (H_1 \vee H_2)) = \frac{1}{4}$$

and $m_0(C) = 0$ for any other event $C \in \langle E, \mathcal{L} \rangle$, is a solution of S_0, giving positive plausibility to both the events H_i.

We give now an example showing that the lower bound of Eq. (7) can be violated when we refer to the plausibilistic framework.

Example 2 Let $\mathcal{L} = \{H_1, H_2\}$ be a partition and E an event logically independent of all the events H_i.

Consider the following aggregated likelihood on $\mathcal{H}$

$$f(E|H_1) = f(E|H_2) = \frac{2}{3}, f(E|H_1 \vee H_2) = \frac{1}{2}.$$

To prove that the assessment is coherent within a conditional plausibility, we consider the following system with unknowns $m_0(C)$, where $C \in \langle E, \mathcal{L} \rangle$

$$(S^0) = \begin{cases} 2/3 \cdot \sum_{H_1 \wedge C \neq \emptyset} m_0(C) = \sum_{H_1 \wedge E \wedge C \neq \emptyset} m_0(C), \\ 2/3 \cdot \sum_{H_2 \wedge C \neq \emptyset} m_0(C) = \sum_{H_2 \wedge E \wedge C \neq \emptyset} m_0(C), \\ 1/2 \cdot \sum_{(H_1 \vee H_2) \wedge C \neq \emptyset} m_0(C) = \sum_{(H_1 \vee H_2) \wedge E \wedge C \neq \emptyset} m_0(C), \\ \sum_{C \subseteq H_1 \vee H_2} m_0(C) = 1 \\ m_0(C) \geq 0, \qquad\qquad\qquad \forall C \in \langle E, \mathcal{L} \rangle \end{cases}$$

The following basic assignment on $\langle E, \mathcal{L} \rangle$:

$$m_0 = (E^c \wedge H_1) = m_0(E^c \wedge H_2) = m_0(E) = m_0(\Omega) = \frac{1}{4}$$

and $m_0(C) = 0$ for any other event $C \in \langle E, \mathcal{L} \rangle$, is a solution of S_0, giving positive plausibility to both the events H_i.

Remark 1 The fact that the lower bound of coherent values of $Pl(E|H_i \vee H_j)$ can be less than $\min\{Pl(E|H_i), Pl(E|H_j)\}$ is an indirect proof that a conditional plausibility (Definition 1) is not an upper envelope of a set of conditional probabilities.

Theorem 5 *If* $f : E \times \mathcal{L} \to [0, 1]$ *is a likelihood, then any coherent conditional plausibility g extending f satisfies, for every* $K \in \mathcal{H}$, *the following inequality:*

$$(L2) \quad 0 \leq g(E|K) \leq \min\{ \sum_{H_i \subseteq K} f(E|H_i), 1 \}.$$

Proof There is a coherent conditional plausibility assessment g on $\mathcal{B} \times \mathcal{H}$ with $\mathcal{B} = \langle \mathcal{H} \cup \{E\} \rangle$, extending f. For every $K \in \mathcal{H}$, Pl satisfies (4) and $g(E|K) \geq 0$ and so $0 \leq g(E|K) \leq \sum_{H_i \subseteq K} f(E|H_i) g(H_i|K)$. Then, the thesis follows. $\qquad\square$

Theorem 5 shows that in plausibility framework there is much more freedom than in both probabilistic and possibilistic ones, where aggregated likelihood functions are monotone, with respect to $\subseteq$, only if the extension is obtained, for every K, as $\max_{H_i \subseteq K} f(E|H_i)$ and they are anti-monotone if and only if their extensions are obtained as $\min_{H_i \subseteq K} f(E|H_i)$.

Since any likelihood is also a coherent conditional probability and in [13, 16] it has been proved that an aggregated likelihood coherent within conditional probability can be obtained by taking the minimum (maximum), this extension can be taken also in the plausibility framework.

In the following proposition we prove that in the plausibilistic framework we could take the sum of likelihoods $Pl(E|H_i)$ as plausibility of $E| \bigvee H_i$.

Proposition 1 *Let f be a likelihood on $\mathscr{L}$ related to an event E. If $\sum_{H_i \in \mathscr{L}} f(E|H_i) \leq 1$, the function g on $\{E\} \times \mathscr{H}$ defined for all $K_1, K_2 \in \mathscr{H}$ with $K_1 \wedge K_2 = \emptyset$ as*

$$g(E|K_1 \vee K_2) = g(E|K_1) + g(E|K_2).$$

is a coherent conditional plausibility extending f.

Proof To prove the result it is enough to consider the following basic assignment m on $\langle E, \mathscr{L} \rangle$:

$$m((E \wedge H_i) \vee \bigvee_{j \neq i}(E^c \wedge H_j)) + m(H_i \vee \bigvee_{j \neq i}(E^c \wedge H_j)) = f(E|H_i)$$

for $H_i \in \mathscr{L}$ and $m(E^c) = 1 - \sum_{H_i \in \mathscr{L}} f(E|H_i)$.

It is easy to show that this basic assignment m is agreeing with g (see Theorem 3) and the plausibility of H_i is positive. $\square$

4 Fuzzy Sets as Coherent Conditional Plausibilities

The aim of this sections is to give an interpretation of the membership of the fuzzy subsets in terms of coherent conditional plausibility ("plausibilistic likelihood"), to study which t-norms and t-conorms can be used under this framework. This problem essentially is based on coherent extensions of a conditional plausibility. Finally in order to apply the results of the previous section we introduce an inferential problem, starting from linguistic information (fuzzy sets) and imprecise prior information.

4.1 Main Definition

For our aim we start with the interpretation of fuzzy sets in terms of coherent conditional plausibility as a function of the conditioning event (plausibilistic likelihood). This interpretation generalises both those given in terms of coherent conditional probability (see for instance [11–13]) and in terms of coherent T-conditional possibility (see [8]), with T the usual product.

Formally speaking: let X be a (not necessarily numerical) variable, with range $\mathscr{C}_X$, and, for any $x \in \mathscr{C}_X$, let us denote by A_x or x the event $\{X = x\}$. The family of events $\{x\}_{x \in \mathscr{C}_X}$ is obviously a *partition* of the certain event Ω.

Let φ be any *property* related to the variable X, and let us refer to the state of information of a real (or fictitious) person that will be denoted by "You".

Let us consider the Boolean event

$$E_\varphi = \text{"You claim that } X \text{ has property } \varphi\text{"}, \tag{8}$$

in order to give the following definition of fuzzy set E_φ^*:

Definition 1 Let X be any variable with range $\mathscr{C}_X$, φ a related property and E_φ the corresponding event. A **fuzzy subset** E_φ^* of $\mathscr{C}_X$ is a pair

$$E_\varphi^* = \{E_\varphi, \mu_\varphi\}, \tag{9}$$

with $\mu_\varphi(x) = f(E_\varphi|x)$, for every $x \in \mathscr{C}_X$, a likelihood function.

Note that, by the Corollary 3, a fuzzy subset E_φ^* is a crisp set when the property φ is such that, for every $x \in \mathscr{C}_X$, either $E_\varphi \wedge A_x = \emptyset$ or $A_x \subseteq E_\varphi$.

Then we can interpret the membership function $f(E_\varphi|x)$, for $x \in \mathscr{C}_X$, as the measure of Your degree of belief in E_φ, when X assumes the different values of its range.

By Theorem 3, this measure can be regarded as a coherent conditional probability as well as a coherent T-conditional possibility or as a coherent conditional plausibility. Obviously the choice of the framework of interpretation of the likelihood function impacts on the resulting fuzzy operations.

4.2 Fuzzy Operations in the Plausibilistic Framework

We study the binary operations of union and intersection and the unary operation of complementation for fuzzy sets. Obviously, these operations depend on the chosen framework of reference.

By following [11–13] for the probabilistic interpretation and [10] for the possibilistic interpretation, the operation of complementation of a fuzzy set E_φ and those of union and intersection between two fuzzy sets E_φ^* and E_ψ^*, can be directly obtained by using the rules of coherent conditional plausibility and the logical independence between E_φ and E_ψ with respect to the partition generated by the relevant variable (or variables).

Definition 2 Let X a random variable generating the partition $\mathscr{C}_X = \{H_i\}_{i \in I}$ and $\mathscr{E} = \{E_j\}_{j=1,\ldots,m}$ a finite set of events. The events in $\mathscr{E}$ are logically independent with respect to $\mathscr{C}_X$ if, denoting with E_j^* either E_j or E_j^c, the following conditions hold:

(i) the events in $\mathcal{E}$ are logically independent, i.e., $\bigwedge_{j=1}^{m} E_j^* \neq \emptyset$;

(ii) for every $i \in I$, $\bigwedge_{j=1}^{m} E_j^* \wedge H_i = \emptyset \Longrightarrow E_j^* \wedge H_i = \emptyset$ for some $j = 1, \ldots, m$.

Notice that the events E_φ and E_ψ are "usually" logically independent, in particular they are logically independent when $\psi = \neg\varphi$: indeed, we can claim both "X has the property φ" and "X has the property $\neg\varphi$", or only one of them or finally neither of them. Similarly, E_φ and E_ψ are logically independent in case ψ is the superlative or a diminutive of φ.

Let us denote by $\varphi \vee \psi$ and $\varphi \wedge \psi$, respectively, the properties "φ or ψ", "φ and ψ" (note that the symbols $\wedge$ and $\vee$ do not indicate Boolean operations, since φ and ψ are not Boolean objects) and define:

$$E_{\varphi\vee\psi} = E_\varphi \vee E_\psi, \quad E_{\varphi\wedge\psi} = E_\varphi \wedge E_\psi. \tag{10}$$

Consider now the problem of complementary which essentially coincides with that discussed in the probabilistic interpretation of fuzzy sets [12, 13]. Denoting by $(E_\varphi^*)' = E_{\neg\varphi}^* = (E_{\neg\varphi}, \mu_{\neg\varphi})$ the complementary fuzzy set of E_φ^*, due to the logical independence of $\{E_\varphi, E_{\neg\varphi}\}$, with respect to X, any value in $[0, 1]$ is coherent for $\mu_{\neg\varphi}(x)$ for any x.

The main remark is related to the fact that the relation $E_{\neg\varphi} \neq E_\varphi^c$ holds. In fact, while $E_\varphi \vee E_\varphi^c = \Omega$, due to the logical independence with respect to X of $\{E_\varphi, E_{\neg\varphi}\}$, we have instead $E_\varphi \vee E_{\neg\varphi} \subseteq \Omega$. Then it is not necessary to require $\mu_{\neg\varphi}(x) = 1$ if $\mu_\varphi(x) < 1$. In particular we can take

$$\mu_{\neg\varphi}(x) = 1 - \mu_\varphi(x) = 1 - Pl(E_\varphi|x) = Pl(E_{\neg\varphi}|x). \tag{11}$$

In fact, the above function $\mu_{\neg\varphi} : \mathcal{L} \to [0, 1]$ is a likelihood function and so a coherent conditional plausibility (as well as a coherent T-conditional possibility and coherent conditional probability).

Let us consider two properties φ and ψ related to the same variable X, such that $\{E_\varphi, E_\psi\}$ are logical independent with respect to the partition generated by the variable X.

Theorem 1 *Given two fuzzy sets E_φ^*, E_ψ^* related to the variable X, with $\mu_\varphi(x) = Pl(E_\varphi|x)$ and $\mu_\psi(x) = P(E_\psi|x)$ (for any $x \in \mathcal{C}_X$) two coherent conditional plausibility. For any given x in $\mathcal{C}_X$, the assessment $Pl(E_\varphi \wedge E_\psi|x) = v$ is a coherent conditional plausibility if and only if*

$$0 \leq v \leq \min\{Pl(E_\varphi|x), Pl(E_\psi|x)\}. \tag{12}$$

Proof First of all we need to prove that the assessment $\{Pl(E_\varphi|x), Pl(E_\psi|x) : x \in \mathcal{C}_X\}$ is a coherent conditional plausibility. Since a likelihood is a coherent conditional probability and the assessment $\{Pl(E_\varphi|x), Pl(E_\psi|x) : x \in \mathcal{C}_X\}$ is a coherent conditional probability [6], then coherence under conditional plausibility follows.

The upper bound follows from monotonicity of capacities, it is a sharp bound, that means it can be assumed, since it is a coherent value under probability.

The lower bound is not coherent under probability. In order to show that the lower bound is sharp under plausibility, consider the following basic assignment, for any $A_x = (X = x)$ with $x \in \mathscr{C}_X$

$$m(E_\varphi \wedge E_\psi \wedge A_x) = 0$$

$$m(E_\varphi \wedge E_\psi^c \wedge A_x) = m(E_\varphi^c \wedge E_\psi \wedge A_x) = \min\{Pl(E_\varphi|x), Pl(E_\psi|x)\}\frac{1}{n}$$

$$m(E_\varphi^c \wedge E_\psi \wedge A_x) = (Pl(E_\psi|x) - \min\{Pl(E_\varphi|x), Pl(E_\psi|x)\})\frac{1}{n}$$

$$m(E_\varphi \wedge E_\psi^c \wedge A_x) = (Pl(E_\varphi|x) - \min\{Pl(E_\varphi|x), Pl(E_\psi|x)\})\frac{1}{n}$$

$$m(E_\varphi^c \wedge E_\psi^c \wedge A_x) = (1 - \max\{Pl(E_\varphi|x), Pl(E_\psi|x)\})\frac{1}{n}$$

and 0 otherwise (with n the cardinality of $\mathscr{C}_X$).

The above basic assignment generates an unconditional plausibility Pl' on $\langle\{E_\varphi E_\psi \wedge, A_x : x \in \mathscr{C}_X\}\rangle$ such that $Pl'(A_x) = \frac{1}{n}$ and $Pl'(E_\varphi|x) = Pl(E_\varphi|x)$, $Pl'(E_\psi|x) = Pl(E_\psi|x)$, for any $x \in \mathscr{C}_X$. $\qquad\square$

Theorem 1 emphasizes a first difference with the probabilistic framework, in fact the upper bounds in the two frameworks coincide while the lower bounds differ, in fact under a probability P the lower bound is not 0, but coincides with Fréchet-Hoeffding lower bound, that is determined by the Lukasiewcz t-norm T_L, so

$$max\{0, P(E_\varphi|x) + P(E_\psi|x) - 1\} \le v \le \min\{P(E_\varphi|x), P(E_\psi|x)\},$$

Obviously, due to coherence of the starting assessment, any value in the interval of coherent values can be accepted, so in particular those obtained by a t-norm (if the obtained values are inside the interval). This is true for two fuzzy events.

The question now is: starting from a set of $\{E_{\varphi_1}, \ldots, E_{\varphi_n}\}$ of logically independent events with respect to $\mathscr{C}_X$ and the relevant $\mu_i = f(E_{\varphi_i}|x)$ is it possible (i.e. coherent with the measure of reference) to compute all the intersections among the fuzzy sets, by using the same t-norm?

The answer is: it depends on the t-norm. If, for instance, we consider the minimum, then the answer is positive in all the considered frameworks of reference:

Theorem 2 *Let $\{E^*_{\varphi_i}\}_I$ be a finite family of fuzzy sets related to a variable X, with $\{E_{\varphi_i}\}_I$ logical independent with respect to the random variable X, and consider the set $\mathscr{F}$ of events obtained as the intersection of a finite set of events in $\{E_{\varphi_i}\}_I$. The assessment*

$$\{f(A|x), : A \in \mathscr{F}\},$$

where $f(A|x) = \min(\mu_{\varphi_i}(x) : A \subseteq E_{\varphi_i})$ is

- *a coherent conditional probability*
- *a coherent conditional plausibility*
- *a coherent conditional possibility*

Proof Since the likelihood $\mu_{\varphi_j}(\cdot)$ is a coherent conditional probability, there is a coherent extension on $\bigwedge_{j \in J} E_{\varphi_j} | A_x$ for any $J \subseteq I$ and $x \in \mathscr{C}_X$.

For a given $x \in \mathscr{C}_X$, assume without loss of generality that $\mu_{\varphi_i}(x) \leq \mu_{\varphi_{i+1}}(x)$ for $i = 1, \dots, 2$. We can define for any $J_1 \subseteq I$ with $1 \in J_1$

$$f_x(\bigwedge_{i \in I} E_{\varphi_i}) = \mu_{\varphi_1}(x)$$

moreover for a set $J \subseteq I$ with $1 \notin J$, let $r = \min\{i : i \in I \bigcap J\}$ and $s = \max\{i : i \in I \bigcap J\}$ $(r < s)$

$$f_x(\bigwedge_{j \geq r} E_{\varphi_j} \bigwedge_{i < r} E^c_{\varphi_i}) = \mu_{\varphi_r}(x) - \mu_{\varphi_{r-1}}(x)$$

and 0 on the other atoms. Any f_x is a probability, so a probability P on the algebra generated by $\{E_{\varphi_i}, A_x : i \in I, x \in \mathscr{C}_X\}$ can be defined as

$$P(B) = \sum_{A_x \wedge B \neq \emptyset} \frac{1}{n} f_x(B \wedge A_x)$$

(with n the cardinality of $\mathscr{C}_x$) and it gives rise to a strictly positive probability and it generates a conditional probability that is an extension of $\{\mu_{\varphi_i}\}_I$.

Then, the assignment f is a coherent conditional probability and then a coherent conditional plausibility.

Furthermore, the above assignment f_x for a given $x \in \mathscr{C}_x$ is a possibilistic distribution and so a possibility Π on the algebra generated by $\{E_{\varphi_i}, A_x : i \in I, x \in \mathscr{C}_X\}$ can be defined as

$$\Pi(B) = \max_{A_x \wedge B \neq \emptyset} f_x(B \wedge A_x)$$

and it gives rise to a strictly positive possibility and it generates a P-conditional possibility that is an extension of $\{\mu_{\varphi_i}\}_I$. $\qquad\square$

The same result can be easily proved for the product t-norm, which implement the case where the events E_{φ_i} are stochastically independent.

On the contrary, if we consider as t-norm T_L, the extension to the intersection computed trough T_L can be not a coherent conditional probability, as the following example shows.

Example 3 Let $\mathscr{H} = \{H, H^c\}$ be a partition, and $\mathscr{E} = \{E_i | H\}_{i=1,2,3}$ be a set of conditional events such that $\bigwedge_{i=1}^3 E_i^* \wedge H \neq \emptyset$ for any $H \in \mathscr{H}$, so the events in $\mathscr{E}$ are logical independent with respect to $\mathscr{H}$.

Suppose that $P(E_1|H) = P(E_2|H) = 0.6$ and $P(E_3|H) = 0.7$, while $P(E_i|H^c) = 0.5$ for $i = 1, 2, 3$.

It is easy to check that the conditional probability P is coherent. Furthermore it is easy to prove that, from Fréchet-Hoeffding bounds, the coherent values for P for an event E obtained as finite intersection of E_i are such that:

$$0 \le P(E_1 \wedge E_2 \wedge E_3|H) \le 0.6; \quad 0.2 \le P(E_1 \wedge E_2|H) \le 0.6;$$
$$0.3 \le P(E_1 \wedge E_3|H) \le 0.6; \quad 0.3 \le P(E_2 \wedge E_3|H) \le 0.6$$

We could show that the function $f(\wedge_I E_i|H)$ $I \subseteq \{1, 2, 3\}$ taking the minimum coherent values is not coherent: in fact the function

$$f(E_1 \wedge E_2 \wedge E_3|H) = 0, \quad f(E_1 \wedge E_2|H) = 0.2,$$
$$f(E_1 \wedge E_3|H) = 0.3, \quad f(E_1|H) = f(E_2|H) = 0.6, \quad f(E_3|H) = 0.7$$

is not a coherent conditional probability.

The next Theorem 6 proves that under a plausibility we can compute, for every $x \in \mathscr{C}_X$, the membership of the intersection of a set of fuzzy sets by using t-norm T_L. This shows that considering coherent conditional plausibility, instead of coherent conditional probability, for measuring the degree of belief of You on the events E_φ we actually capture more parallelism with the classical theory of fuzzy sets, where the inference is made by using t-norms and t-conorms.

Theorem 6 *Let $\mathscr{C} = \{E_{\varphi_i}^*\}_I$ be a finite family of fuzzy sets related to a variable X, with range $\mathscr{C}_X$ such that E_{φ_i} logical independent with respect to the random variable X, and consider the set $\mathscr{F}$ of fuzzy subsets obtained as the intersection of any finite number of these fuzzy sets. The assessment*

$$Pl(\bigwedge_J E_{\varphi_j}|x) = T_L\{\mu_{\varphi_j}(x) \; : \; j \in J\},$$

for any $J \subseteq I$ (with T_L the t-norm of Lukasiewicz) is a coherent conditional plausibility.

Proof Given a finite set $\mathscr{C} = \{E_{\varphi_i} \; : \; i = 1, \dots, m\}$ of events, consider the set $\mathscr{F} \supseteq \mathscr{C}$. Due to the logical independence of the events E_{φ_i} w.r.to X, the assessment $\mathscr{P} = \{Pl(E_{\varphi_i}|x) \; : \; i = 1, \dots, m\}_{x \in \mathscr{C}_X}$ is coherent with a conditional plausibility; we prove that it can be coherently extended to $\mathscr{F}$, by computing, for any $x \in \mathscr{C}_X$, every intersection trough T_L, that is

$$Pl(\bigwedge_{J \subseteq \{1,\dots,m\}} E_{\varphi_i}|x) = T_L\{Pl(E_{\varphi_i}|x) \; : \; i \in J\}, \tag{13}$$

is a coherent conditional plausibility.

Actually we prove the result for $m = 3$ but the basic assignment for any $x \in \mathscr{C}_X$ can be built analogously.

Assume without loss of generality that $Pl(E_{\varphi_i}|x) \le Pl(E_{\varphi_{i+1}}|x)$ for $i = 1, 2$.

Let $m(\wedge_{i=1}^{3} E_{\varphi_i}) = T_L(Pl(E_{\varphi_1}|x), \dots, Pl(E_{\varphi_3}|x))$,

$m(\vee_{j=1}^{3}(E_{\varphi_j}^c \wedge_{i \ne j} E_{\varphi_i})) = T_L(Pl(E_{\varphi_1}|x), Pl(E_{\varphi_2}|x)) - T_L(Pl(E_{\varphi_1}|x), \dots, Pl(E_{\varphi_3}|x))$,

$m(E_{\varphi_1}^c \wedge_{j=2}^{3} E_{\varphi_j}) = T_L(Pl(E_{\varphi_2}|x), Pl(E_{\varphi_3}|x)) - T_L(Pl(E_{\varphi_1}|x), Pl(E_{\varphi_2}|x))$,

$m(E_{\varphi_2}^c \wedge_{j=1,3} E_{\varphi_j}) = T_L(Pl(E_{\varphi_1}|x), P(E_{\varphi_3}|x)) - T_L(Pl(E_{\varphi_1}|x), Pl(E_{\varphi_2}|x))$,

$m(\vee_{i=1}^{3} E_{\varphi_i} \wedge_{j \ne i} E_{\varphi_j}^c) = Pl(E_{\varphi_1}|x) - T_L(Pl(E_{\varphi_1}|x), Pl(E_{\varphi_3}|x))$,

$m(\vee_{i=2}^{3} E_{\varphi_i} \wedge_{j \ne i} E_{\varphi_j}^c) = Pl(E_{\varphi_2}|x) - Pl(E_{\varphi_2}|x)$.

Furthermore $m(\wedge_{i=1}^{3} E_{\varphi_j}^c) = 1 - Pl(E_{\varphi_3}|x)$.

It is easy to check that the function m taking the above values and zero otherwise is a basic assignment generating the function Pl defined by Eq. (13), that therefore is coherent. $\qquad \square$

The aim is to discuss now, under the plausibilistic interpretation, the coherent values for the membership of the union of fuzzy sets.

We first recall that in the probabilistic interpretation, fixed the value for the membership function of the fuzzy intersection, the value for the membership function of the fuzzy union is uniquely determined [12, 13], by the equation

$$\mu_{\varphi \vee \psi}(x) = \mu_\varphi(x) + \mu_\psi(x) - \mu_{\varphi \wedge \psi}(x)$$

so that the only pair of t-norm and t-conorm are those of Frank's class [37].

On the contrary, in the possibilistic interpretation [10], independently of the value of $\mu_{\varphi \wedge \psi}(x) = \Pi(E_\varphi \wedge E_\psi|x)$, we get a unique value for the fuzzy union which is

$$\mu_{\varphi \vee \psi}(x) = \Pi(E_\varphi \vee E_\psi|x) = \max\{\Pi(E_\varphi|x), \Pi(E_\psi|x)\} = \max\{\mu_\varphi(x), \mu\psi(x)\}. \tag{14}$$

In the case of plausibility we have that the value of $f(E_\varphi \vee E_\psi|A_x)$ is not univocally determined but it must satisfies the following constraints

$$\max\{f(E_\varphi|x), f(E_\psi|x)\} \le f(E_\varphi \vee E_\psi|x) \le \min\{f(E_\varphi|x) + f(E_\psi|x) - f(E_\varphi \wedge E_\psi|x), 1\}.$$

Then, we can put

$$E_\varphi^* \cup E_\psi^* = \{E_{\varphi \vee \psi}, \mu_{\varphi \vee \psi}\}; \qquad E_\varphi^* \cap E_\psi^* = \{E_{\varphi \wedge \psi}, \mu_{\varphi \wedge \psi}\},$$

with

$$\mu_{\varphi \vee \psi}(x) = f(E_\varphi \vee E_\psi|A_x) \qquad \mu_{\varphi \wedge \psi}(x) = f(E_\varphi \wedge E_\psi|A_x).$$

Obviously both any pair of t-norm and t-conorm of Frank's class or a pair (max, T) with T any t-norm can be used to compute coherent extension on the union and intersection of fuzzy sets.

Unfortunately no pair in the most famous other classes (Hamacher class [52], Yager class [49], Dubois and Prade class [30]) seems to be apt for computing coherent extension on the union and intersection of fuzzy sets. In fact for any t-norm $\odot$ and dual t-conorm $\oplus$ in these classes there exist $(x', y'), (x'', y'') \in [0, 1]^2$ such that

$$x' \odot y' < x' + y' - x' \oplus y'; \qquad x'' \odot y'' > x'' + y'' - x'' \oplus y'$$

so that both two-alternativity and two-monotonicity could be violated by $f(\cdot|x)$, and so f cannot be a coherent conditional plausibility (moreover it cannot be a coherent conditional belief function).

From the above considerations, it follows that the coherent conditional plausibility $Pl(E_\varphi|\cdot)$ comes out to be a natural interpretation of the membership function $\mu_\varphi(\cdot)$.

5 Plausibility of "Fuzzy Events"

First of all, we recall that the concept of fuzzy event, as introduced by Zadeh [54], in the context of the interpretation of a fuzzy set as a pair whose elements are a (Boolean) event E_φ and a conditional measure $f(E_\varphi|x)$, coincides exactly with the event E_φ = "You claim that X has property φ".

For any "prior" uncertainty measure (probability, possibility and plausibility) on the algebra generated by X the assessment together μ_φ is coherent with respect the relative measure (see Theorem 4) and so coherently extendible to E_φ (Theorem 2 for plausibilities, [18] for conditional probabilities and [8, 10, 18] for conditional possibilities).

Since the variable X has finite range, by taking a probability or a T-possibility as "prior" uncertainty measure of reference, it is easy to see that the only coherent value for the probability or possibility of E_φ is

$$g(E_\varphi) = \bigoplus_{x \in \mathscr{C}_X} \mu_\varphi(x) \odot g(x), \tag{15}$$

where $\oplus$ and $\odot$ are the sum and the product in the case of probability, while they are the maximum and the t-norm T in the case of possibility and g is either a probability or a possibility.

Obviously, only in the case of probability $g(E_\varphi)$ coincides with Zadeh's definition of the probability of a "fuzzy event" [53].

The Eq. (15) is based on the disintegration formula holding for both the two measures; as discussed before it does not hold for plausibility. In fact, for plausibility just a weak form of disintegration holds, (see inequality in (4)). Then, we need to compute plausibility of an event E_φ by means the Choquet integral:

$$Pl(E_\varphi) = \oint \mu_\varphi(x) dPl(x) = \int_0^1 Pl(\mu_\varphi(x) \geq t) dt. \qquad (16)$$

Consider now n fuzzy sets $E^*_{\varphi_i}$, and compute by a suitable t-norm $\odot$ (for instance T_L or min), the memberships of the fuzzy sets $E^*_{\varphi_i} \cap E^*_{\varphi_j}$ and then by using (16) the plausibility of the relevant fuzzy events $E_{\varphi_i} \wedge E_{\varphi_j}$. Obviously the global assessment is coherent with a conditional plausibility and then it can be further extended to any new conditional event $A|B$ where A, B are events of the algebra $\mathscr{B}$ spanned by $\{E_{\varphi_i}\}_{i \in I} \cup \{A_x\}_{x \in \mathscr{C}_x}$, with $B \neq \emptyset$.

This extension is not unique in general but, for the events $A|B$, with $A = E_{\varphi_i}$ and $B = E_{\varphi_j}$, with $Pl(E_{\varphi_j}) > 0$ the only coherent extension and, for $i \neq j$, is

$$Pl_\odot(E_{\varphi_i}|E_{\varphi_j}) = \frac{Pl_\odot(E_{\varphi_i} \wedge E_{\varphi_j})}{Pl_\odot(E_{\varphi_j})} = \cdot \frac{\int_0^1 Pl(\mu_{\varphi_i} \odot \mu_{\varphi_j}(x) \geq t) dt}{\int_0^1 Pl(\mu_{\varphi_j}(x) \geq t) dt} \qquad (17)$$

When $P_\odot(E_{\varphi_j}) = 0$, we obtain in general a not unique extension to the events $E_{\varphi_i}|E_{\varphi_j}$.

Nevertheless we note that one has $P_\odot(E_{\varphi_j}) = 0$ if and only if $Pl(H) = 0$, where $H = \bigvee \{x_k : \mu_{\varphi_i}(x_k) = Pl_\odot(E_{\varphi_i}|x_k)\} > 0$.

In this case to obtain a unique extension we need to have also the conditional plausibility $Pl(\cdot|B)$, where B is the logical sum of the events x_k such that $Pl(E_{\varphi_j}|x_k) = 0$.

Remark 2 As in the probabilistic and possibilistic framework, the values $Pl(E_{\varphi_i}|E_{\varphi_j})$ computed by the formula above are coherent only when the events E_{φ_i} and E_{φ_j} are logically independent, so, for instance, the same formula cannot be used for obtaining the coherent extension of Pl to $E_{\varphi_i}|E_{\varphi_i}$, which is necessarily 1.

6 Conclusion

The first part of the paper is devoted into studying likelihood functions seen as assessment on a set of conditional events $E|H_i$, with E the evidence and H_i varying on a partition. It is shown that likelihood functions are assessments coherent with respect probability, possibility and plausibility. Then, inferential processes, like Bayesian one, is studied in the different setting taking a likelihood function and a prior, that could be a probability, a possibility and a plausibility. In particular we prove that any likelihood function on $E \times \mathscr{L}$ and any plausibility on $\mathscr{L}$, with $\mathscr{L}$ a partition, are globally coherent within a plausibility setting. Then, a comparison of aggregated likelihoods, that are coherent extensions of a likelihood function on $E \times \mathscr{L}$ to $E \times \langle \mathscr{L} \rangle$ is studied in the different setting by showing the common characteristic and the specific features.

These results are applied in order to interpret the membership of a fuzzy event as a coherent conditional plausibility. A comparison under the different interpretations based on coherent conditional probabilities, possibilities and plausibilities of the obtained fuzzy operations is carried out. The syntactical advantages related to the use of a plausibilistic frameworks are shown, in particular by referring to the operations of $n \geq 3$ fuzzy sets.

Acknowledgements This research was partially supported by by GNAMPA of INdAM.

References

1. Baioletti M, Coletti G, Petturiti D, Vantaggi B (2011) Inferential models and relevant algorithms in a possibilistic framework. Int J Approx Reason 52:580–598
2. Bouchon-Meunier B, Coletti G, Marsala C (2002) Independence and possibilistic conditioning. Ann Math Artif Intell 35:107–123
3. Brunelli M, Fedrizzi M (2010) The representation of uncertainty in operational research: considerations on the use of possibility, probability, and fuzzy numbers. In: Proceedings of the 2nd international conference on applied operational research, Uniprint, Turku, pp 24–33. ISBN 9789521224140
4. Capotorti A, Coletti G, Vantaggi B (2014) Standard and nonstandard representability of positive uncertainty orderings. Kybertika 50:189–215
5. Chateauneuf A, Jaffray JY (1989) Some characterizations of lower probabilities and other monotone capacities through the use of Mobius inversion. Math Soc Sci 17(3):263–283
6. Coletti G, Gervasi O, Tasso S, Vantaggi B (2012) Generalized Bayesian inference in a fuzzy context: from theory to a virtual reality application. Comput Stat Data Anal 56:967–980
7. Coletti G, Petturiti D, Vantaggi B (2016) Conditional belief functions as lower envelopes of conditional probabilities in a finite setting. Inf Sci 339:64–84
8. Coletti G, Petturiti D, Vantaggi B (2014) Possibilistic and probabilistic likelihood functions and their extensions: common features and specific characteristics. Fuzzy Sets Syst 250:25–51
9. Coletti G, Petturiti D, Vantaggi B (2014) Possibilistic and probabilistic likelihood functions and their extensions: common features and specific characteristics. Fuzzy Sets Syst 250:25–51
10. Coletti G, Petturiti D, Vantaggi B (in press) Fuzzy memberships as likelihood functions in a possibilistic framework. Int J Approx Reason. doi:10.1016/j.ijar.2016.11.017
11. Coletti G, Scozzafava R (2002) Probabilistic logic in a coherent setting. Trends in logic, no 15. Kluwer, Dordrecht
12. Coletti G, Scozzafava R (2004) Conditional probability, fuzzy sets, and possibility: a unifying view. Fuzzy Sets Syst 144:227–249
13. Coletti G, Scozzafava R (2006) Conditional probability and fuzzy information. Comput Stat Data Anal 51:115–132
14. Coletti G, Scozzafava R, Vantaggi B (2013) Inferential processes leading to possibility and necessity. Inf Sci 245:132–145
15. Coletti G, Scozzafava R (2006) Toward a general theory of conditional beliefs. Int J Intell Syst 21:229–259
16. Coletti G, Scozzafava R, Vantaggi B (2009) Integrated likelihood in a finitely additive setting. Lect Notes Comput Sci: LNAI 5590:554–565
17. Coletti G, Vantaggi B (2006) Possibility theory: conditional independence. Fuzzy Sets Syst 157(11):1491–1513
18. Coletti G, Vantaggi B (2009) T-conditional possibilities: coherence and inference. Fuzzy Set Syst 160:306–324

19. Coletti G, Vantaggi B (2012) Probabilistic reasoning in a fuzzy context. In: Proceedings of the second world conference on soft computing, Baku, Azerbaijan, pp 65–72
20. Coletti G, Vantaggi B (2013) Hybrid models: probabilistic and fuzzy information. Synergies of soft computing and statistics for intelligent data analysis. In: Kruse R, Berthold MR, Moewes C, Gil MA, Grzegorzewski P, Hryniewicz O (eds) Advances in intelligent systems and computing, pp 63–72
21. Coletti G, Vantaggi B (2013) Conditional nonadditive measures and fuzzy sets. In: Proceedings of 8th international symposium on imprecise probability: theories and applications, Compiégne, France, 2013 (ISIPTA 2013)
22. Cuzzolin F (2010) Three alternative combinatorial formulations of the theory of evidence. J Intell Data Anal 14:439–464
23. de Finetti B (1970) Teoria della probabilitá. Einaudi, Torino (Engl. Transl. (1974) Theory of probability vols I, II. Wiley, London)
24. de Finetti B (1949) Sull'impostazione assiomatica del calcolo delle probabilità. Annali Univ. Trieste 19:3–55. (Engl. transl.: Ch. 5 in Probability, induction, statistics. Wiley, London (1972))
25. de Cooman G, Miranda E, Couso I (2005) Lower previsions induced by multi-valued mappings. J Stat Plan Infer 133:173–197
26. Dempster AP (1968) A generalization of Bayesian inference. R Stat Soc B 50:205–247
27. Brüning M, Denneberg D (2002) Max-min (σ-)additive representation of monotone measures. Stat Pap 43(1):23–35
28. Denoeux T, Smets P (2006) Classification using belief functions: the relationship between the case-based and model-based approaches. IEEE Trans Syst Man Cybern B 36(6):1395–1406
29. Dubins L (1975) Finitely additive conditional probabilities, conglomerability and disintegrations. Ann Prob 3(1):89–99
30. Dubois D, Prade H (1986) New results about properties and semantics of fuzzy set-theoretic operators. In: Wang PP, Chang SK (eds) Fuzzy Sets. Plenum Press, New York, pp 59–75
31. Dubois D, Moral S, Prade H (1997) A semantics for possibility theory based on likelihoods. J Math Anal Appl 205:359–380
32. Dubois D, Prade H (1989) Fuzzy sets, probability and measurement. Eur J Oper Res 40:135–154
33. Dubois D, Prade H (1992) When upper probabilities are possibility measures. Fuzzy Sets Syst 49:65–74
34. Fagin R, Halpern JY (1991) A new approach to updating beliefs. In: Bonissone PP, Henrion M, Kanal LN, Lemmer JF (eds) Uncertainty in artificial intelligence, vol 6, pp 347–374
35. Fedrizzi M, Esogbue A, Kacprzyk J (1988) Fuzzy dynamic programming with stochastic systems. In: Kacprzyk J, Fedrizzi M (a cura di) Combining fuzzy imprecision with probabilistic uncertainty in decision making. Lecture notes in economics and mathematical systems. Springer, Berlin, pp 266–285
36. Fedrizzi M, Kacprzyk J (a cura di) (1988) Combining fuzzy imprecision with probabilistic uncertainty in decision making. Lecture notes in economics and mathematical systems, vol 310. Springer, Berlin, 399 pp
37. Frank MJ (1979) On the simultaneous associativity of $F(x, y)$ and $x + y - F(x, y)$. Aequationes Math. 19:194–226
38. Halpern J (2003) Reasoning about uncertainty. The MIT Press, Boston
39. Hisdal E (1988) Are grades of membership probabilities. Fuzzy Sets Syst 25:325–348
40. Jaffray JY (1992) Bayesian updating and belief functions. IEEE Trans Syst Man Cybern 22:1144–1152
41. Krauss PH (1968) Representation of conditional probability measures on Boolean algebras. Acta Mathematica Academiae Scientiarum Hungaricae 19(3–4):229–241
42. Liu Z, Li HX (2005) A probabilistic fuzzy logic system for modeling and control. IEEE Trans Fuzzy Syst 13(6):848–859
43. Loginov VI (1966) Probability treatment of Zadeh membership functions and theris use in patter recognition. Eng Cybern 68–69

44. Lukasiewicz T, Straccia U (2008) Managing uncertainty and vagueness in description logics for the Semantic Web. Web Semant: Sci Serv Agents World Wide Web 6(4):291–308
45. Mastroleo M, Vantaggi B (2007) An independence concept under plausibility function. In: Proceeding of 5th international symposium on imprecise probabilities and their applications, pp 287–296
46. Petturiti D, Vantaggi B (2017) Envelopes of conditional probabilities extending a strategy and a prior probability. Int J Approx Reason 81:160–182
47. Singpurwalla ND, Booker JM (2004) Membership functions and probability measures of fuzzy sets (with discussion). J Am Stat Assoc 99:867–889
48. Tripathy BK, Mohant RK, Sooraj TR (2016) Application of uncertainty models in bioinformatics, chapter 9. In: Dash S, Subudhi B, Dash S (eds) Handbook of research on computational intelligence applications in bioinformatics. IGI Global, pp 169–182
49. Yager RR (1980) On a general class of fuzzy connectives. Fuzzy Sets Syst 4:235–242
50. Walley P (1987) Belief function representations of statistical evidence. Ann Stat 4:1439–1465
51. Walley P (1991) Statistical reasoning with imprecise probabilities. Chapman and Hall, London
52. Weber S (1983) A general concept of fuzzy connectives, negations and implications based on t-norms and t-conorms. Fuzzy Sets Syst 11:115–134
53. Zadeh L (1965) Fuzzy sets. Inf Control 8:338–353
54. Zadeh LA (1968) Probability measures of fuzzy events. J Math Anal Appl 23:421–427
55. Zadeh LA (2002) Toward a perception-based theory of probabilistic reasoning with imprecise probabilities. J Stat Plan Infer 105:233–264
56. Zhao R, Liu B (2004) Redundancy optimization problems with uncertainty of combining randomness and fuzziness. Eur J Oper Res 157(3):716–735
57. Zimmermann H-J (1987) Fuzzy sets, decision making, and expert systems. Kluwer, Boston

Intuitionistic Fuzzy Interpretations of Some Formulas for Estimation of Preference Degree

Krassimir T. Atanassov, Vassia Atanassova, Eulalia Szmidt and Janusz Kacprzyk

Abstract Two intuitionistic fuzzy interpretations of M. Fedrizzi, M. Fedrizzi and R. A. M. Pereira's, and of V. Peneva and I. Popchev's formulas are introduced and some of their properties are discussed.

1 Introduction

Already, there are a lot of research over procedures for obtaining of a consensus in group decision making. One of them—a fuzzy approach, is introduced by Mario Fedrizzi, Michele Fedrizzi, and R. A. Marques Pereira in [3]. Another approach is given by Vanja Peneva and Ivan Popchev in [4]. In [5], the authors introduced an extension of V. Peneva and I. Popchev's formulas, using intuitionistic fuzzy approach.

K.T. Atanassov (✉) · V. Atanassova
Department of Bioinformatics and Mathematical Modelling, Institute of Biophysics
and Biomedical Engineering, Bulgarian Academy of Sciences,
105 Acad. G. Bonchev Str., 1113 Sofia, Bulgaria
e-mail: krat@bas.bg

V. Atanassova
e-mail: vassia.atanassova@gmail.com

K.T. Atanassov
Intelligent Systems Laboratory, Prof. Asen Zlatarov University, 8010 Burgas, Bulgaria

E. Szmidt · J. Kacprzyk
Systems Research Institute, Polish Academy of Sciences,
ul. Newelska 6, 01-447 Warsaw, Poland
e-mail: szmidt@ibspan.waw.pl

E. Szmidt · J. Kacprzyk
WIT – Warsaw School of Information Technology,
ul. Newelska 6, 01-447 Warsaw, Poland
e-mail: kacprzyk@ibspan.waw.pl

© Springer International Publishing AG 2018

153

M. Collan and J. Kacprzyk (eds.), *Soft Computing Applications for Group
Decision-making and Consensus Modeling*, Studies in Fuzziness and Soft
Computing 357, DOI 10.1007/978-3-319-60207-3_10

Here, we continue our research from [5], combining the two previous approaches. Let

$$A = \{a_1, a_2, \ldots, a_n\}$$

be a set of alternatives and let decision makers $D_1, D_2, \ldots, D_d$ must determine their preferences among the alternatives.

Therefore, the following matrix can be constructed:

$$X = \begin{vmatrix} x_{1,1} & \cdots & x_{1,n} \\ \vdots & \cdots & \vdots \\ x_{m,1} & \cdots & x_{m,n} \end{vmatrix},$$

where the real number $x_{i,j}$ is the degree of preference of alternative a_i before alternative a_j. In [3], M. Fedrizzi, M. Fedrizzi, and R. A. M. Pereira put (here "iff" is an abbreviation of "if and only it" and $1 \leq i, j \leq n$):

$$x_{i,j} = \begin{cases} 1, & \text{iff } a_i \text{ is definitely preferred to } a_j \\ 0, & \text{iff } a_j \text{ is definitely preferred to } a_i \\ 0.5, & \text{iff } a_i \text{ and } a_j \text{ are indifferent} \end{cases}$$

and always

$$x_{i,j} + x_{j,i} = 1.$$

Therefore, $x_{i,i} = 0.5$.

In [4], V. Peneva and I. Popchev's introduce the following formula for estimation of preference degree for each $1 \leq k \leq d$:

$$\mu_k(a_i, a_j) = \begin{cases} 1, & \text{if } i = j \\ 0.5 + \dfrac{x_{i,k} - x_{j,k}}{2\left(\max\limits_i x_{i,k} - \min\limits_i x_{i,k}\right)}, & \text{if } i \neq j \end{cases} \tag{1}$$

We must mention that in V. Peneva and I. Popchev's paper is not mentioned especially that numbers $x_{i,j} \geq 0$, but in [4] they suppose this. Really, in the opposite case, e.g., if (let for clearness we use a variable p instead of i in operations max and min)

$$x_{i,k} = \min_p x_{p,k} = -1$$

and

$$x_{j,k} = \max_p x_{p,k} = 1,$$

then for $i \neq j$ we obtain

$$\mu_k(a_i, a_j) = 0.5 + \frac{-2}{2} = -0.5 \notin [0, 1],$$

that is impossible.

It is easy to see that the most important difference of the two approaches is in the evaluation of the case, when some alternative is compared with itself, because in the V. Peneva and I. Popchev's case, $\mu_k(a_i, a_i) = 1$, while in M. Fedrizzi, M. Fedrizzi, and R. A. M. Pereira's case $\mu_k(a_i, a_i) = 0.5$, that we suppose is better.

In [3, 4], the matrix X is ordinary one, but for the aims of our future research we can change it with an Index Matrix (IM, see [1]):

$$X = \begin{array}{c|ccc} & a_1 & \cdots & a_n \\ \hline a_1 & x_{1,1} & \cdots & x_{1,n} \\ \vdots & \vdots & \cdots & \vdots \\ a_n & x_{n,1} & \cdots & x_{n,n} \end{array}.$$

In future, the authors plan to study the properties of the IM-presentation of the problem.

Below, we suppose that there are at least two different number $x_{i,j}$. In this case,

$$\max_{1 \le p \le n} x_{p,k} - \min_{1 \le p \le n} x_{p,k} \equiv \max_p x_{p,k} - \min_p x_{p,k} \ne 0$$

for every k $(1 \le k \le d)$.

2 Short Remarks on Intuitionistic Fuzzy Pairs

The Intuitionistic Fuzzy Pair (IFP; see [2]) is an object with the form $\langle a, b \rangle$, where $a, b \in [0, 1]$ and $a + b \le 1$, that is used as an evaluation of some object or process and which components (a and b) are interpreted as degrees of membership and non-membership, or degrees of validity and non-validity, or degree of correctness and non-correctness, etc.

Let us have two IFPs $x = \langle a, b \rangle$ and $y = \langle c, d \rangle$.

In [2], we defined analogous of operations "conjunction", "disjunction", "implication", "negation", e.g.,

$$\neg x = \langle b, a \rangle,$$

and other operations, e.g.

$$x @ y = \langle \frac{a+c}{2}, \frac{b+d}{2} \rangle,$$

relations and three types of operators from modal, topological and level types.

3   On Intuitionistic Fuzzy Interpretation of M. Fedrizzi, M. Fedrizzi, and R. A. M. Pereira's and of V. Peneva and I. Popchev's Formulas

The intuitionistic fuzzy interpretations that we will introduce below, will have the form of a pair $\langle \mu_k(a_i, a_j), v_k(a_i, a_j)\rangle$, where $\mu_k(a_i, a_j)$ and $v_k(a_i, a_j)$ are the degrees of preference and non-preference between alternatives a_i and a_j.

If we like to write (1) as an intuitionistic fuzzy pair, changing the value for case $i = j$ in M. Fedrizzi, M. Fedrizzi, and R. A. M. Pereira's form, it can obtain the form

$$\mu_k(a_i, a_j) = \begin{cases} \langle 0.5, 0.5\rangle, & \text{if } i = j \\[2mm] \langle 0.5 + \dfrac{x_{i,k} - x_{j,k}}{2\left(\max\limits_i x_{i,k} - \min\limits_i x_{i,k}\right)}, 0.5 + \dfrac{x_{j,k} - x_{i,k}}{2\left(\max\limits_i x_{i,k} - \min\limits_i x_{i,k}\right)}\rangle, & \text{if } i \neq j \end{cases}$$

This pair is really an intuitionistic fuzzy pair, because for the

$$0.5 + \frac{x_{i,k} - x_{j,k}}{2\left(\max\limits_i x_{i,k} - \min\limits_i x_{i,k}\right)} \geq 0.5 + \frac{-x_{j,k}}{2\left(\max\limits_i x_{i,k}\right)} \geq 0,$$

$$0.5 + \frac{x_{j,k} - x_{i,k}}{2\left(\max\limits_i x_{i,k} - \min\limits_i x_{i,k}\right)} \geq 0.5 + \frac{-x_{i,k}}{2\left(\max\limits_i x_{i,k}\right)} \geq 0,$$

$$0.5 + \frac{x_{i,k} - x_{j,k}}{2\left(\max\limits_i x_{i,k} - \min\limits_i x_{i,k}\right)} + 0.5 + \frac{x_{j,k} - x_{i,k}}{2\left(\max\limits_i x_{i,k} - \min\limits_i x_{i,k}\right)} = 1 \leq 1.$$

Now, we will try to generalize the above interpretation.

Let $\alpha, \beta \in [0, 1]$ and $\alpha + \beta \leq 1$. The intuitionistic fuzzy interpretation can have the form

$$\langle \mu_{k,\alpha,\beta}(a_i, a_j), v_{k,\alpha,\beta}(a_i, a_j)\rangle$$

$$= \begin{cases} \langle 0.5, 0.5\rangle, & \text{if } i = j \\[2mm] \langle \alpha + \dfrac{x_{i,k} - x_{j,k}}{2\left(\max\limits_i x_{i,k} - \min\limits_i x_{i,k}\right)}, \beta + \dfrac{x_{j,k} - x_{i,k}}{2\left(\max\limits_i x_{i,k} - \min\limits_i x_{i,k}\right)}\rangle, & \text{if } i \neq j \end{cases} \qquad (2)$$

But, if $n = 2, k = 1, x_{1,k} = 1, x_{2,k} = 0$, then $\max\limits_i x_{i,k} = 1$, $\min\limits_i x_{i,k} = 0$ and if $\beta < 0.5$, then

$$\beta + \frac{x_{2,k} - x_{1,k}}{2\left(\max\limits_i x_{i,k} - \min\limits_i x_{i,k}\right)} = \beta - 0.5 < 0,$$

that is a contradiction. Therefore, β must satisfy $\beta \geq 0.5$. By similar way we see that $\alpha \geq 0.5$, i.e., α and β must be exactly $\alpha = \beta = 0.5$, i.e., the idea for a generalization of (1) in the form of (2) failed. In practice, we obtain only a new (intuitionistic fuzzy) record of the original formula. Therefore, we must change something in the form of (2).

4   First Intuitionistic Fuzzy Interpretation

It has the form

$$\langle \mu_k^1(a_i, a_j), v_k^1(a_i, a_j) \rangle = \left\langle \frac{\max\limits_{p \neq i} x_{p,k} + x_{i,k} - x_{j,k}}{2 \max\limits_{p} x_{p,k}}, \frac{\max\limits_{p \neq j} x_{p,k} + x_{j,k} - x_{i,k}}{2 \max\limits_{p} x_{p,k}} \right\rangle, \qquad (3)$$

where

$$\max_{p \neq i} x_{p,k} = \max_{p \in \{1,\ldots,i-1,i+1,\ldots,n\}} x_{p,k},$$

$$\min_{p \neq i} x_{p,k} = \min_{p \in \{1,\ldots,i-1,i+1,\ldots,n\}} x_{p,k}.$$

First, we check that the pair is an intuitionistic fuzzy pair. Really,

$$\frac{\max\limits_{p \neq i} x_{p,k} + x_{i,k} - x_{j,k}}{2 \max\limits_{p} x_{p,k}} \geq \frac{\max\limits_{p \neq i} x_{p,k} - x_{j,k}}{2 \max\limits_{p} x_{p,k}} \geq 0,$$

$$\frac{\max\limits_{p \neq j} x_{p,k} + x_{j,k} - x_{i,k}}{2 \max\limits_{p} x_{p,k}} \geq \frac{\max\limits_{p \neq j} x_{p,k} - x_{i,k}}{2 \max\limits_{p} x_{p,k}} \geq 0$$

and

$$\frac{\max\limits_{p \neq i} x_{p,k} + x_{i,k} - x_{j,k}}{2 \max\limits_{p} x_{p,k}} + \frac{\max\limits_{p \neq j} x_{p,k} + x_{j,k} - x_{i,k}}{2 \max\limits_{p} x_{p,k}} = \frac{\max\limits_{p \neq i} x_{p,k} + \max\limits_{p \neq j} x_{p,k}}{2 \max\limits_{p} x_{p,k}} \leq 1.$$

Second, the degree of uncertainty is

$$\pi_k^1(a_i, a_j) = 1 - \frac{\max\limits_{p \neq i} x_{p,k} + \max\limits_{p \neq j} x_{p,k}}{2 \max\limits_{p} x_{p,k}} \geq 0.$$

Third, we see that for $i = j$, (3) obtains the form

$$\langle \mu_k^1(a_i, a_i), v_k^1(a_i, a_i) \rangle = \left\langle \frac{\max\limits_{p \neq i} x_{p,k} + x_{i,k} - x_{i,k}}{2 \max\limits_p x_{p,k}}, \frac{\max\limits_{p \neq j} x_{p,k} + x_{i,k} - x_{i,k}}{2 \max\limits_p x_{p,k}} \right\rangle$$

$$= \left\langle \frac{\max\limits_{p \neq i} x_{p,k}}{2 \max\limits_p x_{p,k}}, \frac{\max\limits_{p \neq j} x_{p,k}}{2 \max\limits_p x_{p,k}} \right\rangle,$$

that is an analogous of formula (2), but now, rendering an account the fact that in the intuitionistic fuzzy case there exists a degree of uncertainty.

Fourth, it is valid

Proposition 1 For every two alternatives a_i and a_j,

$$\neg\langle \mu_k^1(a_i, a_j), v_k^1(a_i, a_j) \rangle = \langle \mu_k^1(a_j, a_i), v_k^1(a_j, a_i) \rangle.$$

Proof Using the definition of operation negation, we obtain:

$$\neg\langle \mu_k^1(a_i, a_j), v_k^1(a_i, a_j) \rangle$$

$$\neg \left\langle \frac{\max\limits_{p \neq i} x_{p,k} + x_{i,k} - x_{j,k}}{2 \max\limits_p x_{p,k}}, \frac{\max\limits_{p \neq j} x_{p,k} + x_{j,k} - x_{i,k}}{2 \max\limits_p x_{p,k}} \right\rangle$$

$$= \left\langle \frac{\max\limits_{p \neq j} x_{p,k} + x_{j,k} - x_{i,k}}{2 \max\limits_p x_{p,k}}, \frac{\max\limits_{p \neq i} x_{p,k} + x_{i,k} - x_{j,k}}{2 \max\limits_p x_{p,k}} \right\rangle$$

$$= \langle \mu_k^1(a_j, a_i), v_k^1(a_j, a_i) \rangle.$$

It is easy seen that

$$\langle \mu_k^1(a_i, a_j), v_k^1(a_i, a_j) \rangle$$

$$= \left\langle \frac{\max\limits_{p \neq i} x_{p,k} - x_{j,k}}{2 \max\limits_p x_{p,k}}, \frac{x_{j,k}}{2 \max\limits_p x_{p,k}} \right\rangle @ \left\langle \frac{x_{i,k}}{2 \max\limits_p x_{p,k}}, \frac{\max\limits_{p \neq j} x_{p,k} - x_{i,k}}{2 \max\limits_p x_{p,k}} \right\rangle$$

$$= \left\langle \frac{x_{i,k}}{2 \max\limits_p x_{p,k}}, \frac{\max\limits_{p \neq j} x_{p,k} - x_{i,k}}{2 \max\limits_p x_{p,k}} \right\rangle @ \left\langle \frac{\max\limits_{p \neq i} x_{p,k} - x_{j,k}}{2 \max\limits_p x_{p,k}}, \frac{x_{j,k}}{2 \max\limits_p x_{p,k}} \right\rangle$$

that are intuitionistic fuzzy pairs, because, as we shown above,

$$\max_{p\neq i} x_{p,k} - x_{j,k} \geq 0,$$

$$\max_{p\neq j} x_{p,k} - x_{i,k} \geq 0$$

and above check.

5 Second Intuitionistic Fuzzy Interpretation

Let $\alpha, \beta, \gamma, \delta \in [0, 1]$, $\alpha \geq \gamma, \beta \geq \delta$ and

$$\alpha + \beta + \frac{1}{2}|\gamma - \delta| \leq 1.$$

The second intuitionistic fuzzy interpretation has the form

$$\langle \mu^2_{k,\alpha,\beta,\gamma,\delta}(a_i, a_j), v^2_{k,\alpha,\beta,\gamma,\delta}(a_i, a_j) \rangle$$

$$= \begin{cases} \langle \alpha, \beta \rangle, & \text{if } i = j \\ \langle \alpha + \gamma \frac{x_{i,k}-x_{j,k}}{2\left(\max\limits_i x_{i,k}-\min\limits_i x_{i,k}\right)}, \beta + \delta \frac{x_{j,k}-x_{i,k}}{2\left(\max\limits_i x_{i,k}-\min\limits_i x_{i,k}\right)} \rangle, & \text{if } i \neq j \end{cases} \quad (4)$$

First, we check again that the interpretation is an intuitionistic fuzzy pair. Really,

$$\alpha + \gamma \frac{x_{i,k} - x_{j,k}}{2\left(\max\limits_i x_{i,k} - \min\limits_i x_{i,k}\right)} \geq \alpha - \gamma \frac{x_{j,k}}{2\max\limits_i x_{i,k}} \geq \alpha - \gamma \geq 0,$$

$$\beta + \delta \frac{x_{j,k} - x_{i,k}}{2\left(\max\limits_i x_{i,k} - \min\limits_i x_{i,k}\right)} \geq \beta - \delta \frac{x_{i,k}}{2\max\limits_i x_{i,k}} \geq \beta - \delta \geq 0$$

and

$$\alpha + \gamma \frac{x_{i,k} - x_{j,k}}{2\left(\max\limits_i x_{i,k} - \min\limits_i x_{i,k}\right)} + \beta + \delta \frac{x_{j,k} - x_{i,k}}{2\left(\max\limits_i x_{i,k} - \min\limits_i x_{i,k}\right)}$$

$$= \alpha + \beta + (\gamma - \delta) \frac{x_{i,k} - x_{j,k}}{2\left(\max\limits_i x_{i,k} - \min\limits_i x_{i,k}\right)}$$

$$\leq \alpha + \beta + |\gamma - \delta| \frac{x_{i,k}}{2\max\limits_i x_{i,k} i x_{i,k}}$$

$$\leq \alpha + \beta + |\gamma - \delta| \frac{1}{2} \leq 1.$$

Second, the degree of uncertainty now is

$$\pi_k^1(a_i, a_j) = 1 - \alpha - \gamma \frac{x_{i,k} - x_{j,k}}{2\left(\max_i x_{i,k} - \min_i x_{i,k}\right)} - \beta - \delta \frac{x_{j,k} - x_{i,k}}{2\left(\max_i x_{i,k} - \min_i x_{i,k}\right)}$$

$$= 1 - \alpha - \beta - (\gamma - \delta) \frac{x_{i,k} - x_{j,k}}{2\left(\max_i x_{i,k} - \min_i x_{i,k}\right)} \geq 0.$$

Third, we see that for $i = j$, using the second term of (4), we obtain

$$\langle \mu_k^1(a_i, a_i), v_k^1(a_i, a_i) \rangle = \langle \alpha + \gamma \frac{0}{2\left(\max_i x_{i,k} - \min_i x_{i,k}\right)}, \beta +$$

$$+ \delta \frac{0}{2\left(\max_i x_{i,k} - \min_i x_{i,k}\right)} \rangle = \langle \alpha, \beta \rangle,$$

i.e., the first term of (4).

Fourth, it is valid

Proposition 2 For every two alternatives a_i and a_j, so that $i \neq j$,

$$\neg\langle \mu_{k,\alpha,\beta,\gamma,\delta}^2(a_i, a_j), v_{k,\alpha,\beta,\gamma,\delta}^2(a_i, a_j) \rangle = \langle \mu_{k,\beta,\alpha,\delta,\gamma}^2(a_j, a_i), v_{k,\beta,\alpha,\delta,\gamma}^2(a_j, a_i) \rangle.$$

The proof is similar to the above one.

The present interpretation is a real generalization of the modified formula (2), that is the left components of the intuitionistic fuzzy pair for $\alpha = \beta = 0.5$ and $\gamma = \delta = 1$.

6 Conclusion

As we mentioned above, in future, we plan to research the properties of the IM, generated by the above way.

Acknowledgements The first two authors are thankful for the support provided by the Bulgarian National Science Fund under Grant Ref. No. DFNI-I-02-5 "InterCriteria Analysis: A New Approach to Decision Making".

References

1. Atanassov K (2014) Index matrices: towards an augmented matrix calculus. Springer, Cham
2. Atanassov K, Szmidt E, Kacprzyk J (2013) On intuitionistic fuzzy pairs. Note Intuit Fuzzy Set 19(3):1–13
3. Fedrizzi M, Fedrizzi M, Pereira RAM (1999) Soft consensus and network dynamics in group decision making. Int J Intel Syst 14:63–77
4. Peneva V, Popchev I (2010) Fuzzy multi-criteria decision making algorithms. Comptes rendus de l'Academie Bulgare des Sciences 63(7):979–992
5. Szmidt, E, Kacprzyk J, Atanassov K (2013) Intuitionistic fuzzy modifications of some PenevaP-opchev formulas for estimation of preference degree. Part 1. Issue Intuit Fuzzy Set Generalized Net 10:12–20

Part III
Judgments and Aggregation

Fuzzified Likert Scales in Group Multiple-Criteria Evaluation

Jan Stoklasa, Tomáš Talášek and Pasi Luukka

Abstract Likert scales have been in use since 1930s as tool for attitude expression in many fields of social science. Recently there have even been several attempts for the fuzzification of this instrument. In this chapter we explore the possibility of their use in multiple-criteria multi-expert evaluation. We focus on discrete fuzzy Likert scales, that are a generalization of the standard Likert scales. We propose a methodology that deals with the non-uniformity of the distribution of linguistic labels along the underlying ordinal evaluation scale and also with possible response bias. We also consider the analogy of Likert scales (crisp and fuzzy) on continuous universes. Likert-type evaluations of an alternative with respect to various criteria are represented using histograms. Histograms are also used to aggregate the Likert-type evaluations. A transformation of the multi-expert multiple-criteria evaluation represented by a histogram into a 3-bin histogram to control for the response bias is performed and an ideal-evaluation 3-bin histogram is defined. We propose a distance measure to assess the closeness of the overall evaluation to the ideal and suggest the use of the proposed methodology in multiple-criteria multi-expert evaluation.

J. Stoklasa (✉) · P. Luukka
Lappeenranta University of Technology, School of Business and Management,
Skinnarilankatu 34, 53850 Lappeenranta, Finland
e-mail: jan.stoklasa@lut.fi; jan.stoklasa@upol.cz

P. Luukka
e-mail: pasi.luukka@lut.fi

J. Stoklasa · T. Talášek
Olomouc, Faculty of Arts, Department of Applied Economics, Palacký University,
Křížkovského 8, 771 47 Olomouc, Czech Republic
e-mail: tomas.talasek@upol.cz

© Springer International Publishing AG 2018   165
M. Collan and J. Kacprzyk (eds.), *Soft Computing Applications for Group Decision-making and Consensus Modeling*, Studies in Fuzziness and Soft Computing 357, DOI 10.1007/978-3-319-60207-3_11

1   Introduction

Likert scales were introduced by Likert in 1930s [18] as a tool for the measurement
and assessment of attitudes. Since then Likert scales (and Likert-type scales) have
grown popular in many fields of social science, including management and marketing research [2]. They became a frequently used tool for the extraction of information
from participants concerning not only their attitudes, but also preferences, evaluations etc. Likert scales are an easy-to-use tool for the measurement of attitudes or for
evaluation [31]. There is, however, still an ongoing debate on how to properly deal
with the data acquired via Likert scales, what statistical methods are appropriate (see
e.g. [21]), how to aggregate information obtained through Likert scales and how to
summarize it. Even attempts of fuzzification of Likert scales have been presented
(see e.g. [5, 9, 17] or the chapter by Gil & Gil in [35]). The combination of linguistic and numerical labels for the elements of the Likert scale, however, introduces
several methodological issues that have not been addressed sufficiently in the literature so far. Using linguistic labels for more than the endpoints of the scale and the
middle point (if a discrete scale with an odd number of elements is applied) renders
the scale ordinal, since the meanings of the linguistic terms cannot be considered
equidistant (at least not automatically). Aggregation of information provided by the
Likert scale can therefore be problematic (in the sense of the necessary restriction to
ordinal computational methods). Also the well known response biases in self-report
questionnaires, mainly the extremity response (*leniency*) tendency and the mid-point
response (*central*) tendency of the decision makers (see e.g. [2, 12, 13]) can complicate the aggregation.

   In this paper we aim to discuss Likert scales in the context of multiple-criteria and
multi-expert (MCME) evaluation in the fuzzy context, as opposed to [31], where only
crisp Likert scales were considered. As such, also the area of psychology, management, marketing and economical research, behavioral science, sociology and related
fields are the possible recipients of the presented results. In fact the combination of
the potential of Likert scales with the possibility of reflecting uncertainty and dealing with response-bias can open interesting new doors in social science research.
In this chapter we suggest a way to aggregate individual evaluations (provided by
Likert scales as fuzzy sets on the underlying evaluation universe—crisp or continuous) into an overall evaluation and a way of aggregating these overall evaluations
across experts that reflect the specifics of Likert scales and offer tools for handling
response-bias.

   In line with [31] we suggest to use histogram representation for the summarization and aggregation of Likert-scale answers. The extremity-response and mid-point-
response tendencies are taken into account by joining specific bins of the histograms.
We propose a similarity-to-ideal (or distance-from-ideal) assessment of the fulfillment of a goal (an ideal overall evaluation in the context of the fuzzy evaluations
represented by histograms). Contrary to [31], the proposed approach results in general in a relative-type evaluation, since general (subnormal) fuzzy sets are also considered as answers provided by the decision makers. The presented methodology,

however, constitutes a fuzzy extension of the methodology presented in [31], the latter being a special case of the former. In the case when only normal fuzzy sets are considered, the overall evaluation can be interpreted as a degree of fulfillment of the overall goal and as such is an absolute-type evaluation.

2 Likert Scales

The *n-point Likert scale* [18] can be defined as a discrete bipolar evaluation scale with n integer values (e.g. $[1, 2, 3, 4, \dots, n]$). The values are naturally ordered, the endpoints 1 and n are interpreted as extreme evaluations (1 as completely positive, n as completely negative, or vice versa). If n is odd, then the middle value of the scale is considered as neutral evaluation (or "a cannot decide value"), if n is an even number then there is no neutral evaluation allowed and any evaluation expressed by the user of the scale needs to tend either to the positive or the negative direction with respect to the theoretical center of the scale. Extensive discussions can be found e.g. in the psychometrics and methodological literature concerning the appropriate number of elements of these scales, the choice of an odd or even n and even the selection of a minimum value of the scale (see e.g. [20]). Likert scales in the form of $[0, 1, 2, \dots, n-1]$ or $[-\frac{n}{2} + 0.5, -\frac{n}{2} + 1.5, \dots, 0, \dots, \frac{n}{2} - 1.5, \frac{n}{2} - 0.5]$ for an odd n or $[-\frac{n}{2}, -\frac{n}{2} + 1, \dots, \frac{n}{2} - 1, \frac{n}{2}]$ for an even n can be found in the literature. Despite the widespread use of Likert scales in e.g. self-reporting psychodiagnostics, a preference of only 3, 4 and 5 point scales seems to prevail, although Likert scales with 6 or 7 points are also used. An example of a method utilizing a seven-point Likert scale is the questionnaire "Experiences in Close Relationships—Czech", assessing relationship anxiety and relationship evasiveness of the adult population in the context of the theory of relational cohesion [27]. Without any loss of generality we will be considering the $[1, 2, 3, 4, \dots, n]$ form of the Likert scale in this paper.

The Likert scale can have two interconnected levels of description—the numerical one represented by the integer values and a linguistic level assigning each (or some) of the values a linguistic label (see Fig. 1). In practical applications either

Strong positive	Positive	Neutral	Negative	Strong negative		
(1)	(2)	(3)	(4)	(5)		

Strongly agree	Agree	More or less agree	Undecided	More or less disagree	Disagree	Strongly disagree
(1)	(2)	(3)	(4)	(5)	(6)	(7)

Fig. 1 An example of a 5-point and 7-point Likert scale

both levels, or just one of the levels is provided to the decision makers. Generally speaking, four types of measurement (scales) can be distinguished (see e.g. [23, 28, 31, 36]):

- *nominal*, where the elements of the scale have no numerical interpretation, there is no ordering of these elements. Histograms can be used to summarize the results, but the position of the bins is not fixed—they can be reordered freely without any change to the presented information.
- *ordinal*, where ordering of the values of the scale is known. It is possible to find minima and maxima. The distance between the elements of the scale is not known. As such average values and differences should not be computed (unless e.g. the assumption of equidistance of the elements of the scale is made).
- *interval* and *ratio* (also called *cardinal*), where the ordering and distance of the elements of the scale is known. Averaging operators can be applied.

Even a more detailed classification is possible (see e.g. [23]), but for the purpose of this chapter it is sufficient to distinguish between nominal, ordinal and other types of scales (measurement), where averaging and standard forms of summarization of information (aggregation of evaluations) can be used.

When the linguistic labels are used along with the numerical level, or when the numerical level is not presented within the questionnaire, the values (evaluations) obtained from the decision makers via Likert scales should be considered to originate from an ordinal scale, since the meanings of the linguistic terms used as labels might not be equidistant. Although the distances between the meanings of the linguistic labels are not known, the labels are usually constructed so that at least the symmetry with respect to the middle element is preserved. The calibration of the meanings of linguistic terms (assigning them appropriate position on the evaluation universe) is possible, but requires more time and resources to be done properly—see e.g. [3, 14, 22, 26]. In general, the results of such a calibration might not be transferable to a different context, thus limiting the use of these methods to some extent in the practical real-life use (e.g. in management research, diagnostics etc.); see also [29, 30] for a discussion of the issues related to the calibration of scales. We therefore leave the possibility of using calibrated Likert scales out of the scope of this paper.

When only the numerical values of the Likert scale are presented to the decision-makers, the obtained answers can be considered as values of the interval scale from the mathematical point of view. In theory it is possible to compute an "average evaluation" applying some aggregation operator (e.g. an arithmetic or weighted mean, a sum of the evaluations etc.) on all the Likert scale values that assess different aspects of the same alternative (multiple criteria evaluation). The aggregation of such overall evaluations of alternatives across more decision-makers is theoretically possible in the same way. As pointed out by Stoklasa et al. [31], aggregating the numerical values across more decision-makers may lead to the distortion or loss of information. The differences in answering patterns and answering customs of the decision-makers—possible manifestations of the leniency and central tendency—need to be taken into account. Let us consider the 7-point Likert scale in Fig. 1. If a decision-maker tends to never answer using the extreme values of the scale (central tendency bias), his/her evaluation expressed by 2 can describe exactly the same attitude (evaluation) as the value 1 chosen by someone who has no problems using the extreme values. The point here is that in some situations the difference between a 1 and 2 can lie in the interpretation of the question or in the customs of the decision-maker. This becomes even more apparent when the linguistic level is applied. Some decision-makers can interpret "Strongly agree" as being too categorical and thus choose "Agree" without having any objections to the statement presented in the question. If we are then looking for a consensus of the decision-makers or if we want to define the "ideal evaluation", we should not expect that all the decision-makers provide the evaluation 1. For those with a central tendency in answering the agreement can be expressed by a slightly lower value, but it might mean the same level of conviction as 1 provided by those decision-makers that do not have the central tendency.

It is also possible to approach Likert scale through the fuzzy modelling perspective (see Fig. 2 for an illustration of the fuzzy set representation of crisp Likert scale answers). Once in the fuzzy set context, one can allow more uncertain answers to be provided by the decision makers (see Fig. 3 for an example). This way the capabilities of Likert scales can be extended to reflect the possible uncertainty in the answers/evaluations provided by the decision makers. The transition into the fuzzy domain might remedy some of the above mentioned issues. The core of the central-tendency and leniency effects, however, remain unchanged. If a person selects an extreme answer only rarely and considers the second to most extreme as the "best usual" evaluation, even the possibility of expressing uncertainty or choosing more values of the scale with different membership values might not change this. Hence in the fuzzy versions of Likert scales response-bias can be an issue and should be considered.

Overall there seems to be enough reasons for not considering the values provided by Likert scale items in the questionnaire in their crisp or fuzzy form as interval or ratio scale values in the process of their aggregation. The meaning of the linguistic labels can be context dependent (hence the distances between them expressed in the numerical level can differ). Even more importantly, in multi-expert decision making, the distances between the linguistic labels can be perceived differently by different decision-makers, and e.g. the central tendency and the extreme-answer tendency combined in one set of decision-makers can complicate the interpretability of the aggregated results if the values are treated as cardinal. We therefore suggest the use of histograms to represent the information in its raw and aggregated form in the context of Likert scales.

It is also possible to generalize the situation even further and consider continuous universes instead of discrete ones as suggested in the standard Likert scale. Although this might be consider to be a significant departure from the original idea of Likert scales, even when using fuzzy sets on a pre-defined continuous universe (see Fig. 4) can face the problems of leniency and central tendency. We will therefore briefly focus on this form of obtaining expert evaluations as well in this chapter.

	Strong positive	Positive	More or less positive	Neutral	More or less negative	Negative	Strong negative
	(1)	(2)	(3)	(4)	(5)	(6)	(7)
Item 1	( )	(x)	( )	( )	( )	( )	( )
Item 2	( )	( )	( )	(x)	( )	( )	( )

...

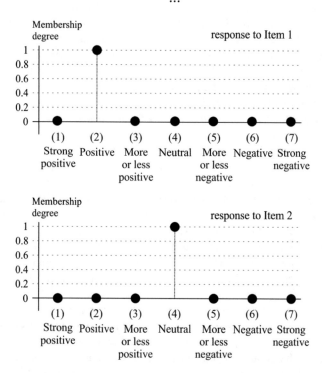

Fig. 2 Representation of two crisp answers in a 7-point Likert-scale questionnaire by fuzzy sets

3   Notation and Basic Concepts Concerning Fuzzy Sets

Let U be a nonempty set (the universe of discourse). A *fuzzy set A* on U is defined by the mapping $A : U \to [0, 1]$. For each $x \in U$ the value $A(x)$ is called a *membership degree* of the element x in the fuzzy set A and $A(.)$ is called a *membership function* of the fuzzy set A. If the universe is a discrete set, i.e. if $U = \{x_1, \ldots, x_n\}$, then a fuzzy set A on U is denoted $A = \{ {}^{A(x_1)}|_{x_1}, \ldots, {}^{A(x_n)}|_{x_n} \}$. The family of all fuzzy sets on U us denoted $\mathscr{F}(U)$. $\mathrm{Ker}(A) = \{x \in U | A(x) = 1\}$ denotes a *kernel* of A, $A_\alpha = \{x \in U | A(x) \geq \alpha\}$ denotes an *α-cut* of A for any $\alpha \in [0, 1]$, $\mathrm{Supp}(A) = \{x \in U | A(x) > 0\}$ denotes a *support* of A.

Fig. 3 Possible generalization of the Likert-scale framework to fuzzy answers. Example of two uncertain answers provided by a decision maker via a 7-point Likert scale. Note, that the top answer is a normal fuzzy set, whereas the bottom one is not

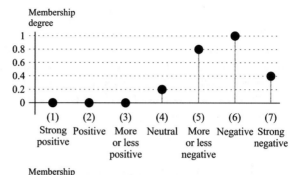

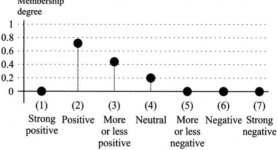

Fig. 4 Evaluations provided as fuzzy sets on the Likert-inspired continuous universal set. A is a representation of a crisp answer, B and C are fuzzy numbers

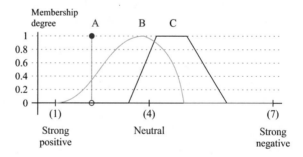

A fuzzy number is a fuzzy set A on the set of real numbers which satisfies the following conditions: (1) $\mathrm{Ker}(A) \neq \emptyset$ (A is *normal*); (2) A_α are closed intervals for all $\alpha \in (0, 1]$ (this implies A is *unimodal*); (3) $\mathrm{Supp}(A)$ is bounded. A family of all fuzzy numbers on $U \subseteq \mathbb{R}$ is denoted by $\mathscr{F}_N(U)$. A fuzzy number A can be characterized by a quaternion of its significant values (a_1, a_2, a_3, a_4), where $a_1, a_2, a_3, a_4 \in U$, $a_1 \leq a_2 \leq a_3 \leq a_4$, $[a_1, a_4] = \mathrm{Cl}(\mathrm{Supp}(A))$ and $[a_2, a_3] = \mathrm{Ker}(A)$. A fuzzy number representing the interval $[c, d] \subseteq \mathbb{R}$ can be defined as (c, c, d, d). An *intersection* of two fuzzy sets A and B on U is a fuzzy set $(A \cap B)$ on U with the membership function defined as follows: $(A \cap B)(x) = \min\{A(x), B(x)\}, \forall x \in U$. Although different t-norms can be used to define the intersection of A and B (see e.g. [10, 15]), we will stick with the minimum t-norm in this paper. Let $A \in \mathscr{F}(U)$, then the cardinality of A (Card(A)) is computed as Card(A) = $\sum_{i=1}^{n} A(x_i)$ in case $U = \{x_1, \ldots, x_n\}$ and as Card(A) = $\int_a^b A(x)dx$ in case $U = [a, b] \subset \mathbb{R}$.

A *fuzzy scale* on $[a, b]$ is defined as a set of fuzzy numbers $T_1, T_2, \ldots, T_s$ on $[a, b]$, that form a Ruspini fuzzy partition (see [25]) of the interval $[a, b]$, i.e. for all $x \in [a, b]$ it holds that $\sum_{i=1}^{s} T_i(x) = 1$, and the T's are indexed according to their ordering. If we weaken the requirements for a fuzzy scale e.g. to $\sum_{i=1}^{s} T_i(x) \geq 1$, we obtain a general fuzzy partition.

4 Histograms and Fuzzy Histograms

Histograms are a frequentistic summary of information concerning a given sample (set of evaluations or answers to items provided by decision-makers). Histograms have been widely used in statistics, pattern recognition and many other fields. The distance of histograms plays an important role in pattern recognition, clustering (see e.g. [8, 11]), time series analysis [34] and virtually in any application field dealing with simulation. Histograms can be used to represent the current state (e.g. prevailing evaluation in the given set of experts expressed via discrete Likert scales, a characteristic pattern) and also the desired evaluation (ideal, aspiration level etc.). They thus seem to be a proper tool for decision making based on Likert-scales outputs. Even in the fuzzy case histograms can be used to deal with some response-bias, such as the central-tendency or leniency effect. Histograms can be utilized to define the ideal evaluation, or even a fuzzy ideal evaluation is such a way that reduces the possible response bias. It is the main idea of the use of crisp Likert-scale in multiple-criteria multi-expert decision making as proposed in [31] and it will be further generalised here to the fuzzy case. First we need to summarize the notation for crisp and fuzzy histograms and recall some relevant distance measures for histograms. We will then utilize these concepts in the next section—introducing a fuzzy-Likert-scale based multiple criteria multi-expert evaluation methodology capable of dealing with some important response biases.

4.1 Crisp Histograms

We will be using the vector representation of a histogram in accordance with e.g. [8] and we will be using the standard graphical representation of histograms in figures. We will also use the notation introduced in [31], where crisp Likert scales were suggested as a tool for multiple-criteria multi-expert evaluation. The histogram can be represented by a vector of fixed dimension. The dimension specifies the number of bins of the histogram, the components of the vector represent the number of observations/evaluations belonging to each bin. More specifically let us consider an n-point Likert scale with the set of possible values $L = \{l_1, \ldots, l_n\}$ that is used to provide

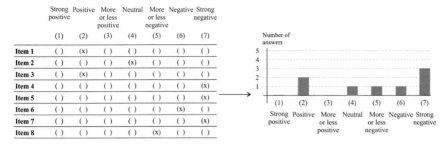

Fig. 5 Representation/summary of an 8-item questionnaire by a histogram. A 7-point crisp Likert scale is used to provide evaluations of each item by the decision-maker

evaluations (measurement) of alternative (feature) x (in the case of the Likert scales presented in Fig. 1 we can write $L = \{1, \ldots, n\}$).

Let us consider a set E of m evaluations (measurements) of the alternative (feature) x, $E = \{e_1, \ldots, e_m\}$, where $e_1, \ldots, e_m \in L$. The histogram of the set E is a vector (an ordered n-tuple) $H(E) = [H_1(E), \ldots, H_n(E)]$, where $H_i(E)$ represents the number of times x has been evaluated l_i, for all $i = 1, \ldots, n$, that is

$$H_i(E) = \sum_{j=1}^{m} c_{ij}, \quad \text{where } c_{ij} = \begin{cases} 1 & \text{if } e_j = l_i, \\ 0 & \text{otherwise.} \end{cases} \tag{1}$$

The histogram presented in Fig. 5 would be represented by the vector $[0, 2, 0, 1, 1, 1, 3]$.

4.2 Fuzzy Histograms on Discrete Universes

Fuzzy histograms (i.e. histogram of fuzzy numbers with crisp classes, histograms of crisp numbers with fuzzy classes or histograms of fuzzy numbers with fuzzy classes) have been studied e.g. in [6, 37]. Let us consider again an n-point Likert scale with the set of possible values $L = \{l_1, \ldots, l_n\}$ and let m evaluations be provided as fuzzy sets on L, i.e. $\widetilde{E} = \{\widetilde{e}_1, \ldots, \widetilde{e}_m\}$, where $\widetilde{e}_j \in \mathscr{F}(L), \widetilde{e}_j = \{\widetilde{e}_j(l_1)|_{l_1}, \ldots, \widetilde{e}_j(l_n)|_{l_n}\}$. The histogram of the set $\widetilde{E}$ is a vector (an ordered n-tuple) $\widetilde{H}(\widetilde{E}) = [\widetilde{H}_1(\widetilde{E}), \ldots, \widetilde{H}_n(\widetilde{E})]$, where $\widetilde{H}_i(\widetilde{E})$ represents the relative amount of support for l_i across all $\widetilde{e}_j$, corrected for subnormality of $\widetilde{e}_j$, for all $i = 1, \ldots, n, j = 1, \ldots, m$, that is

$$\widetilde{H}_i(\widetilde{E}) = \sum_{j=1}^{m} \left(\frac{\widetilde{e}_j(l_i)}{\text{Card}(\widetilde{e}_j)} \cdot \text{hgt}(\widetilde{e}_j) \right). \tag{2}$$

The multiplication by $\mathrm{hgt}(\widetilde{e}_j)$ in (2) serves as a compensation for answers that are represented by subnormal fuzzy sets—these answers are considered less informative (no fully fitting element of l is expressed by the decision maker) and hence contribute less to the overall evaluation. Note, that if the crisp case of Likert scales is considered, where the answer l_i to item j is represented as $\widetilde{e}_j = \{^0|_{l_1}, \ldots, ^0|_{l_{i-1}}, ^1|_{l_i}, ^0|_{l_{i+1}}, \ldots, ^0|_{l_n}\}$, $j = 1, \ldots, m$ then (2) reduces to (1).

4.3 Fuzzy Histograms on Continuous Universes

Let us now depart a bit further from the original idea of an n-point Likert scale and consider the underlying evaluation scale to be continuous, more specifically $L_C = [1, n]$. The evaluations $\widetilde{e}_j, j = 1, \ldots, m$, provided by the m decision makers can now be considered as fuzzy sets on L_C, i.e. $\widetilde{E} = \{\widetilde{e}_1, \ldots, \widetilde{e}_m\}$, where $\widetilde{e}_j \in \mathscr{F}(L_C)$. We can now introduce a partition of L_C to define the bins for the histogram. E.g. if a uniform partition into u subintervals is considered, i.e. $L_C = l_1^I \cup l_2^I \cup \cdots \cup l_u^I$, where $l_s^I = \left[1 + (s-1)\frac{n-1}{u}, 1 + s\frac{n-1}{u}\right]$, $s = 1, \ldots, u$. A fuzzy histogram of the set $\widetilde{E}$ can again be defined as a vector (an ordered n-tuple) $\widetilde{H}(\widetilde{E}) = [\widetilde{H}_{l_1^I}(\widetilde{E}), \ldots, \widetilde{H}_{l_u^I}(\widetilde{E})]$, where $\widetilde{H}_{l_s^I}(\widetilde{E})$ represents the sum of relative cardinalities of all $\left(\widetilde{e}_j \cap \widetilde{l}_s^I\right)$, where $\widetilde{l}_s^I \in \mathscr{F}(L_C)$, such that $\widetilde{l}_s^I(x) = 1$ for all $x \in l_s^I$ and $\widetilde{l}_s^I(x) = 0$ otherwise. Again the correction for subnormality of $\widetilde{e}_j$ is applied, $j = 1, \ldots, m$, that is

$$\widetilde{H}_{l_s^I}(\widetilde{E}) = \sum_{j=1}^{m} \left(\frac{\mathrm{Card}\left(\widetilde{e}_j \cap \widetilde{l}_s^I\right)}{\mathrm{Card}(\widetilde{e}_j)} \cdot \mathrm{hgt}(\widetilde{e}_j) \right). \tag{3}$$

We can also consider the bins to be defined through the introduction of a fuzzy partition of $L_C = [1, n]$. In this case we define u fuzzy numbers $\widetilde{l}_s^{FN} \in \mathscr{F}(L_C)$, $s = 1, \ldots, u$, such that $\sum_{s=1}^{u} \widetilde{l}_s^{FN}(x) \geq 1$ for all $x \in L_C$. Ruspini fuzzy partitions (see [25]) might be a reasonable tool for this purpose. For each $\widetilde{e}_j \in \mathscr{F}(L_C)$ we define u fuzzy sets $\widetilde{e}_{js}, j = 1, \ldots, m, s = 1, \ldots, u$ in the following way

$$\widetilde{e}_{js}(x) = \widetilde{e}_j(x) \cdot \frac{\widetilde{l}_s^{FN}(x)}{\sum_{q=1}^{u} \widetilde{l}_q^{FN}(x)}. \tag{4}$$

These fuzzy sets reflect the distribution of the membership degrees of a given $\widetilde{e}_j$ among all $\widetilde{l}_s^{FN}$, $s = 1, \ldots, u$. Using these fuzzy sets we can now define the fuzzy histogram of the set $\widetilde{E}$ again as a vector (an ordered n-tuple) $\widetilde{H}(\widetilde{E}) = [\widetilde{H}_{\widetilde{l}_1^{FN}}(\widetilde{E}), \ldots, \widetilde{H}_{\widetilde{l}_u^{FN}}(\widetilde{E})]$, where

$$\widetilde{H}_{\widetilde{l}_s^{FN}}(\widetilde{E}) = \sum_{j=1}^{m} \left(\frac{\text{Card}\left(\widetilde{e}_{js}\right)}{\text{Card}\left(\widetilde{e}_j\right)} \cdot \text{hgt}\left(\widetilde{e}_j\right) \right). \tag{5}$$

Again, the correction for subnormality is applied by multiplying the ratio of cardinalities in the sum in (5) by $\text{hgt}\left(\widetilde{e}_j\right)$.

4.4 Histogram Distances

Since we aim to generalize the methodology proposed in [31] for crisp Likert scales to fuzzified versions of Likert scales, we will briefly summarize some vector distance measures for histograms that were considered in [31]. The family of probabilistic distance measures for histograms (distance measures for probability density functions) will not be considered here (we refer the interested readers e.g. to [1, 7]).

Vector distance measures for histograms are defined differently based on the underlying evaluation/measurement scales. We can distinguish between distances for nominal, ordinal and modulo histograms (i.e. histograms where the values of the underlying measurement scale form a circle—see e.g. [8, 19]). Since Likert scales usually do not provide a modulo-type measurement, modulo-type histograms are left out of the scope of this chapter.

Let us consider two sets of evaluations $E = \{e_1, \ldots, e_m\}$ and $F = \{f_1, \ldots, f_m\}$, $e_j, f_j \in L, j = 1, \ldots, m$, and their respective histograms $H(E)$ and $H(F)$. The distance of the individual measurements (see e.g. [8]) can be defined for the nominal, ordinal and modulo types of measurements respectively e.g. by (6), (7) and (8).

$$d_{nom}(e_j, f_j) = \begin{cases} 0 & \text{if } e_j = f_j, \\ 1 & \text{otherwise} \end{cases} \tag{6}$$

$$d_{ord}(e_j, f_j) = |e_j - f_j| \tag{7}$$

$$d_{mod}(e_j, f_j) = \begin{cases} |e_j - f_j| & \text{if } |e_j - f_j| \leq \frac{n}{2}, \\ n - |e_j - f_j| & \text{otherwise.} \end{cases} \tag{8}$$

An intuitive definition of a distance between two histograms is the minimum number of necessary changes of evaluations for the transformation of one histogram into the other (taking into account the nature of the evaluation scale and thus the "magnitude" of the change if it can be assessed). For nominal type histograms $H(E)$ and $H(F)$, their distance $D_{nom}(H(E), H(F))$ can be defined e.g. by (9) or as the number of non-matching answers (evaluations) across all bins. The fact that the underlying evaluation scale is nominal (not even ordinal) implies that distances between bins have no meaning and hence cannot be considered in the computation.

For ordinal type histograms $H(E)$ and $H(F)$, their distance $D_{ord}(H(E), H(F))$ can be defined e.g. by (10) as the minimum number of necessary unit changes of evaluation (under the assumption of the uniformity of the evaluation scale) to transform $H(E)$ into $H(F)$ (see e.g. [8]).

$$D_{nom}(H(E), H(F)) = \frac{\sum_{i=1}^{n} |H_i(E) - H_i(F)|}{2} \tag{9}$$

$$D_{ord}(H(E), H(F)) = \sum_{i=1}^{n} \left| \sum_{j=1}^{i} (H_j(E) - H_j(F)) \right| \tag{10}$$

5 Definition of the Problem

This chapter deals with the use of fuzzified Likert scales in the context of multiple-criteria multi-expert evaluation. In general we aim on problems, where a set of k decision-makers $\{DM_1, \dots, DM_k\}$ provide their evaluations of a certain object or phenomenon via m Likert-type items in a questionnaire. The set of evaluations of the m items by an evaluator r is denoted $E_r = \{e_1^r, \dots, e_m^r\}, r = 1, \dots, k$. That is the phenomenon/object is assessed by each individual using m criteria, points of view or questions. For the sake of simplicity, we will assume that the items have the same weight (descriptive power). We also assume all the decision makers have the same weight. As such we present a fuzzy extension of the analysis and methodology proposed in [31]. We will further distinguish between two settings:

- discrete Likert-scale setting—n-point Likert scales are considered, and the evaluations provided by the decision makers are allowed to be fuzzy sets on $L = \{l_1, \dots, l_n\}$, i.e. $\widetilde{e}_j^r \in \mathscr{F}(L), r = 1, \dots, k, j = 1, \dots, m$. Note, that the standard (crisp) use of Likert scales is a special case of this fuzzy approach, i.e. the crisp case equivalent of an answer to item j by the evaluator r in this fuzzy representation is $\widetilde{e}_j^r = \{\widetilde{e}_j^r(l_1)|_{l_1}, \dots, \widetilde{e}_j^r(l_n)|_{l_n}\}$, where exactly one of the membership degrees $\widetilde{e}_j^r(l_1), \dots, \widetilde{e}_j^r(l_n)$ is equal to 1 and the others are equal to 0.
- a continuous generalization of the Likert scale, where $L_C = [1, n]$ is considered to be the underlying (continuous) evaluation scale and $\widetilde{e}_j^r \in \mathscr{F}(L_C), r = 1, \dots, k,$ $j = 1, \dots, m$.

In both cases we propose a methodology for the aggregation of evaluations of an alternative with respect to multiple criteria obtained by multiple experts into one overall evaluation represented by a histogram. We define a reduced 3-bin histogram representation to address the possible response-bias and define an appropriate 3-bin ideal evaluation. We suggest to compute the distance from this ideal as a measure based on which the alternatives can be ordered and the best alternative chosen subsequently.

6 Proposed Solution—Fuzzified Discrete Likert Scales

We now consider an n-point discrete Likert scale with the set of possible values $L = \{l_1, \ldots, l_n\}$, k decision makers (evaluators) are providing evaluations of an alternative with respect to m criteria. The m evaluations suggested by the decision-maker r are fuzzy sets on L, i.e. $\widetilde{E}_r = \{\widetilde{e}_1^r, \ldots, \widetilde{e}_m^r\}$, where $\widetilde{e}_j^r \in \mathscr{F}(L), \widetilde{e}_j^r = \{\widetilde{e}_j^r{}^{(l_1)}|_{l_1}, \ldots, \widetilde{e}_j^r{}^{(l_n)}|_{l_n}\}$, $r = 1, \ldots, k$. We propose to aggregate the m evaluations provided by the decision-maker r into an overall evaluation $\widehat{\widetilde{E}}_r$ and the subsequent aggregation of these overall evaluations across all the experts onto the group evaluation $\widehat{\widetilde{E}}$, such that the ordinal character of the evaluation scales is respected and the response-bias described in the previous sections can be dealt with in the next step. We therefore suggest to represent the overall evaluations in the form of fuzzy histograms using (2):

$$\widehat{\widetilde{E}}_r = \widetilde{H}(\widetilde{E}_r); \ r = 1, \ldots, k, \tag{11}$$

and the group evaluation analogically also in the form of a histogram:

$$\widehat{\widetilde{E}} = \widetilde{H}(\widetilde{E}_1 \cup \widetilde{E}_2 \cup \cdots \cup \widetilde{E}_k). \tag{12}$$

The histogram representation provides a graphical level for the presentation of aggregated information which is a desired property in mathematical modelling. The histogram is defined in such a way that each normal fuzzy answer is considered to contain a unit of "support" which is distributed among the bins representing $l_1, \ldots, l_n$. A subnormal fuzzy-set answer is considered to contain (1 unit times its height of "support"). The histogram thus summarizes the evaluation in a relative format, but so far no assumption concerning the distance of the bins had to be made. Also notice, that for fuzzy-set representation of crisp answers we obtain exactly the same as was proposed in [31] for the crisp Likert scales. All the $\widehat{\widetilde{E}}_r$ and the $\widehat{\widetilde{E}}$ are now n-bin histograms (Fig. 5 presents an analogy of this process for the crisp case, in our case the height of the bins is defined by (2)). Without any loss of generality we can assume, that 1 is the best evaluation and n is the worst possible evaluation. Since we are going to use the distance-from-ideal based approach, we now need to specify, how an ideal group-evaluation (the best possible evaluation) looks like? As pointed out in [31], in a crisp situation this answer leads to the specification of an absolute-type evaluation ideal. In the fuzzy case this holds only if all the fuzzy answers provided by all the decision-makers are normal. Otherwise, if subnormal answers are also allowed, the definition has to reflect possible subnormality of the fuzzy-set answers, thus becoming a relative-type evaluation ideal.

If we now define the ideal group evaluation as $\widehat{\widetilde{I}}$, such that

$$\widehat{\widetilde{I}} = \widetilde{H}(\widetilde{I}_1 \cup \widetilde{I}_2 \cup \cdots \cup \widetilde{I}_k) = \widetilde{H}(\widetilde{I}), \tag{13}$$

where $\hat{\tilde{I}}_1, \ldots, \hat{\tilde{I}}_k$ are the ideal overall evaluations by each decision-maker represented by histograms, $\tilde{I}_r = \{\tilde{e}_1^r, \ldots, \tilde{e}_m^r\}, \tilde{e}_j^r = \{^{\mathrm{hgt}(\tilde{e}_j^r)}|_{l_1},^0|_{l_2}, \ldots,^0|_{l_n}\}$, $r = 1, \ldots, k$, $j = 1, \ldots, m$. Note, that in the crisp case these would be fuzzy singletons, in the fuzzy case where subnormal fuzzy answers are also allowed, the membership degree of l_1 to each respective fuzzy set $\tilde{e}_j^r$ is $\mathrm{hgt}(\tilde{e}_j^r)$ and the membership degrees of all $l_2, \ldots, l_n$ are assumed to be zero for the definition of such an ideal. The histogram representation of the ideal thus becomes $\hat{\tilde{I}} = [\tilde{H}_1(\tilde{I}), \ldots, \tilde{H}_n(\tilde{I})]$, where $\tilde{H}_1(\tilde{I}) = \sum_{j=1}^{m} \sum_{r=1}^{k} \mathrm{hgt}(\tilde{e}_j^r)$ and $\tilde{H}_i(\tilde{I}) = 0$ for all $i = 2, \ldots, n$. Mathematically speaking this would be a correct definition of the best possible evaluation under the given assumptions.

So far, we have ignored the *central-tendency bias*. Can we really say, that the ideal evaluation is defined correctly by $\hat{\tilde{I}}$? That is can we say that the ideal evaluation is such that all the decision makers provided only support for the best (most extreme) evaluation in all the items that represent the given goal in the questionnaire (or evaluation tool)? As pointed out in [31], when we need to aggregate the evaluations of different decision makers, we can never be sure that there is a significant difference e.g. between an evaluation of 1 provided by one decision maker and an evaluation of 2 provided by a different decision maker, with whom the central tendency bias (very rare selection of extreme evaluations) manifests itself. It may be that these two different evaluations in fact represent the same (or very close) level of satisfaction for both decision makers. We should therefore be careful in considering these values as very different.

The avoidance of extreme answers is to be expected, since Likert scales are frequently presented with labels such as "completely agree", "strongly approve", "strong negative" etc. for the extremes of the evaluation scale, which might be too categorical for the decision makers to use and hence the next, less extreme, label can be chosen instead (i.e. "agree", "approve" or "negative" respectively). In the context of a possible central-tendency bias, it therefore seems illogical to define the ideal as being composed solely of extreme answers—such an output might not be achievable. Note, that defining unachievable ideal evaluations can seriously bias the absolute type evaluation procedure.

The ideas of fuzzy optimization and the concept of sufficiently high evaluation [4, 32, 33] seem to be more feasible for this purpose. Instead of the ideal defined by $\hat{\tilde{I}}$ we can define the ideal in a more central-tendency-friendly way as "*Nonzero membership degrees of values close to the positive extreme point of the scale and zero memberships everywhere else provided by all the decision makers in all the items relevant for the evaluation of the given alternative*". Here "close to the positive extreme point" can be operationally defined as "l_t or lower", where $t < \frac{n}{2}$, and $t \in \{1, \ldots, n\}$ is the pre-specified threshold index (e.g. $t = 2$ for a 6-point or 7-point Likert scale). The introduction of the threshold t reflects the central-tendency bias and makes the evaluations more comparable.

We can thus define reduced 3-bin histograms to represent the overall evaluations of the decision makers by (14).

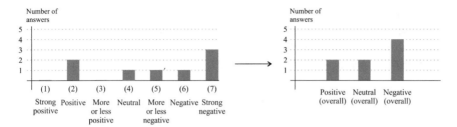

Fig. 6 The construction of a 3-bin reduced histogram (*right figure*) from an overall evaluation provided by one of the decision-makers (*left figure*). A 7-point Likert scale is used to provide evaluations by the decision-maker. A crisp Likert scale is presented here for better clarity, the fuzzified Likert scale would be a direct analogy

$$\hat{\tilde{E}}'_r = \tilde{H}'(\tilde{E}_r) = [\tilde{H}'_+(\tilde{E}_r), \tilde{H}'_0(\tilde{E}_r), \tilde{H}'_-(\tilde{E}_r)], \text{ where}$$

$$\tilde{H}'_+(\tilde{E}_r) = \tilde{H}_1(\tilde{E}_r) + \cdots + \tilde{H}_t(\tilde{E}_r),$$

$$\tilde{H}'_0(\tilde{E}_r) = \tilde{H}_{t+1}(\tilde{E}_r) + \cdots + \tilde{H}_{n-t}(\tilde{E}_r),$$

$$\tilde{H}'_-(\tilde{E}_r) = \tilde{H}_{n-(t-1)}(\tilde{E}_r) + \cdots + \tilde{H}_n(\tilde{E}_r),$$

$$r = 1, \ldots, k \quad (14)$$

Note, that the middle bin (0) of the histogram summarizes the neutral and close-to-neutral evaluations, the (+) bin summarizes all positive evaluations with respect to the given threshold t and defines the "less than t from the best"-bin, the (−) bin summarizes all negative evaluations with respect to the given threshold t and defines the "less than t from the worst"-bin (see Fig. 6 for an analogy in the crisp case). The resulting 3-bin histogram preserves the symmetry with respect to the middle value property.

This way we can consider the underlying 3-point scale to be cardinal in nature (an interval scale to be precise). The difference between a (+) and a (−) evaluation is twice as large as the difference between e.g. (+) and (0). We have thus achieved an aggregation of the information that compensates for the possible manifestation of the central-tendency effect and at the same time we have introduced a cardinal (interval) underlying scale for the reduced representation of the evaluations. At the level of a single decision maker, this takes into account the possible inter-item differences in the decision-maker's tendency to provide extreme evaluations. The correction for the central-tendency effect is however more necessary in the group aggregation (aggregation across all the decision makers). The group evaluation can be defined as the reduced 3-bin histogram (15):

$$\hat{\tilde{E}}' = \tilde{H}'(\tilde{E}_1 \cup \tilde{E}_2 \cup \cdots \cup \tilde{E}_k) = \left[\sum_{r=1}^{k} \tilde{H}'_+(\tilde{E}_r), \sum_{r=1}^{k} \tilde{H}'_0(\tilde{E}_r), \sum_{r=1}^{k} \tilde{H}'_-(\tilde{E}_r) \right]. \quad (15)$$

The ideal evaluation in this case represents such a set of evaluations provided by the k decision-makers to the m items of the questionnaire (evaluation tool) that are all no worse than t. That is the ideal, reflecting the possibility of providing subnormal fuzzy sets as answers, is in this case defined by (16).

$$\hat{\tilde{I}'} = \left[\sum_{j=1}^{m} \sum_{r=1}^{k} \text{hgt}(\tilde{e}_j^r), 0, 0 \right] \tag{16}$$

This can be considered to be a prototype of such evaluation, that represents the complete fulfillment of a given goal. Note, that if only normal fuzzy sets were allowed, then $\sum_{j=1}^{m} \sum_{r=1}^{k} \text{hgt}(\tilde{e}_j^r) = m \cdot k$ and hence $\hat{\tilde{I}'} = [m \cdot k, 0, 0]$, which is the same as in [31]. In case when only normal fuzzy sets are allowed as answers, ordering of alternatives can be easily obtained by means of computing the distance of the group evaluation 3-bin histogram representing the group evaluation of the alternative from the 3-bin ideal $D(\hat{\tilde{I}'}, \hat{\tilde{E}'})$. We just need to choose an appropriate distance measure for the histograms. In case we allow also subnormal answers, computing the distance from ideal solution defined by (16) we obtain a measure of dissensus on the decision that the alternative is ideal.

After closer inspection we can clearly see that the nominal-histogram distance (9) is not applicable, since it considers the difference between $(+)$ and $(-)$ to be the same as between $(+)$ and (0). This is obviously not a desired property (see Fig. 7). We require the distance measure to be able to reflect that the further the evaluations are on the ordinal scale, the higher the distance of the respective histograms have to be (a similar idea as in the earth mover's distance measure, e.g. [24]).

The original symmetry of the distribution of the elements of the Likert scale with respect to the middle of the scale implies the same property in our reduced histogram representation—bins are symmetrically distributed with respect to the middle bin. We can therefore now define a distance between the reduced ideal evaluation $\hat{\tilde{I}'}$ and

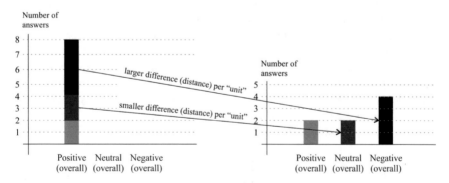

Fig. 7 An intuitive requirement on the distance between ordinal histograms—the difference between the positive $(+)$ and negative $(-)$ bins should be higher than between $(+)$ and the neutral bin (0)

the actual reduced group evaluation $\hat{\tilde{E}}'$ by (17). Let us for the purpose of the notation now consider $[+, 0, -] = [1, 2, 3]$ for the subscripts of $\tilde{H}'$.

$$D_{red}(\hat{\tilde{I}}', \hat{\tilde{E}}') = \sum_{i=1}^{2} i \cdot \left(\tilde{H}'_{i+1}(\tilde{E}) - \tilde{H}'_{i+1}(\tilde{I}) \right) = \tilde{H}'_0(\tilde{E}) + 2\tilde{H}'_-(\tilde{E}) \qquad (17)$$

The *degree of fulfillment of the overall goal* (*GF*) can be defined by (18), that is as a normalized similarity of the group evaluation to the ideal evaluation for those cases, when only normal fuzzy sets are allowed as answers.

$$GF(\hat{\tilde{E}}') = 1 - \frac{D_{red}(\hat{\tilde{I}}', \hat{\tilde{E}}')}{2 \cdot m \cdot k} \qquad (18)$$

We can easily verify, that $GF(\hat{\tilde{E}}') \in [0, 1]$ and $GF(\hat{\tilde{E}}') = 1$ if and only if $\hat{\tilde{I}}'$ and $\hat{\tilde{E}}'$ are identical. If $\hat{\tilde{E}}'$ is the complete opposite of $\hat{\tilde{I}}'$, i.e. $\hat{\tilde{E}}' = [0, 0, m \cdot k]$, then $GF(\hat{\tilde{E}}') = 0$. Note, that (17) is in fact a special case od the ordinal histogram distance (10). The proof can be found in [31]. The distance measure D_{red} is a special case of D_{ord}, since the symmetry of the scale $\{+, 0, -\}$ with respect to 0 implies that the bins of the histogram are equidistant, which is also the underlying assumption for D_{ord}. It is worth noting, that in our case the equidistance does not need to be assumed, it is implied by the symmetry of the Likert scale and by the construction of $\hat{\tilde{E}}'$ introduced by (15).

For those applications, where subnormal fuzzy sets can/must be accepted as answers, we can define the *degree of consensus on the given alternative being ideal* (*CD*) by (19).

$$CD(\hat{\tilde{E}}') = 1 - \frac{D_{red}(\hat{\tilde{I}}', \hat{\tilde{E}}')}{2 \cdot \sum_{j=1}^{m} \sum_{r=1}^{k} \text{hgt}(\tilde{e}_j^r)} \qquad (19)$$

7 Proposed Solution—Continuous Likert-Type Scales

Let us now broaden the scope a bit and consider the continuous evaluation universe $L_C = [1, n]$ instead of the classic n-point discrete Likert scale with the set of possible values $L = \{l_1, \dots, l_n\}$. We will again consider k decision makers (evaluators) that provide evaluations of an alternative with respect to m criteria. The m evaluations expressed by the decision-maker r are fuzzy sets on L_C, i.e. $\tilde{E}_r = \{\tilde{e}_1^r, \dots, \tilde{e}_m^r\}$, where $\tilde{e}_j^r \in \mathscr{F}(L_C)$, $r = 1, \dots, k$. See Fig. 4 for an example of such fuzzy evaluations. The m fuzzy evaluations provided by the decision maker can be summarized e.g. using the method proposed in [16]—i.e. defining a fuzzy set $\tilde{e}^r \in \mathscr{F}(L_C)$ representing the overall evaluation of the alternative by decision-maker r by aggregating the member-ship functions of $\tilde{e}_j^r$ for all $j = 1, \dots, m$. Alternatively this summary can be defined in a more histogram-similar manner just by defining the function $e^r(x) = \sum_{j=1}^{m} \widetilde{e}_j^r(x)$ for all $x \in L_C$. Such defined $e^r(x)$ is, however, no longer a fuzzy set on L_C.

Since we are interested mainly in the overall MCME evaluation of the alternative, that would take into account the possible response bias, we can proceed to directly defining the three-bin fuzzy histogram as a representation of the overall multiple-criteria multi-expert evaluation of the alternative analogically to (14). First we define a threshold $t \in [1, \frac{n+1}{2})$ and thus introduce three subintervals of the evaluation universe: $l_+^l = [1, t]$ representing the values considered to be *sufficiently positive evaluations*, $l_0^l = [t, n-t]$ representing the values of *more or less neutral evaluations* and $l_-^l = [n-t, n]$ representing the values of *sufficiently negative evaluations*. We define the fuzzy number representations of these intervals as $\widetilde{l}_+^l = (1, 1, t, t)$, $\widetilde{0}_+^l = (t, t, n-t, n-t)$ and $\widetilde{l}_-^l = (n-t, n-t, n, n)$ respectively. The three-bin histogram representing the final MCME evaluation can now be defined in accordance with Sect. 4.3, Eq. (3) as

$$\widehat{\widetilde{E}}' = \left[\widetilde{H}_{l_+^l}(\widetilde{E}), \widetilde{H}_{l_0^l}(\widetilde{E}), \widetilde{H}_{l_-^l}(\widetilde{E}) \right], \tag{20}$$

where $\widetilde{E} = \widetilde{E}_r \cup \cdots \cup \widetilde{E}_k$ and $H_{l_g^l}(\widetilde{E}) = \sum_{j=1}^{m} \sum_{r=1}^{k} \left(\frac{\text{Card}\left(\widetilde{e}_j^r \cap \widetilde{l}_g^l\right)}{\text{Card}\left(\widetilde{e}_j^r\right)} \cdot \text{hgt}\left(\widetilde{e}_j^r\right) \right)$ for all $g \in \{+, 0, -\}$. This way each fuzzy set contributes to each of the three bins proportionally to the relative cardinality of its intersection with the fuzzy set representation of the underlying interval. The strength of the contribution is modified by the height of the respective fuzzy set—normal fuzzy sets are considered to provide a unit information, subnormal less than a unit. The definition of the ideal evaluation $\widehat{\widetilde{I}}'$ formulated in (16) applies here as well, i.e. $\widehat{\widetilde{I}}' = \left[\sum_{j=1}^{m} \sum_{r=1}^{k} \text{hgt}(\widetilde{e}_j^r), 0, 0 \right]$. The distance from the ideal evaluation (17) can be used with analogical interpretation as in the previous section.

Alternatively, instead of a crisp partition of L_C by l_+^l, l_0^l and l_-^l, we can introduce a fuzzy partition of L_C. We can again choose a threshold value $t \in [1, \frac{n+1}{2})$ and a reasonable "overlap" $\Delta \in (0, \frac{n+1}{2} - t)$ and define three fuzzy numbers $\widetilde{l}_+^{FN} = (1, 1, t - \frac{\Delta}{2}, t + \frac{\Delta}{2})$, $\widetilde{l}_0^{FN} = (t - \frac{\Delta}{2}, t + \frac{\Delta}{2}, n - t - \frac{\Delta}{2}, n - t + \frac{\Delta}{2})$ and $\widetilde{l}_-^{FN} = (n - t - \frac{\Delta}{2}, n - t + \frac{\Delta}{2}, n, n)$ such that $\widetilde{l}_+^{FN}, \widetilde{l}_0^{FN}$ and $\widetilde{l}_-^{FN}$ form a fuzzy partition of L_C. Note, that the resulting fuzzy partition of L_C needs to be symmetrical with respect to the middle point of L_C. In line with (4) we now define the fuzzy sets $\widetilde{e}_{j+}^r$, $\widetilde{e}_{j0}^r$ and $\widetilde{e}_{j-}^r$ representing the distribution of the membership degree of each $x \in L_C$ to $\widetilde{e}_j^r$ among $\widetilde{l}_+^{FN}, \widetilde{l}_0^{FN}$ and $\widetilde{l}_-^{FN}$ based on the membership degrees of x to $\widetilde{l}_+^{FN}, \widetilde{l}_0^{FN}$ and $\widetilde{l}_-^{FN}$, for all $j = 1, \ldots, m$ and $r = 1, \ldots, k$, by (21).

$$\widetilde{e}_{jg}^r(x) = \widetilde{e}_j^r(x) \cdot \frac{\widetilde{l}_g^{FN}(x)}{\sum_{q \in \{+, 0, -\}} \widetilde{l}_q^{FN}(x)}, \text{ for all } x \in L_C, g \in \{+, 0, -\}. \tag{21}$$

The three-bin histogram representing the final MCME evaluation can now be defined in accordance with Sect. 4.3, equation (5) as

$$\hat{\widetilde{E}}' = \left[\widetilde{H}_{l^{FN}_+}(\widetilde{E}), \widetilde{H}_{l^{FN}_0}(\widetilde{E}), \widetilde{H}_{l^{FN}_-}(\widetilde{E}) \right], \tag{22}$$

where $\widetilde{E} = \widetilde{E}_r \cup \cdots \cup \widetilde{E}_k$ and $H_{l^{FN}_g}(\widetilde{E}) = \sum_{j=1}^m \sum_{r=1}^k \left(\dfrac{\mathrm{Card}\left(\widetilde{e}^r_{jg}\right)}{\mathrm{Card}\left(\widetilde{e}^r_j\right)} \cdot \mathrm{hgt}\left(\widetilde{e}^r_j\right) \right)$ for all $g \in \{+, 0, -\}$. Again this way each fuzzy evaluation contributes to the fuzzy bin g proportionally to the relative cardinality of its $\widetilde{e}^r_{jg}$ function multiplied by the height of the respective fuzzy set. Again normal fuzzy sets are considered to provide a unit information, subnormal less than a unit. The definition of the ideal evaluation $\hat{I}'$ formulated in (16) can be applied here again, i.e. $\hat{I}' = \left[\sum_{j=1}^m \sum_{r=1}^k \mathrm{hgt}(\widetilde{e}^r_j), 0, 0 \right]$. The distance from the ideal evaluation (17) can be used with analogical interpretation too.

8 Conclusion

This chapter extends the methodology for the use of discrete crisp Likert scales in multiple-criteria multi-expert evaluation presented in [31] to the fuzzy domain. Discrete fuzzy Likert scales are considered as well as the generalization of the Likert scales to continuous evaluation universes. The results obtained for the discrete fuzzy Likert scales are a direct generalization of the results obtained in [31]. The proposed methodology suggests a fuzzy histogram representation of fuzzy Likert-type evaluations and their aggregation. Three-bin histograms are proposed to dela with response bias such as central-tendency and leniency effects. In the discrete case the definition of the three bin histogram introduces equidistance of bins and enables the application of the distance-to-ideal based approach to select the best alternative. In the continuous case, the three-bin histograms are used to deal with the response bias mainly, since the underlying evaluation universe is cardinal. It is interesting to note, that under the proposed methodology, it is possible to introduce a fuzzy ideal evaluation (in the three-bin histogram evaluation) that allows for the evaluation of absolute type (interpretation in terms of the "fulfillment of the overall goal"), under the condition that normal fuzzy evaluations are provided. If subnormal fuzzy evaluation can be considered, the definition of an absolute ideal evaluation is more tricky. The suggested definition of ideal evaluation for this case results in evaluations with the interpretation of "consensus on the given alternative being good". The presented methodology extends the findings of [31] to the more general fuzzy domain, allowing for the expression of uncertainty or imprecision in the Likert-type evaluations. As such it provides tools for the aggregation of fuzzy Likert-type evaluations and proposes an extension of the standard use of Likert scales to multiple-criteria multi-expert fuzzy evaluation.

Acknowledgements This research was partially supported by the grant IGA_FF_2017_011 of the Internal Grant Agency of Palacký University, Olomouc, Czech Republic.

References

1. Aherne F, Thacker N, Rockett P (1998) The Bhattacharyya metric as an absolute similarity measure for frequency coded data. Kybernetika 32(4):363–368
2. Albaum G (1997) The Likert scale revisited: an alternate version. J Mark Res Soc 39(2):331–348
3. Arfi B (2010) Linguistic fuzzy logic methods in social sciences. Springer, Berlin, Heidelberg
4. Bellman RE, Zadeh LA (1970) Decision-making in a fuzzy environment. Manage Sci 17(4):141–164
5. Bharadwaj B (2007) Development of a fuzzy Likert scale for the WHO ICF to include categorical definitions on the basis of a continuum. PhD thesis, Wayne State University
6. Bodjanova S (2000) A generalized histogram. Fuzzy Sets Syst 116(2):155–166
7. Cha S (2007) Comprehensive survey on distance/similarity measures between probability density functions. Int J Math Models Methods Appl Sci 1(4):300–307
8. Cha S, Srihari SN (2002) On measuring the distance between histograms. Pattern Recogn 35(6):1355–1370
9. De La Rosa De Saa S, Van Aelst S (2013) Comparing the representativeness of the 1-norm median for Likert and free-response fuzzy scales. In: Borgelt C, Gil M, Sousa J, Verleysen M (eds) Towards advanced data analysis, vol 285. Springer, Berlin, Heidelberg, pp 87–98
10. Dubois D, Prade H (eds) (2000) Fundamentals of fuzzy sets. Kluwer Academic Publishers, Massachusetts
11. Duda RO, Hart PE, Stork DG (2001) Pattern classification, 2nd edn. Wiley-Interscience, New York
12. Furnham A (1986) Response bias, social desirability dissimulation. Pers Individ Differ 7(3):385–400
13. Furnham A, Henderson M (1982) The good, the bad and the mad: response bias in self-report measures. Pers Individ Differ 3(3):311–320
14. Ishizaka A, Nguyen NH (2013) Calibrated fuzzy AHP for current bank account selection. Expert Syst Appl 40(9):3775–3783
15. Klir GJ, Yuan B (1995) Fuzzy sets and fuzzy logic: theory and aplications. Prentice Hall, New Jersey
16. Luukka P, Collan M, Tam F, Lawryshyn Y (2018) Estimating one-off operational risk events with the lossless fuzzy weighted average method. In: Collan M, Kacprzyk J (eds) Soft computing applications for group decision-making and consensus modelling. Springer International Publishing AG
17. Li Q (2013) A novel Likert scale based on fuzzy sets theory. Expert Syst Appl 40(5):1609–1618
18. Likert R (1932) A technique for the measurement of attitudes. Arch Psychol 22(140):1–55. 2731047
19. Luukka P, Collan M (2015) Modulo similarity in comparing histograms. In: Proceedings of the 16th world congress of the international fuzzy systems association (IFSA), pp 393–397
20. Matell MS, Jacoby J (1971) Is there an optimal number of alternatives for Likert scale items? Study I: reliability and validity. Educ Psychol Meas 31(3):657–674
21. Norman G (2010) Likert scales, levels of measurement and the laws of statistics. Adv Health Sci Educ 15(5):625–632
22. Ragin CC (2008) Redesigning social inquiry: fuzzy sets and beyond. University of Chicago Press, Chicago
23. Ramík J, Vlach M (2013) Measuring consistency and inconsistency of pair comparison systems. Kybernetika 49(3):465–486

24. Rubner Y, Tomasi C, Guibas LJ (1998) A metric for distributions with applications to image databases. In: Proceedings of the 1998 IEEE international conference on computer vision, Bombay, India, pp 104–111
25. Ruspini EH (1969) A new approach to clustering. Inform Control 15(1):22–32
26. Schneider CQ, Wageman C (2012) Set-theoretic methods for the social sciences: a guide to qualitative comparative analysis. Cambridge University Press, Cambridge
27. Seitl M, Charvát M, Lečbych M (2016) Psychometrické charakteristiky české verze škály Experiences in Close Relationships (ECR). Československá Psychologie 60(4):351–371
28. Stevens SS (1946) On the theory of scales of measurement. Science 103(2684):677–680
29. Stoklasa J (2014) Linguistic models for decision support. Lappeenranta University of Technology, Lappeenranta
30. Stoklasa J, Talášek T (2016) On the use of linguistic labels in AHP: calibration, consistency and related issues. In: Proceedings of the 34th international conference on mathematical methods in economics, pp 785–790. Technical University of Liberec, Liberec
31. Stoklasa J, Talášek T, Kubátová J, Seitlová K (2017) Likert scales in group multiple-criteria evaluation. J Multiple-Valued Logic Soft Comput (in press)
32. Stoklasa J, Talášek T, Musilová J (2014) Fuzzy approach—a new chapter in the methodology of psychology? Human Affairs 24(2):189–203
33. Stoklasa J, Talašová J, Holeček P (2011) Academic staff performance evaluation—variants of models. Acta Polytechnica Hungarica 8(3):91–111
34. Strelkov VV (2008) A new similarity measure for histogram comparison and its application in time series analysis. Pattern Recogn Lett 29(13):1768–1774
35. Trillas E, Bonissone PP, Magdalena L, Kacprzyk J (eds) (2012) Combining experimentation and theory: a homage to Abe Mamdani. Springer, Berlin, Heidelberg
36. Velleman PF, Wilkinson L (1993) Nominal, ordinal, interval, and ratio typologies are misleading. Am Stat 47(1):65–72
37. Viertl R, Trutschnig W (2006) Fuzzy histograms and fuzzy probability distributions. Technical report, Vienna University of Technology

Maximal Entropy and Minimal Variability OWA Operator Weights: A Short Survey of Recent Developments

Christer Carlsson and Robert Fullér

Abstract The determination of ordered weighted averaging (OWA) operator weights is a very important issue of applying the OWA operator for decision making. One of the first approaches, suggested by O'Hagan, determines a special class of OWA operators having maximal entropy of the OWA weights for a given level of orness; algorithmically it is based on the solution of a constrained optimization problem. In 2001, using the method of Lagrange multipliers, Fullér and Majlender solved this constrained optimization problem analytically and determined the optimal weighting vector. In 2003 Fullér and Majlender computed the exact minimal variability weighting vector for any level of orness using the Karush-Kuhn-Tucker second-order sufficiency conditions for optimality. The problem of maximizing an OWA aggregation of a group of variables that are interrelated and constrained by a collection of linear inequalities was first considered by Yager in 1996, where he showed how this problem can be modeled as a mixed integer linear programming problem. In 2003 Carlsson, Fullér and Majlender derived an algorithm for solving the constrained OWA aggregation problem under a simple linear constraint: the sum of the variables is less than or equal to one. In this paper we give a short survey of numerous later works which extend and develop these models.

1 OWA Operators

The process of information aggregation appears in many applications related to the development of intelligent systems. In 1988 Yager introduced a new aggregation technique based on the ordered weighted averaging operators [44]. The determination of ordered weighted averaging operator weights is a very important issue

C. Carlsson
IAMSR, Åbo Akademi University, Turku, Finland
e-mail: christer.carlsson@abo.fi

R. Fullér (✉)
Széchenyi István University, Győr, Hungary
e-mail: rfuller@sze.hu

© Springer International Publishing AG 2018
M. Collan and J. Kacprzyk (eds.), *Soft Computing Applications for Group Decision-making and Consensus Modeling*, Studies in Fuzziness and Soft Computing 357, DOI 10.1007/978-3-319-60207-3_12

of applying the OWA operator for decision making. One of the first approaches, suggested by O'Hagan [34], determines a special class of OWA operators having maximal entropy of the OWA weights for a given level of *orness*; algorithmically it is based on the solution of a constrained optimization problem. In 2001, using the method of Lagrange multipliers, Fullér and Majlender [12] solved this constrained optimization problem analytically and determined the optimal weighting vector. In 2003 using the Karush-Kuhn-Tucker second-order sufficiency conditions for optimality, Fullér and Majlender [13] computed the exact minimal variability weighting vector for any level of orness. In 2003 Carlsson, Fullér and Majlender [7] derived an algorithm for solving the (nonlinear) constrained OWA aggregation problem. In this work we shall give a short survey of some later works that extend and develop these models.

In a decision process the idea of *trade-offs* corresponds to viewing the global evaluation of an action as lying between the *worst* and the *best* local ratings. This occurs in the presence of conflicting goals, when a compensation between the corresponding compatibilities is allowed. Averaging operators realize trade-offs between objectives, by allowing a positive compensation between ratings. The concept of *ordered weighted averaging* operators was introduced by Yager in 1988 [44] as a way for providing aggregations which lie between the maximum and minimums operators. The structure of this operator involves a nonlinearity in the form of an ordering operation on the elements to be aggregated. The OWA operator provides a new information aggregation technique and has already aroused considerable research interest [49].

Definition 1.1 ([44]) An OWA operator of dimension n is a mapping $F: \mathbb{R}^n \to \mathbb{R}$, that has an associated weighting vector $W = (w_1, w_2, \ldots, w_n)^T$ such as $w_i \in [0, 1]$, $1 \leq i \leq n$, and $w_1 + \cdots + w_n = 1$. Furthermore,

$$F(a_1, \ldots, a_n) = w_1 b_1 + \cdots + w_n b_n = \sum_{j=1}^{n} w_j b_j,$$

where b_j is the jth largest element of the bag $\langle a_1, \ldots, a_n \rangle$.

A fundamental aspect of this operator is the re-ordering step, in particular an aggregate a_i is not associated with a particular weight w_i but rather a weight is associated with a particular ordered position of aggregate. It is noted that different OWA operators are distinguished by their weighting function. In order to classify OWA operators in regard to their location between *and* and *or*, a measure of *orness*, associated with any vector W is introduced by Yager [44] as follows,

$$\text{orness}(W) = \frac{1}{n-1} \sum_{i=1}^{n} (n - i) w_i.$$

It is easy to see that for any W the orness(W) is always in the unit interval Furthermore, note that the nearer W is to an *or*, the closer its measure is to one; while the nearer it is to an *and*, the closer is to zero. It can easily be shown that

$\mathrm{orness}(W^*) = 1, \mathrm{orness}(W_*) = 0$ and $\mathrm{orness}(W_A) = 0.5$. A measure of *andness* is defined as, $\mathrm{andness}(W) = 1 - \mathrm{orness}(W)$. Generally, an OWA operator with much of nonzero weights near the top will be an *orlike* operator, that is, $\mathrm{orness}(W) \geq 0.5$, and when much of the weights are nonzero near the bottom, the OWA operator will be *andlike*, that is, $\mathrm{andness}(W) \geq 0.5$. In [44] Yager defined the measure of dispersion (or entropy) of an OWA vector by,

$$\mathrm{disp}(W) = - \sum_{i=1}^{n} w_i \ln w_i.$$

We can see when using the OWA operator as an averaging operator $\mathrm{disp}(W)$ measures the degree to which we use all the aggregates equally.

2 Obtaining OWA Operator Weights

One important issue in the theory of OWA operators is the determination of the associated weights. One of the first approaches, suggested by O'Hagan, determines a special class of OWA operators having maximal entropy of the OWA weights for a given level of *orness*; algorithmically it is based on the solution of a constrained optimization problem. Another consideration that may be of interest to a decision maker involves the variability associated with a weighting vector. In particular, a decision maker may desire low variability associated with a chosen weighting vector. It is clear that the actual type of aggregation performed by an OWA operator depends upon the form of the weighting vector [45]. A number of approaches have been suggested for obtaining the associated weights, i.e., quantifier guided aggregation [44, 45], exponential smoothing and learning [50]. O'Hagan [34] determined a special class of OWA operators having maximal entropy of the OWA weights for a given level of *orness*. His approach is based on the solution of he following mathematical programming problem,

$$\text{maximize} \qquad \mathrm{disp}(W) = - \sum_{i=1}^{n} w_i \ln w_i$$

$$\text{subject to } \mathrm{orness}(W) = \sum_{i=1}^{n} \frac{n-i}{n-1} \cdot w_i = \alpha, \; 0 \leq \alpha \leq 1 \qquad (1)$$

$$w_1 + \cdots + w_n = 1, \; 0 \leq w_i, \; i = 1, \ldots, n.$$

In 2001, using the method of Lagrange multipliers, Fullér and Majlender [12] transformed constrained optimization problem (1) into a polynomial equation which is then was solved to determine the maximal entropy OWA operator weights. By their method, the associated weighting vector is easily obtained by

$$\ln w_j = \frac{j-1}{n-1} \ln w_n + \frac{n-j}{n-1} \ln w_1 \implies w_j = \sqrt[n-1]{w_1^{n-j} w_n^{j-1}}$$

and

$$w_n = \frac{((n-1)\alpha - n)w_1 + 1}{(n-1)\alpha + 1 - nw_1}$$

then

$$w_1[(n-1)\alpha + 1 - nw_1]^n = ((n-1)\alpha)^{n-1}[((n-1)\alpha - n)w_1 + 1]$$

where $n \geq 3$. For $n = 2$ then from orness$(w_1, w_2) = \alpha$ the optimal weights are uniquely defined as $w_1^* = \alpha$ and $w_2^* = 1 - \alpha$. Furthermore, if $\alpha = 0$ or $\alpha = 1$ then the associated weighting vectors are uniquely defined as $(0, 0, \ldots, 0, 1)^T$ and $(1, 0, \ldots, 0, 0)^T$, respectively.

An interesting question is to determine the minimal variability weighting vector under given level of orness [48]. The variance of a given weighting vector is computed as follows

$$D^2(W) = \sum_{i=1}^n \frac{1}{n}(w_i - E(W))^2 = \frac{1}{n}\sum_{i=1}^n w_i^2 - \left(\frac{1}{n}\sum_{i=1}^n w_i\right)^2 = \frac{1}{n}\sum_{i=1}^n w_i^2 - \frac{1}{n^2}.$$

where $E(W) = (w_1 + \cdots + w_n)/n = 1/n$ stands for the arithmetic mean of weights.

In 2003 Fullér and Majlender [13] suggested a minimum variance method to obtain the minimal variability OWA operator weights. A set of OWA operator weights with minimal variability could then be generated. Their approach requires the solution of the following mathematical programming problem:

$$\text{minimize} \qquad D^2(W) = \frac{1}{n} \cdot \sum_{i=1}^n w_i^2 - \frac{1}{n^2}$$

$$\text{subject to orness}(w) = \sum_{i=1}^n \frac{n-i}{n-1} \cdot w_i = \alpha, \ 0 \leq \alpha \leq 1, \qquad (2)$$

$$w_1 + \cdots + w_n = 1, \ 0 \leq w_i, \ i = 1, \ldots, n.$$

Fullér and Majlender [13] computed the exact minimal variability weighting vector for any level of orness using the Karush-Kuhn-Tucker second-order sufficiency conditions for optimality.

Yager [47] considered the problem of maximizing an OWA aggregation of a group of variables that are interrelated and constrained by a collection of linear inequalities and he showed how this problem can be modeled as a mixed integer linear programming problem. The constrained OWA aggregation problem [47] can be expressed as the following mathematical programming problem

$$\max w^T y$$
$$\text{subject to } Ax \leq b, x \geq 0,$$

where $w^T y = w_1 y_1 + \cdots + w_n y_n$ and y_j denotes the jth largest element of the bag $\langle x_1, \ldots, x_n \rangle$.

In 2003 Carlsson, Fullér and Majlender [7] showed an algorithm for solving the (nonlinear) constrained OWA aggregation problem

$$\max w^T y; \text{ subject to} \{x_1 + \cdots + x_n \leq 1, x \geq 0\}. \tag{3}$$

where y_j denotes the jth largest element of the bag $\langle x_1, \ldots, x_n \rangle$.

3 Recent Advances

In this section we will give a short chronological survey of some later works that extend and develop the maximal entropy, the minimal variability and the constrained OWA operator weights models. We will mention only those works in which the authors extended, improved or used the findings of our original papers [7, 12, 13].

In 2004 Liu and Chen [21] introduced the concept of parametric geometric OWA operator (PGOWA) and a parametric maximum entropy OWA operator (PMEOWA) and showed the equivalence of parametric geometric OWA operator and parametric maximum entropy OWA operator weights. Carlsson et al. [8] showed how to evaluate the quality of elderly care services by OWA operators.

In 2005 Wang and Parkan [39] presented a minimax disparity approach, which minimizes the maximum disparity between two adjacent weights under a given level of orness. Their approach was formulated as

$$\text{minimize } \max_{i=1,2,\ldots,n-1} |w_i - w_{i+1}|$$
$$\text{subject to orness}(w) = \sum_{i=1}^{n} \frac{n-i}{n-1} w_i = \alpha, \ 0 \leq \alpha \leq 1,$$
$$w_1 + \cdots + w_n = 1, \ 0 \leq w_i \leq 1, \ i = 1, \ldots, n.$$

Majlender [32] developed a *maximal Rényi entropy* method for generating a parametric class of OWA operators and the maximal Rényi entropy OWA weights. His approach was formulated as

$$\text{maximize } H_\beta(w) = \frac{1}{1-\beta} \log_2 \sum_{i=1}^{n} w_i^\beta$$

$$\text{subject to orness}(w) = \sum_{i=1}^{n} \frac{n-i}{n-1} w_i = \alpha, \ 0 \leq \alpha \leq 1,$$

$$w_1 + \cdots + w_n = 1, \ 0 \leq w_i \leq 1, \ i = 1, \ldots, n.$$

where $\beta \in \mathbb{R}$ and $H_1(w) = -\sum_{i=1}^{n} w_i \log_2 w_i$. Liu [22] extended the the properties of OWA operator to the RIM (regular increasing monotone) quantifier which is represented with a monotone function instead of the OWA weighting vector. He also introduced a class of parameterized equidifferent RIM quantifier which has minimum variance generating function. This equidifferent RIM quantifier is consistent with its orness level for any aggregated elements, which can be used to represent the decision maker's preference. Troiano and Yager [37] pointed out that OWA weighting vector and the fuzzy quantifiers are strongly related. An intuitive way for shaping a monotonic quantifier, is by means of the threshold that makes a separation between the regions of what is satisfactory and what is not. Therefore, the characteristics of a threshold can be directly related to the OWA weighting vector and to its metrics: the attitudinal character and the entropy. Usually these two metrics are supposed to be independent, although some limitations in their value come when they are considered jointly. They argued that these two metrics are strongly related by the definition of quantifier threshold, and they showed how they can be used jointly to verify and validate a quantifier and its threshold.

In 2006 Xu [43] investigated the dependent OWA operators, and developed a new argument-dependent approach to determining the OWA weights, which can relieve the influence of unfair arguments on the aggregated results. Zadrozny and Kacprzyk [54] discussed the use of the Yager's OWA operators within a flexible querying interface. Their key issue is the adaptation of an OWA operator to the specifics of a user's query. They considered some well-known approaches to the manipulation of the weights vector and proposed a new one that is simple and efficient. They discussed the tuning (selection of weights) of the OWA operators, and proposed an algorithm that is effective and efficient in the context of their FQUERY for Access package. Wang et al. [40] developed the query system of practical hemodialysis database for a regional hospital in Taiwan, which can help the doctors to make more accurate decision in hemodialysis. They built the fuzzy membership function of hemodialysis indices based on experts' interviews. They proposed a fuzzy OWA query method, and let the decision makers (doctors) just need to change the weights of attributes dynamical, then the proposed method can revise the weight of each attributes based on aggregation situation and the system will provide synthetic suggestions to the decision makers. Chang et al. [10] proposed a dynamic fuzzy OWA model to deal with problems of group multiple criteria decision making. Their proposed model can help users to solve MCDM problems under the situation of fuzzy or incomplete information. Amin and Emrouznejad [6] introduced an extended minimax disparity model to determine the OWA operator weights as follows,

$$\text{minimize } \delta$$

$$\text{subject to orness}(w) = \sum_{i=1}^{n} \frac{n-i}{n-1} w_i = \alpha, \; 0 \le \alpha \le 1,$$

$$w_j - w_i + \delta \ge 0, \; i = 1, \ldots, n-1, j = i+1, \ldots, n$$

$$w_i - w_j + \delta \ge 0, \; i = 1, \ldots, n-1, j = i+1, \ldots, n$$

$$w_1 + \cdots + w_n = 1, \; 0 \le w_i \le 1, \; i = 1, \ldots, n.$$

In this model it is assumed that the deviation $|w_i - w_j|$ is always equal to δ, $i \neq j$.

In 2007 Liu [23] proved that the solutions of the minimum variance OWA operator problem under given orness level and the minimax disparity problem for OWA operator are equivalent, both of them have the same form of maximum spread equidifferent OWA operator. He also introduced the concept of maximum spread equidifferent OWA operator and proved its equivalence to the minimum variance OWA operator. Llamazares [30] proposed determining OWA operator weights regarding the class of majority rule that one should want to obtain when individuals do not grade their preferences between the alternatives. Wang et al. [41] introduced two models determining as equally important OWA operator weights as possible for a given orness degree. Their models can be written as

$$\text{minimize } J_1 = \sum_{i=1}^{n-1} (w_i - w_{i+1})^2$$

$$\text{subject to orness}(w) = \sum_{i=1}^{n} \frac{n-i}{n-1} w_i = \alpha, \; 0 \le \alpha \le 1,$$

$$w_1 + \cdots + w_n = 1, \; 0 \le w_i \le 1, \; i = 1, \ldots, n.$$

and

$$\text{minimize } J_2 = \sum_{i=1}^{n-1} \left(\frac{w_i}{w_{i+1}} - \frac{w_{i+1}}{w_i} \right)^2$$

$$\text{subject to orness}(w) = \sum_{i=1}^{n} \frac{n-i}{n-1} w_i = \alpha, \; 0 \le \alpha \le 1,$$

$$w_1 + \cdots + w_n = 1, \; 0 \le w_i \le 1, \; i = 1, \ldots, n.$$

Yager [51] used stress functions to obtain OWA operator weights. With this stress function, a user can "stress" which argument values they want to give more weight in the aggregation. An important feature of this stress function is that it is only required to be nonnegative function on the unit interval. This allows a user to completely focus on the issue of where to put the stress in the aggregation without having to consider satisfaction of any other requirements.

In 2008 Liu [24] proposed a *general optimization model with strictly convex objective function* to obtain the OWA operator under given orness level,

$$\text{minimize } \sum_{i=1}^{n} F(w_i)$$

$$\text{subject to orness}(w) = \sum_{i=1}^{n} \frac{n-i}{n-1} w_i = \alpha, \ 0 \le \alpha \le 1,$$

$$w_1 + \cdots + w_n = 1, \ 0 \le w_i \le 1, \ i = 1, \ldots, n.$$

and where F is a strictly convex function on $[0, 1]$, and it is at least two order differentiable. His approach includes the *maximum entropy* (for $F(x) = x \ln x$) and the *minimum variance* (for $F(x) = x^2$ problems as special cases. More generally, when $F(x) = x^{\alpha}, \alpha > 0$ it becomes the OWA problem of Rényi entropy [32], which includes the maximum entropy and the minimum variance OWA problem as special cases. Liu also included into this general model the solution methods and the properties of maximum entropy and minimum variance problems that were studied separately earlier. The consistent property that the aggregation value for any aggregated set monotonically increases with the given orness value is still kept, which gives more alternatives to represent the preference information in the aggregation of decision making. Then, with the conclusion that the RIM quantifier can be seen as the continuous case of OWA operator with infinite dimension, Liu [25] further suggested a general RIM quantifier determination model, and analytically solved it with the optimal control technique. Ahn [2] developed some new quantifier functions for aiding the quantifier-guided aggregation. They are related to the weighting functions that show properties such that the weights are strictly ranked and that a *value of orness is constant independently of the number of criteria considered*. These new quantifiers show the same properties that the weighting functions do and they can be used for the quantifier-guided aggregation of a multiple-criteria input. The proposed RIM and regular decreasing monotone (RDM) quantifiers produce the same orness as the weighting functions from which each quantifier function originates. the quantifier orness rapidly converges into the value of orness of the weighting functions having a constant value of orness. This result indicates that a quantifier-guided OWA aggregation will result in a similar aggregate in case the number of criteria is not too small.

In 2009 Wu et al. [42] used a linear programming model for determining ordered weighted averaging operator weights with maximal Yager's entropy [46]. By analyzing the desirable properties with this measure of entropy, they proposed a novel approach to determine the weights of the OWA operator. Ahn [3] showed that a closed form of weights, obtained by the least-squared OWA (LSOWA) method, is equivalent to the minimax disparity approach solution when a condition ensuring all positive weights is added into the formulation of minimax disparity approach. Liu [26] presented some methods of OWA determination with different dimension instantiations, that is to get an OWA operator series that can be used to the different dimensional application cases of the same type. He also showed some OWA determination methods that can make the elements distributed in monotonic, symmetric or any function shape cases with different dimensions. Using Yager's stress function method [51] he managed to extend an OWA operator to another dimensional case with the same aggregation properties.

In 2010 Ahn [4] presented a general method for obtaining OWA operator weights via an extreme point approach. The extreme points are identified by the intersection of an attitudinal character constraint and a fundamental ordered weight simplex that is defined as

$$K = \{w \in \mathbb{R}^n \mid w_1 + w_2 + \cdots + w_n = 1, w_j \geq 0, j = 1, \ldots, n\}.$$

The parameterized OWA operator weights, which are located in a convex hull of the identified extreme points, can then be specifically determined by selecting an appropriate parameter. Vergara and Xia [38] proposed a new method to find the weights of an OWA for uncertain information sources. Given a set of uncertainty data, the proposed method finds the combination of weights that reduces aggregated uncertainty for a predetermined orness level. Their approach assures best information quality and precision by reducing uncertainty. Yager [52] introduced a measure of diversity related to the problem of selecting of selecting n objects from a pool of candidates lying in q categories.

In 2011 Liu [27] summarizing the main OWA determination methods (the optimization criteria methods, the sample learning methods, the function based methods, the argument dependent methods and the preference methods) showed some relationships between the methods in the same kind and the relationships between different kinds. Gong [15] generated minimal disparity OWA operator weights by minimizing the combination disparity between any two adjacent weights and its expectation. Ahn [5] showed that the weights generated by the maximum entropy method show equally compatible performance with the rank order centroid weights under certain conditions. Hong [17] proved a relationship between the minimum-variance and minimax disparity RIM quantifier problems.

In 2012 Zhou et al. [55] introduced the concept of generalized ordered weighted logarithmic proportional averaging (GOWLPA) operator and proposed the generalized logarithm chi-square method to obtain GOWLPA operator weights. Zhou et al. [56] presented new aggregation operator called the generalized ordered weighted exponential proportional averaging (GOWEPA) operator and introduced the least exponential squares method to determine GOWEPA operator weights based on its orness measure. Yari and Chaji [53] used maximum Bayesian entropy method for determining ordered weighted averaging operator weights. Liu [28] provided analytical solutions of the maximum entropy and minimum variance problems with given linear medianness values.

In 2013 Cheng et al. [11] proposed a new time series model, which employs the ordered weighted averaging operator to fuse high-order data into the aggregated values of single attributes, a fusion adaptive network-based fuzzy inference system procedure, for forecasting stock price in *Taiwanese stock markets*. Luukka and Kurama

[31] showed how to apply OWA operators to similarity classifier. This newly derived classifier is examined with four different medical data set. Data sets used in this experiment were taken from a UCI-Repository of Machine Learning Database. Liu et al. [29] introduced a new aggregation operator: the induced ordered weighted averaging standardized distance (IOWASD) operator. The IOWASD is an aggregation operator that includes a parameterized family of standardized distance aggregation operators in its formulation that ranges from the minimum to the maximum standardized distance. By using the IOWA operator in the VIKOR method, it is possible to deal with complex attitudinal characters (or complex degrees of orness) of decision maker and provide a more complete picture of the decision making process.

In 2014 Sang and Liu [36] showed an analytic approach to obtain the least square deviation OWA operator weights. Kim and Singh [19] outlined an entropy-based hydrologic alteration assessment of biologically relevant flow regimes using gauged flow data. The maximum entropy ordered weighted averaging method is used to aggregate non-commensurable biologically relevant flow regimes to fit an eco-index such that the harnessed level of the ecosystem is reflected. Kishor et al. [20] introduced orness measures in an axiomatic framework and to propose an alternate definition of orness that is based on these axioms. The proposed orness measure satisfies a more generalized set of axioms than Yager's orness measure.

In 2015 Zhou et al. [57] introduced the generalized least squares method to determine the generalized ordered weighted logarithmic harmonic averaging (GOWLHA) operator weights based on its orness measure. Gao et al. [14] proposed a new operator named as the generalized ordered weighted utility averaging-hyperbolic absolute risk aversion (GOWUA-HARA) operator and constructed a new optimization model to determine its optimal weights. Aggarwal [1] presented a method to learn the criteria weights in multi-criteria decision making by applying emerging learning-to-rank machine learning techniques.

In 2016 Kaur et al. [18] applied minimal variability OWA operator weights to reduce computational complexity of high dimensional data and ANFIS with the fuzzy c-means clustering is used to produce understandable rules for investors. They verified their model through an empirical analysis of the stock data sets, collected from Bombay stock market to forecast the *Bombay Stock Exchange Index*. Gong et al. [16] presented two new disparity models to obtain the associated weights, which is determined by considering the absolute deviation and relative deviation of any distinct pairs of weights. Mohammed [33] demonstrated the application of a Laplace-distribution-based ordered weighted averaging operator to the problem of breast tumor classification.

In 2017 Reimann et al. [35] performed a large-scale empirical study and test whether preferences exhibited by subjects can be represented better by the OWA operator or by a more standard multi-attribute decision model. Chaji [9] presented an analytic approach to obtain maximal Bayesian entropy OWA weights. His approach is based on the solution of he following mathematical programming problem,

$$\text{maximize}\quad W = -\sum_{i=1}^{n} w_i \ln \frac{w_i}{\beta_i / \min\{\beta_1, \ldots, \beta_n\}} = -\sum_{i=1}^{n} w_i \ln \frac{w_i}{\beta_i} - \ln \min\{\beta_1, \ldots, \beta_n\}$$

$$\text{subject to}\qquad \text{orness}(W) = \sum_{i=1}^{n} \frac{n-i}{n-1} \cdot w_i = \alpha,\ 0 \le \alpha \le 1$$

$$w_1 + \cdots + w_n = 1,\ 0 \le w_i,\ i = 1, \ldots, n.$$

where $\beta_1, \ldots, \beta_n$ are given prior OWA weights, such that $\beta_1 + \cdots + \beta_n = 1$, $\beta_i > 0$, $i = 1, \ldots, n$.

References

1. Aggarwal M (2015) On learning of weights through preferences. Inf Sci 321:90–102
2. Ahn BS (2008) Some quantier functions from weighting functions with constant value of orness. IEEE Trans Syst Man Cybern Part B 38:540–546
3. Ahn BS (2009) Some remarks on the LSOWA approach for obtaining OWA operator weights. Int J Intell Syst 24(12):1265–1279
4. Ahn BS (2010) Parameterized OWA operator weights: an extreme point approach. Int J Approx Reason 51:820–831
5. Ahn BS (2011) Compatible weighting method with rank order centroid: maximum entropy ordered weighted averaging approach. Eur J Oper Res 212:552–559
6. Amin GR, Emrouznejad A (2006) An extended minimax disparity to determine the OWA operator weights. Comput Ind Eng 50:312–316
7. Carlsson C, Fullér R, Majlender P (2003) A note on constrained OWA aggregations. Fuzzy Sets Syst 139:543–546
8. Carlsson C, Fedrizzi M, Fullér R (2004) Fuzzy logic in management. International series in operations research and management science, vol 66. Kluwer Academic Publishers, Boston
9. Chaji A (2017) Analytic approach on maximum Bayesian entropy ordered weighted averaging operators. Comput Ind Eng 105:260–264
10. Chang JR, Ho TH, Cheng CH, Chen AP (2006) Dynamic fuzzy OWA model for group multiple criteria decision. Soft Comput 10:543–554
11. Cheng CH, Wei LY, Liu JW, Chen TL (2013) OWA-based ANFIS model for TAIEX forecasting. Econ Model 30:442–448
12. Fullér R, Majlender P (2001) An analytic approach for obtaining maximal entropy OWA operator weights. Fuzzy Sets Syst 124:53–57
13. Fullér R, Majlender P (2003) On obtaining minimal variability OWA operator weights. Fuzzy Sets Syst 136:203–215
14. Gao J, Li M, Liu H (2015) Generalized ordered weighted utility averaging-hyperbolic absolute risk aversion operators and their applications to group decision-making. Eur J Oper Res 243:258–270
15. Gong Y (2011) A combination approach for obtaining the minimize disparity OWA operator weights. Fuzzy Optim and Decis Mak 10:311–321
16. Gong Y, Dai L, Hu N (2016) An extended minimax absolute and relative disparity approach to obtain the OWA operator weights. J Intell Fuzzy Syst 31:1921–1927
17. Hong DH (2011) On proving the extended minimax disparity OWA problem. Fuzzy Sets Syst 168:35–46
18. Kaur Gurbinder, Dhar Joydip, Guha RK (2016) Minimal variability OWA operator combining ANFIS and fuzzy c-means for forecasting BSE index. Math Comput Simul 122:69–80

19. Kim Z, Singh VP (2014) Assessment of environmental flow requirements by entropy-based multi-criteria decision. Water Resour Manage 28:459–474
20. Kishor A, Singh A, Pal N (2014) Orness measure of OWA operators: a new approach. IEEE Trans Fuzzy Syst 22:1039–1045
21. Liu X, Chen L (2004) On the properties of parametric geometric OWA operator. Int J Approx Reason 35:163–178
22. Liu X (2005) On the properties of equidifferent RIM quantifier with generating function. Int J Gener Syst 34:579–594
23. Liu X (2007) The solution equivalence of minimax disparity and minimum variance problems for OWA operators. Int J Approx Reason 45:68–81
24. Liu X (2008) A general model of parameterized OWA aggregation with given orness level. Int J Approx Reason 48:598–627
25. Liu X, Han S (2008) Orness and parameterized RIM quantier aggregation with OWA operators: a summary. Int J Approx Reason 48:77–97
26. Liu X (2009) On the methods of OWA operator determination with different dimensional instantiations. In: Proceedings of the 6th international conference on fuzzy systems and knowledge discovery, FSKD 2009, 14–16 Aug 2009, Tianjin, China, vol 7, pp 200–204. ISBN 978-076953735-1, Article number 5359982
27. Liu X (2011) A review of the OWA determination methods: classification and some extensions. In: Yager RR, Kacprzyk J, Beliakov G (eds) Recent developments in the ordered weighted averaging operators: theory and practice. Studies in fuzziness and soft computing, vol 265. Springer, pp 49–90. ISBN 978-3-642-17909-9
28. Liu X (2012) Models to determine parameterized ordered weighted averaging operators using optimization criteria. Inf Sci 190:27–55
29. Liu HC, Mao LX, Zhang ZY, Li P (2013) Induced aggregation operators in the VIKOR method and its application in material selection. Appl Math Model 37:6325–6338
30. Llamazares B (2007) Choosing OWA operator weights in the field of social choice. Inf Sci 177:4745–4756
31. Luukka P, Kurama O (2013) Similarity classifier with ordered weighted averaging operators. Expert Syst Appl 40:995–1002
32. Majlender P (2005) OWA operators with maximal Renyi entropy. Fuzzy Sets Syst 155:340–360
33. Mohammed EA, Naugler CT, Far BH (2016) Breast tumor classification using a new OWA operator. Expert Syst Appl 61:302–313
34. O'Hagan M (1988) Aggregating template or rule antecedents in real-time expert systems with fuzzy set logic. In: Proceedings of 22nd annual IEEE Asilomar conference signals, systems, computers, Pacific Grove, CA, pp 681-689
35. Reimann O, Schumacher C, Vetschera R (2017) How well does the OWA operator represent real preferences? Eur J Oper Res 258:993–1003
36. Sang X, Liu X (2014) An analytic approach to obtain the least square deviation OWA operator weights. Fuzzy Sets Syst 240:103–116
37. Troiano L, Yager RR (2005) A measure of dispersion for OWA operators. In: Liu Y, Chen G, Ying M (eds) Proceedings of the eleventh international fuzzy systems association world congress, 28–31 July 2005, Beijing, China. Tsinghua University Press and Springer, pp 82–87
38. Vergara VM, Xia S (2010) Minimization of uncertainty for ordered weighted average. Int J Intell Syst 25:581–595
39. Wang Y-M, Parkan C (2005) A minimax disparity approach for obtaining OWA operator weights. Inf Sci 75:20–29
40. Wang JW, Chang JR, Cheng CH (2006) Flexible fuzzy OWA querying method for hemodialysis database. Soft Comput 10:1031–1042
41. Wang YM, Luo Y, Liu XW (2007) Two new models for determining OWA operator weights. Comput Ind Eng 52:203–209
42. Wu J, Sun B-L, Liang C-Y, Yang S-L (2016) A linear programming model for determining ordered weighted averaging operator weights with maximal Yager's entropy. Comput Ind Eng 57(3):742–747

43. Xu ZS (2006) Dependent OWA operators. Lect Notes Comput Sci 3885:172–178
44. Yager RR (1988) Ordered weighted averaging aggregation operators in multi-criteria decision making. IEEE Trans Syst Man Cybern 18:183-190
45. Yager RR (1993) Families of OWA operators. Fuzzy Sets Syst 59:125–148
46. Yager RR (1995) Measures of entropy and fuzziness related to aggregation operators. Inf Sci 82:147–166
47. Yager RR (1996) Constrained OWA aggregation. Fuzzy Sets Syst 81:89–101
48. Yager RR (1995) On the inclusion of variance in decision making under uncertainty. Int J Uncertain Fuzziness Knowl-Based Syst 4:401–419
49. Yager RR, Kacprzyk J (1997) The ordered weighted averaging operators: theory and applications. Kluwer, Norwell
50. Yager RR, Filev D (1999) Induced ordered weighted averaging operators. IEEE Trans Syst Man Cybern—Part B: Cybern 29:141–150
51. Yager RR (2007) Using stress functions to obtain OWA operators. IEEE Trans Fuzzy Syst 15:1122–1129
52. Yager RR (2010) Including a diversity criterion in decision making. Int J Intell Syst 25:958–969
53. Yari G, Chaji AR (2012) Maximum Bayesian entropy method for determining ordered weighted averaging operator weights. Comput Ind Eng 63:338–342
54. Zadrozny S, Kacprzyk J (2006) On tuning OWA operators in a flexible querying interface. Lect Notes Comput Sci 4027:97–108
55. Zhou L, Chen H, Liu J (2012) Generalized logarithmic proportional averaging operators and their applications to group decision making. Knowl-Based Syst 36:268–279
56. Zhou L, Chen H, Liu J (2012) Generalized weighted exponential proportional aggregation operators and their applications to group decision making. Appl Math Model 36:4365–4384
57. Zhou L, Tao Z, Chen H, Liu J (2015) Generalized ordered weighted logarithmic harmonic averaging operators and their applications to group decision making. Soft Comput 19:715–730

A Closer Look at the Relation Between Orness and Entropy of OWA Function

József Mezei and Matteo Brunelli

Abstract Ordered weighted averaging (OWA) functions have been extensively used to model problem of choice and consensus in the presence of multiple experts and decision makers. Since each OWA is associated to a weight vector many scholars have focused on the problem of the determination of this weight vector. In this study, we consider orness and entropy, two characterizing measures of priority vectors, and we study their interplay from a graphical point of view.

1 Introduction

In a world where the amount of data is ever increasing, it becomes more and more important to find meaningful ways to aggregate numerical information into a single representative number. This is the field of investigation related to aggregation functions [2]. The most important domains in which aggregation is extensively applied include multicriteria decision analysis, group decision making or information retrieval.

There is a wide variety of aggregation functions available to be used in various applications and it is a difficult but extremely important task to choose one that captures the requirements that one wants to achieve with the aggregation process.

J. Mezei (✉)
School of Business and Management, Lappeenranta University of Technology,
Lappeenranta, Finland
e-mail: jozsef.mezei@lut.fi

M. Brunelli
Department of Industrial Engineering, University of Trento, Trento, Italy
e-mail: matteo.brunelli@unitn.it

M. Brunelli
Department of Mathematics and Systems Analysis, Aalto University, Espoo, Finland

© Springer International Publishing AG 2018
M. Collan and J. Kacprzyk (eds.), *Soft Computing Applications for Group Decision-making and Consensus Modeling*, Studies in Fuzziness and Soft Computing 357, DOI 10.1007/978-3-319-60207-3_13

If we consider multicriteria decision problems, one can for example require an aggregation function to evaluate alternatives based on: (i) their best performance among all the criteria using the *maximum* operator; (ii) their worst performance among all the criteria *minimum* operator; or (iii) their average performance using the *arithmetic mean* operator. In this article we will consider a widely used parametrized class of aggregation functions that includes the above examples as special cases, namely the Ordered Weighting Averaging (*OWA*) function introduced by Yager [16].

OWA functions have been used in the setting of group decisions to extend the concept of consensus from a crisp to a fuzzy perspective, where consensus becomes a matter of degree [6]. A similar model was extended to work in cases where preferences are expressed in a linguistic form [3]. More recently, *OWA* functions were used to aggregate the agreement between pairs of agents and estimate the level of agreement in larger groups [4].

While maximum, minimum and average offer the right choice in many decision problems, typically one would like to use an aggregation function resulting in a value between the minimum and maximum (i.e., an averaging function) that satisfies a set of predefined criteria. In case of the *OWA* function, the most widely used characterizing measure, *orness*, captures its similarity to the maximum. For a given orness value, typically, there is not a unique corresponding *OWA* function, except for the two extreme cases, i.e. maximum and minimum are the only *OWA* functions with orness 1 and 0, respectively. Another widely used characterizing measure is *entropy* or dispersion. It measures the uniformity of the weights taking its unique maximum value in case of the arithmetic mean.

In this chapter, we will look at the relationship of orness and entropy and provide a visual representation to understand this relationship. Since a typical approach in the literature for choosing an *OWA* function is to find the one that maximizes entropy for a given orness value, an important, but so far in the literature not considered, question would be to what extent the value of orness determines the value that entropy can take. As we will see, the interval containing entropy values that can be achieved by an *OWA* with a given orness value can vary significantly. Additionally, we will also look at this orness-entropy relationship for specific families of *OWA* functions corresponding to various quantifier guided aggregation procedures.

The rest of the chapter is structured as follows. In Sect. 2, we recall the most important definitions regarding *OWA* functions, their characteristic measures and discuss the related contributions from the literature. Section 3 presents the numerical experiments focusing on the visual representation of various *OWA* characteristic measures. The results are discussed and conclusions are offered in Sect. 4.

2 Preliminaries and Literature Review

In this section the definitions and notation used in later parts of the paper are introduced. We will also summarize contributions from the literature considering different problems related to orness and entropy of *OWA* operators.

2.1 OWA Operators

One of the most general definitions of an aggregation function requires two basic properties to be satisfied.

Definition 1 (*Aggregation function* [9]) A function $f : \mathbb{I}^n \to \mathbb{I}$, where $\mathbb{I}$ is a nonempty interval of the extended real number system, is an *aggregation function* of n arguments if it satisfies the following properties:

1. boundary conditions:

$$\inf_{x \in \mathbb{I}^n} f(x) = \inf \mathbb{I}, \qquad \sup_{x \in \mathbb{I}^n} f(x) = \sup \mathbb{I};$$

2. monotonicity: f is non-decreasing in each variable.

In most of the applications and a large part of the literature, a specific case of aggregation functions, namely averaging functions, are used. An aggregation function, f, is an *averaging function* if it satisfies the following property:

$$\min(x_1, \ldots, x_n) \leq f(x_1, \ldots, x_n) \leq \max(x_1, \ldots, x_n) \ \forall (x_1, \ldots, x_n) \in \mathbb{I}^n.$$

An especially relevant and studied family of aggregation functions is represented by the Ordered Weighted Averaging (*OWA*) functions, originally introduced by Yager [16]. Before the definition, we specify the general concept of a weight vector.

Definition 2 (*Weight vector*) A real valued vector **w** is a weight vector if and only if $w_i \geq 0 \ \forall i$ and $\sum_{i=1}^{n} w_i = 1$. We call $\mathscr{W}_n$ the set of all the weight vectors of length n, i.e.

$$\mathscr{W}_n = \left\{ (w_1, \ldots, w_n) \,\middle|\, w_i \geq 0 \ \forall i, \ \sum_{i=1}^{n} w_i = 1 \right\}$$

Using an associated weight vector, an *OWA* function can be defined as follows.

Definition 3 (*OWA function* [16]) An Ordered Weighted Averaging (*OWA*) function is a mapping $OWA_{\mathbf{w}} : \mathbb{R}^n \to \mathbb{R}$ with an associated weight vector **w**, such that

$$OWA_{\mathbf{w}}(x_1, \ldots, x_n) = \sum_{i=1}^{n} w_i y_i$$

where y_i is the ith largest element of the multiset $\{x_1, \ldots, x_n\}$.

It is clear that the attitudinal character of an *OWA* function depends on its associated weight vector $\mathbf{w} = (w_1, \ldots, w_n)$. Some indices have been proposed in the literature to give a numerical estimation of various characteristics of a weight vector and the associated *OWA* operator. The most widely used one, proposed in [16], quantifies the degree to which the *OWA* function is similar to the maximum operator.

$$O_y = orness(\mathbf{w}) = \frac{1}{n-1} \sum_{i=1}^{n} w_i(n-i) \tag{1}$$

The value of *orness*(**w**) is equal to 1 if and only if the *OWA* function represents the maximum. Similarly, it is equal to 0, if and only if the *OWA* collapses to the minimum. Values between 0 and 1 represent *OWA* functions which are trade-offs between maximum and minimum. While the above definition is the most widely adopted, there are alternative definitions of orness in the literature that can be characterized by different sets of reasonable properties required for a well-defined orness measure. A novel definition was proposed by Kishor et al. [10] as follows:

$$O_p = \frac{1}{2n'} \sum_{i=1}^{n} \left(2n' + 1 - i - \left\lfloor \frac{2i-1}{n} \right\rfloor - \left\lfloor \frac{2n'-1}{n} \right\rfloor \left\lfloor \frac{i-1}{n'} \right\rfloor \right) w_i, \tag{2}$$

where $n' = \left\lfloor \frac{n+1}{2} \right\rfloor$.

The second main characterizing measure of an *OWA* function is the entropy (or dispersion) [16], measuring the degree to which the information in the arguments is taken into consideration during the aggregation process.

$$E_y = entropy(\mathbf{w}) = - \sum_{i=1}^{n} w_i \log w_i \tag{3}$$

The value of *entropy*(**w**) characterizes the dispersion of the weights in the same way entropy is used to estimate the level of uncertainty in a probability distribution. A different measure of dispersion, introduced by Yager [17], can be defined as follows:

$$E_p = 1 - \max_i w_i. \tag{4}$$

For both entropy measures, the maximum (minimum) values are attained when all the weights are equal (one of the weights is equal to 1). Figure 1 depicts the points in three dimensions corresponding to weights associated to *OWA* functions with given O_y and E_y values.

A crucial problem when using *OWA* functions is to determine the weights to be used in the aggregation [15]. A widely used procedure is to utilize a linguistic quantifier. While there are different types of quantifiers, the most widely used ones belong to the family of Regular Increasing Monotone (RIM) quantifiers.

Definition 4 (*RIM quantifier* [18]) A RIM linguistic quantifier Q is a fuzzy subset of the real line that satisfies the following properties:

- $Q(0) = 0$ and $Q(1) = 1$;
- $Q(x) \geq Q(y)$ if $x > y$.

Using RIM quantifiers, the weight vector associated to the *OWA* operator can be defined in the following way:

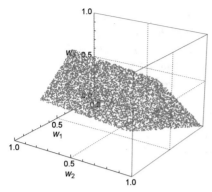
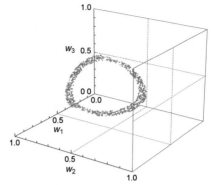
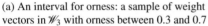

(a) An interval for orness: a sample of weight vectors in $\mathscr{W}_3$ with orness between 0.3 and 0.7

(b) An interval for entropy: a sample of weight vectors in $\mathscr{W}_3$ with entropy between 0.95 and 1

Fig. 1 Graphical interpretations of orness and entropy

$$w_i = Q\left(\frac{i}{n}\right) - Q\left(\frac{i-1}{n}\right).$$ (5)

The two most widely used parametric families of RIM quantifiers are the following:

- $Q_1(r) = r^\alpha$ with $\alpha \geq 0$;
- $Q_2(r) = 1 - (1-r)^\alpha$ with $\alpha \geq 0$.

The associated weight vectors can be specified as follows:

- $w_{1,i} = \left(\frac{i}{n}\right)^\alpha - \left(\frac{i-1}{n}\right)^\alpha$;
- $w_{2,i} = \left(\frac{n+1-i}{n}\right)^\alpha - \left(\frac{n-i}{n}\right)^\alpha$.

Additionally to RIM quantifiers, there exist other classes, namely Regular Decreasing Monotone (RDM) and Regular UniModal (RUM) quantifiers. In the subsequent analysis, we will utilize the following functions:

- $Q_3(r) = e^{-\alpha r}$;
- $Q_4(r) = \dfrac{2}{1 + e^{-\lambda}\left(\frac{r}{1-r}\right)} - 1.$

with the corresponding weight vectors.

2.2 Related Literature

A typical starting point for generating the weights of an *OWA* operator to be used in the aggregation process is to specify the required attitudinal character in terms of the orness value. As we noted before, except for the two extreme cases of orness 0 and 1, there is no unique weight vector corresponding to a given value in the unit interval. For this reason, additional properties need to be specified in order to arrive to a unique weight vector. A widely used approach first proposed in [13] is to find the weight vector that maximizes entropy for a given orness level, hereafter called β. This can be obtained by solving the following optimization problem:

$$\begin{aligned} \underset{(w_1,\ldots,w_n)}{\text{maximize}} \quad & E_y(\mathbf{w}) \\ \text{subject to} \quad & O_y(\mathbf{w}) = \beta \\ & w_1 + \cdots + w_n = 1 \\ & w_1, \ldots, w_n \geq 0. \end{aligned} \tag{6}$$

Fullér and Majlender [7] found an analytical solution to the above optimization problem. Yager [17] considered the same problem with the dispersion measure E_p. We note here that there are alternative measures for capturing the dispersion of the weight vector other than the ones used in this article. For example, Majlender [12] formulated a similar optimization problem with the objective of maximizing Rényi entropy for a given orness value and derived an analytical solution to the problem.

Additionally to various dispersion measures, other characteristics of the *OWA* function have also been used to determine the weights as the solution to an optimization problem similar to the one presented above. Again Fullér and Majlender [8] derived an analytical solution to the optimization problem that, for a given orness level, minimizes the variability of the weight vector:

$$D^2(w) = \frac{1}{n} \sum_{i=1}^{n} w_i^2 - \frac{1}{n^2}.$$

Motivated by the previous studies, Wang and Parkan [14] identified the weight vector minimizing the maximal disparity for a given orness level, where *disparity* is defined as

$$\max_{i \in \{1,2,\ldots,n-1\}} |w_i - w_{i+1}|. \tag{7}$$

The disparity model was further investigated and extended by Amin and Emrouznejad [1, 5]. Furthermore, Liu [11] proved that the optimal solution for the minimal variability and disparity problem is the maximum spread equidifferent *OWA*.

3 Numerical Experiments

In this section we present a visual representation of the relationship between various characteristic measures of *OWA* functions. Particularly, we will look at the efficient frontier of the various optimization problems that have been presented in the literature to determine weights of *OWA* functions. We will also look at how different weights resulting from linguistic quantifiers compare to those optimal frontiers.

In the experiments, we generated 1,00,000 random weight vectors of order n for various n values, calculated the corresponding characteristic measures, and created scatterplots for specific pairs of measures.

3.1 Orness and Entropy

First we look at the relationship between orness O_y and entropy E_y. Figure 2 depicts the points that represent orness and entropy of the 10,000 randomly generated weight vectors. As we mentioned before, the optimal frontier from the perspective of the discussed optimization problem can be analytically identified using the solution from [7].

Especially in the subfigures with $n = 3, 4$, it is evident that for each n there are n low points where the entropy is zero. For instance, with $n = 4$, these minimums are represented by the vectors $(0, 0, 0, 1)$, $(0, 0, 1, 0)$, $(0, 1, 0, 0)$ and $(1, 0, 0, 0)$. Conversely, the maximum is reached, in each case, by the vector $(1/n, \ldots, 1/n)$. This

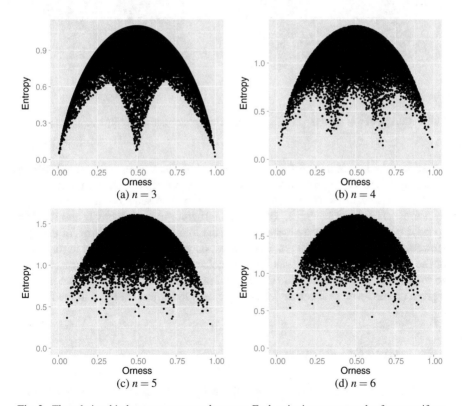

Fig. 2 The relationship between orness and entropy. Each point is vector samples from a uniform distribution on $\mathscr{W}_n$

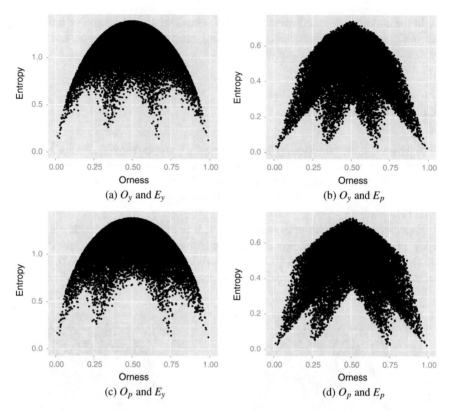

(a) O_y and E_y

(b) O_y and E_p

(c) O_p and E_y

(d) O_p and E_p

Fig. 3 Comparison of the relationship between various orness and dispersity measures

behaviour can be confirmed for larger n values, in the figures a larger number of randomly generated weights would be necessary to visualize these minimum points.

Next, we look at the relationship between the two presented orness and entropy measures as depicted in Fig. 3. The overall structure of the graphs in the solution space is quite similar, as we always observe the four low points in terms of entropy. However, we can observe several differences, particularly with respect to the shape of the efficient frontier. Additionally, the choice of the characteristic measures has an important effect on how wide the interval of values that the entropy can take for a given orness level, as it can be for example seen for orness level 0.5 in the figures.

3.2 Linguistic Quantifiers

As presented in the previous sections, the weights of the OWA functions can be found by means of linguistic quantifiers. Now, by noting that these quantifiers are parametric we focus on the subset of vectors in $\mathscr{W}_n$ which can be obtained through

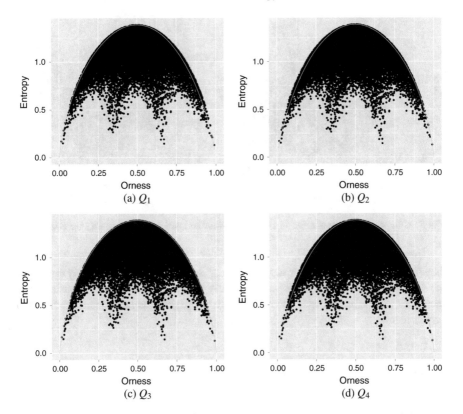

Fig. 4 The bright curves represent the sets of weights obtained with Q_1, Q_2, Q_3 and Q_4, respectively

the quantifiers. In particular we want to discover how the entropy of these vectors changes in relation to its orness. We shall consider the four quantifiers introduced in the previous section and look at the orness and the entropy which can be obtained using different values of $\alpha \in]0, \infty[$. The results are reported in Fig. 4 and show that in all cases the obtained vectors are remarkably close to the maximum entropy vectors, at least for $\mathscr{W}_4$. While the weights generated by this procedures are clearly not the optimal ones derived in [7], the presented figures show that relying on linguistic quantifiers will result in close to optimal weights in terms of entropy.

3.3 Variability and Disparity

Variance and disparity have been proposed in the literature as indices alternative to the entropy. The definition of disparity was recalled in (7). In contrast to the previous problems, in these cases one is interested in the minimal variability and minimal

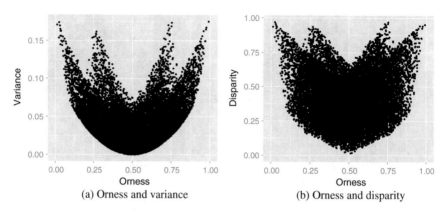

(a) Orness and variance   (b) Orness and disparity

Fig. 5 Comparison of the relationship between orness and variance and disparity measures

dispersion *OWA* functions for a given orness level. Figure 5 depicts the results of the simulation. The main difference, similarly to the behaviours of various entropy measures, can be identified in the wider range of possible disparity values for a given orness level.

4 Discussion and Conclusion

Studying various families of aggregation functions is an important problem from both the theoretical and the practical standpoints. In this chapter, we have studied *OWA* functions, one of the most widely used families of aggregation functions in the literature and in real life applications. A typical question related the use of *OWA* operators is the selection of the aggregation weights. A common strategy is to select weights based on a predefined value of one or several characterizing measures, such as orness and entropy. In this chapter, instead of focusing on the approach of determining weights that maximize or minimize one of the characterizing measures why keeping another fixed, we offer a visual representation of the problem.

We looked at the relationship of different orness and entropy measures, and additionally to visualizing the efficient frontier of the typically considered optimization problems, we depicted the complete objective space. Additionally to the traditional entropy and orness definitions, we looked at more recent proposals, and the visual analysis illustrated some differences between the behaviour of the different characteristics. As an interesting finding of the analysis, we found that by using linguistic quantifiers, the generated weights have entropy value close to the optimal value for a given orness level.

The presented research is exploratory by nature, as we only present and discuss visual representation of the relationship between various characteristic measures of *OWA* functions. The natural future research direction would be to derive analytical formulas for the efficient frontiers in case of different entropy and orness measures, and also to offer a quantitative estimation on how far the solutions based on linguistic quantifiers are from the frontier.

Acknowledgements The research of Matteo Brunelli was funded by the Academy of Finland (decision no. 277135).

References

1. Amin GR, Emrouznejad A (2006) An extended minimax disparity to determine the OWA operator weights. Comput Ind Eng 50(3):312–316
2. Beliakov G, Pradera A, Calvo T (2007) Aggregation functions: a guide for practicioners. In: Studies in fuzziness and soft computing, vol 221. Springer
3. Bordogna G, Fedrizzi M, Pasi G (1997) A linguistic modeling of consensus in group decision making based on OWA operators. IEEE Trans Syst Man Cybern Part A Syst Hum 27(1):126–133
4. Brunelli M, Fedrizzi M, Fedrizzi M (2014) Fuzzy m-ary adjacency relations in social network analysis: optimization and consensus evaluation. Inf Fusion 17:36–45
5. Emrouznejad A, Amin GR (2010) Improving minimax disparity model to determine the OWA operator weights. Inf Sci 180(8):1477–1485
6. Fedrizzi M, Kacprzyk J, Nurmi H (1993) Consensus degrees under fuzzy majorities and fuzzy preferences using OWA (ordered weighted average) operators. Control Cybern 22(4):71–80
7. Fullér R, Majlender P (2001) An analytic approach for obtaining maximal entropy OWA operator weights. Fuzzy Sets Syst 124(1):53–57
8. Fullér R, Majlender P (2003) On obtaining minimal variability OWA operator weights. Fuzzy Sets Syst 136(2):203–215
9. Grabisch M, Marichal JL, Mesiar R, Pap E (2009) Aggregation functions. Cambridge University Press
10. Kishor A, Singh AK, Pal NR (2014) Orness measure of OWA operators: a new approach. IEEE Trans Fuzzy Syst 22(4):1039–1045
11. Liu X (2007) The solution equivalence of minimax disparity and minimum variance problems for OWA operators. Int J Approx Reason 45(1):68–81
12. Majlender P (2005) OWA operators with maximal Rényi entropy. Fuzzy Sets Syst 155(3):340–360
13. O'Hagan M (1988) Aggregating template or rule antecedents in real-time expert systems with fuzzy set logic. In: Twenty-second Asilomar conference on signals, systems and computers
14. Wang YM, Parkan C (2005) A minimax disparity approach for obtaining OWA operator weights. Inf Sci 175(1):20–29
15. Xu Z (2005) An overview of methods for determining OWA weights. Int J Intell Syst 20(8):843–865
16. Yager RR (1988) Ordered weighted averaging operators in multicriteria decision making. IEEE Trans Syst Man Cybern 18(1):183–190
17. Yager RR (1993) Families of OWA operators. Fuzzy Sets Syst 59(2):125–148
18. Yager RR (1996) Quantifier guided aggregation using OWA operators. Int J Intell Syst 11(1):49–73

Rank Reversal in the AHP with Consistent Judgements: A Numerical Study in Single and Group Decision Making

Michele Fedrizzi, Silvio Giove and Nicolas Predella

Abstract In this paper we study, by means of numerical simulations, the influence of some relevant factors on the Rank Reversal phenomenon in the Analytic Hierarchy Process, AHP. We consider both the case of a single decision maker and the case of group decision making. The idea is to focus on a condition which preserves Rank Reversal, RR in the following, and progressively relax it. First, we study how the estimated probability of RR depends on the distribution of the criteria weights and, more precisely, on the entropy of this distribution. In fact, it is known that RR does'nt occur if all the weights are concentrated in a single criterion, i.e. the zero entropy case. We derive an interesting increasing behavior of the RR estimated probability as a function of weights entropy. Second, we focus on the aggregation method of the local weight vectors. Barzilay and Golany proved that the weighted geometric mean preserves from RR. By using the usual weighted arithmetic mean suggested in AHP, on the contrary, RR may occur. Therefore, we use the more general aggregation rule based on the weighted power mean, where the weighted geometric mean and the weighted arithmetic mean are particular cases obtained for the values $p \to 0$ and $p = 1$ of the power parameter p respectively. By studying the RR probability as a function of parameter p, we again obtain a monotonic behavior. Finally, we repeat our study in the case of a group decision making problem and we observe that the estimated probability of RR decreases by aggregating the DMs' preferences. This fact suggests an inverse relationship between consensus and rank reversal. Note that we assume that all judgements are totally consistent, so that the effect of inconsistency is avoided.

Keywords Rank reversal · Analytic hierarchy process · Consistency · Group decision making

M. Fedrizzi (✉)
Department of Industrial Engineering, University of Trento, Via Sommarive 9, 38123 Trento, Italy
e-mail: michele.fedrizzi@unitn.it

S. Giove
Department of Economics, University of Venezia, Cannareggio 873, 30121 Venice, Italy

N. Predella
University of Trento, Via Inama, 5, 38122 Trento, Italy

© Springer International Publishing AG 2018
M. Collan and J. Kacprzyk (eds.), *Soft Computing Applications for Group Decision-making and Consensus Modeling*, Studies in Fuzziness and Soft Computing 357, DOI 10.1007/978-3-319-60207-3_14

1   Introduction

Despite its popularity, Saaty's Analytic Hierarchy Process, AHP in the following, has been criticized by many authors since its introduction in 1977 [16]. Some researchers pointed out single drawbacks or weaknesses, whereas other researchers criticized and rejected the very foundations of the method. Among these criticisms, we mention only few popular ones. The interested reader may refer to [7] and [19] for a more extended debate. Barzilai [4], as an example, considers the eigenvector method and the normalization procedure in AHP as mathematical errors. Well-known criticisms came also from Bana e Costa and Vansnick [3] and concern the meaning of the priority vector derived from the principal eigenvalue method and the Saaty's consistency ratio. Nevertheless, the best known and most cited drawback of AHP is certainly the rank reversal (RR) phenomenon. Rank reversal is the change of ranking of alternatives as a consequence, for example, of the addition or deletion of an alternative. This clearly contradicts the principle of the independence of irrelevant alternatives. Rank reversal in AHP was firstly evidenced in 1983 by Belton and Gear [6]. After that, numerous paper were published on this subject [5, 10, 15, 20]. Saaty regularly answered to the criticisms on AHP and, in particular, on RR, arguing on the validity of his method and on the legitimacy of RR [17]. A survey on this topic can be found in [13]. Numerous authors argued on the main causes/factors influencing RR, mainly focusing on the role of vector normalization, aggregation rule and inconsistency. The aim of this paper is not to enter the debate in favor or against AHP and/or RR. Our scope is, instead, to contribute to the understanding on how some relevant factors influence the probability of the RR phenomenon. The paper is organized as follows. In Sect. 2 we set the necessary notation and definitions in order to specify the framework of pairwise comparison and AHP. In the same section we also define the main issues on RR. In Sect. 3 we describe the plan and the results of our numerical study, both in the case of a single decision maker and in the case of group decision making. Finally, we discuss and comment our results.

2   Preliminaries

2.1   Pairwise Comparisons and AHP

We assume that the reader is familiar with AHP [12, 16, 18], so that we only briefly recall the main steps of the method. Let us consider a set of n alternatives $X = \{x_1, \ldots, x_n\}$. We first recall the definition of *pairwise comparison matrix*, PCM in the following, which is a positive and reciprocal square matrix $\mathbf{A}$ of order n obtained by pairwise comparing the n alternatives. More precisely, it is $\mathbf{A} = (a_{ij})_{n \times n}$, with $a_{ij} > 0$, $a_{ij}a_{ji} = 1$, $\forall i, j$, where a_{ij} is an estimation of the degree of preference of x_i over x_j. Given a PCM, the most relevant task is to derive a weight vector, that is a vector $\mathbf{w} = (w_1, \ldots, w_n)$, where w_j is a numerical value quantifying the priority of alternative x_j as estimated on the basis of the pairwise comparisons. Vector **w** is defined up to a positive factor. Normalization is often applied to **w**, so that the components sum up to one.

In order to derive the vector **w**, Saaty [16] proposes the eigenvector method. Namely, **w** is the solution of the following equation

$$\mathbf{A}\mathbf{w} = \lambda_{max}\mathbf{w}. \tag{1}$$

Note that the maximum eigenvalue of **A**, here denoted by λ_{max}, refers to the Perron–Frobenius theorem. Another popular method to derive $\mathbf{w} = (w_1, \ldots, w_n)$ is the geometric mean method [2, 9], where

$$w_i = \left(\prod_{j=1}^{n} a_{ij} \right)^{\frac{1}{n}} \quad \forall i. \tag{2}$$

Several other methods were proposed for obtaining a weight vector **w** from a PCM [8], but we do not consider them in this paper. In general, given a PCM, different methods lead to different vectors **w**, but in particular cases the result is the same, as pointed out below.

Beside reciprocity, which is a property required in the definition of a PCM, *consistency* is another relevant property. A pairwise comparison matrix is called *consistent* if and only if the following condition holds:

$$a_{ik} = a_{ij}a_{jk} \quad \forall i, j, k. \tag{3}$$

Property (3) can be considered as a cardinal transitivity condition and means that preferences are fully coherent. If and only if **A** is consistent, then there exists a weight vector $\mathbf{w} = (w_1, \ldots, w_n)$ such that

$$a_{ij} = \frac{w_i}{w_j} \quad \forall i, j. \tag{4}$$

If **A** is consistent, then the Saaty's eigenvector method and the geometric mean method lead to the same weight vector **w**.

Having defined a PCM and the corresponding weight vector **w**, we can now apply these basic concepts to a hierarchy, which is the characterizing structure of AHP.

Let us consider a set of m criteria $C = \{c_1, \ldots, c_m\}$ and require that the n alternatives $x_1, \ldots, x_n$ must be evaluated on the basis of all criteria. Similarly to the case of a single PCM, the aim of AHP is to provide a weight vector $\mathbf{w} = (w_1, \ldots, w_n)$, where w_j a numerical value quantifying the global priority of alternative x_j as estimated on the basis of all the m criteria. For each fixed criterion c_k, Saaty proposed to construct a PCM, say $\mathbf{A}_k$ where the n alternatives are pairwise compared on the basis of c_k, $k = 1, \ldots, m$. In such a way, m PCMs of order n are obtained. Then, the corresponding 'local weight vector', say $\mathbf{w}_k$, is derived for each $\mathbf{A}_k$ by using the eigenvector

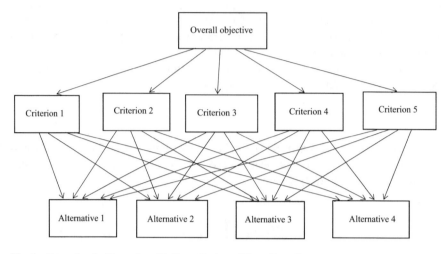

Fig. 1 Example of a hierarchy with five criteria and four alternatives

method. Vectors $\mathbf{w}_k$ are normalized so that for each vector the sum of the components equals one. Finally, by means of the so-called 'hierarchical composition', the m local weight vectors $\mathbf{w}_1, \ldots, \mathbf{w}_m$ are aggregated in order to obtain the global weight vector $\mathbf{w}$. The AHP prescribes that the aggregation of the local weight vectors is done through the weighted arithmetic mean,

$$\mathbf{w} = v_1 \mathbf{w}_1 + \cdots + v_m \mathbf{w}_m = \sum_{i=1}^{m} v_i \mathbf{w}_i, \tag{5}$$

where $v_1, \ldots, v_m$ are the weights, or priorities, of the criteria $c_1, \ldots, c_m$ respectively. AHP requires that also the weights $v_1, \ldots, v_m$ are computed as the components of the maximal eigenvector of the $m \times m$ PCM obtained by pairwise comparing the m criteria. Nevertheless, this is not relevant for our study, so that, in the next section, we will determine $v_1, \ldots, v_m$ more directly. In Fig. 1 we give an example of a hierarchy where the four alternatives constitute the lowest layer and the five criteria the layer immediately above.

2.2 Rank Reversal

Few years after the introduction of AHP by T. Saaty in 1977 [16], Belton and Gear proved in 1983 [6] that AHP may suffer of a drawback that was then named Rank Reversal. They considered an example with three alternatives and three criteria. The three PCMs where consistent and the ranking of the alternatives were computed by means of AHP as described in the preceding subsection. Belton and Gear showed that, by adding a fourth alternative, the ranking of the original three alternatives changed even if the preferences among them remained unchanged and consistency was preserved for the new 4×4 PCMs. Other types of RR were studied in the following years, evidencing, for example, that the phenomenon may occur also by replacing an alternative with a similar one. Other authors evidenced the role of the weight vectors normalization. It was also proved that RR is avoided by aggregating the local weight vectors using the weighted geometric mean instead of the weighted arithmetic meas prescribed by the original AHP. Moreover, it was proved by Saaty itself that RR may not occur in the case of a single PCM, that is for a single criterion, provided that the PCM is consistent. On the contrary, if consistency assumption is removed, RR may occur even for a single PCM. In the next section, we describe our numerical simulations in the case of multiple criteria and consistent PCMs. Our study is aimed at investigating the influence on RR of some relevant factors, as the aggregation method for local weight vectors and the entropy of the criteria weight distribution.

3 Numerical Study on Rank Reversal

3.1 The Effect of Criteria Weights Distribution on Rank Reversal

In this subsection, we study how the distribution of the normalized criteria weights $(v_1, \ldots, v_m)$, $\sum_{i=1}^{m} v_i = 1$ can influence the probability of RR. We start by observing that if $m - 1$ weights are null, being 1 the remaining weight, as for example in $(v_1, \ldots, v_m) = (1, 0, \ldots, 0)$, then RR doesn't occur. The simple reason is that this case leads back to the single criterion consistent case. As mentioned above, Saaty proved that this latter case is RR-free. By drifting away from this polarized case, RR can arise. In particular, our assumption was to consider the uniform case $(v_1, \ldots, v_m) = (\frac{1}{m}, \frac{1}{m}, \ldots, \frac{1}{m})$ as an 'opposite' case with respect to the previous polarized one. In our opinion, the most suitable quantity to describe the range between these two extreme cases is the entropy of $(v_1, \ldots, v_m)$,

$$H(v_1, \ldots, v_m) = - \sum_{i=1}^{m} v_i \ln(v_i). \tag{6}$$

Note that entropy of $(v_1, \ldots, v_m)$ measures the 'closeness' to the uniform case $(\frac{1}{m}, \frac{1}{m}, \ldots, \frac{1}{m})$. In fact, the minimum value of entropy (6) is reached in the fully polarized case, for example, $H(1, 0, \ldots, 0) = 0$, whereas the maximum value of entropy (6) is reached in the case of uniformly distributed weights, $H(\frac{1}{m}, \frac{1}{m}, \ldots, \frac{1}{m}) = \ln m$. We assume that, in the case of $v_i = 0$, the value of the corresponding term in (6) is taken to be 0, according to the limit $\lim_{v_i \to 0^+} (v_i \ln(v_i)) = 0$. We performed numerous

simulations with different values of the number of alternatives and criteria. We antic-
ipate that, for the sake of simplicity, Fig. 2a reports a single example, corresponding
to $n = 4$ alternatives and $m = 5$ criteria. In the following, we briefly describe the
plan of our numerical study.

1. We construct m consistent PCMs $\mathbf{A} = (a_{ij})_{n \times n}$ by setting $a_{ij} = \frac{w_i}{w_j}$, where $(w_1, \dots,$
   $w_n)$ is a randomly generated vector by uniformly sampling in the set $\{1, 2, 3, 4, 5,$
   $6, 7, 8, 9\}$.
2. We associate to each PCM constructed in the previous point the corresponding
   normalized weight vector. Given that all the PCMs are consistent, it is clearly
   irrelevant wether to use the eigenvector method or the geometric mean method.
   Moreover, the obtained normalized vector will be proportional to that used for
   constructing the PCM. At this point, we have the m normalized local weight vectors $\mathbf{w}_1, \dots, \mathbf{w}_m$.
3. The m local weight vectors $\mathbf{w}_1, \dots, \mathbf{w}_m$ are aggregated in order to obtain the global
   weight vector $\mathbf{w}$. The aggregation is performed using the standard weighted arithmetic mean (5) and weights $v_1, \dots, v_m$ of the criteria are randomly generated.
4. For each one of the m consistent PCMs $\mathbf{A} = (a_{ij})_{n \times n}$ constructed at the point 1, we
   remove the last row and the last column, thus obtaining a new set of m consistent
   PCMs of order $n - 1$.
5. We repeat on the new set of m consistent PCMs of order $n - 1$ exactly the same
   computations as described at points 2 and 3 with the same weights $v_1, \dots, v_m$
   of the criteria. So, we obtain the corresponding global weight vector with $n - 1$
   components.
6. We compare, for each instance of m consistent PCMs, the rank obtained in the
   n alternatives case with that obtained in the $(n - 1)$ alternatives case, in order to
   verify whether the RR occurred.

We repeat 100.000 times the points from 1 to 6, thus obtaining the output data set
of our study. We can now better describe the graphical results shown in Fig. 2a,
where the case with 4 alternatives and 5 criteria is reported. Each point in the plot
is determined as follows. We report on the horizontal axis the entropy values, ranging from 0 to its maximum value $\ln m = \ln 5$. This interval is then partitioned in k
equally spaced subintervals. For each subinterval, we consider all the weight vectors
$(v_1, v_2, v_3, v_4, v_5)$ with entropy value belonging to this subinterval. Correspondingly,
we compute the percentage of occurrence of RR for all 4×4 PCMs, constructed as
described above, provided that the aggregation is made using the criteria weight vectors with entropy values belonging to the fixed subinterval. Finally, this percentage
is reported on the vertical axis, thus obtaining the second coordinate of the point.
The first coordinate of the point is the center of the subinterval. As synthesized
in Fig. 2a, it is apparent that the probability of RR increases when the entropy of
$(v_1, v_2, v_3, v_4, v_5)$ increases, thus evidencing the role of criteria weights distribution
on RR. Figure 2b reports the outcome of a study which is very similar to the one
reported in Fig. 2a. The only difference is that, instead of measuring the closeness of
a criteria weight vector $(v_1, v_2, v_3, v_4, v_5)$ to the uniform case $(\frac{1}{5}, \frac{1}{5}, \frac{1}{5}, \frac{1}{5}, \frac{1}{5})$ with the

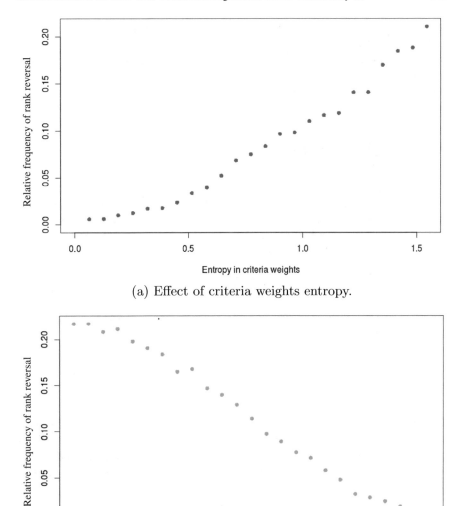

(a) Effect of criteria weights entropy.

(b) Effect of criteria weights standard deviation.

Fig. 2 Estimated probability of rank reversal for a single DM

entropy of $(v_1, v_2, v_3, v_4, v_5)$, we use its standard deviation. Note that, actually, the standard deviation measures the distance from $(\frac{1}{5}, \frac{1}{5}, \frac{1}{5}, \frac{1}{5}, \frac{1}{5})$, rather than the closeness. As pointed out above, the interpretation of Fig. 2a is straightforward. The percentage of cases in which RR occurred increases when the entropy (6) increases.

In the case of maximum entropy, i.e. $H(v_1, v_2, v_3, v_4, v_5) = H(\frac{1}{5}, \frac{1}{5}, \frac{1}{5}, \frac{1}{5}, \frac{1}{5}) = \ln 5 \approx$ 1.6, we found that RR occurred approximately in 23% of cases. Figure 2b represents the same outcome as in Fig. 2a, but referring to the standard deviation of criteria weights. Coherently with the outcome in Figs. 2a, 2b shows that the percentage of cases in which RR occurred decreases when the standard deviation increases.

3.2   The Effect of Aggregation Methods on Rank Reversal

Let us consider again the question of the aggregation of the local weight vectors $\mathbf{w}_1, \ldots, \mathbf{w}_m$. As it is known, Saaty's AHP states that the aggregation is made using the weighted arithmetic mean (5). On the other hand, Barzilai and Golany [5] proposed to use the weighted geometric mean,

$$\mathbf{w} = \prod_{i=1}^{m} \mathbf{w}_i^{v_i}. \tag{7}$$

Note that the exponents v_i in (7) act componentwise. This means that components w_j of vector $\mathbf{w}$ are computed as

$$w_j = \prod_{i=1}^{m} (\mathbf{w}_i)_j^{v_i}. \tag{8}$$

Barzilai and Golany proved that if the weighted geometric mean aggregation (7) is applied, RR cannot occur, thus evidencing that the weighted arithmetic mean can be considered as a relevant factor determining RR. In our study, we consider the weighted arithmetic mean and the weighted geometric mean as particular cases of the weighted power mean,

$$\mathbf{w} = \left(\sum_{i=1}^{m} v_i (\mathbf{w}_i)^p \right)^{\frac{1}{p}}, \tag{9}$$

where, again, the exponent p in (9) acts componentwise.

   More precisely, it is known that the weighted arithmetic mean (5) is obtained from (9) for $p = 1$ and the weighted geometric mean (7) is obtained from (9) for $p \to 0$. Justified by this latter result, definition (9) is often completed assuming that for $p = 0$ the weighted power mean is defined to be the weighted geometric mean. We assume this extended definition in the following. Clearly, each value of p in $[0, 1]$ is associated with a different aggregation method which acts between the two extreme cases of the weighted geometric and arithmetic mean. By means of numerical simulations, we study how the relative frequency of RR varies by moving from the RR-free case of the geometric mean aggregation, corresponding to $p = 0$, to the weighted arithmetic mean case, corresponding to $p = 1$. As it might be expected, the outcome is a monotonically increasing behavior, as showed in Fig. 3. Similarly to the study described in the Sect. 3.1, we performed numerous simulations with different values of the number of alternatives and criteria but, in Fig. 3, for the sake of simplicity, we report a single example, corresponding to $n = 4$ alternatives and $m = 5$ criteria. The following description of the plan of our numerical study is quite similar to the one described in the previous subsection, as it differs only for what concerns the aggregation method at point 3.

1. We construct m consistent PCMs $\mathbf{A} = (a_{ij})_{n \times n}$ by setting $a_{ij} = \frac{w_i}{w_j}$, where $(w_1, \ldots, w_n)$ is a randomly generated vector by uniformly sampling in the set $\{1, 2, 3, 4, 5, 6, 7, 8, 9\}$.
2. We associate to each PCM constructed in the previous point the corresponding normalized weight vector. Given that all the PCMs are consistent, it is clearly irrelevant whether to use the eigenvector method or the geometric mean method. Moreover, the obtained normalized vector will be proportional to that used for constructing the PCM. At this point, we have the m normalized local weight vectors $\mathbf{w}_1, \ldots, \mathbf{w}_m$.
3. The m local weight vectors $\mathbf{w}_1, \ldots, \mathbf{w}_m$ are aggregated in order to obtain the global weight vector $\mathbf{w}$. The weights $v_1, \ldots, v_m$ of the criteria are randomly generated and the aggregation is performed using the weighted power mean (9) with 100 different values of p uniformly spaced in $[0, 1]$.
4. For each one of the m consistent PCMs $\mathbf{A} = (a_{ij})_{n \times n}$ constructed at the point 1, we remove the last row and the last column, thus obtaining a new set of m consistent PCMs of order $n - 1$.
5. We repeat on the new set of m consistent PCMs of order $n - 1$ exactly the same computations as described at points 2 and 3 with the same weights $v_1, \ldots, v_m$ of the criteria and the same 100 different values of p used in the $n \times n$ case.
6. We compare, for each instance, the rank obtained in the $n \times n$ case with that obtained in the $(n - 1) \times (n - 1)$ case, in order to verify whether the RR occurred.

Let us describe how the plot in Fig. 3 is obtained. Each one of the 100 points in the plot corresponds to a fixed value of p on the horizontal axis. The second coordinate of the point is obtained by reporting on the vertical axis the relative frequency of RR, as computed by performing the numerical simulations described above. This relative frequency was computed on 5.000 different instances of five 4×4 PCMs. We use the same set of instances for generating all points in the plot, using different values of p in the aggregation method (9). As synthesized in Fig. 3, it is apparent that the probability of RR increases with p, thus evidencing how the RR is influenced by the aggregation method (9). For example, it can be noted that points corresponding to values of p close to zero represent aggregation methods close to weighted geometric mean (7). In these cases, RR occurs very rarely. Although we reported only few examples in Figs. 2 and 3, we can send the interested readers other outcomes corresponding to different values of n and m.

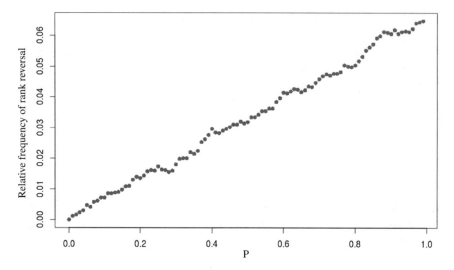

Fig. 3 Estimated probability of rank reversal as influenced by the aggregation method in the case of a single DM

3.3 Group AHP and Rank Reversal

As mentioned in the introduction, we extended our study to the case of a group of Decision Makers, DMs in the following. We assume that N DMs express their preferences on n alternatives $\{x_1, \ldots, x_n\}$ through AHP exactly as described in the previous subsections. Then, m PCMs of order n are associated to each DM, corresponding to the m criteria. We denote by $\mathbf{A}_k^l$ the PCM of DM l referring to criterion k, for $l = 1, \ldots, N$ and $k = 1, \ldots, m$. In order to derive a final weight vector $\mathbf{w} = (w_1, \ldots, w_n)$ quantifying the group priorities on the n alternatives, one has to aggregate the data provided by the N DMs. As it is known, there are two main aggregation procedures within AHP, i.e. the aggregation of individual priorities (AIP) and the aggregation of individual judgments (AIJ). In AIP procedure, each DM derives independently his/her individual priorities. Then, the N priority vectors are aggregated to a group priority vector using either a (weighted) arithmetic (WAMM) or a (weighted) geometric mean method. Conversely, in AIJ procedure, the group PCMs are first determined and the group priority vector is computer after that [11, 21]. In the of AIJ procedure, each entry of the group PCMs $\mathbf{A}_k^G$ is obtained using the geometric mean method on the corresponding entries of the PCMs of all DMs,

$$\left(\mathbf{A}_k^G\right)_{ij} = \left[\prod_{l=1}^{N} \left(\mathbf{A}_k^l\right)_{ij}\right]^{\frac{1}{N}} \quad k = 1, \ldots, m \ \ i, j = 1, \ldots, n. \tag{10}$$

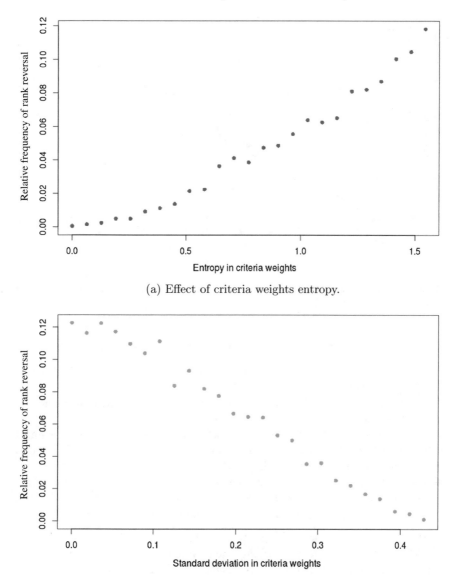

(a) Effect of criteria weights entropy.

(b) Effect of criteria weights standard deviation.

Fig. 4 Estimated probability of rank reversal in a group of 5 DMs

As pointed out by Aczél and Saaty [1], the geometric mean method must be used in AIJ procedure in order to preserve the reciprocity of the group PCMs (10). In the following, we use AIJ procedure, since we consider it as more relevant for our study. The interested reader can refer to [14] for an updated review on group aggregation techniques for AHP.

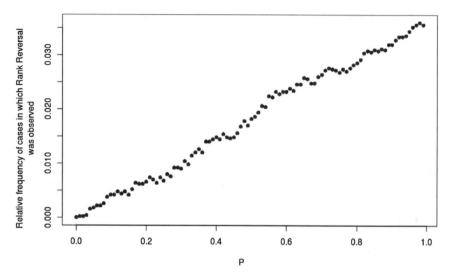

Fig. 5 Estimated probability of group rank reversal as influenced by the aggregation method. Case with 4 DMs

Our numerical study on group RR is then performed as follows.

1. The number N of DMs constituting the group is fixed and the PCMs of the N DMs are constructed as described in Sect. 3.1.
2. The PCMs of the N DMs are aggregated using componentwise the geometric mean method, as in (10), in order to form the the group PCMs.
3. The same study and graphical representation described in Sect. 3.1 for a single DM is performed for the group. An illustrative example of the obtained outcomes is reported in Fig. 4, corresponding to the case with $N = 5$.
4. The study described in the preceding points is performed on the aggregation method too, thus extending to the group case the investigation carried out in Sect. 3.2. Again, we report an illustrative example with $N = 4$ in Fig. 5.

Similarly to the study described in Sects. 3.1 and 3.2, we considered different values of n, m and N. For the sake of simplicity, we reported only the illustrative examples in Fig. 4 and in Fig. 5. A first remark is that the monotone behavior of the estimated RR probability is confirmed for the group case too. Although Fig. 4 resembles Fig. 2, it can be observed that the values reported in the latter are approximately double that the values in the former. For different values of the parameters n, m and N we obtain results that are coherent with this last observation. We can therefore conclude that outcomes of our simulations support the conjecture that RR probability decreases when the number of experts in a group increases. The impact of the number of experts on rank reversal was studied from a different point of view in [10]. Comparison between Figs. 3 and 5 leads to very similar conclusions for the aggregation method too.

References

1. Aczél J, Saaty TL (1983) Procedures for synthesizing ratio judgments. J Math Psychol 27:93–102
2. Aguaròn J, Moreno-Jimènez JM (2003) The geometric consistency index: approximated threshold. Eur J Oper Res 147:137–145
3. Bana e Costa CA, Vansnick J-C (2008) A critical analysis of the eigenvalue method used to derive priorities in AHP. Eur J Oper Res 187:1422–1428
4. Barzilai J (2010) Preference function modelling: the mathematical foundations of decision theory. In: Ehrgott M, Figueira JR, Greco S (eds) Trends in multiple criteria decision analysis. Springer
5. Barzilai J, Golany B (1994) AHP rank reversal, normalization and aggregation rules. INFOR: Inf Syst Oper Res 32(2):57–64
6. Belton V, Gear AE (1983) On a short-coming of Saaty's method of analytic hierarchies. Omega 11:228–230
7. Belton V, Stewart T (2002) Multiple criteria decision analysis: an integrated approach. Springer
8. Choo EU, Wedley WC (2004) A common framework for deriving preference values from pairwise comparison matrices. Comput Oper Res 31:893–908
9. Crawford G, Williams C (1985) A note on the analysis of subjective judgement matrices. J Math Psychol 29:25–40
10. Dede G, Kamalakis T, Sphicopoulos T (2015) Convergence properties and practical estimation of the probability of rank reversal in pairwise comparisons for multi-criteria decision making problems. Eur J Oper Res 241:458–468
11. Forman E, Peniwati K (1998) Aggregating individual judgments and priorities with the analytic hierarchy process. Eur J Oper Res 108:165–169
12. Forman EH, Gass SI (2001) The analytic hierarchy process-an exposition. Oper Res 49:469–486
13. Maleki H, Zahir S (2013) A comprehensive literature review of the rank reversal phenomenon in the analytic hierarchy process. J Multi-Criteria Decis Anal 20:141–155
14. Ossadnik W, Schinke S, Kaspar RH (2016) Group aggregation techniques for analytic hierarchy process and analytic network process: a comparative analysis. Group Decis Negot 25(2):421–457
15. Raharjo H, Endah D (2006) Evaluating relationship of consistency ratio and number of alternatives on rank reversal in the AHP. Qual Eng 18(1):39–46
16. Saaty TL (1977) A scaling method for priorities in hierarchical structures. J Math Psychol 15:234–281
17. Saaty TL, Vargas LG (1984) The legitimacy of rank reversal. Omega 12:513–516
18. Saaty TL (1994) Highlights and critical points in the theory and application of the analytic hierarchy process. Eur J Oper Res 74:426–447
19. Smith JE, von Winterfeldt D (2004) Decision analysis in management science. Manage Sci 50:561–574
20. Triantaphyllou E (2001) Two new cases of rank reversals when the AHP and some of its additive variants are used that do not occur with the multiplicative AHP. J Multi-Criteria Decis Anal 10:11–25
21. Van Den Honert RC, Lootsma FA (1996) Group preference aggregation in the multiplicative AHP The model of the group decision process and Pareto optimality. Eur J Oper Res 96:363–370

Estimating One-Off Operational Risk Events with the Lossless Fuzzy Weighted Average Method

Pasi Luukka, Mikael Collan, Fai Tam and Yuri Lawryshyn

Abstract Banks are required by the Basel II Accord to report on their operational risks, including reporting an estimate for the size of possible one-off negative operational events. The typical way to produce these estimates is to use a quantitative value at risk methodology that is based on a limited amount of data, but also the use of qualitative, expert estimate-based methodologies is sanctioned by the regulations. The final estimations are most often reached by fusing the input from multiple experts. In this chapter we propose and introduce a new lossless fuzzy weighted averaging method and show how and why it is a usable tool for the aggregation of expert estimates in the context of estimating the unlikely one-off operational losses originating from single risks. The method is simple to use, intuitive to understand, and does not suffer from the loss of information associated with using many other weighted averaging methods.

Keywords Risk management · Operational risk · One-off events · Information fusion · Consensus · Multi-expert decision-making

P. Luukka (✉) · M. Collan
School of Business and Management, Lappeenranta University of Technology,
Lappeenranta, Finland
e-mail: pasi.luukka@lut.fi

M. Collan
e-mail: mikael.collan@lut.fi

F. Tam · Y. Lawryshyn
Centre for Management of Technology and Entrepreneurship, University of Toronto,
Toronto, Canada
e-mail: yuri.lawryshyn@utoronto.ca

© Springer International Publishing AG 2018   227
M. Collan and J. Kacprzyk (eds.), *Soft Computing Applications for Group Decision-making and Consensus Modeling*, Studies in Fuzziness and Soft Computing 357, DOI 10.1007/978-3-319-60207-3_15

1 Introduction

Under current regulations, set by Basel II [2, 3], banks are required to set aside capital to ensure solvency to account for risk events related to credit, market and operational risk. Currently, our client, "the Bank", uses two methods to quantify operational risk, as part of the Basel II guideline: through a quantitative statistical approach, and a qualitative scenario analysis approach. In the statistical approach, losses are assumed to follow a statistical distribution, and the Value-at-Risk VaR measure is calculated based on the distribution tail. This value can be difficult to quantify given that typically, a limited amount of internal operational risk incident data is available. In the scenario analysis approach, subject-matter experts (SMEs) review historical operational risk incidents and corroborate on potential future losses. Our objective is to create a methodology to aggregate expert opinions within the scenario analysis approach, utilizing fuzzy set theory, to estimate the expected loss from a single severe and infrequent event.

The Basel II framework defines operational risk as "the risk of loss resulting from inadequate or failed internal processes, people and systems or from external events". The framework outlines three methods for calculating operational risk, namely (i) the Basic Indicator Approach, (ii) the Standard Approach, and (iii) the Advanced Measurement Approach (AMA), each requiring a greater level of sophistication. Under the AMA [4], besides a quantitative analysis, banks must also use scenario analysis based on expert opinion, to evaluate exposure to high-severity events. The reason for the qualitative scenario analysis is to augment quantitative analyses, which typically lack sufficient data to provide dependable results.

A number of authors have highlighted the challenges of utilizing internal/ external data to quantify operational loss, and recommend the use of expert judgment based scenario analysis to augment the analysis see, e.g., [1, 10, 15, 16]. There are a few papers which attempt to formalize the scenario analysis approach based on qualitative subjective input from subject matter experts in the operational risk context. Dutta and Babbel [10] make the claim that there has been little focus on the development of scenario-based methodology for the operational risk measurement, citing one exception, namely [13], where a Bayesian inference method is proposed to incorporate expert opinion with available data. Arguably, the approach used by Lambrigger et al. is mathematically complex to the point that managers may not accept the methodology due to a lack of familiarity with the formulation.

Stepanek et al. [17] utilize a semantic linguistic approach to rank risk scenarios within the insurance industry. The authors make a strong case for their method's effectiveness in ranking the scenarios, but a monetary quantification of the scenarios is not provided. Amin [1] presents a list of methodological challenges associated with quantifying operational risk based on historical data. He proposes an analytical approach for quantifying operational risk based on expert judgment by utilizing a survey to quantify information on the quality of risk management, the business environment, and the effectiveness of internal control factors using a four point scale. Amin uses a uniform distribution to estimate the potential financial impact associated with the four point scale. Through Monte Carlo simulation, VaR estimates are obtained. According to our experience, we feel that SMEs are more inclined to provide low, medium, and high estimated losses. From previous literature on profitability analysis [7] and strategic project selection [8] we know that such estimations can be used in formulating fuzzy numbers from these estimates that represent, in this context, the losses. We choose to utilize fuzzy numbers and fuzzy set theory as the basis for our work.

Fuzzy set theory provides a methodology to utilize linguistic information to estimate numerical outcomes [18], and thus, there appears to be a natural fit for fuzzy logic to be utilized in the scenario analysis of the AMA. Furthermore, we find that fuzzy representations of information are easy to understand for managers, when presented in a graphical form and require no simulation. According to Reveiz and León Rincon [14] while there are several studies that support the use of fuzzy logic in operational risk management, there is very little literature on the application of the fuzzy models in modeling operational risk.

Previously in the academic literature the use of fuzzy methods in capturing one-off events in the banking context is scarce, however, a contribution by Durfee and Tselykh [9] provides one such example. Their method is based on summing up fuzzy numbers originating from expert estimates of different operational risks in order to arrive at a quantification of the total operational risk. This approach is different from the one that we take in this chapter, as we consider the estimation of one such risk. Outside of banking, operational risks have been previously quantified by utilizing fuzzy numbers in, e.g., the telecom business [6]. It is rather clear that there is a lot of room for new methods and for research on using fuzzy modeling in framing operational risks generally, and in the banking context specifically.

An important issue that is connected to risk management and to estimation of losses, and particularly important in the banking risk estimation context presented here, is the ability to aggregate information contained in the estimates of multiple SMEs in a way that does not lose information contained in these estimates. This observation is not a trivial one, as commonly used aggregation methods such as the weighted average and the fuzzy weighted average may lose information about the estimated minima and maxima in the aggregation process that are important in determining the expected losses. For this reason, in this chapter we present a new lossless fuzzy weighted averaging (LFWA) method that is designed to overcome the problem of information loss that is connected to the use of fuzzy weighted averaging.

We illustrate the use of the new method in the banking context with a situation where multiple subject matter experts are asked to estimate the size of a single one-off risk with three scenarios. After creating triangular fuzzy loss estimates from these scenarios they are aggregated to yield an overall loss estimate. For this aggregation we use the fuzzy weighted average and the proposed lossless fuzzy weighted average method and study the difference between the results. We show that the way the aggregation is done has a remarkable effect on the result, which indicates that there are "gains" to be made by introducing smarter aggregation methods in this context.

This chapter continues with an introduction of the mathematical background needed and then with the presentation of the new proposed lossless fuzzy weighted averaging method. The use of the method in the context of banking is illustrated numerically and the chapter is closed with some discussion and conclusions.

2 Preliminaries

In this section we briefly go through the mathematical background starting from the definition of a fuzzy set used in this chapter, the basic mathematical operations on fuzzy sets, and then go into how we present a triangular fuzzy number.

Definition 1 Let X be an nonempty set (a universe of discourse). A fuzzy set A on X is defined by a mapping: $\mu_A: X \to [0, 1]$, where μ_A is called a membership function of A. The set of all fuzzy sets on X is denoted by $F(X)$. From the membership function, instead of $\mu_A(x)$, often simply $A(x)$ is used, see, e.g., [12].

Some properties of fuzzy sets that are required later are given in Definition 2.

Definition 2 Let $A \in F(X)$, then the core of A is a crisp set $core(A) = \{x \in X | A(x) = 1\}$ the support of A is crisp set $Supp(A) = \{x \in X | A(x))0\}$.
Let $Supp(A) = (a, b)$ then $cl(Supp(A)) = [a, b]$.

In the proposed method we consider only fuzzy numbers of the triangular type that are of the form given in Definition 3.

Definition 3 A triangular fuzzy number A can be defined by a triplet $A = (a_1, a_2, a_3)$. The membership function $A(x)$ is defined as [11]:

$$A(x) = \begin{cases} \frac{x - a_1}{a_2 - a_1}, & a_1 \leq x \leq a_2 \\ \frac{x - a_3}{a_2 - a_3}, & a_2 \leq x \leq a_3 \\ 0, & otherwise \end{cases} \tag{1}$$

For arithmetic operations for triangular fuzzy numbers we refer to [11].

One commonly used way to aggregate fuzzy numbers is by computing their weighted average. This can be done, for example, by using the methodology described in Definition 4.

Definition 4 Let U be the set of fuzzy numbers. A Fuzzy Weighted Averaging (FWA) operator of dimension n is a mapping FWA: $U^n \to U$ that has an associated weighting vector W, of dimension n, such that $\sum_{i=1}^{n} w_i = 1$ and $w_i \in [0, 1]$, then:

$$FWA(A_1, \ldots, A_n) = \sum_{i=1}^{n} w_i A_i \tag{2}$$

where $(A_1, \ldots, A_n)$, are now fuzzy triangular numbers of the form given in Definition 3.

The FWA and other fuzzy weighted aggregation operators that are based on the FWA are often applied, for example, in connection with cases where one needs to aggregate the evaluations of many decision makers, see, for example, [5, 8] for more discussion about fuzzy weighted averaging operators. The advantage of the FWA is that it is simple to understand and easy to use.

When one takes a fuzzy weighted average from n fuzzy numbers provided typically by experts, an aspect that can in some situations prove to be problematic is that the aggregation procedure causes one to lose a part of the original information provided.

Say, for example that one has triangular fuzzy evaluations (A) from three experts: $A_1 = (2, 4, 10)$, $A_2 = (3, 4, 6)$ and $A_3 = (1, 3, 4)$, and if these evaluations are assumed to carry equal weights, then the fuzzy weighted average will be $A_{ave} = \left(\frac{2+3+1}{3}, \frac{4+4+3}{3}, \frac{10+6+4}{3}\right) = (2, 3.67, 6.67)$.

The aggregation procedure, as noted above, faces the problem that information regarding the full range of the expert identified possible states of the world may be lost. In the above example, expert A_3 provided the estimate of 1 for the minimum possible outcome, in other words a possible by the expert identified "state of the world" is 1, however, this value is not included in the outcome of the fuzzy weighted average resulting in a loss of information provided by expert evaluation.

If the example presented above were a risk management situation, such as one found in the banks that are the context of this chapter, then the lowest identified possible outcome, certainly relevant to the risk profile, would be omitted. It is clear that such loss of relevant information is not good for decision-making. In fact, when experts with different experience are used, they may have different information that may not be optimally aggregated by FWA and methods that are derivatives of FWA, because of this observed possible loss of information.

To avoid this loss of information we propose, as an alternative, a lossless new way of aggregating expert information into a fuzzy number that preserves the information about the managerially identified estimated possible minimum and maximum values. If this kind of a new method is used as the basis of calculating a single number representation for the evaluations of the group of experts involved, it seems to have a different "attitude", than using FWA for the same purpose, because it will not lose the estimated extreme values.

3 The Proposed New Lossless Fuzzy Weighted Average and Computation of Single Number Consensus

The new proposed method is based on the idea that the lossless fuzzy weighted average (LFWA) is constructed by calculating the "contribution" from each fuzzy number A_i that contributes to the LFWA, for each possible point x_j of the LFWA in a way that individual contributing weights w_i are provided for each contributing fuzzy number. The procedure is described as follows:

First, n fuzzy numbers, where n is the number of decision makers giving the aggregated estimates, are typically elicited.

Second, the decision makers evaluate credibility scores (CS) for each fuzzy number used in the creation of the LFWA. These credibility scores are then transformed into individual weights for each fuzzy number, by the following formula:

$$w_i = \frac{CS_i}{\sum_{i=1}^{n} CS_i} \tag{3}$$

Third, with n fuzzy numbers and n weights, an average for each point x_j of the resulting LFWA is calculated. This is done by computing the weighted average A_c from n membership values of the contributing fuzzy numbers for each particular point of the LFWA. This is formulated as:

$$A_c(x_j) = \sum_{i=1}^{n} w_i A_i(x_j) \tag{4}$$

To set the universe or range, for which this procedure is applied, an interval, where non-zero averages are reached is defined. This is done by taking the union of the fuzzy numbers contributing to the LFWA and by computing the support area of this union:

$$X = cl(supp(A_1 \cup A_2 \cup \cdots \cup A_n)) = [x_1, x_2] \tag{5}$$

where $supp(A) = \{x \in X | A(x) > 0\}$, used union is the standard max operator. With the closure used denoting that the support is bounded and leads to a closed interval $[x_1, x_2]$. This way we, for example, get $X = [1, 10]$ for the previously presented example and the information from both, the lowest and the highest expert estimates is included.

Fourth, the constructed LFWA or A_c is used as the basis for computing an expected value that can be also interpreted as a consensus estimate of the evaluations from the involved experts. For the purpose of deriving a single number expected value we use the center of gravity (COG):

$$EX(A_c) = \frac{\int_{x_1}^{x_2} xA_c(x)dx}{\int_{x_1}^{x_2} A_c(x)dx}. \tag{6}$$

The resulting expected value (consensus estimate) is a representation that considers the original extremes of the expert estimates, which is an important issue in terms of the applicability of such representations in risk management (or for other purposes), where it is paramount not to exclude the extreme values.

4 Case: Estimating One-Off Operational Loss Events by Using the LFWA

The case presented here is based on a realistic situation of an international bank that must annually report the risks of one-off operational losses to the regulators. The bank has decided to use multiple-expert evaluation in order to estimate the size of possible one-off operational loss events and for this reason needs a method that will not lose relevant information that is coming from the experts and that pertains to the minimum and the maximum possible sizes of the said one-off operational losses.

In Table 1 we can find estimates of three scenarios that have been evaluated for minimum possible (low), expected and maximum possible (high) losses. It is simply assumed that the expected loss fully belongs to the set of possible outcomes and that the minimum possible and the maximum possible losses belong to the set of possible outcomes with a membership degree that is limited to zero; this is similar to what is presented in [7].

Also, a credibility score for each expert opinion has been estimated. The credibility score can be understood as the credibility of the expert giving the evaluation, e.g., such that senior experts or risk specialists get a higher credibility score, while junior experts get a lower score. The credibility scores are transformed into normalized (within the unit interval) weights resulting in a weighting vector $= (0.2857, 0.5714, 0.1429)$, also visible in Table 1.

The commonly used way to aggregate the information would be to use the fuzzy weighted average [2], the result of which is for this information would be $A_{ave} = (27.14, 40, 67.14)$. The resulting triangular fuzzy number is provided in Fig. 1.

Calculation of a single expected value with the COG from the result of the fuzzy weighted average of the expert estimates with [6] gives us $EX(A_{ave}) = 44.76$.

This result has the observed and visible flaw of disregarding the identified minimum and maximum possible values of the losses and for this reason the bank wants to use the proposed lossless fuzzy weighted average method for determination of a single one-off operational loss estimate.

This can be done by using the estimated scenarios represented as fuzzy numbers $A_1 = (20, 30, 100)$, $A_2 = (35, 50, 60)$, and $A_3 = (10, 20, 30)$ and the weight vector of the normalized credibility scores $W = (0.2857, 0.5714, 0.1429)$ by computing the

Table 1 Expert estimates, credibility scores, and normalized weights for the three estimates

Expert	Low estimate ($MM)	Expected loss ($MM)	High estimate ($MM)	Credibility score	Normalized weight
1	20	30	100	2	$\frac{2}{2+4+1} = 0.2857$
2	35	50	60	4	$\frac{4}{2+4+1} = 0.5714$
3	10	20	30	1	$\frac{1}{2+4+1} = 0.1429$

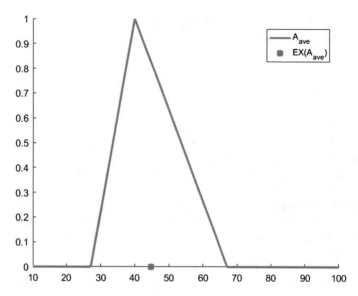

Fig. 1 The fuzzy number resulting from using the FWA to aggregate the scenarios

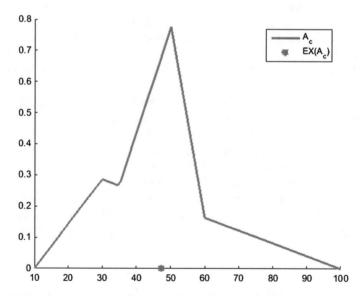

Fig. 2 The fuzzy number A_c resulting from using LFWA to aggregate the scenarios

consensus for each individual point x_j, by using the estimated scenarios and the credibility scores as described above in [4].

The resulting fuzzy number A_c from using the lossless fuzzy weighted averaging in aggregating the expert given scenarios can be seen in Fig. 2.

When a single expected value representing the fuzzy number A_c is calculated by using [6] we get $EX(A_c) = 47.26$. This value is remarkably different from $EX(A_{ave})$. The difference of this magnitude put in the context of banking is significant and reflects the importance of having a lossless aggregation method available in aggregating SME estimates in this context.

5 Conclusions

In this chapter we have discussed the evaluation and estimation of possible one-off operational losses in the banking context, a task that is required from banking institutions and based on the Basel II framework. We have learned that this requirement of estimation requirement can be partially fulfilled by using subject-matter expert evaluation that can in turn be transferred into fuzzy number loss estimates. Typically multiple experts are used in creating the overall estimate of these losses, which in practice means that there is a need to aggregate the estimates.

In the framework of risk management that we are discussing here in the banking context, it is important that the information elicited from the experts is aggregated in a way that the extreme (minimum and maximum) situations that are deemed possible by the experts are not lost in the aggregation process. This means that commonly used aggregation methods that are based on the procedure used in the fuzzy weighted averaging are not a good choice in this context as the process may lose the information about the estimated minima and the maxima. It has been for this reason that we have introduced a new lossless fuzzy weighted averaging method that is able to fuse the information contained in the multiple estimates without loss and is thus better for the purposes of the overall estimation of possible one-off operational losses in the banking context.

The use of the new method was illustrated numerically in the banking context and the gained results show that the difference to using the fuzzy weighted averaging are remarkable and important from the point of view of operational risk estimation in this context. The method proposed is generally usable, intuitively understandable and the aggregation result can be visually presented.

There are many interesting future research directions opened up by this research such as the possible extensions of the new lossless fuzzy weighted averaging method and utilizing the method in framing the evaluation of multiple risks simultaneously in the context of one-off operational loss estimation in the banking business.

References

1. Amin Z (2016) Quantification of operational risk: a scenario-based approach. North Am Actuarial J 20(3):97–286
2. BIS (2016) Consultative document, standardised measurement approach for operational risk. Basel Committee on Banking Supervision, Bank for International Settlements. http://www.bis.org
3. BIS (2006) International convergence of capital measurement and capital standards, Basel Committee on Banking Supervision, A Revised Framework, Comprehensive Version, Bank for International Settlements. http://www.bis.org
4. BIS (2011) Operational risk—supervisory guidelines for the advanced measurement approaches, Basel Committee on Banking Supervision, Bank for International Settlements. http://www.bis.org
5. Bojadziev G, Bojadziev M (2007) Fuzzy logic for business, finance, and management, vol 23. World Scientific, Washington, D.C.
6. Cerchiello P, Giudici P (2012/13) Fuzzy methods for variable selection in operational risk management. J Oper Risk 7(4):25–41
7. Collan M, Fullér R, Mézei J (2009) Fuzzy pay-off method for real option valuation. J Appl Math Decis Syst
8. Collan M, Luukka P (2016) Strategic R&D project analysis: keeping it simple and smart. In: Collan M, Kacprzyk J, Fedrizzi M (eds) Fuzzy technology, vol 335, Springer International Publishing, Heidelberg, pp 169–91
9. Durfee A, Tselykh A (2011) Evaluating operational risk exposure using fuzzy number approach to scenario analysis. In: Galichet S, Montero J, Mauris G (eds) 7th conference of the European Society for fuzzy logic and technology (EUSFLAT-2011). Atlantis Press, Aix-les-Bains, France, pp 1045–1051
10. Dutta K, Babbel D (2013) Scenario analysis in the measurement of operational risk capital: a change of measure approach. J Risk Insurance 81(2):34–303
11. Kaufmann M, Gupta M (1985) Introduction to fuzzy arithmetics: theory and applications. Van Nostrand Reinhold, New York, NY
12. Klir GJ, Yuan B (1995) Fuzzy sets and fuzzy logic—theory and applications. Prentice Hall, Upper Saddle River, NJ
13. Lambrigger D, Shevchenko P, Wûthrich M (2007) The quantification of operational risk using internal data, relevant external data and expert opinions. J Oper Risk 2(3):3–27
14. Reveiz HA, León Rincon CE (2009) Operational risk management using a fuzzy logic inference system. Borradores de economia, vol 574. Banco de la republica Colombia, Bogota, pp 1–30
15. Rosengren E (2006) Scenario Analysis and the AMA, Federal Reserve Bank of Boston, Power Point presentation. https://www.boj.or.jp/en/announcements/release_2006/data/fsc0608be9.pdf. Accessed 6 Jan 2017
16. Segal S (2011) Corporate value of enterprise risk management. Wiley, New York
17. Stepanek L, Urban R, Urban R (2013) A new operational risk assessment technique: the CASTL method. J Oper Risk 8(3):101–117
18. Stoklasa J (2014) Linguistic models for decision support, vol 604. Yliopistopaino, Lappeenranta

Fuzzy Signature Based Methods for Modelling the Structural Condition of Residential Buildings

Ádám Bukovics, István Á. Harmati and László T. Kóczy

Abstract Conservation, extension or renovation of residential buildings is a task that requires intensive attention, where it must be ensured that design and construction works are carried out in proper quality. Priority is given to the proper use of the available financial resources. Incorrect assessment of renovation or reconstruction needs might cause considerable financial loss without implementing necessary interventions (which could eliminate eventual deteriorations, or hinder their reoccurrence).

1 Introduction

Conservation, extension or renovation of residential buildings is a task that requires intensive attention, where it must be ensured that design and construction works are carried out in proper quality. Priority is given to the proper use of the available financial resources. Incorrect assessment of renovation or reconstruction needs might cause considerable financial loss without implementing necessary interventions (which could eliminate eventual deteriorations, or hinder their reoccurrence).

Á. Bukovics
Department of Structural and Geotechnical Engineering,
Széchenyi István University, Győr, Hungary
e-mail: bukovics@sze.hu

I.Á. Harmati (✉)
Department of Mathematics and Computational Sciences,
Széchenyi István University, Győr, Hungary
e-mail: harmati@sze.hu

L.T. Kóczy
Department of Informatics, Széchenyi István University, Győr, Hungary
e-mail: koczy@tmit.bme.hu; koczy@sze.hu

L.T. Kóczy
Department of Telecommunications and Media Informatics,
Budapest University of Technology and Economics, Budapest, Hungary

© Springer International Publishing AG 2018
M. Collan and J. Kacprzyk (eds.), *Soft Computing Applications for Group Decision-making and Consensus Modeling*, Studies in Fuzziness and Soft Computing 357, DOI 10.1007/978-3-319-60207-3_16

In this chapter through the survey of a real stock of buildings it will be introduced what kind of examinations and researches precede the preparation of a decision support and ranking method, related to the renovation (or other utilization) of residential buildings, and what factors influence the reliability thereof. Thereafter three fuzzy signature based methods are introduced shortly which—in different depth—are suitable for determining the condition of a bigger stock of buildings and for ranking them. These methods are suitable for aggregate handling of expert evaluations of different detail and depth. Finally the sensitivity analysis of the method will also be introduced.

2 Characteristic of Status Assessments

2.1 Role of Building Diagnostics and Building Pathology in Status Assessments

In Hungary, and in many other historical countries of the world buildings and engineering constructions provide a significant part of the national property. Hospitals, schools, residential buildings, bridges, roads, and many other constructions could be listed here, where conscious protection and maintenance is in the public interest.

Construction pathology analyses the processes of deteriorations in building structures, and reveals the reasons and consequences of these processes. Building diagnostics is a process of determining the condition of buildings and rate of defects in buildings. These days there is a tendency to recognise the significance of both, since a considerable part of buildings is in very bad condition, and this professional renovation is inevitable. It is enough to think of the stock of rented flats of the central districts of Budapest (the capital of Hungary), where the systematic renovation of more than 100 years old buildings cannot be further postponed for a long time.

The significance of building pathology is appreciated these days. Many international organisations and experts are involved in the study of the building defects and the reasons thereof, and are documenting the obtained results and details in a professional manner.

It often occurs that a bigger stock of buildings is examined at the same time, where advantage is that lots of data are available for drawing more precise conclusions.

Often such examinations were initiated by the owner or the building maintainer, because many defects can be avoided with the help of experience gained in the course of examinations, thereby making renewal or construction works more cost-effective.

2.2   Uncertainties in the Status Assessment

In the next the causes of uncertainties and inaccuracy are introduced which occur while preparing the expert reports.

It has an influence on the correctness of the evaluation of the condition how detailed assessment of the given building element has been completed. The quality of building structures and building materials cannot be clearly identified in case of covered structural elements. Sometimes it is even difficult to identify the structural systems themselves.

The hidden structural defects, which can be discovered only by using destructive examination is hindering an accurate condition assessment. Usually there is no sign of such defects, and the defect is noticed only when the structure is damaged.

In most cases when examining an inhabited building no destruction check can be carried out. So it is not possible (or only possible to a restricted extent) to complete laboratory investigations either. In such a case material quality and the properties, and often the structural systems themselves can be determined by estimation only. In this situation the expert can rely on building diagnostics by visual check only.

When preparing an expert report it is not sure whether two well prepared experts would give the same evaluation of the condition of building structures. Professional experience and skill of the experts completing the status evaluation analysis may be different to a significant extent, so the condition of the examined building structures, defined by certain experts, may include significant inaccuracy.

There are construction materials which were accepted and prevailing at the time of the construction of the residential building but later adverse features were detected and so these are not used nowadays. These construction materials are unreliable, often dangerous. Usually it is necessary to replace the structure, or at least these are taken into consideration as a negative feature when qualifying the structure, even in the case when seemingly the structure is in good condition as a whole. Bauxite concrete and slag concrete are the most well-known construction materials of this type.

It may be of a help with carrying out the work, if the original design documentation of the building or a part of it is still available, or the discovered conditions have been documented during a later renovation or assessment. In case of residential buildings constructed at the end of the 19th century and the beginning of the 20th century usually no original design documentations are any more available, partly due to the sad historical events of the 20th century. As a consequence of the two world-wars, the Hungarian revolution in 1956, and also as a result of the closing down the big planning companies at the beginning of the 1990s—including the liquidation of the archives of architectural and structural plans, the design documentations of lots of residential buildings disappeared or were destroyed.

All factors, influencing the evaluation of the building structure are included, which influence the quality of the structure and at the same time make uncertain the expert evaluation.

Uncertainties of expert evaluations			
GU		Partial subjectivity of expert evaluation	ES
		Professional preparedness of the expert	EE
		Features of the expert behaviour	ERE
		Elaborateness of the expert evaluation	EB
SU	SUP	Method of checking	VO1
		Quality of visual observability	VO2
		Presence of active deteriorations	AD
		Quality of building materials	BMQ1
		Extent to which the construction materials are accepted	BMQ2
	SUT	Active vibration effects	AV
		Former design documentations	ADD
		Date of construction	DC

Fig. 1 General and special uncertainties of expert evaluation

Expert uncertainties are sorted into two groups (Fig. 1). Uncertainties in the first group (general uncertainties) have impact on the expert evaluation in each case, and here the expert's subjectivity is integrated into the system. (It depends on the professional skill and time investment). The second group are those uncertainties, which depend on the quality and quantity of data defining the circumstances of status assessment, are called special uncertainties (SU). At this point two further sub-groups are defined, subject to whether the given uncertainty relates to the whole building or to a specific structure within the building.

Special uncertainties of expert evaluation featuring certain building structures (SUP), takes into consideration the circumstances and methods of the examination of certain building structures as well as the quality and acceptance of the examined building materials.

The special uncertainties of the expert evaluation, featuring the total building (SUT) takes into consideration the impact of the direct environment of the building exerted on the whole building, as well as the circumstances of the construction of the building, and the documentations of later interventions (renovation, reconstruction, strengthening).

3   Generating the Database and Knowledge Base Used for Modelling the Condition of Buildings

3.1   Presentation of the Historic Residential Building Stock of Budapest

The city of Budapest had grown into a metropolis over several decades therefore the development of this city significantly differed from other big cities in many respects.

Budapest became one of the most dynamically developing cities of the world at the end of the 19th century and in the beginning of the 20th century. The following data well indicate the extent of its development: while in 1869 280349 inhabitants lived in 52583 flats, in 1900 733358 inhabitants lived in 182214 flats. Thus over 31 years the number of inhabitants increased by 450,000, and more than 129,000 flats were built. During this time period new residential districts were built, using the well-known construction technology and structural solutions of the given era (Figs. 2 and 3). A large amount of these buildings still exist and are a significant element of the cityscape. The condition of a part of these buildings is often degraded. Static, functional and social deterioration can be noticed in case of a significant part of these old buildings. To solve this problem is one of the most pressing issues of Budapest.

The development of Budapest city was the most dynamical over the period between the Compromise between Hungary and Austria of 1867 and the outbreak of World War I in 1914. In this time period Budapest became a world city. A considerable part of the tenement houses and the network of public institutions were established at that time. Even nowadays the public and residential buildings of this time period dominate the image of the city (Fig. 4).

Many buildings were declared protected as National Heritage thus their existence in the future seems to be ensured. Some parts of the city are a part of World Heritage (Buda Castle Quarter, Riverside of the Danube, Andrássy Avenue), thereby guaranteeing the preservation of the image of the cityscape in addition to the protected buildings (Figs. 5 and 6).

At that time an unprecedented development process started in the Hungarian capital. Rapid growth of the city was enhanced by the milling and the distillery industries, as well as the boom in the heavy industry, connected with the railway construction, too. The most dynamical development was between 1890 and 1900

Fig. 2 Typical urban structure of inner Budapest

Fig. 3 Typical streetview of inner Budapest

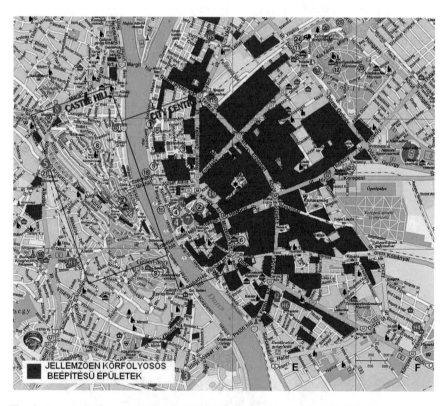

Fig. 4 Areas in Budapest including buildings, similar to those under review

Fig. 5 Andrássy Avenue at the end of the 19th century

Fig. 6 The hungarian Parliament at the riverside of the Danube

when the number of the population of the city increased by 23,000 heads per year. There was a time period when it was considered the most dynamically developing city of the world, thus overtaking even Chicago. The intensive growth was largely promoted also by the gigantic transport infrastructure and cultural projects, connecting to the year of the Millenium, the 1000-year anniversary of the Hungarian state in 1896 (the first continental underground railway was open here).

The growth of the number of flats was connected to the growth of the stock of tenement houses. In the last three decades of the 19th century about 7000 new tenement houses were built, typically multi-storey residential buildings.

3.2   The Available Expert Reports and Their Analysis

Old residential buildings are typical in the 13th district of the city, too. Layout arrangements and building structures resemble buildings from the same age in other districts of the city. In this district expert reports were prepared about 340 residential buildings altogether (Fig. 7). The majority of the buildings were built between 1880 and 1950.

The buildings were studied on the basis of building structure, building diagnostic, social and functional aspects. The knowledge and conclusions obtained from the structural and diagnostic research of these residential buildings can be efficiently utilized on a important part of the residential buildings of Budapest, since a large part (especially in the inner districts) were built at the end of the 19th century, and at the beginning of the 20th century. Therefore their age, structural arrangement and conditions are similar to those of the buildings under review.

The study may be helpful in the optimum allocation of economic resources which are available for renovations and for rehabilitation.

Conclusions can be drawn on the typical construction methods of various ages and their respective characteristic construction deficiencies. Significant financial savings can be achieved by the elimination of construction and operational

Fig. 7 Location of stock of buildings under review

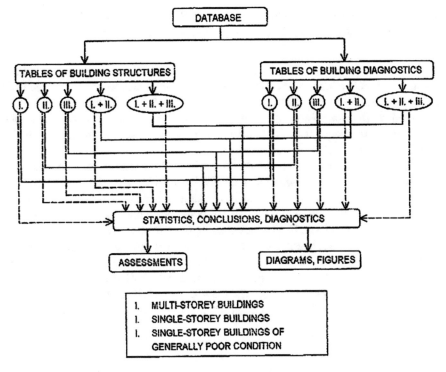

Fig. 8 The processing flow-chart

deficiencies. The knowledge of the building systems and the construction materials makes it easy to recognise the frequency of the typical defects.

The buildings were examined separately on the basis of the aspects of building structures and building diagnostics.

Relying on the database the buildings were arranged into 3 groups. Group 1 includes multi-storey buildings, with number of storeys ranging from 2–5. Group 2 includes higher quality single-storey buildings, and Group 3 includes single-storey residential buildings of generally poor conditions.

Various groups were studied separately and altogether, as well, and the connections between the diagnosed building defects and the systems and materials of the structures were identified. The most important statistics and connections will be represented on diagrams and figures. The flow-chart of the process can be seen on Fig. 8.

3.3 Examples of the Building Structural and Building Diagnostic Analysis of the Examined Buildings

Over the past decades—often because of financial reasons—the necessary main-tenance and renovation works of a lot of residential buildings were not fulfilled this lead unfortunately to fast deterioration of the status of older buildings.

Based on the supervision of buildings we can draw conclusions with regard to the characteristic construction defaults and the characteristic construction methods of the given time period.

The conclusion and resorts of the analysis of the database may facilitate to implement the ideas of rehabilitation of the given parts of the city. It can be used in the course of the renovation and utilisation of buildings which are in bad condition and it may be helpful in the optimum utilisation of economical opportunities. Considerable financial savings can be realised by eliminating the errors during operation.

Primary main load bearing structures (foundation structures, wall structures, cellar floor, intermediate floor, cover floor, side corridor structures, step structures and roof structures) and those secondary structures which play an important role in protecting the main load bearing structures (roof covering, facade, footing, tin structures and insulation against ground water and soil moisture) were in the focus of my studies.

In this chapter some examples of the results of the studies of side corridor structures and floor structures is described. These results are from the building structural and building diagnostic analysis of the available expert reports [1].

3.3.1 Building Structural Analysis

At the end of the 19th century side corridors became dominant in Hungary, when lots of multi-storey residential buildings were built.

Although these days the central corridor design was applied in case of a remarkable part of multi-storey residential buildings to access the flats, in certain periods of the 19th and 20th centuries the side corridor design was nearly exclu-sively applied to access the flats. Since a considerable part of these residential buildings still exists, it is unavoidable to analyse them for building diagnostics and building structure in the interest of economical sustainability of such structures.

Side corridors were usually built with three different types of structural systems: with steel brackets (58%), stone brackets (24%), and monolithic reinforced concrete plates. Side corridors built with steel brackets were made with reinforced concrete plates (51%), Prussian vault (26%), cinder concrete plates (MATRAI) (16%) and natural stone plates (7%) alike. Side corridors built with stone brackets are covered by natural stone plates (46%), monolithic reinforced concrete plates (47%) and precast concrete plates (7%) (See Fig. 9).

The plate structure of steel and stone bracket type side corridors is nearly exclusively of the single span support design in case of cinder concrete, natural stone

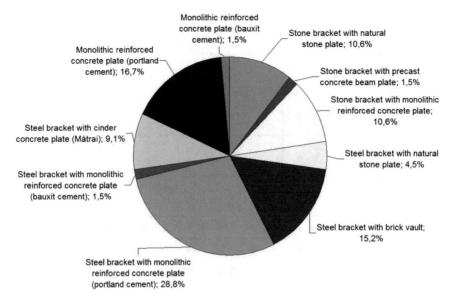

Monolithic reinforced concrete plate (bauxit cement); 1,5%

Stone bracket with natural stone plate; 10,6%

Monolithic reinforced concrete plate (portland cement); 16,7%

Stone bracket with precast concrete beam plate; 1,5%

Stone bracket with monolithic reinforced concrete plate; 10,6%

Steel bracket with cinder concrete plate (Mátrai); 9,1%

Steel bracket with natural stone plate; 4,5%

Steel bracket with monolithic reinforced concrete plate (bauxit cement); 1,5%

Steel bracket with brick vault; 15,2%

Steel bracket with monolithic reinforced concrete plate (portland cement); 28,8%

Fig. 9 Apportionment of structures of side corridors in case of buildings under review

and prefabricated reinforced concrete plates. If monolithic reinforced concrete materials are used for the plate structure, it can be of single span plate or continuous plate according to its static frame. In case of single span design the reinforced concrete plate is arranged between the bottom and the top flange of the steel beams, and the plate is supported by the bottom member. In case of the continuous slab design two general structures were used. In the first case the bottom reinforcement rod is of the single span plate design, and is arranged between the steel beams, and the top reinforcement rod is of the continuous plate design arranged over the steel beam. In the second case the reinforced concrete plate is placed over the steel beams, and the bottom and the top reinforcement rod is of the continuous design alike.

In case of the majority of side corridors constructed with monolithic reinforced concrete one side of the slab is restrained into the reinforced concrete ring beam in its entire width. The stability is ensured by the weight of the brickwork over the slab.

Steel bracket type suspension corridors were usually built between 1890 and 1920, the monolithic reinforced concrete plates became popular from the 1930s. In case of the examined building stock stone brackets were used in all of the construction periods (Fig. 10).

Floors are one of the most significant type of load bearing structure of a residential building. It is intended to transfer the forces to the walls and the lintel beams. Based on their location the floors can differ within one building, too. So the cellar floor, the intermediate floor and the cover floor were separately studied.

In case of intermediate floors the most frequently used (70%) type of floors is the steel beam. Typically brick vault (54%), reinforced concrete or Horcsik slab (9%) was used between the steel beams. 12% of the floors was made of bottom or top

Á. Bukovics et al.

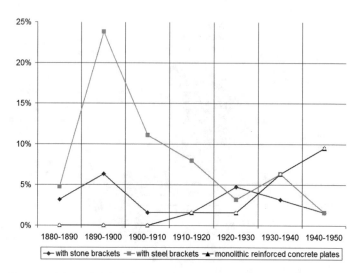

Fig. 10 Changes of the material of side corridors in term of time

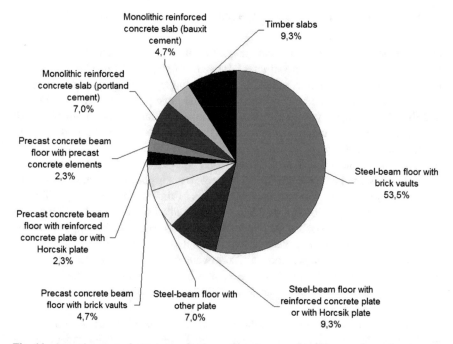

Fig. 11 Apportionment of structures of side corridors in case of buildings under review

ribbed monolithic reinforced concrete (using Portland cement or bauxite cement), and 9–9% of floors was made of precast reinforced concrete beams (with brick vault, reinforced concrete or Horcsik slab, prefabricated concrete or reinforced concrete element), and of timber (Fig. 11).

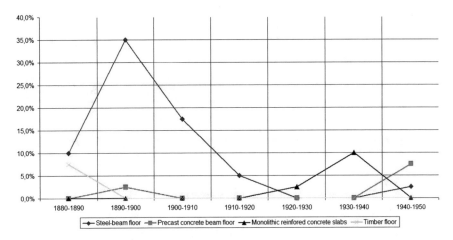

Fig. 12 Changes of the material of intermediate floor in term of time

Wooden floors were only made up to 1890. Steel-beam floors became popular around the turn of the century, Monolithic reinforced concrete plates were basically built from 1930 (Fig. 12).

3.4 Building Diagnostic Analysis

In case of the monolithic reinforced concrete plate type side corridors the most frequently occurring defects are the frost damage (43%), cracks of the slab plate (17%) and corrosion of the reinforcement rod (13%) (See Fig. 13). In case of steel brackets the corrosion of steel beams is often occurring.

Fig. 13 Corrosion of the reinforcement rod

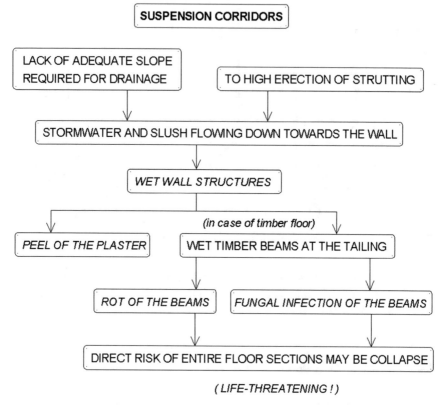

Fig. 14 Flowchart of structural damages of side corridors caused by inadequate slope

The lack of providing an adequate slope required for drainage was a common mistake made during the construction of side corridors, thereby causing constant soaking of flats in the line of perimeter walls. A part of the side corridors in the older residential buildings is in bad condition.

24% of the analysed side corridors were strutted in order to ensure safe usage. Inefficient strutting deteriorated wall and floor structures in many cases. In many cases strutting was erected too high, as a consequence whereof slush and storm-water was flowing down towards the wall, making them wet, thereby causing deterioration, many often the peel off of the plaster. In case of timber floor wetting occurs at the place where the joists are the most sensitive to wetting, that is at the tailing thereof. Water-soaked beams may get rotted, as a result whereof entire floor sections may be collapse (Fig. 14).

In case of side corridors made with Prussian vault brickwork between the steel beams it was a frequently made mistake that the Prussian vault brickwork was made with irregular brick bonds.

In case of cellar floors and intermediate floors built with steel beams the most frequently occurring defect is the corrosion of steel beams, while in the case of reinforced concrete plates it is the corrosion of the reinforcement rod occurring due to the imperfect concrete cover for reinforcement.

42% of bottom or top ribbed monolithic reinforced concrete slabs was made of bauxite cement. This material is exposed to considerable decrease of strength in an environment with high moisture content and due to high temperature. The cinder concrete slab structure (Matrai-slab) between steel beams is very sensitive to moisture, too. If exposed to moisture, sulfuric and the derivatives thereof may be produced, which may cause corrosion of the steel beams and the reinforcement rod.

It is a frequent damage of steel beam floor with reinforced concrete slabs that the reinforced concrete slab elements get loosened from the flange of the beam. Usually it happens as a result of the movement of buildings. As a result of further movements of buildings the floor may become life-threatening due to smaller support of the slab sections.

4 The Applicability of the Fuzzy Approach on the Problem

In many cases the traditional two valued logic is not suitable for modelling or handling a given occurrence. For example, when describing the condition of a load-bearing building structure the linguistic characteristics of "proper condition" cannot be handled by Boolean-logic, because it cannot be sharply determined where the border between the "proper" and "improper" condition is. In the case of decisive majority of this type of linguistic variables a certain type of joint attribute of inaccuracy and uncertainty can be well observed. Fuzzy type inaccuracy is somehow linked to human thinking.

Fuzzy sets were introduced in 1965, and is considered now as one of the main areas of the Soft Computing methods [2]. Shortly after its introduction several significant results were achieved in the field of modelling and control as well. Systems for drawing conclusions were elaborated, which were successfully used in many fields of engineering. Binary logic can be considered a special case of fuzzy logic, since fuzzy logic retains one of the basic axioms of two-valued logic. Interval of fuzzy logic variables [0, 1], and within this any optional value can be taken. In case of 0 the applied linguistic variable is "totally false" while in case 1 it is "totally true". In case of 0.5 the statement is, half true, while, for example, in the case of 0.1 it is "nearly false".

Modelling of the condition of load bearing structures of residential buildings is a complex task, the components of which are well structured and a hierarchical structure can be built up from them. Thereby certain components of the structure are determined by a partial tree of components on a higher level. Also the built-up structure may provide significant additional information on the problem. Building structures can be analysed and compared also on various levels.

Fuzzy signatures can be well used for modelling problems what may be modelled by a hierarchical structure [3, 4]. They are also suitable for handling cases where some components of data are missing. It can be well used, too if there is partial difference in the structure of two data elements.

With the help of the methods created in the course of our work it is possible to draw conclusions on the condition of a residential building or on the stock of residential buildings, modelled by fuzzy signatures.

The set of data which belongs to the problem to be modelled has a joint basic structure.

Structure of data may be different to a small extent, because some of the components of set of data may be missing. (For example, a building with or without side corridor). In order to be able to evaluate and compare the data an aggregator operator is assigned to each tip of the base structure with the aim of modifying the structure. Relevance weight is also specified, by which further information can be buffered on the relationship of components.

In a mathematical sense, fuzzy signatures are hierarchical representations of data structured into vectors of fuzzy values. A fuzzy signature can be illustrated by a nested vector valued fuzzy sets and by a tree graph.

An aggregated value can be computed from a set of aggregator operator values. The operation of the aggregation can be specified by an n-operand function [1]. h: $[0, 1]^n \rightarrow [0, 1]$.

5 Methods Developed for Modelling Status Surveys

We have worked out methods for ranking the buildings based on various viewpoints which are modelling and utilizing the results of status surveys of residential buildings, promoting optimal utilization and priority of buildings, supporting interventions (their rankings and priorities) in the course of renovations and transformation of buildings.

Some of these methods do not take into account the expert's uncertainties, and some of them take into consideration the circumstances of construction and survey, as well as the competence of and also the work invested by the expert into the survey.

The results, characterising the condition of a building can be different depending on the applied method, too. The condition of a building can be characterised by a singleton value, an interval valued fuzzy set, triangular or trapezoidal shaped fuzzy membership function with two parts (which represent the lower and upper bound of the uncertain membership grade).

In the next three methods will described which are suitable for modelling the condition of building structures.

5.1 Fuzzy Singleton Signature Based Model for Qualification and Ranking of Residential Buildings

5.1.1 The Basic Structure of Fuzzy Signatures

In order to determine the conditions of buildings, a fuzzy singleton signature based model was proposed.

In order to model the goals, initially the basic structure of fuzzy singleton signature, characterizing the problem was set up on the basis of the data available from the prepared data base.

Fuzzy membership degrees, which can be used in the system, were assigned to the individual data elements by experts, from the interval [0, 1].

When applying the fuzzy signature based qualifying and ranking method, the examined main load bearing structures will be jointly named as "primary" structures. These are the following:

- foundation structures,
- wall structures,
- cellar floor structures,
- intermediate floor structures,
- cover floor structures,
- side corridor structures,
- step structures
- roof structures.

Those not primary load-bearing structures, which however play an important role in the protection of the main load bearing structures themselves will be called "secondary structures". These are the following:

- roof covering,
- surface formation,
- tin structures,
- insulation against soil moisture and ground water.

The applicability of the elaborated method is demonstrated by using a database, created from 340 real expert assessment reports which contains multi-storey residential buildings.

The basic structure to apply here is a four-level fuzzy signature, because in the course of the examination of these old residential buildings this depth was found to be the most appropriate to achieve sufficient accuracy in describing the condition of buildings.

Because the database was prepared on the basis of the examination of numerous residential buildings in district 13, characteristic of Budapest, so the results obtained, express well the actual general conditions. The set up of the fuzzy singleton signature structure in the form of a tree structure is shown in Fig. 15.

The information available about the conditions of the load bearing structure of the buildings, providing the basis for the database, was classified into two main groups. The first level of the fuzzy signature tree splits the information into "primary structures" and "secondary structures" groups (μ_{11}, μ_{12}).

The same set-up of the fuzzy singleton signature in vector format is shown in Fig. 16. The vector format illustrates the structure in the line of vectors embedded in one another.

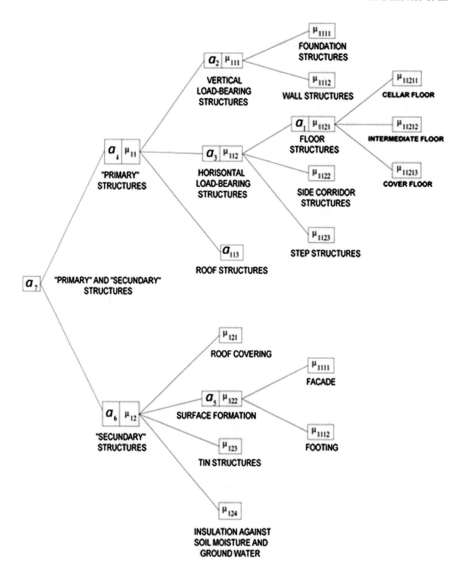

Fig. 15 Set up of fuzzy singleton signature structure

5.1.2 The Applied Aggregation Operators and Relevance Weights

It may be necessary to modify the structures and the values appearing on the leaves of the modified structure depend on the aggregation operators. Therefore the aggregation operators play thereby significant role in the evaluation, and then in the comparison of two signatures. A loss of information will unavoidably occur, when reducing the sub-trees, since the aggregated status descriptors may be equal to one

Fig. 16 Complete set-up of
the fuzzy singleton signature
structure in vector format

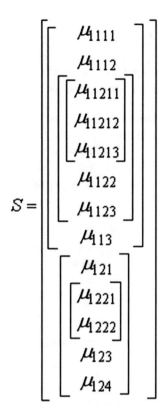

another, while the original signatures were different. It occurs even if the sub-trees
have different arrangement and information contents. With the use of weighted
aggregations it is possible to take further expert knowledge into consideration. The
relevance weight shows the importance of each node, related to the root of the
sub-tree. The sub-groups of variables together specify a component on a higher
level. Therefore the components within the sub-trees of the structure may relate to
the roots of the sub-trees in a way unlike the components of other sub-trees related
to their respective roots. Thus different aggregation operators should be assigned to
each node of the fuzzy signature structure.

The aggregation operators and the relevance weights were determined based on
the official expert reports, and on the basis of evaluating the importance of the
structure related to the whole residential building.

In this case for aggregations the weighted average was used which is a special
case of the weighted generalized mean aggregation operation [5] where p = 1.

$$a(\mu_1, \mu_2, \ldots, \mu_n) = \left(\sum_{i=1}^{n} w_i \cdot \mu_i^p \right)^{\frac{1}{p}}, \tag{1}$$

The notations used in the formula: a: (aggregation function), μ_i: (membership degree of descendant i), w_i: (relevance weight of descendant i), n: (number of successors in the node to be aggregated), p: (aggregation factor ($p \in R$, $p \neq 0$)).

After the evaluation of the related literature and the expert reports the proper relevance weights were chosen.

With the use of the aggregation operators we can modify the structure of the fuzzy signatures. Then a sub-tree of the variables is reduced to the root of the sub-tree.

In Fig. 17 one of the recursive processes of the model can be seen. In order to reduce the sub-tree with the root corresponding to μ_1, in the first instance the sub-tree marked μ_{12} must be reduced. After that the root, marked by μ_1 can be reduced, too.

The applied aggregation operators were determined by using the values on the root of the structure and the relevance weights.

The applied aggregation operators are the following:

Floor structures:

$$a_1 = \frac{0.35 \cdot m \cdot \mu_{11211} + 0.45 \cdot (n-1) \cdot \mu_{11212} + 0.20 \cdot \mu_{11213}}{0.2 + 0.45 \cdot (n-1) + 0.35 \cdot m} \tag{2}$$

Vertical load bearing structures:

$$a_2 = (0.50 - 0.05 \cdot (n-1)) \cdot \mu_{1111} + (0.50 + 0.05 \cdot (n-1)) \cdot \mu_{1112} \tag{3}$$

Horizontal load bearing structures:

$$a_3 = \frac{0.65 \cdot \mu_{1121} + 0.20 \cdot f \cdot \mu_{1112} + 0.15 \cdot \mu_{1112}}{0.80 + 0.20 \cdot f} \tag{4}$$

Primary structures:

$$a_4 = 0.40 \cdot \mu_{111} + \frac{0.60 \cdot n}{n+1} \cdot \mu_{112} + \frac{0.60}{n+1} \cdot \mu_{113}$$
$$= 0.40 \cdot \mu_{111} + \frac{0.60}{n+1} \cdot (n \cdot \mu_{112} + \mu_{113}) \tag{5}$$

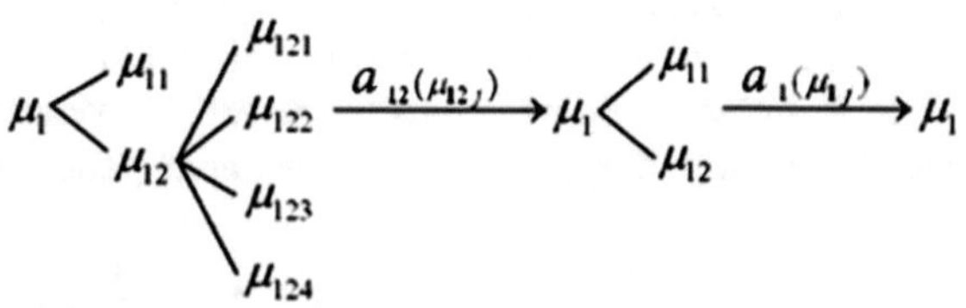

Fig. 17 A recursive process of the evaluation

Surface formation:

$$a_5 = \frac{n - 0.50}{n} \cdot \mu_{1221} + \frac{0.50}{n} \cdot \mu_{1222} \tag{6}$$

Secondary structures:

$$a_6 = \frac{0.40 \cdot \mu_{121} + 0.20 \cdot n \cdot \mu_{122} + 0.20 \cdot \mu_{123} + 0.20 \cdot \mu_{124}}{0.80 + 0.20 \cdot n} \tag{7}$$

Primary and secondary structures:

$$a_7 = 0.75 \cdot \mu_{11} + 0.25 \cdot \mu_{12} \tag{8}$$

Possible values of the parameters are:
$n = 2, 3, 4, 5$ (number of the storeys of the building)
$0 \le m \le 1$ (extent of the cellar)
$f = 0$ or 1 (building with or without side corridors).

5.1.3 Software

In order to determine the status and do the ranking of multi-storey residential buildings of similar formation and age a simple software was prepared on the basis of the fuzzy singleton signature.

The input data may be constant or it may change in terms of time. Constant input data are for example the applied materials, the applied structures and the formation of the building. Input data which change in terms of time are for example the extent of cracks the appearance of cracks and the extent of the corrosion.

For each member of the examined residential buildings a value in [0, 1] can be computed, which express the overall quality of the residential building.

Defining the present-day condition of the buildings and the future utilisation ("suggested for demolition"; "suggested for renewal"; "renewal is necessary, but not immediately") is the aim of the ranking.

With the ranking a more sophisticated decision making is possible as to what can and should be done with the given residential building.

5.1.4 Application of the Model

In order to study and analyze the database a fuzzy singleton signature based software was created. With the help of this software valuable information can be gained on the status of the examined building structures and their respective links to each another. The overall condition of the residential building is characterised by the aggregated status descriptor μ_1 (which was calculated with the help of the fuzzy signature based software). The condition of the buildings and the whole set of buildings were examined in two ways, which were called "normal tuning" and "fine tuning" method. Input data are less detailed in the case of the "normal tuning method", than in the case of the "fine-tuning method", and the output results give only approximation classification.

The results of the computation in the case of the "normal tuning method" are illustrated in Fig. 18.

Studying the diagram the following observation can be done:

In 17.65% of the tested cases the value of μ_1 summarised aggregated state descriptor does not exceed 0.3. This means that the status of the tested structures of the building would be qualified "inappropriate". More than 80% of the examined buildings are qualified as "appropriate" ($0.3 \leq \mu_1 \leq 0.7$). Only 1.47% of the buildings are qualified as "good condition" ($\mu_1 > 0.7$). It can be established that a large part of the examined residential buildings is in need of renovation. It is partly because of the omission of former preservation works.

The stock of buildings was tested with the normal and the fine tuning methods. It was determined that the value of the summarised aggregated state descriptor (μ_1), describing the condition of the buildings was different only to a minimum extent in the case of the two methods.

It may be concluded that in the tests it is usually enough the use the normal tuning method. The application of the fine-tuning method is necessary only in special cases.

With the use of the model it is possible to take a suggestion for prompt interventions in order to avoid further rapid deterioration of the condition. It is worth to begin the preservation process immediately when the condition of the main load-bearing structures is relatively good, however the condition of the secondary

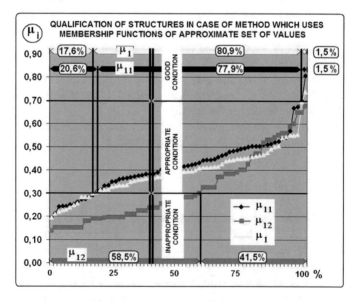

Fig. 18 Qualification of structures in case of the "normal tuning method"

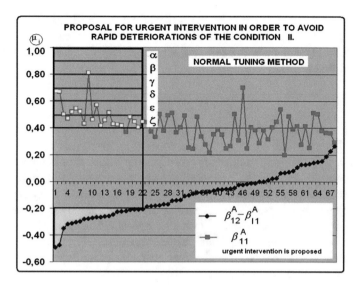

Fig. 19 Suggestion for urgent interventions

structures, protecting the same, is in deteriorated condition. In this case the "secondary" structures can not protect the main load bearing structures from deteriorative effects any more.

In the next an urgency ranking (suggested ranking of restoration) on the basis of the database is specified. It must be defined in which case of the summarised aggregated state descriptor, related to μ_{11} (main load bearing structure) it is advisable to check the summarised condition of the "secondary" structures. After studying the database it is suggested that in the case of $\mu_{11} \geq 0.4$ summarised aggregated status descriptor values should be checked if it is necessary to perform quick intervention.

In such a case the condition of the "primary" structures is good enough for an intervention and taking into account the financial aspects, it is necessary to do so in order to prevent any accidents later. It is also advisable to determine the range of μ_{12} summarised aggregated status descriptor value, when it is advisable to start renovation. When $\mu_{11} \geq 0.4$ and $\mu_{12} \leq 0.3$ or $\mu_{12} - \mu_{11} \leq -0.2$, quick intervention of the "secondary" structures is recommended (Fig. 19).

5.2 Modelling the Condition of Residential Buildings by Real Fuzzy Sets

5.2.1 Introduction of Real Fuzzy Sets and the Applied Operations

According to the original definition of fuzzy sets the elements of the basic set (universe of discourse X) are mapped into the [0, 1] unit interval by the membership function, thereby providing membership values for the given elements of the set [4] m: $X \rightarrow [0, 1]$. Later on, this definition was generalised in many ways, and the one, best suiting our present goals is the membership function extended to the whole real line [6]:

$$\mu_A : X \rightarrow R^1 \tag{9}$$

Negative and greater than one membership values have no physical meaning, however this extended set of values enable us to make certain technical computations in a way that such "imaginary" membership values should be generated as intermediate results (while the final result is a true membership value in [0, 1]). This extension is primarily justified by the fact that in the so-called I-fuzzy and R-fuzzy axiomatic systems it is possible to determine the inverse of the set intersection and the union under certain conditions. In many aspects the features of these inverse operations are similar to the algebraic inverse operations [7].

The simplest and most commonly used generally prevailing fuzzy operations, satisfying the R-fuzzy axioms are as follows.

- Disjunction (algebraic sum):

$$\mu_{A \vee B} = \mu_A + \mu_B - \mu_A \cdot \mu_B \tag{10}$$

- Conjunction (algebraic product):

$$\mu_{A \wedge B} = \mu_A \cdot \mu_B \tag{11}$$

Based on the axiomatics it is possible to introduce fuzzy inverse disjunction and fuzzy inverse conjunction.

- Inverse disjunction:

$$\alpha_{A \vee -B} = \frac{\alpha_A - \alpha_B}{1 - \alpha_B} \quad \alpha_B \neq 1 \tag{12}$$

If $\alpha_B > \alpha_A$, then $\alpha_{A \vee -B} < 0$;

- Inverse conjunction:

$$\alpha_{A \wedge -B} = \frac{\alpha_A}{\alpha_B} \quad \alpha_B \neq 0 \tag{13}$$

If $\alpha_A > \alpha_B$, then $\alpha_{A \wedge -B} > 1$.

5.2.2 Determining the Membership Values

For the determination of the membership values of the examined attributes, the number of the factors influencing the condition of the examined building element must be specified. These factors describe the intensity and the expansion of the investigated effect exerted on the status of the examined building element.

We introduced "status improving factors" that modify the condition of the building element in a positive direction. They will be denoted by b_i (i = 1,..., n, n being the number of status improving factors taken into account). Similarly "status deterioration factors" modify the condition of the building in a negative direction. This will be denoted by α_i (i = 1,..., m, m being the number of status deterioration factors taken into account).

In the examined case the application of 6 (3 status improving and 3 status deterioration) factors were justified. (In the general case there may be any number of such factors).

It is supposed that each factor is effecting the condition in each case the examination must be restricted to strictly monotonously behaving fuzzy operators. Fuzzy disjunction interpreted as a strictly monotonous fuzzy operator, is suitable for taking into consideration all the factors. The so called "result of status improving factors" includes the effect of all status improving factors. It is denoted with b_P.

The "result of status deterioration factors" includes the effects of all status deterioration factors. It is denoted with α_N. These two factor was determined with the help of the fuzzy disjunction (1).

In case of any n positive, and m negative factors effecting the membership value, using the fuzzy disjunction, the expression of factors β_P and α_N are given as the next formulas:

$$\beta_P = \beta_{\bigcup_{i=1}^{n} A_i} = \sum_{i=1}^{n} \beta_i - \sum_{\substack{i \neq j \\ i=1}}^{n} \beta_i \cdot \beta_j \pm \cdots \pm (-1)^{n-1} \cdot \prod_{i=1}^{n} \beta_i = 1 - \prod_{i=1}^{n} (1 - \beta_i) \quad (14)$$

$$\alpha_N = \alpha_{\bigcup_{i=1}^{m} A_i} = \sum_{i=1}^{m} \alpha_i - \sum_{\substack{i \neq j \\ i=1}}^{m} \alpha_i \cdot \alpha_j \pm \cdots \pm (-1)^{m-1} \cdot \prod_{i=1}^{m} \alpha_i = 1 - \prod_{i=1}^{m} (1 - \alpha_i) \quad (15)$$

These so called sieve formulas could be converted into a closed form, using De Morgan's equality [8]. In case of three status improving and three status deterioration factors (as in the case of our investigation), the value of β_P and α_N can be calculated by the following:

$$\beta_P = \beta_{Af \vee Bf \vee Cf} = \beta_{Af} + \beta_{Bf} + \beta_{Cf}$$
$$- \beta_{Af} \cdot \beta_{Bf} - \beta_{Af} \cdot \beta_{Cf} - \beta_{Bf} \cdot \beta_{Cf} + \beta_{Af} \cdot \beta_{Bf} \cdot \beta_{Cf} \tag{16}$$

$$\alpha_N = \alpha_{Af \vee Bf \vee Cf} = \alpha_{Af} + \alpha_{Bf} + \alpha_{Cf} - \alpha_{Af} \cdot \alpha_{Bf}$$
$$- \alpha_{Af} \cdot \alpha_{Cf} - \alpha_{Bf} \cdot \alpha_{Cf} + \alpha_{Af} \cdot \alpha_{Bf} \cdot \alpha_{Cf} \tag{17}$$

In order to describe the condition of the building structure in two different ways we propose two extreme values. These two together determine an interval valued fuzzy set (Fig. 20). The two extreme values of the interval are defined by the membership values calculated on the basis of what we call "optimistic" and "pessimistic" estimation.

The "optimistic" estimation (λ_{SUP}) means the expected upper limit of the membership value of the examined structure. The result of the status improving factors is multiplied by the complement of the status deterioration factors and this value is divided by the complement of the relative deterioration ($\beta_P \cdot \alpha_N$).

"Optimistic" estimation:

$$\lambda_{SUP} = \frac{\beta_P \cdot (1 - \alpha_N)}{1 - \beta_P \cdot \alpha_N} \tag{18}$$

The "pessimistic" estimation (λ_{INF}) means the lower limit of the membership value of the tested building structure. The result of the status improving factors is multiplied by the complement of the resultant of status deterioration factors.

"Pessimistic" estimation:

$$\lambda_{INF} = \beta_P \cdot (1 - \alpha_N) \tag{19}$$

The received membership values (the results of the two estimations) were compared with the results of expert evaluation. It is established that this latter one was well approximated by the average of the two values ($\lambda_{INF}, \lambda_{SUP}$). Therefore this interval is recommended for describing the estimated condition of the examined building structure (λ_{AVG}).

The estimated average condition:

$$\lambda_{AVG} = \frac{\lambda_{INF} + \lambda_{SUP}}{2} \tag{20}$$

In the next it is considered as the overall membership value connected with the status of the examined building structure. These values can be applied when using the fuzzy signature-based status determining and ranking model.

Fig. 20 Interval valued fuzzy set

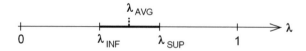

5.2.3 The Applicability of the R-Fuzzy Algebraic Structure

As a result of the methodological research we found that in each case the upper value, (with the application of fuzzy inverse disjunction) was greater than the lower one (obtained by the application of fuzzy conjunction) ($\lambda_{SUP} > \lambda_{INF}$).

We investigated which values may be supposed by the upper and lower membership functions and the estimated condition depending on the resultant of the status improving factors (b_P) and the status deterioration factors (α_N).

In our method (described in the following section) the smallest value of the resultant of status improving factors may be 0.65, since the minimum value of b_{A1}, cannot be smaller than this. The value of μ_P was examined in the interval between 0.65 and 1 with 0.05 accuracy.

The resultant of the α_N status deterioration factors was examined and plotted in the interval [0, 1] with 0.05 accuracy. The value and the character of the examined fuzzy algebraic operators are influenced by status deterioration effects.

The alteration of the value of fuzzy algebraic operators was investigated in connection with the value of the status deterioration factors in case of various characterising values of the status improving factors. In Fig. 21 the resulting diagram can be seen In case of $b_P = 0.85$.

In this diagram the blue line shows the "optimistic" estimation, the yellow one shows the "pessimistic" value, while the purple one shows the estimated average condition of the membership value of the examined building structure.

Usually the difference between the "optimistic" and "pessimistic" values of the membership function (the "open scissors") is great if the value of the resultant of status improving factors and that of the status deterioration factors is great alike. Than the difference between the "optimistic" and "pessimistic" value of the membership function may exceed 0.60. The value of λ_{SUP}-λ_{INF} decreases in the case when the resultant of the status deterioration factors has an extraordinary small or large value. If the difference between the "optimistic" and "pessimistic" values of the membership function exceeds 0.45, a detailed examination of the reasons of deterioration is recommended.

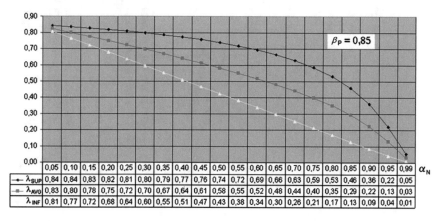

α_N	0,05	0,10	0,15	0,20	0,25	0,30	0,35	0,40	0,45	0,50	0,55	0,60	0,65	0,70	0,75	0,80	0,85	0,90	0,95	0,99
λ_{SUP}	0,84	0,84	0,83	0,82	0,81	0,80	0,79	0,77	0,76	0,74	0,72	0,69	0,66	0,63	0,59	0,53	0,46	0,36	0,22	0,05
λ_{AVG}	0,83	0,80	0,78	0,75	0,72	0,70	0,67	0,64	0,61	0,58	0,55	0,52	0,48	0,44	0,40	0,35	0,29	0,22	0,13	0,03
λ_{INF}	0,81	0,77	0,72	0,68	0,64	0,60	0,55	0,51	0,47	0,43	0,38	0,34	0,30	0,26	0,21	0,17	0,13	0,09	0,04	0,01

Fig. 21 Possible values of λ_{SUP}, λ_{INF} and λ_{AVG} subject to α_N in case of $\beta_P = 0.85$

5.2.4 Investigating the Model on Foundation Structures

In order to investigate the created model with the help of the examining building stock, the membership values of foundation structures were determined. The applicability of the method based on imaginary fuzzy values was examined. It was investigated how the method could be applied from the viewpoint of the qualification and ranking of the load bearing structures. In Fig. 22 the schematic model of the definition of membership values, featuring the condition of foundation structures can be seen.

The membership value, featuring the foundation structure (in our case strip foundation), is influenced by the following factors.

Positive factors:

Material of the strip foundation (β_{Af})
    Its value was selected depending on the material of strip foundation.

Conformance of the width of the strip foundation (positive direction) (β_{Bf})
    First the ratio of the width of the actual strip foundation (w_f) and the idealized, estimated width of the strip foundation (w_{fi}) was calculated. In order to define the width of the idealized foundation it is necessary to calculate the loads exerted on the strip foundation, as well as to know the soil quality under it. These values are available in the expert reports. After that the expected width of the strip foundations of the examined buildings was determined. Comparing the expected width with the actual width, the factor was obtained. If the actual width is smaller than the idealized width, this factor has no impact.

Fig. 22 Schematic model of the definition of μ_{1111}

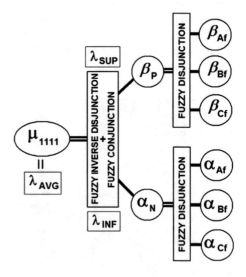

Year of construction (β_{Cf})

It takes two viewpoints into account. One is the number of years that have passed after the construction because of the process of the natural weakening of the status of building structures in terms of time. The second part is whether in the year of the construction the city had already its regulations in effect or has not.

Negative factors:

Detected deteriorations of the strip foundation (α_{Af})

This factor is defined by using the component featuring the status of the formerly determined membership value of foundation structure (μ_{1111}). The bigger the extent of the detected deteriorations is the closer the value of μ_{1111} is to 0.

Conformance of the width of the strip foundation (α_{Bf})

If the width of strip foundation is smaller than the ideal width, the value of α_{Bf} depends on the relation to w_f/w_{fi}.

Other effects deteriorating the quality of the structure (α_{Cf})

It takes into account the effect of vibrations impairing the status of the foundation structures. The effect of vibration is ifferent in case of buildings situated in side streets, in case of buildings situated by roads with high traffic, and in case of buildings situated over the subway train.

5.2.5 Results

In Fig. 23 the values of all three membership functions (λ_{SUP}, λ_{INF}, λ_{AVG}) can be seen. The examined residential buildings were arranged in monotonically increasing sequence on the basis of the estimated value of overall membership functions. It is interesting to observe that in the case of building structures in extremely good and in extremely bad status, the difference between the upper and the lower membership value is generally rather small. The scissors between upper and lower values open up for medium status residential buildings. In the case of the examined buildings the difference between the upper and the lower value of the membership value is 0.25 as an average.

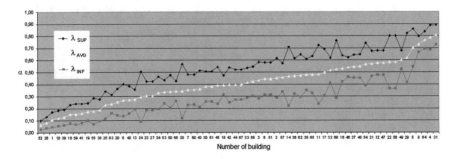

Fig. 23 Values of the membership functions in case of ranked buildings

5.3 Modelling the Uncertainty in the Condition Assessment Using Fuzzy Signature Sets

5.3.1 Effects of Uncertainties Exerted on Membership Functions

When modelling the reliability of expert evaluations uncertainties are assigned to the expert estimation. Then the singleton membership value is transformed into a membership function. In our case triangular- and trapezoidal-shaped membership functions are suggested to model the uncertainties, because they can be easily handled. The steps of determining the membership functions are as follows (Fig. 24). The expert defines a singleton value, which will be first normalised to [0, 1]. The form of the membership function will be changed step by step, while the uncertainties of the expert report are taken into consideration. Every uncertainty modifies the shape of the membership functions.

5.3.2 The Structure of the Fuzzy Signature

In this case a four-level fuzzy signature structure was proposed (see Fig. 25). The membership function at the leaves of the structure is related to the following building structures: foundation structures (A_1), wall structures (A_2), cellar floor (A_3), intermediate floor (A_4), cover floor (A_5), side corridor structures (A_6), step structures (A_7), roof structures (A_8), roof covering (A_9), facade (A_{10}), footing (A_{11}), tin structures (A_{12}), insulation against soil moisture and ground water (A_{13}).

In this approach the applied aggregation operators are piecewise combinations of the min and weighted mean operators, which depend on the special features of the building. For example to aggregation operator h_2, related to the status of the vertical load bearing structures and aggregation operator h_7, related to the status of the building, were determined as follows.

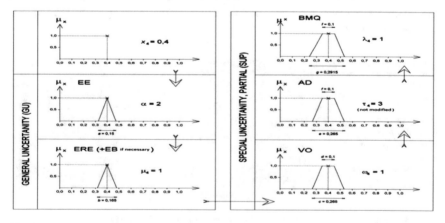

Fig. 24 Modelling the uncertainty with membership function

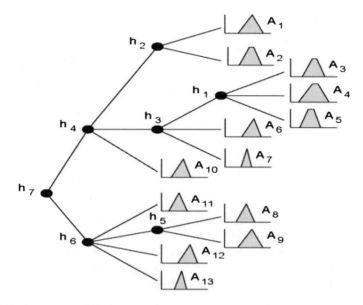

Fig. 25 Basic structure of the fuzzy set signature

$$
h_2 = \left\{
\begin{array}{l}
(0.50 - 0.05 \cdot (n-1)) \cdot x_1 + (0.50 + 0.05 \cdot (n-1)) \cdot x_2 \quad \text{if} \quad x_1 > 0.40 \\
(0.50 - 0.05 \cdot (n-1)) \cdot x_1 + (0.50 + 0.05 \cdot (n-1)) \cdot x_2 \quad \text{if} \quad 0.40 \geq x_1 \geq 0.20 \,\text{and} \,\, x_2 < x_1 \\
(0.50 - 0.05 \cdot n) \cdot x_1 + (0.50 + 0.05 \cdot n) \cdot x_2 \quad \text{if} \quad 0.40 \geq x_1 \geq 0.20 \,\text{and} \,\, x_2 \geq x_1 \\
\min(x_1 ; x_2) \quad \text{if} \quad x_1 < 0.20
\end{array}
\right\}
$$

(21)

$$
h_7 = \left\{
\begin{array}{l}
0.75 \cdot h_4 + 0.25 \cdot h_6 \quad \text{if} \quad h_4 > 0.4 \\
\quad \text{or} \quad 0.2 \leq h_4 \leq 0.4 \,\text{and}\, h_6 < h_4 \\
0.85 \cdot h_4 + 0.15 \cdot h_6 \quad \text{if} \quad 0.2 \leq h_4 \leq 0.4 \,\text{and}\, h_6 \geq h_4 \\
\min(h_4 ; h_2) \quad \text{if} \quad h_4 < 0.2
\end{array}
\right\}
$$

(22)

n is the number of the storeys of the building.

5.3.3 Investigating the Model

For investigating the model database introduced earlier was used. The membership function, which takes into account the general and special uncertainties, is fine-tuned also with this database.

As an example we have chosen an old building which was built in 1894. The human expert's report was modelled by triangular shaped fuzzy membership functions, according to the previous sections. The values of the shape parameters are shown in the table (see Fig. 26).

Fig. 26 Triangular shaped
fuzzy numbers

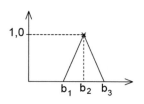

	b_1	b_2	b_3
A_1	0,295	0,350	0,405
A_2	0,200	0,250	0,300
A_3	0,600	0,650	0,700
A_4	0,600	0,650	0,700
A_5	0,202	0,250	0,298
A_6	0,350	0,400	0,450
A_7	0,345	0,400	0,455
A_8	0,190	0,250	0,310
A_9	0,140	0,200	0,260
A_{10}	0,300	0,350	0,400
A_{11}	0,600	0,650	0,700
A_{12}	0,190	0,250	0,310
A_{13}	0,150	0,200	0,250

The fuzzy number, which describes the condition of the whole residential building, is obtained after having taken into consideration the uncertainties in case of each examined building structure in the same way as above. After that the triangular membership function, gained by using the aggregation operators, is further modified, thus taking into account also the special uncertainties characterizing the whole building.

The output of the fuzzy signature is a triangular shaped membership function.

The shape of the final fuzzy set offers lot of information about the uncertainty of the condition assessment, which should be taken into account before the decision making. It is possible to get a crisp conclusion applying some defuzzification methods.

6 Sensitivity and Validity of the Fuzzy Signature Based Evaluation of Residential Building Condition

In the next a few comments will be given as results to the overall sensitivity of the assessment in terms of the uncertainty in the expert report.

6.1 Sensitivity of the Fuzzy Singleton Signature

From the mathematical point of view important questions are the sensitivity and stability of the fuzzy signature-based decision support method. It must be examined whether a small change in the input variables may result a large change of the final membership function, or may not. The answer depends on the structure of the signature and on the applied aggregation operators [9].

The examined fuzzy signature creates the weighted arithmetic mean of the input values in every node, so the final output is the weighted arithmetic mean of the original input values, where the weights can be determined from the tree structure and form the functions used in the nodes. In general, if the input values are $x_1,..., x_n$ ($0 \leq x_i \leq 1$ for all i) $p_1,..., p_n$, where $p_1 + \cdots + p_n = 1$ $p_i \geq 0$ for all i, than the output value is

$$f(x_1, \ldots, x_n) = p_1 \cdot x_1 + \cdots + p_n \cdot x_n \tag{23}$$

Let us change the input values to $x_1^*, \ldots, x_2^*$. Then the absolute value of the change in the output is

$$|\Delta f| = \left| f(x_1^*, \ldots, x_n^*) - f(x_1, \ldots, x_n) \right| = \left| \sum_{i=1}^{n} p_i \cdot \Delta x_i \right| \tag{24}$$

From the well-known Cauchy-Schwarz inequality [10] we get that

$$|\Delta f| \leq \left\| \underline{p} \right\|_2 \cdot \|\Delta \underline{x}\|_2 \tag{25}$$

where $\|*\|_2$ stands for the Euclidean vector norm. From the triangular inequality we get that

$$|\Delta f| = \max_i (p_i) \cdot \|\Delta \underline{x}\|_1 \tag{26}$$

(where $\|*\|_1$ denotes the sum norm (taxicab norm) [11]).

In order to examine the sensitivity of the applied fuzzy singleton signature the input values are numbered form top to bottom. In the final decision their respective weights are the following:

$$p_1 = 0.16500 - 0.01500n \tag{27}$$

$$p_2 = 0.13500 + 0.01500n \tag{28}$$

$$p_3 = 0.1023750 \frac{nm}{(-0.25 + 0.45n + 0.35m)(0.8 + 0.2f)(n+1)} \tag{29}$$

$$p_4 = 0.1316250 \frac{(n-1)n}{(-0.25 + 0.45n + 0.35m)(0.8 + 0.2f)(n+1)} \tag{30}$$

$$p_5 = 0.058500 \frac{n}{(-0.25 + 0.45n + 0.35m)(0.8 + 0.2f)(n+1)} \tag{31}$$

$$p_6 = 0.0900 \frac{fn}{(0.8 + 0.2f)(n+1)} \tag{32}$$

$$p_7 = 0.06750 \frac{n}{(0.8 + 0.2f)(n+1)} \tag{33}$$

$$p_8 = \frac{0.45}{n+1} \tag{34}$$

$$p_9 = \frac{0.1}{0.8 + 0.2n} \tag{35}$$

$$p_{10} = \frac{0.05n - 0.025}{0.8 + 0.2n} \tag{36}$$

$$p_{11} = \frac{0.025}{0.8 + 0.2n} \tag{37}$$

$$p_{12} = \frac{0.005}{0.8 + 0.2n} \tag{38}$$

$$p_{13} = \frac{0.005}{0.8 + 0.2n} \tag{39}$$

Analysing the above expressions numerically we get the maximal value of $\sum_{i=1}^{13} p_i^2$ (see Table 1).

With the results of Table 1 we get an upper estimation of the change of the output.

$$|\Delta f| \le 0.4 \cdot \|\Delta \underline{x}\|_2 \tag{40}$$

Based on the maximal values of $|p_i|$ (see Table 2) we can conclude that

$$|\Delta f| \le 0.28 \cdot \|\Delta \underline{x}\|_1 \tag{41}$$

From those results it follows that the change of the output is bounded in the following sense: a *small* change of the input values can not yield a *large* change of the output.

Table 1 maximal values of $\sum_{i=1}^{13} p_i^2$

f/n	2	3	4	5
0	0.12	0.14	0.15	0.16
1	0.12	0.12	0.13	0.14

Table 2 Maximal values of $|p_i|$

P_1	P_2	P_3	P_4	P_5	P_6	P_7
0.135	0.21	0.0853	0.2743	0.075	0.075	0.0703
P_8	P_9	P_{10}	P_{11}	P_{12}	P_{13}	
0.15	0.0833	0.125	0.0208	0.0417	0.0417	

6.2 Sensitivity of the Applied Operations Defined by the Fuzzy Sets of Real Values

In the pessimistic case

$$|\Delta\lambda_{\inf}| \leq |\Delta\beta_P| + |\Delta\alpha_N| \tag{42}$$

And in the optimistic case

$$|\Delta\lambda_{\sup}| \leq (\Delta^2\alpha_N + \Delta^2\beta_P) \cdot \sqrt{\frac{(1-a)^2 + b^2 \cdot (b-1)^2}{(1-a\cdot b)^4}} \tag{43}$$

Here a is the maximal value of α_N and β is the maximal value of b_P.

Sensitivity of the whole R-fuzzy signature
    "Optimistic case"
    Let us assume that we use $3 + 3$ values to determine the values of α_N and β_P. Applying our results described in the previous sections we can conclude that the upper estimation of $|\Delta f|$ with the Euclidean norm is the following:

$$|\Delta f| \leq \sqrt{0.48} \cdot K_{2\max}(a,b)\sqrt{\sum_{j=1}^{13}\sum_{i=1}^{3}(\Delta^2\beta_{ij} + \Delta^2\alpha_{ij})} \tag{44}$$

where $K_{2\max}(a,b) = \max_j\{K_{2j}(a,b)\}$.

The estimation using the sum norm is the following:

$$|\Delta f| \leq 0.28 \cdot K_{1\max}(a,b)\sum_{j=1}^{13}\sum_{i=1}^{3}(|\Delta\beta_{ij}| + |\Delta\alpha_{ij}|) \tag{45}$$

where $K_{2\max}(a,b) = \max_j\{K_{1j}(a,b)\}$.

"Pessimistic case"

We assume again that we have 3 + 3 values to determine the value of α_N and β_P. Then the estimation with the Euclidean norm of the input values is

$$|\Delta f| \le \sqrt{0.96} \cdot \sqrt{\sum_{j=1}^{13} \sum_{i=1}^{3} (\Delta^2\beta_{ij} + \Delta^2\alpha_{ij})} \qquad (46)$$

while the estimation with the sum norm of the input values is

$$|\Delta f| \le 0.28 \cdot \sum_{j=1}^{13} \sum_{i=1}^{3} (|\Delta\beta_{ij}| + |\Delta\alpha_{ij}|) \qquad (47)$$

7 Conclusion

Based on the above it can be stated that fuzzy signatures are essentially suitable for modelling the available expert evaluation reports.

It is also reasonable to draw all evaluations of different depth and details to a joint platform and to determine which are eventually defective. Also in case of buildings of different structure modelling by fuzzy signatures may present the evaluations on a joint platform.

It was established that the inaccurateness and uncertainties inevitably occurring in expert reports do not increase the uncertainty of resulting evaluation to such an extent that would make the results invalid.

Since the uncertainty in expert evaluations is multiple, in addition to normal fuzzy signatures an extended (rich) toolbox is offered which includes fuzzy sets of real values, fuzzy set signatures, as well as type 2 fuzzy signatures.

Acknowledgements This work was supported by National Research, Development and Innovation Office (NKFIH) K105529, K108405. The authors would like to thank to the National Research, Development and Innovation Office (NKFIH) K105529, K108405 grant for the support of the research.

References

1. Bukovics Á (2009) Building diagnostic and pathological analysis of residantial buildings of budapest. In: PATORREB 2009, 3° Encontro Sobre Patologia E Reabilitacao de Edificios, Porto, pp 1025–1030
2. Molnárka G, Kóczy LT (2011) Decision support system for evaluating existing apartment buildings based on fuzzy signatures. Int J Comput Commun Control VI(3):442–457
3. Kóczy LT, Vámos T, Biró G (1999) Fuzzy signatures. In: Proceedings of EUROFUSE-SIC'99, Budapest, pp 25–28

4. Bukovics Á, Kóczy LT (2012) Fuzzy signature-based model for qualification and ranking of residential buildings. In: XXXVIII. IAHS world congress on housing, Istanbul, Turkey, pp 290–297
5. Bullen PS (2003) Handbook of means and their inequalities. Kluwer Academic Publishers
6. Kóczy LT, Hajnal MA (1977) New attempt to axiomatize fuzzy algebra with an application example. Probl Control Inf Theor 6(1):47–66, Budapest
7. Bukovics Á, Kóczy LT (2012) The application of fuzzy connectives and inverse operations on r-fuzzy descriptors of the static conditions of residential buildings. In: Szakál A (szerk.) Proceedings of the 13th IEEE international symposium on computational intelligence and informatics: CINTI 2012. Konferencia helye, ideje: Budapest, Magyarország, 2012.11.20–2012.11.22. IEEE Hungary Section, Budapest, pp 433–437. ISBN:978-1-4673-5204-8
8. Bíró GY, Kóczy LT, Vámos T (2001) Fuzzy signatures in data mining. In: Joint 9th IFSA world congress and 20th NAFIPS international conference, pp 2842–2846
9. Harmati IÁ, Bukovics Á, Kóczy LT (2014) Sensitivity analysis of the weighted generalized mean aggregation operator and its applications to fuzzy signatures. In: Proceedings of the WCCI 2014 (FUZZ-IEEE 2014), July 6–11, Beijing, China, pp. 1327–1332
10. Hardy H, Littlewood JE, P´olya G (1952) Inequalities. Cambridge University Press
11. Golub GH, van Loane CF (1996) Matrix computations. John Hopkins University Press

Retrieval from Uncertain Data Bases

Ronald R. Yager

Abstract We investigate tools that can enrich the process of querying databases. We show how to include soft conditions with the use of fuzzy sets. We describe some techniques for aggregating the satisfactions of the individual conditions based on the inclusion of importance and the use of the OWA operator. We discuss a method for aggregating the individual satisfactions that can model a lexicographic relation between the individual requirements. We look at querying databases in which the information in the database can have some probabilistic uncertainty.

1 Introduction

Database structures play a pervasive role in underlying many websites, as a result the intelligent retrieval of information from databases is an important task. The use of soft concepts often allows us to model human cognitive concepts. Here we focus on providing a framework for retrieving information from database structures via soft querying [1–4]. A query can be seen as a collection of requirements and an imperative for combining an object's satisfaction to the individual requirements to get the object's overall satisfaction to the query. We describe some fuzzy set based methods that enable the inclusion of human focused concepts in the construction of a query [5–8]. Here we also describe techniques for aggregating the satisfactions of the individual conditions to obtain an object's overall satisfaction to a query. The inclusion of importance information and the use of the OWA operator are some tools that provide this facility. Here we also discuss a method for aggregating the individual satisfaction's that can model a lexicographic relation between the individual requirements. In addition we look at databases in which the information can have some probabilistic uncertainty.

R.R. Yager (✉)
Machine Intelligence Institute, Iona College, New Rochelle, NY 10801, USA
e-mail: yager@panix.com

© Springer International Publishing AG 2018
M. Collan and J. Kacprzyk (eds.), *Soft Computing Applications for Group Decision-making and Consensus Modeling*, Studies in Fuzziness and Soft Computing 357, DOI 10.1007/978-3-319-60207-3_17

2   Standard Querying of Data Bases

A database consists of a collection of attributes, V_j, $j = 1$ to r and a domain X_j for each attribute. Associated with a database is a collection of objects Di, the objects in the database. Each D_i is an r tuple, $(d_{i1}, d_{i2}, \ldots, d_{ir})$. Here $d_{ij} \in X_j$ is the value of attribute V_j for object D_i.

A query Q is of a collection of q pairs. Each pair $P_k = (V_{Q(k)}, F_{Q(k)})$ consists of an attribute $V_{Q(k)}$ and an associated property for that attribute $F_{Q(k)}$. Here Q(k) indicates the index of the attribute in the kth pair. Thus if $Q(k) = i$ then the attribute in the kth pair is V_i. In addition to the collection of pairs a query contains information regarding the relationship between the pairs. We denote this information as Agg.

The querying process consists of testing each object D_i to see if it satisfies the query. The process of testing an object consists of two steps. The first is the determination of the satisfaction of each of the pairs by the object. We denote this as $T_r(P_k|D_i)$ or more simple as t_{ik}. The second step is to calculate the overall satisfaction of D_i to the query Q, $Sat(Q/D_i)$, by combining the individual t_{ik}, guided by the instructions contained in Agg. Here then $Sat(Q/D_i) = Agg(t_{i1}, \ldots, t_{iq}) = T_i$.

The calculation of t_{ik}, $T_r[P_k|D_i]$ is based on determining whether the condition $F_{Q(k)}$ is satisfied based on $d_{iQ(k)}$, the value of $V_{Q(k)}$ for the ith object D_i. In the standard environment $F_{Q(k)}$ is represented as a crisp subset of $X_{Q(k)}$ indicating the desired values of $V_{Q(k)}$. The determination of satisfaction, $T_r[P_k|D_i]$, is based on whether $d_{iQ(k)} \in F_{Q(k)}$. The system returns the value $t_{ik} = 1$ if $d_{iQ(k)} \in F_{Q(k)}$ and $t_{ik} = 0$ if $d_{iQ(k)} \notin F_{Q(k)}$. Using the notation $F_{Q(k)}(d_{iQ(k)})$ to indicate membership of $d_{iQ(k)}$ in $F_{Q(k)}$ we get $t_{ik} = F_{Q(k)}(d_{iQ(k)})$. Thus in this case each of the values $t_{ik} \in \{0, 1\}$ where 0 indicates false, not satisfied, and 1 indicates true, satisfied.

In this binary environment the aggregation process is constructed as a well-formed logical statement consisting of conjunctions, disjunctions, negations and implications. The overall satisfaction $Sat(Q|D_i)$ is equal to the truth-value of this well-formed formula. The calculation of this truth-value involves the use of Min, Max and Negation. For example the requirement that all pairs in Q are satisfied is expressed as $T_i = Min_{k = 1 \text{ to } r}[t_{ik}]$. The requirement that any of the conditions need be satisfied is expressed as $T_i = Max_{k = 1 \text{ to } r}[t_{ik}]$. We note this is a binary environment, $T_i \in \{0, 1\}$, and hence we have two categories of objects, those which satisfy the query and those that do not.

3   Soft Database Querying

An extension of the querying process involves the use of soft or flexible queries [6, 8]. This idea extends the process of querying standard databases in a number of ways. The most fundamental is to allow the search criteria, the $F_{Q(k)}$, to be expressed as fuzzy subsets of the domain $X_{Q(k)}$. This allows one to represent imprecise non-crisply bounded concepts. In this case the determination of the satisfaction of a criteria $P_k = (V_{Q(k)}, F_{Q(k)})$ by an object D_i, t_{ik}, again equals the membership grade of $d_{iQ(k)}$ in $F_{Q(k)}$, $t_{ik} = F_{Q(k)}(d_{iQ(k)})$. However in this case $t_{ik} \in [0, 1]$, it takes a value in the unit interval rather than simply in the binary set $\{0, 1\}$. An important implication of this is that the overall satisfaction of the element D_i will also lie in the unit interval. This will allow for a richer and more discriminating ordering of the elements in regard to their satisfaction to the query then simply satisfied or not, here we get a degree of satisfaction. We note the evaluation of the overall satisfaction to a query Q whose Agg operator is based on a well-formed logical formula can be implemented in this framework. In particular conjuncting the satisfactions to two criteria pairs P_{k1} and P_{k2} is implemented by $Min(t_{ik1}, t_{ik2})$. disjuncting the satisfactions to two criteria pairs P_{k1} and P_{k2} is implemented by $Max(t_{ik1}, t_{ik2})$. The negation of the satisfaction to P_{ki} is implemented as $1 - t_{ik1}$.

A second benefit obtained by using flexible queries is the allowance for more sophisticated formulations for the operator Agg used to determine the overall satisfaction to the query [6]. Let us look at some of these extensions.

A first class of extensions is a generalization of the operations used for implementing the "anding" and "oring" operator. Assume P_1 and P_2 are two conditions whose satisfaction for D_i are t_{i1} and t_{i2}. Classically the "anding" of the satisfactions to these two criteria has been implemented using the Min operator: $P_1 \text{ and } P_2 \Rightarrow Min(t_{i1}, t_{i2})$. The "oring" of the satisfactions to these two criteria has been implemented using the Max operator: $P_1 \text{ or } P_2 \Rightarrow Max(t_{i1}, t_{i2})$. The t-norm operator is a generalization of the "anding" operator [9], it provides a class of cointensive operators that can be used to implement the conjunction. We recall a t-norm is a binary operator $T: [0, 1] \times [0, 1] \rightarrow [0, 1]$ having the properties: symmetry, associativity, monotonicity and has one as the identity. Notable examples of t-norm in addition to the Min are $T_P(a, b) = ab$ and $T_L(a, b) = Max(0, a + b - 1)$. Since for any t-norm T we have $T(a, b) \leq Min(a, b)$ then using other t-norms increases the penalty for not completely satisfying the conditions.

The t-conorm operator provides a generalization of the *oring* operator. A t-conorm is a mapping $S: [0, 1] \times [0, 1] \rightarrow [0, 1]$ which has the same first three properties of the t-norm binary, condition four is replaced by $S(0, a) = a$, zero as Identity. In addition to the Max other notable examples of t-conorm are $S_P(a, b) = a + b - ab$ and $T_L(a, b) = Min(a + b, 1)$. We note for any t–norm S we have $S(a, b) \geq Max(a, b)$.

Another extension to the standard situation is the inclusion of importance associated with the conditions [10, 11]. Here we assume each of the q conditions P_k has an associated importance weight $w_k \in [0, 1]$. The methodology for including importance depends on the aggregation relationship between the criteria, it is different for "anding" and "oring".

Assume for object D_i we have that $t_{ik} \in [0, 1]$ as its degree of satisfaction to P_k. Consider first the case in which we desire an "anding" of the P_k. That is our aggregation imperative is expressed as

$$(P1, \ w1) \ \text{and} \ (P2, \ w2) \ \ldots \ \text{and} \ (Pq, \ wq).$$

In this case the weighted aggregation results in an overall the satisfaction Min $(h_1, \ldots, h_q)$ where $h_k = \text{Max}((1 - w_k), t_{ik})$ [12].

It is interesting to observe the standard situation is obtained when all $w_k = 1$. In this case we get $\text{Min}_k[t_{ik}]$. We can also observe that since $\text{Max}((1 - w_k), t_{ik}) \geq t_{ik}$ then $\text{Min}_k[\text{Max}((1 - w_k), t_{ik})] \geq \text{Min}_k[t_{ik}]$. Thus the use of importances is effectively to reduce the requirements, we are easing the difficulty. We also observe that more generally we can replace Min with any t-norm T and Max with any t-conorm S thus we can use $\mathbf{T}^q_{k=1}[S((1 - w_K), t_{iK})]$ [10].

Consider now the case where we have an "oring" of the criteria. Thus here our aggregation imperative is (P_1, w_1) or (P_2, w_2) or (P_3, w_3) or, $\ldots$, (P_q, w_q). In this case the weighted aggregation and the satisfaction is $\text{Max}(g_1, g_2, \ldots, g_q)$ where $g_k = \text{Min}(w_k, t_{ik})$ [10, 13].

We observe the standard situation is obtained when all $w_k = 1$. In this case we get $\text{Max}_k[t_{ik}]$. We can also observe that since $\text{Min}(w_k, t_{ik}) \leq t_{ik}$ then $\text{Max}_k[\text{Min}(w_k, t_{ik})] \leq \text{Max}_k[t_{ik}]$. Thus the use of importances reduces the influence of individual criteria. We also observe that more generally we can replace Min with any t-norm T and Max with any t-conorm S thus we can use $\mathbf{S}^q_{k=1}[T(w_K, t_{ik})]$.

An interesting observation can be made with respect to the weighted "anding" and "oring" observations. In the case of the weighted *oring*, increasing the weights associated with the criteria can only result in an increase in the overall satisfaction while in the case of the "anding" increasing the weights can only result in a decrease in the satisfaction. More formally if we let w_k and $\tilde{w}_k$ be two pairs of weights such that $\tilde{w}_k \geq w_k$ and let $g_k = T(w_k, t_{ik})$ and let $\tilde{g}_k = T(\tilde{w}_k, t_{ik})$ then for any t-conorm S, we have $S(g_1, \ldots, g_q) \leq S(\tilde{g}_1, \ldots, \tilde{g}_q)$. On the other hand if we let $h_k = S((1 - w_k), t_{ik})$ and $\tilde{h}_k = S((1 - \tilde{w}_k), t_{ik})$ then for any t-norm T, $T(h_1, \ldots, h_q) \geq T(\tilde{h}_1, \ldots, \tilde{h}_q)$.

We can make the following observation about the weighted "oring" aggregation in the case of the using $S = \text{Max}$. In this case our overall aggregation is $\text{Max}(g_1, \ldots, g_q)$ where $g_k = T(w_k, t_{ik})$ for a t-norm T. We further observe that if $w_k < 1$ then for any T and t_{ik} we have $g_k < 1$. From this we can conclude that in the case of the weighted "or" aggregation using $S = \text{Max}$ we can never get an overall satisfaction of one unless at least one at the criteria conditions has importance of one. We note this is a necessary but not sufficient condition for getting an overall satisfaction of one.

A corresponding result can be obtained in the case of the weighted "anding" aggregation in the case of using $T = \text{Min}$. Recall the definition $h_k = S(\tilde{w}_k, t_{ik})$ for any t-conorm S, from this we can observe that if $w_k < 1$, then $\tilde{w}_k > 0$ and hence for any S and t_{ik} we have $h_k \geq \tilde{w}_k > 0$. Since our overall aggregation is $\text{Min}(h_1, \ldots, h_q)$ we can conclude that in this case of weighted "anding" aggregation we can not get an overall satisfaction of zero unless at least one of the criteria conditions has importance weight of one.

An interesting formulation for the weighted "oring" aggregation occurs when we use $S(a, b) = Min(a + b, 1)$ and $T(a, b) = ab$. In this case (P_1, w_1) or (P_2, w_2) or ... or (P_q, w_q) evaluates to $Min\left(\sum_{k=1}^{q} w_k t_{ik}, 1\right) Min\left(\sum_{k=1}^{q} w_k t_{ik,1}\right)$. Thus here we take a simple weighted sum and then bound it by the value one.

A correspondingly interesting formulation for the weighted "anding" can be had if we use $T(a, b) = Max(a + b - 1, 0)$ and $S(\bar{a}, b) = a + b - ab$. In this case

$$T\left(g1, \ \ldots, \ g_q\right) = 1 - Min\left[\sum_{k=1}^{q} w_k \bar{t}_{ik}, 1\right]$$

The OWA operator [14] can provide a useful formulation for the aggregation process in flexible querying of databases. We recall the OWA operator is aggregation operator $F: I^n \rightarrow I$ so that $F(a_1, \ldots, a_n) = \sum_j w_j b_j$ where b_j is the jth largest of the a_i and w_j are collection of n weights such that $w_j \in [0, 1]$ and $\sum_j w_j = 1$. We note that if π is an index function so $\pi(j)$ is the index of the jth largest a_i, then $b_j = a_{\pi(j)}$ and hence $F(a_1, \ldots, a_n) = \sum_j w_j a_{\pi(j)}$. It is common to refer to the collection of w_j using an n vector W whose jth component is w_j, in this case we refer to W as the OWA weighting vector.

We observe that if $w_1 = 1$ and $w_j = 0$ for $j \neq 1$ then $F(a_1, \ldots, a_n) = Max_i[a_i]$. If $w_n = 1$ and $w_j = 0$ for $j \neq n$ then $F(a_1, \ldots, a_n) = Min_i[a]$. If $w_j = 1/n$ for all the j then $F(a_1, \ldots, a_n) = \frac{1}{n} \sum_i a_i$. The OWA operator is a mean operator, it is monotonic, symmetric and bounded, It is also idempotent

The OWA operator can be very directly used in flexible querying. Assume we have a query with n component pairs, $P_k = (V_{Q(k)}, F_{Q(k)})$ where $F_{Q(k)}$ is a fuzzy subset over the domain of $V_{Q(k)}$. As in the preceding for any object D_i we can obtain $t_{ik} = F_{Q(k)}(d_{iQ(k)})$. We can then provide an aggregation of these using the OWA operator, $F(t_{i1}, t_{i2}, \ldots, t_{in})$. Essentially this provides a kind an average of the individual satisfactions. We can denote this $OWA_W(D_i)$. Different choices of the OWA weights will result in different types of aggregation. If we use the weights such that $w_1 = 1$ then we get $Max_k[t_{ik}]$ which is essentially an "oring" of the components. If we use a weighting vector W such that $w_n = 1$ we get $Min_k[t_{ik}]$ which is an "anding" of the conditions. If $w_j = 1/n$ then we are taking a simple average of the criteria satisfactions.

Using quantifiers and particularly linguistic quantifiers we can use the OWA operator to provide a very rich class of aggregation operators [15]. A quantifier or proportion can be seen as a value in the unit interval. In [16] Zadeh generalized the concept of quantifier by introducing the concept of linguistic quantifiers. Examples of these linguistic quantifiers are: *most, about half, all, few* and *more then α percent*. Zadeh [16] suggested that one can represented these linguistic quantifiers as fuzzy subsets of the unit interval. This if R is a linguistic quantifier we can represent it as a fuzzy subset R of the unit interval so that for any $y \in I$, $R(y)$ is the degree to which the proportion y satisfies the concept R. An important class of quantifiers are regular monotonic quantifiers. These quantifiers have the properties: $R(0) = 0$, $R(1) = 1$ and $R(x) \geq R(y)$ for $x > y$. Thus for these quantifiers the satisfaction increases as the proportion increases. Examples of these are "at least p", "most", "some", "all" and "at least one". In [15] Yager showed how to use the fuzzy subsets associated with these quantifiers to generate the weights of an OWA operator. In particular he suggested we obtain the weights as

$$w_j = R(\frac{j}{n}) - R(\frac{j-1}{n}).$$

It can be shown if R is regular monotonic then the weights sum to one and lie in the unit interval.

Using these ideas we can obtain a quantifier guided approach to flexible querying. We can express a query as collection of n pairs, $P_k = (V_{Q(k)}, F_{Q(k)})$ and an aggregation imperative in terms of a linguistic quantifier R describing the proportion of these pairs we require to be satisfied. We then express this quantifier R as a fuzzy subset of the unit interval, R. Using this R we obtain a collection of OWA weights, $w_j = R(\frac{j}{n}) - R(\frac{j-1}{n})$ for j = 1 to n. Once having these weights we can evaluate the satisfaction of each element D_i as $OWA_R(D_i)$. In [17] we showed how to extend this to the case importance weighted criteria.

4   Lexicographic Formulated Queries

In the database query structure presented here we impose a priority ordering on the conditions composing the query. In this approach conditions higher in priority ordering play a more important role in determining the overall satisfaction of an object to the query in the same way that letters in earlier positions a play in determining the alphabetical order. The technique we shall develop will be called a **LEX**icographic Aggregation (LEXA) query and it will make use of the prioritized aggregation operator introduced in [18–20]. This approach will allow us to model different query imperatives than in the preceding.

Here again we shall consider a query Q to consist of a collection of q pairs P_k and an aggregation imperative. Each pair as in the preceding is of the form $(V_{Q(k)}, F_{Q(k)})$ where $V_{Q(k)}$ indicates an attribute name and $F_{Q(k)}$ is a fuzzy subset of the domain of $V_{Q(k)}$ denoting the desired values for $V_{Q(k)}$. Again for each object $D_i = (d_{i1}, \ldots, d_{iq})$ we can obtain the values $t_{ik} = F_{Q(k)}(d_{ik})$, the satisfaction of the pair P_k by D_i. Here we will consider an aggregation imperative in the spirit of a lexicographic ordering. This aggregation imperative requires an ordering of the P_k indicating their priority in formulating the overall satisfaction to the query. While our lexicographic aggregation query framework will work in the flexible environment we shall initially consider its performance in the binary situation.

We now describe the basic binary LEXA query. In this case it is assumed that the t_{ik} are binary either 1 or 0, P_k is satisfied by D_i or not. In the following we shall assume there are m elements in the database, D_i for i = 1 to m. As we subsequently see the output of a LEXA query is an ordering of the elements in D_i according to their satisfaction to lexicographic query.

A lexicographic aggregation query assumes a linear priority among the criteria, the P_k, with respect to their importance. For simplicity we shall assume that the P_k have been indexed according to this priority ordering:

$$P_1 > P_2 > \cdots > P_q.$$

The following algorithm describes the process for forming the ordered list L

(1) Initiate: $k = 1$, $D = \{D_i | i = 1 \text{ to } m\}$, list L as empty
(2) Set $F = \emptyset$
(3) Test all elements $D_i \in D$ with respect P_k and place all those for which $t_{ik} = 0$ in F.
(4) Place all elements in F tied at the top available level of L.
(5) Set $D = D - F$
(6) If $D = \emptyset$ Stop
(7) Set $k = k + 1$
(8) If $k > q$

    (a) Place all remaining elements in D tied as top level of L
    (b) Stop

(9) Go to step 2

The result of this is the list L which is an ordered list of the satisfactions of the elements in D to the lexicographic query Q. We note the higher up the list the better the satisfaction.

There is another way we can generate the satisfaction ordering of the elements in D under the priority ordering $P_1 > \cdots > P_q$. For each D_i, starting from P_1 and preceding in increasing order find the first index k for which $t_{ik} = 0$. Assign D_i a score S_i where

$$S_i = q \qquad \text{if } D_i \text{ meets no failures}$$
$$S_i = k - 1 \qquad \text{otherwise}$$

Using this method each D_i gets a score $S_i \in \{0, 1, \ldots, q\}$. If we order the elements by their value for S_i we get L. We note that S_i is the number of criteria D_i satisfied before it meets failure. This method for evaluating the score of satisfaction of an element D_i to a LEXA query is more useful then the preceding as it provides a score is addition to an ordering.

It is interesting to note there is only one situation in which an element D_i can attain the maximal score $S_i = q$, if all $t_{ik} = 1$. On the other hand there are many ways D_i can get the minimal score of $S_i = 0$. In particular any D_i which has $t_{i1} = 0$ will have $S_i = 0$ regardless of the values of t_{ik} for $k \neq 1$.

We further observe that since $0 \leq S_i \leq q$ we can provide a kind of normalization by defining $G_i = \frac{S_i}{q}$. Here we are getting a value in the unit interval.

**We shall introduce a more general implementation of the lexicographic aggregation type query that will be appropriate for the case where the t_{ik} are values from $[0, 1]$ rather then being binary values from $\{0, 1\}$. In this environment for each element D_i we shall obtain a score G_i indicating its satisfaction to our LEXA query.

Again here we have a query Q consisting of q condition pairs, $P_k = (V_{Q(k)}, F_{Q(k)})$. In addition we have a priority ordering over the pairs guiding the lexicographic aggregation,

$$P_1 > P_2 > P_3 \cdots > P_q.$$

Assume for object D_i we have $t_{ik} = F_{Q(k)}(d_{ik})$ as the degree to which D_i satisfies P_k. Our procedure for determining G_i is to calculate

$$G_i = \sum_{k=1}^{q} w_{ik} t_{ik}$$

where the w_{ik} are determined as follows.

(1) Set $u_{i1} = 1$

(2) $u_{ik} = \prod_{j=2}^{k} t_{i(j-1)}$ for $k = 2$ to q

(3) $w_{ik} = \frac{1}{q} u_{ik}$

The important observation here is that the w_{ik} is proportional to the product of the satisfactions of the higher priority conditions. We note that we can express $u_{ik} = u_{ik-1} H_{ik-1}$.

We can make some observations. For $k_1 < k_2$ we always have $w_{ik1} \geq w_{ik2}$. Thus a lower priority condition can't have a bigger weight than one that is higher.

We observe that if $t_{ik1} = 0$ then $w_{ik2} = 0$ for all $k_2 > k_1$.

We can show that $G_i = 1$ if and only $t_{ik} = 1$ for all k.

We also observe that $G_i = 0$ iff $t_{i1} = 0$. It is important to emphasize that this is independent of the satisfaction to the other criteria. Zero satisfaction to the highest priority criterion means zero overall satisfaction, there is no possibility for compensation by other criteria.

Let us look at this process for the binary case to see that it correctly generalizes the evaluation in the binary case we presented earlier. Consider the list of satisfactions: $t_{i1}, t_{i2}, \ldots, t_{iq}$ which here we assume are either one or zero. Assume t_{ik} is the first one of these that is zero, hence $t_{ij} = 1$ for $j < k$. In this case we get that $u_{ij} = 1$ for $j \leq k$ and $u_{jk} = 0$ for $j > k$.

Hence

$$G_i = \sum_{j=1}^{q} w_{ij} t_{ij} = \frac{1}{q} \sum_{j=1}^{q} u_{ij} t_{ij} = \frac{1}{q} \sum_{j=1}^{q} t_{ij}$$

Furthermore since $t_{ij} = 1$ for $j < k$ and $t_{ij} = 0$ for $j = k$ we get

$$G_i = \frac{1}{q}\sum_{j=1}^{k} t_{ij} = \frac{1}{q}\sum_{j=1}^{k-1} 1 = \frac{k-1}{q}$$

as desired. In the special case where all $t_{ij} = 1$ then all $u_{ij} = 1$ and all $w_{ij} = \frac{1}{q}$ and hence $G_i = \frac{q}{q} = 1$.

As discussed in [18] the approach we suggested can be extended to the case when the ordering among the pairs is a weak ordering, we allow ties among the conditions pairs with respect to their priority. Here we shall denote a condition pair as $P_{(k, j)}$ where all pairs with the same k value are tied in the priority ordering. The j value is just an indexing distinguishing among the tied pairs. Thus here we are assuming the priority ordering is such that for $k_1 < k_2$ that $P_{(k1, j)} > P_{(k2, i)}$ and $P_{(k1, j)} = P_{(k1, i)}$ for all j and i. We let n_k denote the number of pairs with k in the first term, thus n_1 is the number of conditions with the highest priority. We shall also let $n = \sum_{k=1}^{q} n_k$.

Here we let $t_{i(k, j)}$ denote the satisfaction of the object D_i to the condition pair $P_{(k, j)}$. In this case we obtain the value G_i as

$$G_i = \sum_{k=1}^{q} w_{ik}\left(\sum_{r=1}^{n_k} t_{i(j, r)}\right)$$

Again it is a weighted sum of the satisfaction to each of the pairs. We emphasize that each pair with the same k has the same weight w_{ik}. The procedure we use to obtain the weights is similar to the earlier one except in the first step:

1. Calculate: $H_{ij} = \underset{r=1 \text{ to } n_j}{\text{Max}} [t_{i(j, r)}]$

   (It is the satisfaction value of the least satisfied pair at the j level)

   $u_{i1} = 1$

2. $u_{ik} = \prod_{j=1}^{k} H_{i(j-1)}$ for $k = 2$ to q

3. $w_{ik} = \frac{1}{n} u_{ik}$

Let us see how this plays out in the pure binary case where all $t_{i(k, r)} \in \{0, 1\}$. Let b be the k value at which we meet the first condition for which $t_{i(k, r)} = 0$. In this case we have that $H_{ij} = 1$ for $j < b$ and $H_{ij} = 0$ for $j = 0$. Here then while $u_{i1} = 1$ we have for $k = 2$ to q that $u_{ik} = \prod_{j=1}^{n} H_{i(j-1)}$. From the above we get that

$$u_{ik} = 1 \quad 2 \le k \le b$$
$$u_{ik} = 0 \quad k > b$$

Thus we get $w_{ik} = \frac{1}{n}$ for $k = 1$ to b and $w_{ik} = 0$ for $k > b$. From this we get
$G_i = \sum_{k=1}^{q} w_{ik} \left(\sum_{r=1}^{n_k} t_{i(k,r)} \right) = \sum_{k=1}^{b} \frac{1}{n} \left(\sum_{r=1}^{n_k} t_{i(k,r)} \right)$.
For all $k < b$ all the elements have $t_{i(k,r)} = 1$ hence

$$G_i = \frac{1}{n} \left(\sum_{k=1}^{b-1} n_k + \sum_{r=1}^{n_b} t_{i(b,r)} \right)$$

We note at least one element in the second term is zero. We see that the numerator of G_i is equal to the number of elements in the priority classes higher than the first class where we meet failure plus all the pairs satisfied in the class that we meet failure. An alternative expression of this numerator is the sum of all satisfied pairs in priority classes up to and including the class when we meet our first failure.

We here point to related work by Chomicki [21–24] on preferences in databases and the on bipolar queries by Dubois and Prade [25, 26] and Zadrozny and Kacprzyk [27, 28].

5 Querying Probabilistic Databases

In the preceding we have discussed several different paradigms for formulating questions to databases. In particular we considered a query to be a collection of q condition pairs, $P_k = (V_{Q(k)}, F_{Q(k)})$, and an agenda for aggregating the satisfactions to these pairs by a database element D_i. We added flexibility by allowing $F_{Q(k)}$ to be a fuzzy subset of the domain $X_{Q(k)}$ of $V_{Q(k)}$. In this flexible case we obtained the degree of satisfaction of P_k by D_i to be $t_{ik} = F_{Q(k)}(x_{iQ(k)}) \in [0, 1]$ where $x_{iQ(k)}$ is the value of $V_{Q(k)}$ for D_i. Thus t_{ik} is the degree to which $F_{Q(k)}$ is true given the value of $V_{Q(k)}$ for D_i is $x_{iQ(k)}$. We now shall consider the situation in which the value of $V_{Q(k)}$ for D_i, $x_{iQ(k)}$, is random. More specifically if the domain $X_{Q(k)} = \{y_{Q(k)1}, y_{Q(k)2}, \ldots, y_{Q(k)rQ(k)}\}$ then the knowledge of the value of $V_{Q(k)}$ for object D_i, $x_{iQ(k)}$, is best expressed as a probability distribution $\mathbf{P}_{iQ(k)}$ where $\mathrm{Prob}(x_{iQ(k)} = y_{Q(k)j}) = p_{j/iQ(k)}$. In the special case where $P_{j/qQ(k)} = 1$ for some j then we have the preceding situation in which $x_{iQ(k)}$ is exactly $y_{Q(k)j}$.

Initially in the following discussion we shall assume all the $Q(k)$ are distinct. We shall subsequently consider the case where a query can involve multiple occurrences of the same attribute.

The approach we shall follow in this probabilistic situation is in the spirit of the possible words approach used in [29–37]. We shall associate with every D_i a collection Z of q tuples. In particular $Z = X_{Q(1)} \times X_{Q(2)} \times \cdots \times X_{Q(q)}$. That is each element $z \in Z$ is of the form $z = (z_1, z_2, \ldots, z_q)$ where $z_k \in X_{Q(k)}$. For each object D_i in the database we now associate a probability distribution PD_i over the space Z so that the probability of z, $PD_i(z) = \prod_{k=1}^{q} \mathrm{Prob}(x_{iQ(k)} = z_k)$.

We want to make one comment here. In the preceding we assumed all the $V_{Q(k)}$ in a query where distinct. This is not a necessary requirement. However in the case where the we have a multiple occurrences of the same attribute in the query care must be taken in the formulation of the possible worlds. In the preceding we formulated $Z = X_{Q(1)} \times X_{Q(2)} \times X_{Q(q)}$, here each element $z \in Z$ is of the form $z = (z_1, z_2, z_q)$ where each $z_k \in X_{Q(k)}$. In order to understand which happens when all $V_{Q(k)}$ are not distinct we consider the situation where $Q(1) = Q(2)$, here then $V_{Q(1)} = V_{Q(2)}$. Here the associated Z space has the additional required that for all $z \in Z$, $z_1 = z_2$. Thus any element in Z must have the same value of $V_{Q(1)}$ and $V_{Q(2)}$. In addition we must make some modifications in the calculation of $PD_i(z)$. In particular we must note duplicate the probabilities and avoid using them twice, thus $PD_i(z)$ must be based on the product of the distinct probabilities and hence we have

$$PD_i(z) = \prod_{\substack{k=1 \text{ for all} \\ \text{distinct } Q(k)}}^{q} Prob(x_{iQ(k)} = z_k)$$

Another comment we want to make is regarding the relationship between distinct attribute values. Implicit in the preceding has been an assumption of independence between the probability distribution of distinct attributes. In some cases this may not be true. For example some values of V_1 may not be allowable under V_2 while some values for V_1 may mandate a particular value for V_2. Such relationships require us to modify to possible elements in Z and condition the probability distribution in known ways. Nevertheless when including these special conditions we still end up with a subset of tuples from Z with associated probabilities, as a result in the following we shall neglect any special relationships between the attributes as they don't effect the subsequent discussion nor the basic ideas of the approach introduced.

We now recall that our query consists of a collection of q constraints $P_k = (V_{Q(k)}, F_{Q(k)})$ where each $F_{Q(k)}$ is a fuzzy subset over the space $X_{Q(k)}$. We now apply these constraints to the space of possible worlds Z. In particular we transform the space Z to F so that each tuple $z = (z_1, \ldots, z_q)$ is transformed to new tuple

$$F(z) = (F_{Q(1)}(z_1), F_{Q(2)}(x_2), \ldots, F_{Q(q)}(z_q))$$

We note each element $F_{Q(k)}(x_k) \in [0, 1]$. Thus each term in F(z) is a tuple of q values drawn from the unit interval, F(z) is a subset the space I^q. In addition for each element D_i in the database we can associate a probability distribution over the space F. In particular for each D_i we associate $PD_i(z)$ with the tuple F(z). Thus now for any element D_i we have a probability distribution on the space F of satisfactions to the query components. We note that if two z tuples, z_1 and z_2, transform into the same value, $F(z_1) = F(z_2)$, we represent these in F with just one, $F(z_1)$, and use the sum of the probabilities z_1 and z_2.

In order to provide an intuitive understanding of the discussion to follow we shall use the following database to illustrate our ideas.

Example 1 We consider a database with three attributes V_1, V_2 and V_3. The domain of these are respectfully:

$$X_1 = \{a, b, c\}, X_2 = \{red, \ blue\}, X_3 = \{10, 20\}$$

Let D be one object in the database and let x_1, x_2, x_3 denote the values of V_1, V_2, V_3 for this object. In particular for each of these, x_1, x_2, x_3, we have a probability distribution

x_1: Prob(a) = 0.4, Prob(b) = 0.5 and Prob(c) = 0.1
x_2: Prob(red) = 0.7 and Prob(blue) = 0.3
x_3: Prob(10) = 0.6 and Prob(20) = 0.4.

In this example the set of Z of possible worlds are Z = {(a, red, 10), (a, red, 20), (a, blue, 10), (a, blue, 20), (b, red, 10), (b, red, 20), (b, blue, 10), (b, blue, 20), (c, red, 10), (c, red, 20), (c, blue, 10), (c, blue, 20)}.

For the element D we get the probability of each of these components as shown in Table 1.

Before preceding we want to point out that for any other object D* in the database the set of possible worlds Z will be the same, the difference between D and D* will be in the probabilities associated with the elements in Z.

We now consider our query as consisting of three components: $P_1 = (V_1, F_1)$, $P_2 = (V_2, F_2)$ and $P_3 = (V_3, F_3)$ where each F_k is a fuzzy set over the space X_k defining the required condition. Below are the associated fuzzy subsets

$$F_1 = \left\{\frac{1}{a}, \frac{0.6}{b}, \frac{0.2}{c}\right\}, F_2 = \left\{\frac{0.8}{red}, \frac{1}{blue}\right\}, F_3 = \left\{\frac{1}{10}, \frac{0.2}{20}\right\}$$

Table 1 Probabilities of elements in Z

Element in Z	Probability
(a, red, 10)	(0.4) (0.7) (0.6) = 0.167
(a, red, 20)	(0.4) (0.7) (0.4) = 0.112
(a, blue, 10)	(0.4) (0.3) (0.6) = 0.072
(a, blue, 20)	(0.4) (0.3) (0.4) = 0.048
(b, red, 10)	(0.5) (0.7) (0.6) = 0.21
(b, red, 20)	(0.5) (0.7) (0.4) = 0.14
(b, blue, 10)	(0.5) (0.3) (0.6) = 0.09
(b, blue, 20)	(0.5) (0.3) (0.4) = 0.06
(c, red, 10)	(0.1) (0.7) (0.6) = 0.042
(c, red, 20)	(0.1) (0.7) (0.4) = 0.028
(c, blue, 10)	(0.1) (0.3) (0.6) = 0.018
(c, blue, 20)	(0.1) (0.3) (0.4) = 0.012

Table 2 Probabiities of
transformed elements

Element in Z	Transform F(z)	Probability
(a, red, 10)	(1, 0.8, 1)	0.167
(a, red, 20)	(1, 0.8, 0.2)	0.112
(a, blue, 10)	(1, 1, 1)	0.072
(a, blue, 20)	(1, 1, 0.2)	0.048
(b, red, 10)	(0.6, 0.8, 1)	0.21
(b, red, 20)	(0.6, 0.8, 0.2)	0.14
(b, blue, 10)	(0.6, 1, 1)	0.09
(b, blue, 20)	(0.6, 1, 0.2)	0.06
(c, red, 10)	(0.2, 0.8, 1)	0.042
(c, red, 20)	(0.2, 0.8, 0.2)	0.028
(c, blue, 10)	(0.2, 1, 1)	0.018
(c, blue, 20)	(0.2, 1, 0.2)	0.012

We now can apply these constraints to our set Z of possibilities and also use the
probabilities for D and shown the results in Table 2.

We now consider the issue of evaluating the query Q for the object D in the
database. Let us recapitulate the situation. Associated with D we have a space F of
tuples $T_j = (F_1(z_1), \ldots, F_q(z_q))$, each tuple is an element from the space I^q. That is
each tuple is a collection of q values that the unit interval. In addition associated
with each tuple in F we have a probability. Parenthetically we note that the space F
of tuples is the same for each element D_i in the database, the only distinction
between the elements in the database are the probabilities associated with each of
the tuples.

At this point we can reformulate more simply. We have a subset $F \subseteq I^q$ with
arbitrary element $T_j = (t_{j1}, t_{j2}, \ldots, t_{jq})$. Furthermore for any database element D_i we
have a probability distribution over the space F where p_{ij} is the probability associated with the tuple T_j for the database object D_i. We note that $\sum_{j=1}^{n_z} p_{ij} = 1$ where
n_z is the cardinality of the space Z.

In addition associated with a query is an aggregation imperative that dictates
how we aggregate the satisfactions to the individual components, in particular Agg
$(T_j) = Agg(t_{j1}, t_{j2}, \ldots, t_{jq})$. After applying the Agg operation to the tuples in the
space F we end up with a collection of scalar values, Agg(F), consisting of the
elements $Agg(T_j)$. For each D_i we have a probability distribution over the set of
$Agg(T_j)$. In particular for each D_i and each $Agg(T_j)$ we have p_{ij} as its probability.
We now illustrate the preceding with our earlier example and consider the application of different aggregation imperatives.

Example 2 We have the collection of 12 tuples and their associated probabilities. In
Table 3 we consider three aggregation imperatives the first imperative is an
"anding" of all conditions, here $Agg(T_j) = Min_k(t_{jk})$, the second imperative is an

Table 3 Aggregated value of elements in F

Tuple in F	Probability	And	Average	Oring
$T_1 = (1, 0.8, 1)$	0.167	0.8	0.933	1
$T_2 = (1, 0.8, 0.2)$	0.112	0.2	0.66	1
$T_3 = (1, 1, 1)$	0.072	1	1	1
$T_4 = (1, 1, 0.2)$	0.048	0.2	0.73	1
$T_5 = (0.6, 0.8, 1)$	0.21	0.6	0.9	1
$T_6 = (0.6, 0.8, 0.2)$	0.14	0.2	0.53	0.8
$T_7 = (0.6, 1, 1)$	0.09	0.6	0.866	1
$T_8 = (0.6, 1, 0.2)$	0.06	0.2	0.6	1
$T_9 = (0.2, 0.8, 1)$	0.042	0.2	0.66	1
$T_{10} = (0.2, 0.8, 0.2)$		0.028	0.2	0.40.8
$T_{11} = (0.2, 1, 1)$	0.018	0.2	0.73	1
$T_{12} = (0.2, 1, 0.2)$	0.012	0.2	0.466	1

Table 4 Probabilities of "anded" value

Value	Prob
0.2	0.461
0.6	0.3
0.8	0.167
1	0.072

Table 5 Probabilities of "ored" value

Value	Prob
1	0.82
0.8	0.168

average of all conditions, $Agg(T_j) = \frac{1}{q}\sum_{j=1}^{q} t_{jk}$ and the third imperative is an "oring" of all conditions, here $Agg(T_j) = Max_k(t_{jk})$..

In a similar way we can implement any aggregation imperative.

Combining the probabilities of tuples with the same value for their **anding** we get Table 4.

Combining the probabilities of the tuples with the same value for their **oring** we get Table 5

Let us now summarize our situation for a query we obtained a collection T_j of tuples. Associated with the query is an aggregation operation that converts T_j into a single value $Agg(T_j)$ indicating the overall satisfaction of the tuple. For simplicity we denote these as $Agg(T_j) = a_j$. Essentially we have a collection of degrees of satisfaction, a_j, of each possible world to the query. Finally for each element D_i in the database we have a probability distribution over the a_j. Thus the satisfaction of the query by the database object D_i is a probability distribution

$$a_1 \ p_{i1}$$
$$a_2 \ p_{i2}$$
$$\vdots$$
$$a_r \ p_{ir}$$

We emphasize that the set a_j, which are the satisfactions of a possible world to the query, is the same for all D_i, the difference between the D_i is reflected in the probability distribution over the a_j, the probabilities they assign to a possible world.

A natural question that arises is how to order the D_i with respect to their satisfaction to the query. Here we look to [38] for some ideas. If D and $\hat{D}$ are two database objects then for each of these we have a probability distribution over the set $A = \{a_1, \ldots, a_r\}$.

$$
\begin{array}{ccc}
 & \mathbf{P} & \mathbf{\hat{P}} \\
a_1 & p_1 & \hat{P}_1 \\
a_2 & P_2 & \hat{P}_2 \\
a_j & p_j & \hat{P}_j \\
a_r & p_r & \hat{P}_r \\
\end{array}
$$

In the following for simplicity we shall assume the a_j have been indexed so that they are in increasing order, $a_j > a_k$ if $j > k$. In the following we shall use the notation $D > \hat{D}$ if D is a more satisfying alternative than $\hat{D}$. What is clear is that if $p_j = 1$ and $\hat{P}_k = 1$ and $j > k$ than $D > \hat{D}$.

One approach to ordering the P and $\hat{P}$ is to use the cumulative distribution function, CDF, and the idea of stochastic dominance [39, 40].

We define CDF $p(a_j) = \sum_{i=1}^{j} p_i$, this the probability that for object D the actual satisfaction will be less or equal to a_j. Similarly we define CDF$\hat{p}(a_j) = \sum_{i=1}^{j} \hat{p}_i$. Using this we shall say $D > \hat{D}$ if

$$CDF_P(aj) \leq CDF\hat{p}(a_j) \quad \text{for all } j$$
$$CDF_P(aj) \leq CDF\hat{p}(a_j) \quad \text{for at least one } j$$

We note that if $CDF_P(a_j) \leq CDF\ \hat{p}(a_j)$ then $\sum_{i=1}^{j} p_i \leq \sum_{i=1}^{j} \hat{p}_i$ this implies

$$1 - \sum_{i=j+1}^{r} p_i \leq 1 - \sum_{i=j+1}^{r} \hat{p}_i$$

which further implies that $\sum_{i=j+1}^{r} p_i \geq \sum_{i=j+1}^{r} \hat{p}_i$. Thus we see that under this condition object D never has less probability associated with the higher satisfaction then $\hat{D}$. This provides a reasonable justification for asserting that $D > \hat{D}$.

While the CDF provides a way for ordering the elements if the conditions are met, usually the condition $CDF_P(a_j) \leq CDF\ \hat{p}(a_j)$ is not satisfied for all j. This effectively means that the use of stochastic dominance does not provide the kind of general approach necessary to cover all cases.

Another approach to comparing the two database elements is a scalarization. Here we associate with each element a distinct value and then compare these values. Since these are scalar values we are able to order them. One approach to scalarization is to use the expected value of satisfaction of each the elements. Here $EV(D) = \sum_{j=1}^{r} a_j p_j$ and $EV(\hat{D}) = \sum_{j=1}^{r} a_j \hat{p}_j$. We then say $D > \hat{D}$ if $EV(D) > EV(\hat{D})$. If $EV(D) = EV(D)$ then we say they are tied.

It is interesting to show that if $CDF_P(a_j) \leq CDF\ \hat{p}(a_j)$ for all j then $EV(D)$ $EV(\hat{D})$. We shall illustrate this for the case when r = 4, the extension to the more general case of r will be obvious. Since we assumed $a_i > a_j$ for i > j then we can express

$$a_2 = a_1 + \Delta_2 \quad \Delta_2 > 0$$
$$a_3 = a_2 + \Delta_3 \quad \Delta_3 > 0$$
$$a_4 = a_3 + \Delta_4 \quad \Delta_4 > 0$$

We see

$$EV(D) = \sum_{j=1}^{4} a_j p_j = a_1 p_1 + (a_1 + \Delta_2)p_2 + (a_1 + \Delta_2 + \Delta_3)p_3 + (a_1 + \Delta_2 + \Delta_3 + \Delta_4)p_4$$

$$EV(D) = a_1 \sum_{j=1}^{4} p_j + \Delta_2 \sum_{j=2}^{4} p_j + \Delta_3 \sum_{j=3}^{4} p_j + \Delta_4 \sum_{j=4}^{4} p_j$$

As we have already shown if $CDF_p(a_i) \leq CDF\ \hat{p}(a_i)$ for all i then $\sum_{j=4}^{4} p_j \geq \sum_{j=4}^{4} \hat{p}_j$ for all i. From this we see that if $CDF_P(a_i) \leq CDF\ \hat{p}(a_i)$ for all i then $EV(D) \geq EV(\hat{D})$.

A more general approach to scalarization of the probabilities of the database elements can be had if we use some ideas from the OWA aggregation operator. Here we let $\alpha \in [0, 1]$ be a measure our optimism. The more optimistic the more we are anticipating the higher valued satisfactions to occur. Here then if $\alpha = 1$ we are always anticipating that a_r will occur. While if $\alpha = 0$ we are always anticipating a_1 will occur.

We now associate with a given α a function $f(x) = x^q$ where $q = \frac{1-\alpha}{\alpha}$. We see that if $\alpha \to 0$ then $q \to \infty$ and if $\alpha \to 1$ then $q \to 0$ and if $\alpha = 1/2$ then $q = 1$.

Using this function we obtain a set of OWA weights as follows. Letting $S_j = \sum_{k=i}^{r} p_k$ we get

$$w_j = f(S_j) - f(S_j + 1) \quad \text{for } j = 1 \text{ to } r_r$$

Using these weights we obtain $EV_\alpha(D) = \sum_{j=1}^{r} w_j a_j$. We first see that if $\alpha = 1/2$ then $q = 1$ and hence $w_j = \sum_{k=j}^{r} p_k - \sum_{k=j+1}^{r} p_k = p_j$. Here we get the usual expected value. We see that if $\alpha \to 0$, $q = \infty$, since $S_j < 1$ for all $j > 1$ we have $(S_j)^\infty \to 0$ for $j < 1$ and $(S_i)^\alpha = 1$ hence in this case

$$w_1 = 1$$
$$w_j = 0 \text{ for } j > 1$$

From this we obtain $EV_0(D) = a_1$, it is the least satisfaction. If $\alpha \to 1$, $q = 0$, $w_r = 1$, if $p_r > 0$ and hence $ED_1(D) = a_r$. Here we see that in using EV_α for the extreme optimistic and pessimistic of α all database elements D_k will have the same value for $EV_\alpha(D_k)$ however for other values $0 < \alpha < 1$ as in the case of $\alpha = 0.5$ each of the D_k will get its own unique value for $EV_\alpha(D_k)$. Here then choosing α will determine the ordering of the D_k.

6 Conclusion

Here we provided a framework for soft querying of databases. We described a soft query as a collection of required conditions and an imperative for combining an objects satisfaction to individual conditions to get its overall satisfaction. We investigated tools that can enrich this process by enabling the inclusion of more human focused considerations. We described some more sophisticated techniques for aggregating the satisfactions of the individual conditions based on the inclusion of importances and the use of the OWA operator. In addition to considering more human focused aspects of the query we looked at databases in which the information in the database can have some uncertainty. We particularly considered probabilistic.

References

1. Kacprzyk J, Ziolkowsi A (1986) Database queries with fuzzy linguistic quantifiers. IEEE Trans Syst Man Cybern 16:474–479; Prade H, Negoita CV (eds) Verlag TUV Rheinland, Cologne, pp 46–57
2. Bosc P, Galibourg M, Hamon G (1988) querying. Fuzzy Sets Syst 28:333–349
3. Bosc P, Pivert O (1995) SQLf: a relational database language for fuzzy querying. IEEE Trans Fuzzy Syst 3:1–17
4. Kraft DH, Bordogna G, Pasi G (1999) Fuzzy set techniques in information retrieval. In: Bezdek JC, Dubois D, Prade H (eds) Fuzzy sets in approximate reasoning and information systems. Kluwer Academic Publishers, Norwell, MA, pp 469–510

5. Galindo J (2008) Handbook of research on fuzzy information processing in databases. Information Science Reference, Hershey, PA
6. Zadrozny S, de Tré G, de Caluwe R, Kacprzyk J (2008) An overview of fuzzy approaches to flexible database querying. In: Galindo J (ed) Handbook of research on fuzzy information processing in databases, vol 1. Information Science Reference, Hershey, PA, pp 34–54
7. Petry FE (1996) Fuzzy databases principles and applications. Kluwer, Boston
8. Pivert O, Bosc P (2012) Fuzzy preference queries to relational databases. World Scientific, Singapore
9. Beliakov G, Pradera A, Calvo T (2007) Aggregation functions: a guide for practitioners. Springer, Heidelberg
10. Yager RR (1987) A note on weighted queries in information retrieval systems. J Am Soc Inf Sci 38:23–24
11. Dubois D, Prade H, Testemale C (1988) Weighted fuzzy pattern matching. Fuzzy Sets Syst 28:313–331
12. Yager RR (1981) A new methodology for ordinal multiple aspect decisions based on fuzzy sets. Decis Sci 12:589–600
13. Sanchez E (1989) Importance in knowledge systems. Inf Syst 14:455–464
14. Yager RR (1988) On ordered weighted averaging aggregation operators in multi-criteria decision making. IEEE Trans Syst Man Cybern 18:183–190
15. Yager RR (1996) Quantifier guided aggregation using OWA operators. Int J Intell Syst 11:49–73
16. Zadeh LA (1983) A computational approach to fuzzy quantifiers in natural languages. Comput Math Appl 9:149–184
17. Yager RR (1998) Including importances in OWA aggregations using fuzzy systems modeling. IEEE Trans Fuzzy Syst 6:286–294
18. Yager RR (2008) Prioritized aggregation operators. Int J Approx Reason 48:263–274
19. Yager RR (2010) Lexicographic ordinal OWA aggregation of multiple criteria. Inf Fusion 11:374–380
20. Yager RR, Reformat M, Ly C (2011) Using a web personal evaluation tool—PET for lexicographic multi-criteria service selection. Knowl Based Syst 24:929–942
21. Chomicki J (2007) Database querying under changing preferences. Ann Math Artif Intell 50:79–109
22. Mindolin D, Chomicki J (2011) Preference elicitation in prioritized skyline queries. Very Large Data Base J 20:157–182
23. Mindolin D, Chomicki J (2011) Contracting preference relations for database applications. Artif Intell J 175:1092–1121
24. Staworko S, Chomicki J, Marcinkowski J (2012) Prioritized repairing and consistent query answering in relational databases. Ann Math Artif Intell 64(2–3):209–246, 2012
25. Dubois D, Prade H (2002) Bipolarity in flexible querying. In: Proceedings of the 5th international conference on flexible query answering systems, pp 174–182
26. Dubois D, Prade H (2008) Handling bipolar queries in fuzzy information processing. In: Galindo J (ed) Handbook of research on fuzzy information processing in databases, vol 1. Information Science Reference, Hershey, PA, pp 99–114
27. Zadrozny S (2005) Bipolar queries revisited. In: Modeling decisions for artificial intelligence. LNCE 0302-9743. Springer, Heidelberg, pp 387–398
28. Zadrozny SL, Kacprzyk J (2006) Bipolar queries and queries with preferences. In: Proceedings of the 17th international conference on database and expert systems applications, pp 415–419
29. Cavallo R, Pittarelli M (1987) The theory of probabilistic databases. In: Proceedings of the 13th international conference very large databases (VLDB), pp 71–81
30. Barbará D, Garcia-Molina H, Porter D (1992) The management of probabilistic data. IEEE Trans Knowl Data Eng 4:487–502
31. Re C, Dalvi N, Suciu D (2006) Query evaluation on probabilistic data bases. IEEE Data Eng Bull 29:25–31

32. Suciu D (2008) Probabilistic databases. SIGACT News 39:111–124
33. Re C, Suciu D (2008) Management of data with uncertainties. In: CIKM, Lisbon, pp 3–8
34. Dalvi N, Re C, Suciu D (2009) Probabilistic databases: diamonds in the dirt. J Assoc Comput Mach 52(7):86–94
35. Dalvi N, Re C, Suciu D (2011) Queries and materialized views on probabilistic databases. J Comput Syst Sci 77(3):473–490
36. Suciu D, Olteanu D, Re C, Koch C (2011) Probabilistic databases. In: Synthesis lectures on data management. Morgan & Claypool Publishers, San Rafael, CA
37. Dalvi N, Suciu D (2012) The dichotomy of probabilistic inference for unions of conjunctive queries. J Assoc Comput Mach 59(6):30
38. Yager RR (2001) Ordering ordinal probability distributions. Fuzzy Econ Rev 6:3–18
39. Hadar J, Russell W (1969) Rules for ordering uncertain prospects. Am Econ Rev 59:25–34
40. Bawa VS (1975) Optimal rules for ordering uncertain prospects. J Financial Econ 2:95–121

Part IV
Decision Making and Optimization

An Ordinal Multi-criteria Decision-Making Procedure in the Context of Uniform Qualitative Scales

José Luis García-Lapresta and Raquel González del Pozo

Abstract In this contribution, we propose a multi-criteria decision-making procedure that has been devised in a purely ordinal way. Agents evaluate the alternatives regarding several criteria by assigning one or two consecutive terms of a uniform ordered qualitative scale to each alternative in each criterion. Weights assigned to criteria are managed through replications of the corresponding ratings, and alternatives are ranked according to the medians of their ratings after the replications.

Keywords Multi-criteria decision-making · Group decision-making · Qualitative scales · Majority Judgment

1 Introduction

Majority Judgment (MJ) is a recent voting system introduced and analyzed by Balinski and Laraki [2, 3]. Under MJ, agents evaluate each alternative with a linguistic term of a fixed ordered qualitative scale (the authors consider six linguistic terms for evaluating candidates in political elections: 'to reject', 'poor', 'acceptable', 'good', 'very good' and 'excellent'). The alternatives are ranked according to the medians of the obtained ratings. The authors also propose two different tie-breaking processes for obtaining a final ranking on the set of alternatives.

MJ does not care whether the qualitative scale is or not uniform (the psychological distance between consecutive terms of the scale could be or not the same). Additionally, when the number of ratings is even MJ only considers one of the medians, the lower median (as shown in Felsenthal and Machover [8, Example 3.7], if the upper

J.L. García-Lapresta (✉)
PRESAD Research Group, BORDA Research Unit, IMUVA, Departamento de Economía Aplicada, Universidad de Valladolid, Valladolid, Spain
e-mail: lapresta@eco.uva.es

R. González del Pozo
PRESAD Research Group, IMUVA, Departamento de Economía Aplicada, Universidad de Valladolid, Valladolid, Spain

© Springer International Publishing AG 2018
M. Collan and J. Kacprzyk (eds.), *Soft Computing Applications for Group Decision-making and Consensus Modeling*, Studies in Fuzziness and Soft Computing 357, DOI 10.1007/978-3-319-60207-3_18

297

median is chosen the outcome could be different to the one obtained when choosing the lower median). This asymmetry and loss of information could be relevant when the number of ratings is low.

As all voting systems, MJ may produce some paradoxes and inconsistences (some of them can be found in Felsenthal and Machover [8]). Some problems of MJ have been solved by using different techniques (see García-Lapresta and Martínez-Panero [10] and Falcó and García-Lapresta [5]).

In this contribution, we propose an alternative and extended procedure of MJ by allowing agents to assign one or two consecutive terms of the qualitative scale, when they hesitate. Moreover, we consider different criteria that can be weighted in a different way, but by using an ordinal treatment. Additionally, we take into account the two medians of the corresponding ratings, avoiding a loss of information. This richer information requires to consider an appropriate linear order on the set of feasible pairs of medians.

We note that the possibility of using more than one linguistic term for assessing alternatives has been considered by Travé-Massuyès and Piera [14], Roselló et al. [13], Agell et al. [1], Falcó et al. [6, 7] and García-Lapresta et al. [9], among others.

The proposed multi-criteria decision-making procedure is shown by taking into account some data obtained in a case study (García-Lapresta et al. [9]).

The rest of the contribution is organized as follows. Section 2 is devoted to introduce the proposed multi-criteria decision-making procedure. Section 3 includes the case study. Finally, Sect. 4 concludes with some remarks.

2   The Decision Procedure

In this section we establish the multi-criteria decision-making procedure. First, we introduce the notation and basic notions.

2.1   Notation and Basic Notions

Let $A = \{1, \dots, m\}$, with $m \geq 2$, be a set of agents and let $X = \{x_1, \dots, x_n\}$, with $n \geq 2$, be the set of alternatives which have to be evaluated by the agents regarding a set of different criteria $C = \{c_1, \dots, c_q\}$. Initially, each agent may assign a linguistic term to every alternative in each criterion within an ordered qualitative scale $\mathcal{L} = \{l_1, \dots, l_g\}$, arranged from the lowest to the highest linguistic terms, i.e., $l_1 < l_2 < \dots < l_g$, where the granularity of $\mathcal{L}$ is at least 3 ($g \geq 3$). It is assumed that the linguistic scale is uniform: the psychological distance between every pair of consecutive terms of the scale is the same.

Since agents could hesitate on which linguistic term is the more appropriate to assign in each case, agents are allowed to assign two consecutive linguistic terms of the scale. Thus, we consider the set of these intervals:

$$\mathcal{L}_2 = \{[l_r, l_s] \mid r, s \in \{1, \ldots, g\}, s \in \{r, r+1\}\}.$$

Taking into account that $[l_r, l_r] = \{l_r\}$, we will identify the linguistic term $l_r \in \mathcal{L}$ with the interval $[l_r, l_r] \in \mathcal{L}_2$ and, then, $\mathcal{L} \subset \mathcal{L}_2$. Notice that the granularity of $\mathcal{L}_2$ is $2g - 1$.

We extend the original order on $\mathcal{L}$ $(l_1 < l_2 < \cdots < l_g)$ to $\mathcal{L}_2$ in the natural way: $l_r < [l_r, l_{r+1}] < l_{r+1}$, for every $r \in \{1, \ldots, g-1\}$.

The opinions of all the agents over all the alternatives regarding the criterion $c_k \in C$ are collected in a *profile* V^k, that is a matrix of m rows and n columns with coefficients in $\mathcal{L}_2$

$$V^k = \begin{pmatrix} v_1^{1,k} & \cdots & v_i^{1,k} & \cdots & v_n^{1,k} \\ \cdots & \cdots & \cdots & \cdots & \cdots \\ v_1^{a,k} & \cdots & v_i^{a,k} & \cdots & v_n^{a,k} \\ \cdots & \cdots & \cdots & \cdots & \cdots \\ v_1^{m,k} & \cdots & v_i^{m,k} & \cdots & v_n^{m,k} \end{pmatrix} = \left(v_i^{a,k}\right),$$

where $v_i^{a,k}$ is the rating given by the agent a to the alternative x_i with respect to the criterion c_k.

Since each criterion may have different importance in the decision, we consider a weighting vector $w = (w_1, \ldots, w_q) \in [0, 1]^q$, with $w_1 + \cdots + w_q = 1$. For practical reasons, we assume that these weights have at most two decimals, i.e., the percentages $100 \cdot w_1, \ldots, 100 \cdot w_q$ are integer numbers.

A binary relation $\geq$ on a set $Z \neq \emptyset$ is a *linear order* if it is complete ($x \geq y$ or $y \geq x$, for all $x, y \in Z$), transitive (if $x \geq y$ and $y \geq z$, then $x \geq z$, for all $x, y, z \in Z$) and anti-symmetric (if $x \geq y$ and $y \geq x$, then $x = y$, for all $x, y \in Z$). As usual, the asymmetric and symmetric parts of a linear order $\geq$ are denoted by $>$ and $\sim$, respectively.

In the next section we will need to compare some pairs of intervals of linguistic terms,

$$P = \{([l_r, l_s], [l_t, l_u]) \in \mathcal{L}_2 \times \mathcal{L}_2 \mid r + s \leq t + u\}.$$

We will use the following binary relation on P

$$([l_r, l_s], [l_t, l_u]) \geq_P ([l_{r'}, l_{s'}], [l_{t'}, l_{u'}]) \iff$$

$$\begin{cases} r + s + t + u > r' + s' + t' + u' \\ \text{or} \\ r + s + t + u = r' + s' + t' + u' \text{ and } t + u - r - s \leq t' + u' - r' - s', \end{cases} \quad (1)$$

for all $[l_r, l_s], [l_t, l_u], [l_{r'}, l_{s'}], [l_{t'}, l_{u'}] \in P$.

It is easy to check that $\succeq_P$ is a linear order on P and that (1) is equivalent to

$$
\begin{cases}
r + s + t + u > r' + s' + t' + u' \\
\text{or} \\
r + s + t + u = r' + s' + t' + u' \quad \text{and} \quad t + u \leq t' + u'
\end{cases}
\tag{2}
$$

and

$$
\begin{cases}
r + s + t + u > r' + s' + t' + u' \\
\text{or} \\
r + s + t + u = r' + s' + t' + u' \quad \text{and} \quad r + s \geq r' + s'.
\end{cases}
\tag{3}
$$

Notice that

$$
\frac{r + s + t + u}{4} = \frac{\frac{r+s}{2} + \frac{t+u}{2}}{2}
$$

is the average of the midpoints of the intervals $[r, s]$ and $[t, u]$.

Moreover,

$$
\frac{t + u - r - s}{4} = \frac{\frac{t+u}{2} - \frac{r+s}{2}}{2}
$$

is the difference between the midpoints of the intervals $[t, u]$ and $[r, s]$.

Then, the two conditions appearing in the lexicographic order $\succeq_P$ can be interpreted through the midpoints of the corresponding intervals. First, the bigger the average of midpoints, the better. In the case of a tie, the smaller the difference between midpoints, the better.

Example 1 If $g = 3$, then P is ordered in the following way

$$(l_3, l_3) \succ_P ([l_2, l_3], [l_2, l_3]) \succ_P (l_2, l_3) \succ_P (l_2, [l_2, l_3]) \succ_P ([l_1, l_2], l_3) \succ_P$$

$$(l_2, l_2) \succ_P ([l_1, l_2], [l_2, l_3]) \succ_P (l_1, l_3) \succ_P ([l_1, l_2], l_2) \succ_P (l_1, [l_2, l_3]) \succ_P$$

$$([l_1, l_2], [l_1, l_2]) \succ_P (l_1, l_2) \succ_P (l_1, [l_1, l_2]) \succ_P (l_1, l_1).$$

2.2 The Procedure

The proposed multi-criteria decision-making procedure is divided in the following steps.

- *Step 1.* Gather the ratings given by the agents in the corresponding profiles $V^1, \ldots, V^q$.
- *Step 2.* Replicate the previous profiles according to the corresponding percentages $100 \cdot w_1, \ldots, 100 \cdot w_q$. In practice, calculate the greatest common divisor (gcd) of

percentages associated with the weights, and divide each percentage by the gcd. Then, the minimum number of replications of each profile is obtained.

For instance, if there are four criteria and the weights are $w_1 = 0.15$, $w_2 = 0.20$, $w_3 = 0.25$ and $w_4 = 0.40$, then $\gcd(15, 20, 25, 40) = 5$ and the profiles V^1, V^2, V^3 and V^4 should be replicated $15/5 = 3$, $20/5 = 4$, $25/5 = 5$ and $40/5 = 8$ times, respectively.

- *Step 3.* For each alternative $x_i \in X$, arrange the obtained ratings (taking into account the corresponding replications) in an increasing fashion.
- *Step 4.* Select the medians $M_i \in P$ of the ratings for each alternative $x_i \in X$ in the following way:

  1. If the number of ratings is odd, then duplicate the median. Thus, $M_i = ([l_r, l_s], [l_r, l_s])$ for some $[l_r, l_s] \in \mathcal{L}_2$.
  2. If the number of ratings is even, then take into account the two medians. Thus, $M_i = ([l_r, l_s], [l_t, l_u])$ for some $([l_r, l_s], [l_t, l_u]) \in P$.

- *Step 5.* Rank order the alternatives through the linear order $\succeq_P$, defined in (1), (2) and (3), by applying it to the corresponding medians:

$$x_i \succeq_X x_j \Leftrightarrow M_i \succeq_P M_j,$$

for all $x_i, x_j \in X$.
- *Step 6.* Since some alternatives can share the same median(s), it is necessary to devise a tie-breaking process for ordering the alternatives. We propose to use a sequential procedure based on Balinski and Laraki [2] (see Balinski and Laraki [4] for practical examples). It consists of withdrawing the median(s) of the ratings associated with the alternatives that are in a tie, and then selecting the new median(s) of the remaining ratings for the corresponding alternatives and applying Steps 4 and 5. The process continues until the ties are broken. It is important to note that alternatives with different ratings never are in a final tie.

The proposed multi-criteria decision-making procedure inherits (even enhances) good properties from MJ. We now pay attention on some of them (see Felsenthal and Machover [8]).

1. Voter-expressivity: agents not only rank order the alternatives (as in the preferential approach), but they assign grades in a enriched qualitative scale, $\mathcal{L}_2$. In fact, there are $(2g - 1)(g - 1)$ different ways in which $v_i^{a,k} > v_j^{a,k}$.
2. Anonymity: all agents are treated equally.
3. Neutrality: all alternatives are treated equally.
4. Unanimity: if all voters award alternative x_i a higher rating than to every other alternative in each criterion, then x_i wins.
5. Transitive ordering: alternatives are ranked in a linear order; one alternative is necessarily ranked ahead or behind another, unless they have identical ratings in all the criteria.

6. Independence of irrelevant alternatives: if alternative x_i wins, then x_i would still win if another alternative, x_j, is removed, *ceteris paribus*.
7. Monotonicity: if alternative x_i wins, then x_i would still win if one of the x_i's ratings $v_i^{a,k}$ increases, *ceteris paribus*.

3 An Illustrative Case Study

We now show how the proposed multi-criteria decision-making procedure works taking into account some data obtained in a case study carried out in *Trigo* restaurant in Valladolid (November 30th, 2013), under appropriate conditions of temperature, light and service.

A total of six judges (agents) trained in the sensory analysis of wild mushrooms were recruited through the *Gastronomy and Food Academy of Castilla y León*. When the test was being carried out there was no communication among judges, and the samples were given without any identification. For more details, see García-Lapresta et al. [9].

All six judges assessed the wild mushrooms included in Table 1 through the five linguistic terms of Table 2 (or the corresponding intervals of two consecutive linguistic terms, when they hesitated) under three criteria: appearance, smell and taste.

Notice that the granularity of $\mathcal{L}_2$ is 9:

$$\mathcal{L}_2 = \left\{ l_1, [l_1, l_2], l_2, [l_2, l_3], l_3, [l_3, l_4], l_4, [l_4, l_5], l_5 \right\}.$$

Table 1 Alternatives

x_1	Raw *Boletus pinophilus*
x_2	Raw *Tricholoma portentosum*
x_3	Cooked *Boletus pinophilus*
x_4	Cooked *Tricholoma portentosum*

Table 2 Linguistic terms in $\mathcal{L}$

l_1	I don't like it at all
l_2	I don't like it
l_3	I like it
l_4	I rather like it
l_5	I like it so much

Table 3 Ratings given by the judges

		1	2	3	4	5	6
x_1	Appearance	l_4	l_4	l_5	l_5	l_3	l_4
	Smell	$[l_4, l_5]$	l_3	$[l_4, l_5]$	l_5	l_3	l_5
	Taste	l_5	l_5	$[l_4, l_5]$	l_5	l_4	$[l_4, l_5]$
x_2	Appearance	l_4	l_3	l_4	l_5	l_3	$[l_2, l_3]$
	Smell	l_3	$[l_2, l_3]$	l_4	l_3	l_2	$[l_2, l_3]$
	Taste	l_3	l_3	$[l_4, l_5]$	l_4	l_2	l_4
x_3	Appearance	l_5	l_4	l_4	l_4	l_3	l_4
	Smell	l_5	l_5	l_5	l_4	l_3	$[l_4, l_5]$
	Taste	$[l_4, l_5]$	l_5	l_5	l_5	l_4	$[l_4, l_5]$
x_4	Appearance	l_5	l_3	$[l_4, l_5]$	l_5	$[l_2, l_3]$	l_4
	Smell	l_5	l_3	l_5	l_3	$[l_3, l_4]$	l_4
	Taste	l_4	l_4	l_5	l_3	l_4	$[l_3, l_4]$

Table 3 includes the ratings provided to the alternatives in all the criteria.

Notice that only l_1 and $[l_1, l_2]$ of $\mathcal{L}_2$ were not used for the judges, and five of the six judges used at least once two consecutive linguistic terms; in overall 20.83% of the ratings had two consecutive linguistic terms (less than in the tasting described in Agell et al. [1], where 40% of the ratings had two or more linguistic terms).

We considered the weights $w_1 = 0.2$ for appearance, $w_2 = 0.3$ for smell and $w_3 = 0.5$ for taste, i.e., $w = (0.2, 0.3, 0.5)$ (these weights are usual in this kind of tasting). After replicating the profiles V^1, V^2 and V^3 2, 3 and 5 times, respectively, and applying the steps 3, 4 and 5, we obtain the following medians:

$$M_1 = \left([l_4, l_5], [l_4, l_5]\right), \ M_2 = \left(l_3, l_3\right), \ M_3 = \left([l_4, l_5], [l_4, l_5]\right), \ M_4 = \left(l_4, l_4\right).$$

Then, $x_1 \sim_X x_3 \succ_X x_4 \succ_X x_2$. In order to break the tie between x_1 and x_3, we apply step 6 and, after four tiebreakers, we obtain the following medians $M_1 = \left([l_4, l_5], [l_4, l_5]\right)$ and $M_3 = \left([l_4, l_5], l_5\right)$. Consequently, the final ranking is $x_3 \succ_X x_1 \succ_X x_4 \succ_X x_2$. We note that this result coincides with the one obtained in García-Lapresta et al. [9] following a very different procedure based on geodesic distances and a penalization of the imprecision.

4 Concluding Remarks

The multi-criteria decision-making we have proposed in this contribution has been devised within the framework of uniform qualitative scales. In that setting, makes sense to consider the medians of the corresponding ratings as the collective rating. It is not the case of non-uniform qualitative scales, where medians could not capture the collective opinion. As further research, we plan to devise an appropriate proposal which takes into account the ordinal proximities among linguistic terms of the scale, following García-Lapresta and Pérez-Román [11, 12].

Acknowledgements This contribution is dedicated to Mario Fedrizzi in occasion of his retirement. The authors gratefully acknowledge the funding support of the Spanish *Ministerio de Economía y Competitividad* (project ECO2016-77900-P) and ERDF.

References

1. Agell N, Sánchez G, Sánchez M, Ruiz FJ (2013) Selecting the best taste: a group decision-making application to chocolates design. In: Proceedings of the 2013 IFSA-NAFIPS joint congress. Edmonton, pp 939–943
2. Balinski M, Laraki R (2007) A theory of measuring, electing and ranking. Proc Natl Acad Sci USA 104:8720–8725
3. Balinski M, Laraki R (2011) Majority Judgment. Measuring, ranking, and electing. The MIT Press, Cambridge
4. Balinski M, Laraki R (2013) How best to rank wines: Majority Judgment. In: Wine economics: quantitative studies and empirical observations. Palgrave-MacMillan, pp 149–172
5. Falcó E, García-Lapresta JL (2011) A distance-based extension of the Majority Judgement voting system. Acta Universitatis Matthiae Belii, Ser Math 18:17–27
6. Falcó E, García-Lapresta JL, Roselló L (2014) Aggregating imprecise linguistic expressions. In: Guo P, Pedrycz W (eds) Human-centric decision-making models for social sciences. Springer, Berlin, pp 97–113
7. Falcó E, García-Lapresta JL, Roselló L (2014) Allowing agents to be imprecise: a proposal using multiple linguistic terms. Inf Sci 258:249–265
8. Felsenthal DS, Machover M (2008) The Majority Judgement voting procedure: a critical evaluation. Homo Oeconomicus 25:319–334
9. García-Lapresta JL, Aldavero C, de Castro S (2014) A linguistic approach to multi-criteria and multi-expert sensory analysis. In: Laurent A et al (eds) IPMU 2014, Part II, CCIS 443. Springer International Publishing Switzerland, pp 586–595
10. García-Lapresta JL, Martínez-Panero M (2009) Linguistic-based voting through centered OWA operators. Fuzzy Optim Decis Making 8:381–393
11. García-Lapresta JL, Pérez-Román D (2015) Ordinal proximity measures in the context of unbalanced qualitative scales and some applications to consensus and clustering. Appl Soft Comput 35:864–872
12. García-Lapresta JL, Pérez-Román D (2015) Aggregating opinions in non-uniform qualitative scales. In: USB proceedings of the 12th international conference on modeling decisions for artificial intelligence (MDAI 2015). pp 152–163
13. Roselló L, Prats F, Agell N, Sánchez M (2010) Measuring consensus in group decisions by means of qualitative reasoning. Int J Approx Reason 51:441–452
14. Travé-Massuyès L, Piera N (1989) The orders of magnitude models as qualitative algebras. In: Proceedings of the 11th international joint conference on artificial intelligence. Detroit, pp 1261–1266

FRIM—Fuzzy Reference Ideal Method in Multicriteria Decision Making

E. Cables, M.T. Lamata and J.L. Verdegay

To Mario, Professor Mario Fedrizzi, whose friendship, high scientific and academic level, have influenced us along the years.

Abstract There are numerous compensatory multicriteria decision methods that are used for decision making. Among them, we consider the TOPSIS method for its rationality and easy applicability. This method is based on the concept that the alternative chosen should be the one whose distance to the positive ideal solution is smaller and simultaneously, the distance to the negative ideal solution is as large as possible. Based on this idea, the Reference Ideal Method (RIM) can be considered as an extension of the TOPSIS method when considering that the ideal solution does not have to be the maximum or minimum value, but may be a value between them. RIM gives good solutions but does not always obtain the solution when operating with fuzzy numbers. In this paper its extension is proposed to work with vagueness and uncertainty, resulting in the Fuzzy Reference Ideal Method (FRIM), with its applicability being illustrated through an example built frequency from a real practical problem.

1 Introduction

Humans are faced with situations each and every day where they have to choose among a set of options. In general, decision-makers make their choices following a set of rules and heuristic associated with their experience level, their independence degree, the type of information available, etc. In every situation it is necessary to consider that there are not unique criteria for making decisions, but that the decision-maker has to decide taking into account different decision criteria; thus, let us focus on the so-called Multi Criteria Decision Making (MCDM) problems and methods. In these circumstances, to select the most favourable alternative of the set,

E. Cables
Facultad de Ingeniería de Sistemas, Universidad Antonio Nariño, Bogotá, Colombia

M.T. Lamata · J.L. Verdegay (✉)
Depto de Ciencias de la Computación e Inteligencia Artificial, Universidad de Granada, Granada, Spain
e-mail: verdegay@ugr.es; verdegay@decsai.ugr.es

© Springer International Publishing AG 2018
M. Collan and J. Kacprzyk (eds.), *Soft Computing Applications for Group Decision-making and Consensus Modeling*, Studies in Fuzziness and Soft Computing 357, DOI 10.1007/978-3-319-60207-3_19

we must resort to operating methods that assess alternatives objectively and rationally with respect to previously established criteria and they may have associated weights reflecting their value, intensity, importance, etc.

The quality of our decisions depends directly on these methods, which must be able to synthesize large amount of information, often from different sources and therefore with different natures and different meanings. The MCDM problem resolution depends on the effectiveness, efficiency and functionality of these methods which are of great importance.

Among the wide variety of MCDM methods, in this paper we will consider those associated with compensatory strategy which take into account that the chosen alternative is superior to the other alternatives in the sum of the weighted utilities of all the criteria considered; by selecting, at the end of the process, the alternative with a higher score. In other words, we will consider methods that permit trade-offs between criteria [17]. It is important to mention that in these methods, a negative value on one attribute can be compensated by an equal or higher value on another attribute.

- Among the compensatory methods we can consider the methods using:
- A utility function, as is the case of the Analytical Hierarchy Process (AHP) [25], the Analytical Network Process (ANP) [26], or the SMART method [11], among others.
- An outranking relation between alternatives, for example: the ELECTRE method [27] and the PROMETHEE method [3]. The identification of the ideal solution to perform the aggregation of information, for example: the TOPSIS method [33], the VIKOR method [22], and the Reference Ideal Method (RIM) [5].
- Moreover, from the high level of imprecision that is reflected in the information collected in real decision problems, different MCDM methods have been extended or combined, so as to operate with fuzzy information [20], for example: The AHP method, [4, 7, 15, 28, 32], the ANP method, [16, 19, 29], the ELECTRE method and their respective variants and applications such as ELECTRE III [24], ELECTRE TRI [14, 23, 27, 30], the PROMETHEE method [2, 13], the TOPSIS method, with different variants, [1, 5, 9, 12] and the VIKOR method [8, 10, 18, 21, 31] among many others.

As can be observed, there are different MCDM methods using fuzzy numbers. These methods have as their purpose to resolve the high levels of imprecision that the information presents to confront decision making problems in different areas with a high degree of objectivity. In general, these methods consider that the best is the maximum value when it comes to profits and the worst when we consider losses.

It is also necessary to consider problems, both in the crisp as well as in the fuzzy case, where for a given criterion, "the best" should not be the maximum (profit case) or the minimum (losses case), but "the optimum" may be a value between the minimum and the maximum value. Such will be the case of the pH of a cosmetic, the temperature of a wine, the fat content in food, the age of a person to access a specific job, etc. In order to address these situations where the best is not the maximum neither the minimum value and to reach operational solutions, RIM was presented [6]. RIM is a new method based on the concept of "ideal solution value", where this concept will be any value between the maximum and the minimum value, with this ideal solution being the main difference with the VIKOR and TOPSIS methods in which this value is the extreme value.

But RIM, as with other MCDM methods, cannot be applied directly to problems where the information is expressed imprecisely, or in other words to problems in which there is vagueness or imprecision in data and therefore will be expressed as fuzzy numbers. Therefore, the aim of this paper is to present the Fuzzy Reference Ideal Method (FRIM) that modifies, broadens and extends the original RIM. Thus, FRIM can operate in situations where the best for a particular criterion could be the maximum value, the minimum value or an intermediate value among them, it can operate with fuzzy numbers and therefore solve situations that until now had not been addressed.

Consequently, the problem to be solved has already been raised in this section and we have performed a discussion of some of the most recognized compensatory MCDM methods related to the problem, as well as the extension to operate with imprecise information. In the next section, RIM and the problems derived from operating with fuzzy numbers are presented. The third section is dedicated to developing FRIM itself, to finish by presenting a real illustrative example of the new formulation that we extract from the results of a research project that we are currently developing for a consumer organisation in Spain.

2 Background: The Reference Ideal Method (RIM)

Different MCDM reported in the literature, require a valuation matrix M, where its elements x_{ij} represent the evaluation of all alternatives A_i, $i = 1, 2, \ldots, m$ for each one of the criteria C_j, $j = 1, 2, \ldots, n$ and w_j is the weight associated with each criterion.

$$
M = \begin{matrix}
 & & w_1 & w_2 & & w_n \\
 & & C_1 & C_2 & \cdots & C_n \\
 & A_1 & \begin{pmatrix} x_{11} & x_{12} & \cdots & x_{1n} \\
 A_2 & x_{21} & x_{22} & \cdots & x_{2n} \\
 \vdots & \vdots & \vdots & \ddots & \vdots \\
 A_m & x_{m1} & x_{m2} & \cdots & x_{mn} \end{pmatrix}
\end{matrix}
$$

In this case, RIM also uses a valuation or judgments matrix and from it the calculations are performed to rank the alternatives involved in the decision-making process. Therefore, supposing the decision matrix M is known, RIM is based on identifying for each criterion C_j, $j = 1, 2, \ldots, n$ the concepts of Range and Reference Ideal:

The *Range* $R_j = [A_j, B_j]$ indicates any interval, ordered set of labels or ordered set of values that identify a domain of discourse and that is associated with each one of the criteria.

The *Reference Ideal* $RI_j = [C_j, D_j]$, is an interval, an ordered set of labels, labels or simple values, which represent the optimal value, the maximum importance or relevance of the criterion C_j in a given Range.

Then, from the abovementioned concepts, RIM is based on determining the shortest distance to the *Reference Ideal*, considering the distance of a given rating x_{ij}, to their respective $[C_j, D_j]$, as follows:

$$d_{\min}\left(x_{ij}, [C_j, D_j]\right) = \min\left(\left|x_{ij} - C_j\right|, \left|x_{ij} - D_j\right|\right) \tag{1}$$

Once the distance matrix has been obtained, it is necessary to normalize it. This operation is performed with the aim of transforming all values to the same scale, because these values can usually represent different magnitudes and different meanings. Thus, we see that the TOPSIS, VIKOR and RIM methods have different metrics for the process.

Particularly, RIM performs *normalization* of any $x_{ij} \in [A_j, B_j]$ value through the following function [12].

$$f : x_{ij} \otimes [A, B] \otimes [C, D] \to [0, 1]$$

$$f\left(x_{ij}, R_j, RI_j\right) = \begin{cases} 1 & \text{if } x_{ij} \in [C_j, D_j] \\ 1 - \dfrac{d_{\min}\left(x_{ij}, RI_j\right)}{dist\left(A_j, C_j\right)} & \text{if } x_{ij} \in [A_j, C_j] \wedge A_j \neq C_j \\ 1 - \dfrac{d_{\min}\left(x_{ij}, RI_j\right)}{dist\left(D_j, B_j\right)} & \text{if } x_{ij} \in [D_j, B_j] \wedge D_j \neq B_j \\ 0 & \text{in other case} \end{cases} \tag{2}$$

Example 1 To show the behaviour of the f function, we will consider Fig. 1,

Let us suppose $R = [A, B] = [0, 10]$, $[C, D] = [5, 7]$, and the three possibilities $x = 2$, $y = 6$ and $z = 8$. When calculating the f function image for the x, y, z values, we obtain:

Fig. 1 Representation of values A, B, C, D, x, y, and z in the real line

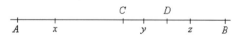

$$f(x, R, RI) = 1 - \frac{\min(|x - C|, |x - D|)}{dist(A, C)} = 1 - \frac{\min(|2 - 5|, |2 - 7|)}{5} = 1 - \frac{3}{5} = \frac{2}{5}$$

$$f(y, R, RI) = 1$$

$$f(z, R, RI) = 1 - \frac{\min(|z - C|, |z - D|)}{dist(D, B)} = 1 - \frac{\min(|8 - 5|, |8 - 7|)}{3} = 1 - \frac{1}{3} = \frac{2}{3}$$

As we see, RIM has been able to solve a decision problem, where "the best" can be any value $v \in [A, B]$, (not just the extremes A or B). However, when fuzzy numbers are considered, RIM presents problems; therefore in the next paragraph we will see what these problems are and how to solve them.

3 The Fuzzy Reference Ideal Method (FRIM)

Until now we have worked on a set of real numbers. However, if the values are not real ones, but fuzzy numbers, it becomes necessary to reformulate the expression (1) and therefore (2). When it is necessary to operate with fuzzy numbers the distance between two fuzzy numbers $\tilde{X}_{ij}, \tilde{D}_{ij}$, will be given by the vertex method distance defined in (3):

$$dist: \tilde{X} \times \tilde{Y} \to \mathbb{R}$$

$$dist(\tilde{X}_{ij}, \tilde{D}_{ij}) = \sqrt{\frac{1}{3}\left((x_1 - d_1)^2 + (x_2 - d_2)^2 + (x_3 - d_3)^2\right)} \qquad (3)$$

Furthermore, and as we have seen before, when there is vagueness in the data, the formulation used by RIM cannot be applied directly and we will thus need to reformulate the distance measure to a fuzzy interval.

3.1 Minimal Distance to a Fuzzy Interval

As it has arisen, RIM is based on determining the shortest distance to the reference ideal, and in this case, when operated with fuzzy numbers it is possible to observe that it is not sufficient with the Euclidean distance. For this, we define the minimal distance of a fuzzy number to an interval bounded by fuzzy numbers (or a fuzzy number) through the following definition.

Definition 1 Let $\tilde{X}, \tilde{C}, \tilde{D}$ be positive fuzzy numbers such that, $\tilde{X} = (x_1, x_2, x_3)$, $\tilde{C} = (c_1, c_2, c_3)$, $\tilde{D} = (d_1, d_2, d_3)$, then the minimal distance of the value $\tilde{X}_{ij}$ to the interval $I\tilde{R}_j = [\tilde{C}_j, \tilde{D}_j]$, is given by the function $d^*_{\min}$, where:

$$d^*_{\min}: \tilde{X}_{ij} \otimes \left[I\tilde{R}_j\right] \to \mathbb{R}$$

$$d^*_{\min}\left(\tilde{X}_{ij}, \left[I\tilde{R}_j\right]\right) = \min\left(dist\left(\tilde{X}_{ij}, \tilde{C}_j\right), dist\left(\tilde{X}_{ij}, \tilde{D}_j\right)\right) \tag{4}$$

where, the functions $dist\left(\tilde{X}_{ij}, \tilde{C}_j\right)$ and $dist\left(\tilde{X}_{ij}, \tilde{D}_j\right)$ are calculated using the expression (3).

3.2 Normalization in FRIM

As such, RIM carries out the normalization process of the decision matrix through expression 2, which should not be used when we operate with fuzzy numbers. It may be the case that the value assigned to the variable $\tilde{X}$ is not completely included in the Reference Ideal interval $\left[\tilde{C}_j, \tilde{D}_j\right]$. In this case $\tilde{X}_{ij} \cap \left[\tilde{C}_j, \tilde{D}_j\right] \neq \emptyset$. This would be the case of the $\tilde{Y}$ value on the interval $\left[\tilde{C}_j, \tilde{D}_j\right]$, as shown in Fig. 2.

Thus, the reformulation of (2) is expressed in the following definition.

Definition 2 Let $\tilde{X}, \tilde{A}, \tilde{B}, \tilde{C}, \tilde{D}$ be positive fuzzy numbers such that, $\tilde{X} = (x_1, x_2, x_3)$, $\tilde{A} = (a_1, a_2, a_3)$, $\tilde{B} = (b_1, b_2, b_3)$, $\tilde{C} = (c_1, c_2, c_3)$, $\tilde{D} = (d_1, d_2, d_3)$, where the interval $\tilde{R}_j = \left[\tilde{A}_j, \tilde{B}_j\right]$ represents the range, the interval $I\tilde{R}_j = \left[\tilde{C}_j, \tilde{D}_j\right]$ represents the Reference Ideal and $\left[\tilde{C}_j, \tilde{D}_j\right] \subseteq \left[\tilde{A}_j, \tilde{B}_j\right]$ for each criterion $\tilde{C}_j$, then the normalization function f^*, is given by:

$$f^*\left(\tilde{X}_{ij}, \left[\tilde{R}_j\right], \left[I\tilde{R}_j\right]\right) = \begin{cases} 1 & if \quad \tilde{X}_{ij} \in \left[I\tilde{R}_j\right] \\ 1 - \dfrac{d^*_{\min}\left(\tilde{X}_{ij}, \left[I\tilde{R}_j\right]\right)}{dist\left(\tilde{A}_j, \tilde{C}_j\right)} & if \quad \tilde{X}_{ij} \in \left[\tilde{A}_j, \tilde{C}_j\right] \wedge \tilde{X}_{ij} \notin \left[I\tilde{R}_j\right] \wedge dist\left(\tilde{A}_j, \tilde{C}_j\right) \neq 0 \\ 1 - \dfrac{d^*_{\min}\left(\tilde{X}_{ij}, \left[I\tilde{R}_j\right]\right)}{dist\left(\tilde{D}_j, \tilde{B}_j\right)} & if \quad \tilde{X}_{ij} \in \left[\tilde{D}_j, \tilde{B}_j\right] \wedge \tilde{X}_{ij} \notin \left[I\tilde{R}_j\right] \wedge dist\left(\tilde{D}_j, \tilde{B}_j\right) \neq 0 \\ 0 & in\ other\ case \end{cases}$$

$$\tag{5}$$

where:

$d^*_{\min}\left(\tilde{X}_{ij}, \left[I\tilde{R}_j\right]\right)$ is obtained by applying (4).

Fig. 2 Representation of the fuzzy Reference ideal and the fuzzy numbers $\tilde{X}$ and $\tilde{Y}$

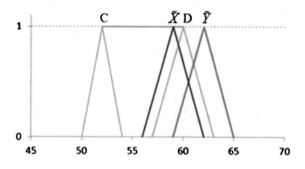

$dist\left(\tilde{A}_j, \tilde{C}_j\right)$ and $dist\left(\tilde{D}_j, \tilde{B}_j\right)$ are obtained by applying (3).

Example   2 Let   us   suppose $\left[\tilde{A}_j, \ \tilde{B}_j\right] = \left[(0, \ 15, \ 35), \ (110, \ 135, \ 150)\right]$, $\left[\tilde{C}_j, \ \tilde{D}_j\right] = \left[(50, \ 52, \ 54), \ (57, \ 60, \ 63)\right]$ and we wish to normalize the values $\tilde{X} = (56, \ 59, \ 62)$ and $\tilde{Y} = (59, \ 62, \ 65)$ then:

$$f^*\left(\tilde{X}, \tilde{R}_j, I\tilde{R}_j\right) = 1$$

In this case the fuzzy number $\tilde{X}$, is totally included in the interval that represents the *Reference Ideal.*

$$f^*\left(\tilde{Y}, \tilde{R}_j, I\tilde{R}_j\right) = 1 - \frac{d^*_{\min}\left(\tilde{Y}, \left[\tilde{C}_j, \tilde{D}_j\right]\right)}{\max\left(dist\left(\tilde{A}_j, \tilde{C}_j\right), dist\left(\tilde{D}_j, \tilde{B}_j\right)\right)} = 1 - \frac{2}{73} = 0.9726$$

In this case $\tilde{Y}$ is not completely included in the *Reference Ideal* with the result being near to 1.

4   Fuzzy RIM Algorithm

From the formulations showed previously, it can be considered that the FRIM algorithm stays similar to the RIM algorithm, because only step 2 changes.

Therefore, the algorithm FRIM steps are described below:

Step 1. Definition of the work context.

First, the conditions in the work context are established, and for each criterion C_j the following aspects are defined:

- The Range $\left[\tilde{A}_j, \tilde{B}_j\right]$, that from now will be denoted by $\tilde{R}_j$.
- The Reference Ideal $\left[\tilde{C}_j, \tilde{D}_j\right]$, that from now will be denoted by $I\tilde{R}_j$.
- The weight w_j associated to the criterion.

Step 2. Obtain the decision matrix V in correspondence with the defined criteria. In this case, the $\tilde{v}_{ij}$ elements represent triangular fuzzy numbers.

$$V = \begin{pmatrix} \tilde{v}_{11} & \tilde{v}_{12} & \cdots & \tilde{v}_{1n} \\ \tilde{v}_{21} & \tilde{v}_{22} & \cdots & \tilde{v}_{2n} \\ \vdots & \vdots & \ddots & \vdots \\ \tilde{v}_{m1} & \tilde{v}_{m2} & \cdots & \tilde{v}_{mn} \end{pmatrix}$$

Step 3. Normalize the decision matrix V, depending on the reference ideal.

$$N = \begin{pmatrix} f^*(\tilde{v}_{11}, \tilde{R}_1, I\tilde{R}_1) & f^*(\tilde{v}_{12}, \tilde{R}_2, I\tilde{R}_2) & \cdots & f^*(\tilde{v}_{1n}, \tilde{R}_n, I\tilde{R}_n) \\ f^*(\tilde{v}_{21}, \tilde{R}_1, I\tilde{R}_1) & f^*(\tilde{v}_{22}, \tilde{R}_2, I\tilde{R}_2) & \cdots & f^*(\tilde{v}_{2n}, \tilde{R}_n, I\tilde{R}_n) \\ \vdots & \vdots & \ddots & \vdots \\ f^*(\tilde{v}_{m1}, \tilde{R}_1, I\tilde{R}_1) & f^*(\tilde{v}_{m2}, \tilde{R}_2, I\tilde{R}_2) & \cdots & f^*(\tilde{v}_{mn}, \tilde{R}_n, I\tilde{R}_n) \end{pmatrix}$$

where, the f^* function is that considered in (5).

Step 4. Calculate the weighted normalized matrix P, through:

$$P = N \otimes W = \begin{pmatrix} n_{11} \cdot w_1 & n_{12} \cdot w_2 & \cdots & n_{1n} \cdot w_n \\ n_{21} \cdot w_1 & n_{22} \cdot w_2 & \cdots & n_{2n} \cdot w_n \\ \vdots & \vdots & \ddots & \vdots \\ n_{m1} \cdot w_1 & n_{m2} \cdot w_2 & \cdots & n_{mn} \cdot w_n \end{pmatrix}$$

Step 5. Calculate the distance to the ideal and non-ideal alternative.

$$A_i^+ = \sqrt{\sum_{j=1}^{n} (p_{ij} - w_j)^2} \text{ and } A_i^- = \sqrt{\sum_{j=1}^{n} (p_{ij})^2}$$

Step 6. Calculate the relative index to the reference ideal of each alternative by the following expression:

$$I_i = \frac{A_i^-}{A_i^+ + A_i^-}, \text{ where, } 0 \le I_i \le 1, i = 1, 2, \ldots, m$$

Step 7. Rank the alternatives in descending order from the relative index I_i. In this case, if the alternative has a relative index I near to the value 1, this indicates that it is very good. However, if this value approaches the value 0, we will interpret that the alternative should be rejected.

5   A Real Illustrative Example

The rationality of FRIM, as well as its practical importance, may be illustrated by the following real example extracted from a much broader project, to classify the different trademarks of olive oil that we are carrying out for a consumer organisation in Spain.

Olive oil is how we refer to the oil obtained from the fruit of olive trees. People have been eating olive oil for thousands of years and it is now more popular than ever, thanks to its many proven health benefits and its culinary usefulness. It is good to understand the different types or grades of olive oil to help decision makers select the appropriate uses for this healthful and flavoursome type of fat.

The basic types of olive oil are: extra virgin olive oil, virgin olive oil and pure olive oil. There are other forms, but these are blends and are not part of the formal grading process. Extra virgin is the highest quality and most expensive olive oil classification. But, as it is evident, not all trademarks of extra virgin olive oil are equal.

There are hundreds of trademarks of virgin olive oil and classifying them is an important problem, both from the economic as well as the methodological point of view, to which a large amount of resources are dedicated (http://www.bestoliveoils. com/). In situations in which the conventional methodologies (TOPSIS, VIKOR) present dysfunctions because they are unable to provide correct solutions, FRIM is shown as a rigorous methodology which perfectly resolves these solutions that are unapproachable for the other methods. For this reason FRIM is being applied to carry out the classification of olive oils that we are working on. Herein we only present this small-sized example for purely illustrative purposes.

Let us consider 8 trademarks of extra virgin olive oil that are available in supermarkets. We wish to know which the best is, considering the price, acidity, wax and qualification of tasting experts.

For each trademark measures have been taken several for each criteria, the minimum, average and maximum of the values, except those relating to the tasting of which only the final values are known, and which correspond to the mean value.

The data are collected in Table 1 and it is considered that the four criteria are equally important. The values of A are determined by minimum values for prices, while for acidity and waxes the minimum values are those given by the experience. The values of B indicate the maximum for prices and value ceilings imposed by the law for acidity and waxes. While the interval [C, D] represents the greater or lesser slack that a decision-maker is willing to admit (Tables 2 and 3).

As we can see, this case cannot be resolved by TOPSIS or VIKOR methods. We detail the reasons why it is not possible through the different criteria that have been taken into account in the case proposed to illustrate the method.

Table 1 The decision matrix

Trademarks	Prices	Acidity	Waxes	Oil tastings
M1	(2.99, 3.29, 3.75)	(0.17, 0.19, 0.21)	(64, 67, 70)	(6.9, 6.9, 6.9)
M2	(2.69, 3.27, 3.85)	(0.18, 0.2, 0.22)	(54, 57, 60)	(5.8, 5.8, 5.8)
M3	(2.93, 3.04, 3.8)	(0.28, 0.31, 0.34)	(59, 62, 65)	(6.3, 6.3, 6.3)
M4	(2.8, 3.24, 3.69)	(0.23, 0.26, 0.29)	(66, 69, 72)	(6.8, 6.8, 6.8)
M5	(2.95, 3.14, 3.46)	(0.22, 0.25, 0.28)	(53, 56, 59)	(6, 6, 6)
M6	(2.74, 2.96, 3.99)	(0.33, 0.37, 0.4)	(46, 49, 52)	(6.4, 6.4, 6.4)
M7	(2.89, 3.08, 3.8)	(0.18, 0.2, 0.22)	(57, 60, 63)	(6.6, 6.6, 6.6)
M8	(3, 3.19, 3.45)	(0.23, 0.26, 0.29)	(56, 59, 62)	(6.4, 6.4, 6.4)

Table 2 The values of the range

Trademarks	Prices	Acidity	Waxes	Oil tastings
A	(2.69, 2.96, 3.45)	(0.15, 0.16, 0.17)	(0, 15, 35)	(1, 1, 1)
B	(3, 3.29, 3.99)	(0.55, 0.70, 0.80)	(110, 135, 150)	(9, 9, 9)

Table 3 The values of the Reference Ideal

Trademarks	Prices	Acidity	Waxes	Oil tastings
C	(2.69, 2.96, 3.04)	(0.16, 0.16, 0.17)	(50, 52, 54)	(8, 9, 9)
D	(2.69, 2.96, 3.04)	(0.2, 0.22, 0.26)	(57, 60, 63)	(8, 9, 9)

Table 4 Normalized valuation matrix

Trademarks	Prices	Acidity	Waxes	Oil tastings
M_1	0.1947	1.0000	0.9042	0.7620
M_2	0.1699	1.0000	1.0000	0.6218
M_3	0.2456	0.8199	0.9726	0.6858
M_4	0.3085	0.9274	0.8768	0.7494
M_5	0.4880	0.9486	1.0000	0.6474
M_6	0.1114	0.6974	0.9172	0.6986
M_7	0.2484	1.0000	1.0000	0.7240
M_8	0.4498	0.9274	1.0000	0.6986

- The prices: It is logical to think that we would seek to pay as little as possible, therefore both TOPSIS and VIKOR, or indeed any other MCDM, could be applied.
- The acidity: In this case we consider any oil as being good if its acidity is within the interval [0.16, 0.26] although by law this may reach 0.80.
- The waxes: For this criterion, neither of the methods (TOPSIS, VIKOR) can give a solution because the optimal is the interval [50, 63], but the range possible goes from a minimum of 0 to a maximum of 150. RIM cannot be applied either because it does not work with fuzzy evaluations. Thus it must be resolved with a method such as FRIM and comparisons with other methods cannot be made since they are inapplicable.
- The oil taste: It is easy to understand that the ideal it to achieve the maximum score by the experts, which means that both TOPSIS and VIKOR could be applied.

The following tables show the different steps of the algorithm (Tables 4, 5 and 6).

Concluding that under these criteria, the trademark with the best quality price ratio is M_5 although M_8 is very close to it.

Table 5 Weighted normalized matrix

Trademarks	Prices	Acidity	Waxes	Oil tastings
M_1	0.0487	0.2500	0.2260	0.1905
M_2	0.0425	0.2500	0.2500	0.1554
M_3	0.0614	0.2050	0.2432	0.1715
M_4	0.0771	0.2318	0.2192	0.1873
M_5	0.1220	0.2372	0.2500	0.1619
M_6	0.0279	0.1744	0.2293	0.1746
M_7	0.0621	0.2500	0.2500	0.1810
M_8	0.1125	0.2318	0.2500	0.1746

Table 6 Variation to the positive and negative reference ideal. Indexes calculation

Trademarks	d_i^+	d_i^-	R_i
M_1	0.2113	0.3902	0.6487
M_2	0.2281	0.3885	0.6301
M_3	0.2093	0.3665	0.6365
M_4	0.1873	0.3779	0.6686
M_5	0.1559	0.3998	**0.7194**
M_6	0.2473	0.3380	0.5775
M_7	0.2002	0.4020	0.6676
M_8	0.1579	0.3992	**0.7166**

6 Final Remarks

Since the measuring instruments are imprecise, it is necessary to work with methods that counteract this problem. In this sense, the fuzzy theory and its arithmetic give good results. On the other hand there are many problems where the best decision is not associated to the maximun or to the minimun but that the best value correspond to intermediate best value correspond to intermediate values as it the case that concerns us. Thus, to assess the quality of Virgin olive oil one of the components to consider are waxes, where the best is neither 0 nor 150, extreme values that take it. We have seen that the optimum would be a value comprised between 50 and 63.

In this paper, from the study of the Reference Ideal Method, a modification thereof is performed if the operation uses fuzzy numbers. Therefore, it has been necessary to modify RIM, because it was not possible to work directly with this method, when the fuzzy number has non-empty intersection with the Reference Ideal.

Given that RIM does not give a solution when the fuzzy number intersects with the reference ideal, a new distance has been defined and from it, the normalization function.

Triangular fuzzy numbers have been considered, but by extension the Fuzzy Reference Ideal Method (FRIM) can work with any other type of fuzzy numbers.

Acknowledgements This work has been partially funded by projects TIN2014-55024-P from the Spanish Ministry of Economy and Competitiveness, P11-TIC-8001 from the Andalusian Government, and both with FEDER funds.

References

1. Arslan M, Cunkas M (2012) Performance evaluation of sugar plants by fuzzy technique for order performance by similarity to ideal solution (TOPSIS). Cybern Syst 43:529–548
2. Behzadian M, Kazemzadeh RB, Albadvi A, Aghdasi M (2010) PROMETHEE: a comprehensive literature review on methodologies and applications. Eur J Oper Res 200:198–215
3. Brans JP, Vincke P, Mareschal B (1986) How to select and how to rank projects: the PROMETHEE method. Eur J Oper Res 24:228–238
4. Buyukozkan G, Cifci G, Guleryuz S (2011) Strategic analysis of healthcare service quality using fuzzy AHP methodology. Expert Syst Appl 38:9407–9424
5. Cables E, Garcia-Cascales MS, Lamata MT (2012) The LTOPSIS: an alternative to TOPSIS decision-making approach for linguistic variables. Expert Syst Appl 39:2119–2126
6. Cables E, Lamata MT, Verdegay JL (2016) RIM-reference ideal method in multicriteria decision making. Inf Sci 337:1–10
7. Calabrese A, Costa R, Menichini T (2013) Using fuzzy AHP to manage intellectual capital assets: an application to the ICT service industry. Expert Syst Appl 40:3747–3755
8. Chang TH (2014) Fuzzy VIKOR method: a case study of the hospital service evaluation in Taiwan. Inf Sci 271:196–212
9. Dymova L, Sevastjanov P, Tikhonenko A (2013) An approach to generalization of fuzzy TOPSIS method. Inf Sci 238:149–162
10. Dincer H, Hacioglu U (2013) Performance evaluation with fuzzy VIKOR and AHP method based on customer satisfaction in Turkish banking sector. Kybernetes 42:1072–1085
11. Edwards W, Barron FH (1994) SMARTS and SMARTER: improves simple methods for multiattibute utility measurement. Organ Behav Hum Decis Process 60:306–325
12. García-Cascales MS, Lamata MT (2011) Multi-criteria analysis for a maintenance management problem in an engine factory: rational choice. J Intell Manuf 22:779–788
13. Gupta R, Sachdeva A, Bhardwaj A (2012) Selection of logistic service provider using fuzzy PROMETHEE for a cement industry. J Manuf Tech Manag 23:899–921
14. Hatami-Marbini A, Tavana M, Moradi M, Kangi F (2013) A fuzzy group Electre method for safety and health assessment in hazardous waste recycling facilities. Saf Sci 51:414–426
15. Ishizaka A, Nguyen NH (2013) Calibrated fuzzy AHP for current bank account selection. Expert Syst Appl 40:3775–3783
16. Kang HY, Lee AH, Yang CY (2012) A fuzzy ANP model for supplier selection as applied to IC packaging. J Intell Manuf 23:1477–1488
17. Keeney RL, Raiffa H (1976) Decisions with multiple objectives: preferences and value tradeoffs. Wiley, New York
18. Kim Y, Chung ES (2013) Fuzzy VIKOR approach for assessing the vulnerability of the water supply to climate change and variability in South Korea. Appl Math Model 37:9419–9430
19. Liou JH, Tzeng G-H, Tsai C-Y, Hsu C-C (2011) A hybrid ANP model in fuzzy environments for strategic alliance partner selection in the airline industry. Appl Soft Comput 11:3515–3524
20. Mardani A, Jusoh A, Zavadskas EK (2015) Fuzzy multiple criteria decision-making techniques and applications—two decades review from 1994 to 2014. Expert Syst Appl 42: 4126–4148
21. Mokhtarian MN, Sadi-Nezhad S, Makui A (2014) A new flexible and reliable interval valued fuzzy VIKOR method based on uncertainty risk reduction in decision making process: an

application for determining a suitable location for digging some pits for municipal wet waste landfill. Comput Ind Eng 78:213–233

22. Opricovic S (1998) Multi-criteria optimization of civil engineering systems. Faculty of Civil Engineering, Belgrade
23. Rouyendegh BD, Erkan TE (2013) An application of the fuzzy electre method for academic staff selection. Hum Factor Ergon Manuf Serv Ind 23:107–115
24. Roy B, Skalka J (1985) ELECTRE IS, aspects méthodologiques et guide d´utilisation. Cahier du LAMSADE. Université Paris-Dauphine, Paris, p 30
25. Saaty TL (1980) The analytic hierarchy process. McGraw-Hill, New York
26. SaatyTL (1999) Fundamentals of the analytic network process. In: ISAHP 1999. Kobe, Japan
27. Sánchez-Lozano M, García-Cascales MS, Lamata MT (2014) Identification and selection of potential sites for onshore wind farms in Region of Murcia, Spain. Energy 73:311–324
28. Sánchez-Lozano M, García-Cascales MS, Lamata MT (2015) Evaluation of optimal sites to implant solar thermoelectric power plants: case study of the coast of the Region of Murcia, Spain. Comput Ind Eng 87:343–355
29. Vahdani B, Hadipour H, Tavakkoli-Moghaddam R (2012) Soft computing based on interval valued fuzzy ANP—A novel methodology. J Intell Manuf 23:1529–1544
30. Vahdani B, Mousavi SM, Tavakkoli-Moghaddam R, Hashemi H (2013) A new design of the elimination and choice translating reality method for multicriteria group decision-making in an intuitionistic fuzzy environment. Appl Math Model 37:1781–1799
31. Wan SP, Wang OY, Dong J-Y (2013) The extended VIKOR method for multi-attribute group decision making with triangular intuitionistic fuzzy numbers. Knowl-Based Syst 52:65–77
32. Wang YJ (2014) A fuzzy multi-criteria decision-making model by associating technique for order preference by similarity to ideal solution with relative preference relation. Inf Sci 268:169–184
33. Yoon K (1980) Systems selection by multiple attribute decision making. PhD dissertation, Kansas State University, Manhattan

A New Approach for Solving CCR Data Envelopment Analysis Model Under Uncertainty

Bindu Bhardwaj, Jagdeep Kaur and Amit Kumar

Abstract Wang and Chin (Expert Syst Appl, 38:11678–11685, 2011 [25]) proposed an optimistic as well as pessimistic fuzzy CCR data envelopment analysis (DEA) model and an approach for solving it to evaluate the best relative fuzzy efficiency as well as worst relative fuzzy efficiency and hence, relative geometric crisp efficiency of decision making units (DMUs). In this chapter, it is shown that the fuzzy CCR models, proposed by Wang and Chin, are not valid and hence cannot be used to evaluate the best relative fuzzy efficiency as well as worst relative fuzzy efficiency and hence, relative geometric crisp efficiency of DMUs. To resolve the flaws of the fuzzy CCR DEA models, proposed by Wang and Chin, new fuzzy CCR DEA models are proposed. Also, a new approach is proposed to solve the proposed fuzzy CCR DEA models for evaluating the relative geometric crisp efficiency of DMUs.

Keywords Data envelopment analysis · Fuzzy input and fuzzy output data · Fuzzy efficiency

1 Introduction

DEA is a non-parametric approach for measuring the relative efficiency of DMUs when the production process presents a structure of multiple inputs and outputs. DEA has found surprising development due to its wide range of applications to real world problems. The conventional CCR and BCC DEA models [1, 2] require accurate measurement of both the inputs and outputs.

In conventional DEA models, all the data is assumed to be exactly known. However, inputs and outputs of DMUs in real world problems may be imprecise. Imprecise evaluations may be the result of unquantifiable, incomplete and non-obtainable information. In recent years, fuzzy set theory has been proven to be useful as a way to quantify imprecise and vague data in DEA models. The DEA

B. Bhardwaj · J. Kaur (✉) · A. Kumar
Thapar University, Patiala 147004, Punjab, India
e-mail: sidhu.deepi87@gmail.com

model with fuzzy data, called "fuzzy DEA" models, can more realistically represent real world problems than the conventional DEA models.

Several authors [1, 3–24, 27, 28] have proposed methods to solve fuzzy DEA models and to solve fuzzy DEA models two approaches are used: (1) by transforming the fuzzy fractional programming model into a fuzzy linear programming model (2) by transforming fuzzy DEA models into two respective pessimistic and optimistic crisp DEA models using α-cut technique. Wang et al. [26] pointed out that the former ignores the fact that a fuzzy fractional programming cannot be transformed into a linear programming model as we do for a crisp fractional programming; while the latter requires the solution of a series of linear programming models based on different α-level sets and therefore it requires much computational efforts to get the fuzzy efficiencies of DMUs. Thus Wang et al. [26] proposed methods to solve fuzzy DEA models without using the aforementioned transformations (1) and (2) and later extended their existing method by considering uncertain weights [25]. In this chapter, the shortcomings of the existing method [25] are pointed out and a new method is proposed for the same.

The rest of the chapter is organized as follows. In Sect. 2, some basic definitions and arithmetic operations on fuzzy numbers are presented. In Sect. 3, the existing method [25] for solving fuzzy DEA problems is reviewed. In Sect. 4, the flaws of the existing method [25] are pointed out. To overcome these flaws, a new method is proposed in Sect. 5 for solving the proposed fuzzy DEA models. The proposed method is illustrated with the help of a real world problem in Sect. 6 and the obtained results are discussed in Sect. 7. Finally, the conclusions are discussed in Sect. 8.

2   Basic Definitions and Arithmetic Operations

In this section, some basic definitions and arithmetic operations on fuzzy numbers are reviewed [29].

2.1   Basic Definitions

In this section, some basic definitions are reviewed [29].

Definition 1 A fuzzy number $\tilde{A} = (a^L, a^M, a^R)$ is said to be a triangular fuzzy number if its membership function is given by

$$\mu_{\tilde{A}}(x) = \begin{cases} \dfrac{(x-a^L)}{(a^M-a^L)}, & a^L \le x < a^M \\ 1 & x = a^M \\ \dfrac{(x-a^R)}{(a^M-a^R)}, & a^M < x \le a^R \\ 0, & \text{otherwise.} \end{cases}$$

Definition 2 A triangular fuzzy number $\tilde{A} = (a^L, a^M, a^R)$ is said to be non-negative triangular fuzzy number if and only if $a^L \geq 0$.

Definition 3 A triangular fuzzy number $\tilde{A} = (a^L, a^M, a^R)$ is said to be positive triangular fuzzy number if and only if $a^L > 0$.

Definition 4 A ranking function is a function $\mathfrak{R}: F(R) \to R$, where $F(R)$ is a set of fuzzy numbers defined on the set of real numbers, which maps each fuzzy number into real line where a natural order exists.

Let $\tilde{A} = (a^L, a^M, a^R)$ and $\tilde{B} = (b^L, b^M, b^R)$ be two triangular fuzzy numbers. Then,

(i) $\tilde{A} \succeq \tilde{B}$ iff $\mathfrak{R}(\tilde{A}) \geq \mathfrak{R}(\tilde{B})$

(ii) $\tilde{A} \approx \tilde{B}$ iff $\mathfrak{R}(\tilde{A}) = \mathfrak{R}(\tilde{B})$

where, $\mathfrak{R}(\tilde{A}) = \dfrac{a^L + 2a^M + a^R}{4}$ and $\mathfrak{R}(\tilde{B}) = \dfrac{b^L + 2b^M + b^R}{4}$.

2.2 Arithmetic Operations on Triangular Fuzzy Numbers

Let $\tilde{A} = (a^L, a^M, a^R)$ and $\tilde{B} = (b^L, b^M, b^R)$ be two arbitrary triangular fuzzy numbers. Then,

(i) $\tilde{A} + \tilde{B} = (a^L + b^L, a^M + b^M, a^R + b^R)$

(ii) $\tilde{A} - \tilde{B} = (a^L - b^R, a^M - b^M, a^R - b^L)$

(iii) $\tilde{A}\tilde{B} = (a^L b^L, a^M b^M, a^R b^R)$, where $\tilde{A}$ and $\tilde{B}$ are non-negative triangular fuzzy numbers.

(iv) $\dfrac{\tilde{A}}{\tilde{B}} = \left(\dfrac{a^L}{b^R}, \dfrac{a^M}{b^M}, \dfrac{a^R}{b^L}\right)$, where $\tilde{A}$ is a non-negative triangular fuzzy number and $\tilde{B}$ is a positive triangular fuzzy number.

3 An Overview of the Existing Fuzzy DEA Approach

Wang and Chin [25] proposed the optimistic fuzzy CCR DEA model (1) and pessimistic fuzzy CCR DEA model (2) to evaluate the best relative fuzzy efficiency $\left(\tilde{E}_p^B\right)$ and worst relative fuzzy efficiency $\left(\tilde{E}_p^W\right)$ respectively of pth DMU by considering input data, output data and weights as trapezoidal fuzzy numbers.

$$\text{Maximize}\left[\tilde{E}_p^B \approx \left(E_{p1}^B, E_{p2}^B, E_{p3}^B, E_{p4}^B\right) \approx \frac{\sum_{r=1}^{s}\left(u_r^L, u_r^M, u_r^N, u_r^U\right)\left(y_{rp}^L, y_{rp}^M, y_{rp}^N, y_{rp}^U\right)}{\sum_{i=1}^{m}\left(v_i^L, v_i^M, v_i^N, v_i^U\right)\left(x_{ip}^L, x_{ip}^M, x_{ip}^N, x_{ip}^U\right)]}\right]$$

Subject to $\qquad\qquad\qquad\qquad\qquad\qquad\qquad\qquad\qquad\qquad$ (1)

$$\frac{\sum_{r=1}^{s}\left(u_r^L, u_r^M, u_r^N, u_r^U\right)\left(y_{rj}^L, y_{rj}^M, y_{rj}^N, y_{rj}^U\right)}{\sum_{i=1}^{m}\left(v_i^L, v_i^M, v_i^N, v_i^U\right)\left(x_{ij}^L, x_{ij}^M, x_{ij}^N, x_{ij}^U\right)]} \leqslant (1,1,1,1), \forall j$$

$\left(u_r^L, u_r^M, u_r^N, u_r^U\right)$ and $\left(v_i^L, v_i^M, v_i^N, v_i^U\right), i = 1, \ldots, m, r = 1, \ldots, s$, are non-negative trapezoidal fuzzy numbers.

$$\text{Minimize}\left[\tilde{E}_p^W \approx \left(E_{p1}^W, E_{p2}^W, E_{p3}^W, E_{p4}^W\right) \approx \frac{\sum_{r=1}^{s}\left(u_r^L, u_r^M, u_r^N, u_r^U\right)\left(y_{rp}^L, y_{rp}^M, y_{rp}^N, y_{rp}^U\right)}{\sum_{i=1}^{m}\left(v_i^L, v_i^M, v_i^N, v_i^U\right)\left(x_{ip}^L, x_{ip}^M, x_{ip}^N, x_{ip}^U\right)]}\right]$$

Subject to

$$\frac{\sum_{r=1}^{s}\left(u_r^L, u_r^M, u_r^N, u_r^U\right)\left(y_{rj}^L, y_{rj}^M, y_{rj}^N, y_{rj}^U\right)}{\sum_{i=1}^{m}\left(v_i^L, v_i^M, v_i^N, v_i^U\right)\left(x_{ij}^L, x_{ij}^M, x_{ij}^N, x_{ij}^U\right)]} \geqslant (1,1,1,1), \forall j$$

$\qquad\qquad\qquad\qquad\qquad\qquad\qquad\qquad\qquad\qquad\qquad\qquad$ (2)

$\left(u_r^L, u_r^M, u_r^N, u_r^U\right)$ and $\left(v_i^L, v_i^M, v_i^N, v_i^U\right), i = 1, \ldots, m, r = 1, \ldots, s$, are non-negative trapezoidal fuzzy numbers.

Wang and Chin [25] proposed the following method to evaluate best relative fuzzy efficiency as well as worst relative fuzzy efficiency and hence, geometric crisp efficiency of DMUs.

Step 1: Using the product of trapezoidal fuzzy numbers, defined in Sect. 2.2, the optimistic fuzzy CCR DEA model (1) and the pessimistic fuzzy CCR DEA model (2) can be transformed into optimistic fuzzy CCR DEA model (3) and the pessimistic fuzzy CCR DEA model (4) respectively.

$$\text{Maximize}\left[\tilde{E}_p^B \approx \left(E_{p1}^B, E_{p2}^B, E_{p3}^B, E_{p4}^B\right) \approx \frac{\sum_{r=1}^{s}\left(u_r^L y_{rp}^L, u_r^M y_{rp}^M, u_r^N y_{rp}^N, u_r^U y_{rp}^U\right)}{\sum_{i=1}^{m}\left(v_i^L x_{ip}^L, v_i^M x_{ip}^M, v_i^N x_{ip}^N, v_i^U x_{ip}^U\right)}\right]$$

Subject to $\qquad\qquad\qquad\qquad\qquad\qquad\qquad\qquad\qquad\qquad$ (3)

$$\frac{\sum_{r=1}^{s}\left(u_r^L y_{rj}^L, u_r^M y_{rj}^M, u_r^N y_{rj}^N, u_r^U y_{rj}^U\right)}{\sum_{i=1}^{m}\left(v_i^L x_{ij}^L, v_i^M x_{ij}^M, v_i^N x_{ij}^N, v_i^U x_{ij}^U\right)} \leqslant (1,1,1,1), \forall j$$

$\left(u_r^L, u_r^M, u_r^N, u_r^U\right)$ and $\left(v_i^L, v_i^M, v_i^N, v_i^U\right), i = 1, \ldots, m, r = 1, \ldots, s$, are non-negative trapezoidal fuzzy numbers.

$$\text{Minimize}\left[\tilde{E}_p^W \approx \left(E_{p1}^W, E_{p2}^W, E_{p3}^W, E_{p4}^W\right) \approx \frac{\sum_{r=1}^{s}\left(u_r^L y_{rp}^L, u_r^M y_{rp}^M, u_r^N y_{rp}^N, u_r^U y_{rp}^U\right)}{\sum_{i=1}^{m}\left(v_i^L x_{ip}^L, v_i^M x_{ip}^M, v_i^N x_{ip}^N, v_i^U x_{ip}^U\right)}\right]$$

Subject to

$$\frac{\sum_{r=1}^{s}\left(u_r^L y_{rj}^L, u_r^M y_{rj}^M, u_r^N y_{rj}^N, u_r^U y_{rj}^U\right)}{\sum_{i=1}^{m}\left(v_i^L x_{ij}^L, v_i^M x_{ij}^M, v_i^N x_{ij}^N, v_i^U x_{ij}^U\right)} \succcurlyeq (1,1,1,1), \forall j$$

(4)

$\left(u_r^L, u_r^M, u_r^N, u_r^U\right)$ and $\left(v_i^L, v_i^M, v_i^N, v_i^U\right), i = 1, \ldots, m, r = 1, \ldots, s$, are non-negative trapezoidal fuzzy numbers.

Step 2: The optimistic fuzzy CCR DEA model (3) and the pessimistic fuzzy CCR DEA model (4) can be transformed into optimistic fuzzy CCR DEA model (5) and the pessimistic fuzzy CCR DEA model (6) respectively.

$$\text{Maximize}\left[\tilde{E}_p^B \approx \left(E_{p1}^B, E_{p2}^B, E_{p3}^B, E_{p4}^B\right) \approx \frac{\left(\sum_{r=1}^{s} u_r^L y_{rp}^L, \sum_{r=1}^{s} u_r^M y_{rp}^M, \sum_{r=1}^{s} u_r^N y_{rp}^N, \sum_{r=1}^{s} u_r^U y_{rp}^U\right)}{\left(\sum_{i=1}^{m} v_i^L x_{ip}^L, \sum_{i=1}^{m} v_i^M x_{ip}^M, \sum_{i=1}^{m} v_i^N x_{ip}^N, \sum_{i=1}^{m} v_i^U x_{ip}^U\right)}\right]$$

Subject to

$$\frac{\left(\sum_{r=1}^{s} u_r^L y_{rj}^L, \sum_{r=1}^{s} u_r^M y_{rj}^M, \sum_{r=1}^{s} u_r^N y_{rj}^N, \sum_{r=1}^{s} u_r^U y_{rj}^U\right)}{\left(\sum_{i=1}^{m} v_i^L x_{ij}^L, \sum_{i=1}^{m} v_i^M x_{ij}^M, \sum_{i=1}^{m} v_i^N x_{ij}^N, \sum_{i=1}^{m} v_i^U x_{ij}^U\right)} \preccurlyeq (1,1,1,1), \forall j$$

(5)

$\left(u_r^L, u_r^M, u_r^N, u_r^U\right)$ and $\left(v_i^L, v_i^M, v_i^N, v_i^U\right), i = 1, \ldots, m, r = 1, \ldots, s$, are non-negative trapezoidal fuzzy numbers.

$$\text{Minimize}\left[\tilde{E}_p^W \approx \left(E_{p1}^W, E_{p2}^W, E_{p3}^W, E_{p4}^W\right) \approx \frac{\left(\sum_{r=1}^{s} u_r^L y_{rp}^L, \sum_{r=1}^{s} u_r^M y_{rp}^M, \sum_{r=1}^{s} u_r^N y_{rp}^N, \sum_{r=1}^{s} u_r^U y_{rp}^U\right)}{\left(\sum_{i=1}^{m} v_i^L x_{ip}^L, \sum_{i=1}^{m} v_i^M x_{ip}^M, \sum_{i=1}^{m} v_i^N x_{ip}^N, \sum_{i=1}^{m} v_i^U x_{ip}^U\right)}\right]$$

Subject to

$$\frac{\left(\sum_{r=1}^{s} u_r^L y_{rj}^L, \sum_{r=1}^{s} u_r^M y_{rj}^M, \sum_{r=1}^{s} u_r^N y_{rj}^N, \sum_{r=1}^{s} u_r^U y_{rj}^U\right)}{\left(\sum_{i=1}^{m} v_i^L x_{ij}^L, \sum_{i=1}^{m} v_i^M x_{ij}^M, \sum_{i=1}^{m} v_i^N x_{ij}^N, \sum_{i=1}^{m} v_i^U x_{ij}^U\right)} \succcurlyeq (1,1,1,1), \forall j$$

(6)

$\left(u_r^L, u_r^M, u_r^N, u_r^U\right)$ and $\left(v_i^L, v_i^M, v_i^N, v_i^U\right), i = 1, \ldots, m, r = 1, \ldots, s$, are non-negative trapezoidal fuzzy numbers.

Step 3: The optimistic fuzzy CCR DEA model (5) and the pessimistic fuzzy CCR DEA model (6) can be transformed into optimistic crisp CCR DEA model (7) and the pessimistic crisp CCR DEA model (8) respectively.

$$\text{Minimize}\left[\Re\left(\tilde{E}_p^B\right) = \Re\left(E_{p1}^B, E_{p2}^B, E_{p3}^B, E_{p4}^B\right) = \Re\left[\frac{\left(\sum_{r=1}^{s} u_r^L y_{rp}^L, \sum_{r=1}^{s} u_r^M y_{rp}^M, \sum_{r=1}^{s} u_r^N y_{rp}^N, \sum_{r=1}^{s} u_r^U y_{rp}^U\right)}{\left(\sum_{i=1}^{m} v_i^L x_{ip}^L, \sum_{i=1}^{m} v_i^M x_{ip}^M, \sum_{i=1}^{m} v_i^N x_{ip}^N, \sum_{i=1}^{m} v_i^U x_{ip}^U\right)}\right]\right]$$

Subject to

$$\Re\left[\frac{\left(\sum_{r=1}^{s} u_r^L y_{rj}^L, \sum_{r=1}^{s} u_r^M y_{rj}^M, \sum_{r=1}^{s} u_r^N y_{rj}^N, \sum_{r=1}^{s} u_r^U y_{rj}^U\right)}{\left(\sum_{i=1}^{m} v_i^L x_{ij}^L, \sum_{i=1}^{m} v_i^M x_{ij}^M, \sum_{i=1}^{m} v_i^N x_{ij}^N, \sum_{i=1}^{m} v_i^U x_{ij}^U\right)}\right] \leq \Re(1,1,1,1), \forall j$$

$$0 \leq u_r^L \leq u_r^M \leq u_r^N \leq u_r^U; 0 \leq v_i^L \leq v_i^M \leq v_i^N \leq v_i^U, i = 1, \ldots, m, r = 1, \ldots, s.$$

(7)

$$\text{Minimize}\left[\Re\left(\tilde{E}_p^W\right) = \Re\left(E_{p1}^W, E_{p2}^W, E_{p3}^W, E_{p4}^W\right) = \Re\left[\frac{\left(\sum_{r=1}^s u_r^L y_{rp}^L, \ \sum_{r=1}^s u_r^M y_{rp}^M, \ \sum_{r=1}^s u_r^N y_{rp}^N, \ \sum_{r=1}^s u_r^U y_{rp}^U\right)}{\left(\sum_{i=1}^m v_i^U x_{ip}^U, \ \sum_{i=1}^m v_i^M x_{ip}^M, \ \sum_{i=1}^m v_i^N x_{ip}^N, \ \sum_{i=1}^m v_i^L x_{ip}^L\right)}\right]\right]$$

Subject to

$$\Re\left[\frac{\left(\sum_{r=1}^s u_r^L y_{rj}^L, \ \sum_{r=1}^s u_r^M y_{rj}^M, \ \sum_{r=1}^s u_r^N y_{rj}^N, \ \sum_{r=1}^s u_r^U y_{rj}^U\right)}{\left(\sum_{i=1}^m v_i^L x_{ij}^L, \ \sum_{i=1}^m v_i^M x_{ij}^M, \ \sum_{i=1}^m v_i^N x_{ij}^N, \ \sum_{i=1}^m v_i^U x_{ij}^U\right)}\right] \geq \Re(1,1,1,1), \forall j \qquad (8)$$

$$0 \leq u_r^L \leq u_r^M \leq u_r^N \leq u_r^U; 0 \leq v_i^L \leq v_i^M \leq v_i^N \leq v_i^U, i = 1, \ldots, m, r = 1, \ldots, s.$$

Step 4: The optimistic crisp CCR DEA model (7) and the pessimistic crisp CCR DEA model (8) can be transformed into optimistic crisp CCR DEA model (9) and the pessimistic crisp CCR DEA model (10) respectively.

$$\text{Maximize}\left[\Re\left(\tilde{E}_p^B\right) = \Re\left(\left(E_{p1}^B, E_{p2}^B, E_{p3}^B, E_{p4}^B\right)\right) = \left[\frac{\Re\left(\sum_{r=1}^s u_r^L y_{rp}^L, \ \sum_{r=1}^s u_r^M y_{rp}^M, \ \sum_{r=1}^s u_r^N y_{rp}^N, \ \sum_{r=1}^s u_r^U y_{rp}^U\right)}{\Re\left(\sum_{i=1}^m v_i^L x_{ip}^L, \ \sum_{i=1}^m v_i^M x_{ip}^M, \ \sum_{i=1}^m v_i^N x_{ip}^N, \ \sum_{i=1}^m v_i^U x_{ip}^U\right)}\right]\right]$$

Subject to

$$\frac{\Re\left(\sum_{r=1}^s u_r^L y_{rj}^L, \ \sum_{r=1}^s u_r^M y_{rj}^M, \ \sum_{r=1}^s u_r^N y_{rj}^N, \ \sum_{r=1}^s u_r^U y_{rj}^U\right)}{\Re\left(\sum_{i=1}^m v_i^L x_{ij}^L, \ \sum_{i=1}^m v_i^M x_{ij}^M, \ \sum_{i=1}^m v_i^N x_{ij}^N, \ \sum_{i=1}^m v_i^U x_{ij}^U\right)} \leq \Re(1,1,1,1), \forall j \qquad (9)$$

$$0 \leq u_r^L \leq u_r^M \leq u_r^N \leq u_r^U; 0 \leq v_i^L \leq v_i^M \leq v_i^N \leq v_i^U, i = 1, \ldots, m, r = 1, \ldots, s.$$

$$\text{Minimize}\left[\Re\left(\tilde{E}_p^W\right) = \Re\left(E_{p1}^W, E_{p2}^W, E_{p3}^W, E_{p4}^W\right) = \frac{\Re\left(\sum_{r=1}^s u_r^L y_{rp}^L, \ \sum_{r=1}^s u_r^M y_{rp}^M, \ \sum_{r=1}^s u_r^N y_{rp}^N, \ \sum_{r=1}^s u_r^U y_{rp}^U\right)}{\Re\left(\sum_{i=1}^m v_i^L x_{ip}^L, \ \sum_{i=1}^m v_i^M x_{ip}^M, \ \sum_{i=1}^m v_i^N x_{ip}^N, \ \sum_{i=1}^m v_i^U x_{ip}^U\right)}\right]$$

Subject to

$$\frac{\Re\left(\sum_{r=1}^s u_r^L y_{rj}^L, \ \sum_{r=1}^s u_r^M y_{rj}^M, \ \sum_{r=1}^s u_r^N y_{rj}^N, \ \sum_{r=1}^s u_r^U y_{rj}^U\right)}{\Re\left(\sum_{i=1}^m v_i^L x_{ij}^L, \ \sum_{i=1}^m v_i^M x_{ij}^M, \ \sum_{i=1}^m v_i^N x_{ij}^N, \ \sum_{i=1}^m v_i^U x_{ij}^U\right)} \geq \Re(1,1,1,1), \forall j \qquad (10)$$

$$0 \leq u_r^L \leq u_r^M \leq u_r^N \leq u_r^U; 0 \leq v_i^L \leq v_i^M \leq v_i^N \leq v_i^U, i = 1, \ldots, m, r = 1, \ldots, s.$$

Step 5: The optimistic crisp CCR DEA model (9) and pessimistic crisp CCR DEA model (10) can be transformed into optimistic crisp CCR DEA model (11) and pessimistic crisp CCR DEA model (12) respectively.

$$\text{Maximize}\left[\Re\left(\tilde{E}_p^B\right) = \frac{\left(\sum_{r=1}^s u_r^L y_{rp}^L + \sum_{r=1}^s u_r^M y_{rp}^M + \sum_{r=1}^s u_r^N y_{rp}^N + \sum_{r=1}^s u_r^U y_{rp}^U\right)}{\left(\sum_{i=1}^m v_i^L x_{ip}^L + \sum_{i=1}^m v_i^M x_{ip}^M + \sum_{i=1}^m v_i^N x_{ip}^N + \sum_{i=1}^m v_i^U x_{ip}^U\right)}\right]$$

Subject to

$$\frac{\left(\sum_{r=1}^s u_r^L y_{rj}^L + \sum_{r=1}^s u_r^M y_{rj}^M + \sum_{r=1}^s u_r^N y_{rj}^N + \sum_{r=1}^s u_r^U y_{rj}^U\right)}{\left(\sum_{i=1}^m v_i^L x_{ij}^L + \sum_{i=1}^m v_i^M x_{ij}^M + \sum_{i=1}^m v_i^N x_{ij}^N + \sum_{i=1}^m v_i^U x_{ij}^U\right)} \leq 1, \forall j \qquad (11)$$

$$0 \leq u_r^L \leq u_r^M \leq u_r^N \leq u_r^U; 0 \leq v_i^L \leq v_i^M \leq v_i^N \leq v_i^U, i = 1, \ldots, m, r = 1, \ldots, s.$$

$$\text{Minimize}\left[\Re\left(\tilde{E}_p^W\right) = \frac{\left(\sum_{r=1}^s u_r^L y_{rp}^L + \sum_{r=1}^s u_r^M y_{rp}^M + \sum_{r=1}^s u_r^N y_{rp}^N + \sum_{r=1}^s u_r^U y_{rp}^U\right)}{\left(\sum_{i=1}^m v_i^L x_{ip}^L + \sum_{i=1}^m v_i^M x_{ip}^M + \sum_{i=1}^m v_i^N x_{ip}^N + \sum_{i=1}^m v_i^U x_{ip}^U\right)}\right]$$

Subject to

$$\frac{\left(\sum_{r=1}^s u_r^L y_{rj}^L + \sum_{r=1}^s u_r^M y_{rj}^M + \sum_{r=1}^s u_r^N y_{rj}^N + \sum_{r=1}^s u_r^U y_{rj}^U\right)}{\left(\sum_{i=1}^m v_i^L x_{ij}^L + \sum_{i=1}^m v_i^M x_{ij}^M + \sum_{i=1}^m v_i^N x_{ij}^N + \sum_{i=1}^m v_i^U x_{ij}^U\right)} \geq 1, \forall j \qquad (12)$$

$$0 \leq u_r^L \leq u_r^M \leq u_r^N \leq u_r^U; 0 \leq v_i^L \leq v_i^M \leq v_i^N \leq v_i^U, i = 1, \ldots, m, r = 1, \ldots, s.$$

Step 6: The optimistic crisp CCR DEA model (11) and pessimistic crisp CCR DEA model (12) can be transformed into optimistic crisp CCR DEA model (13) and pessimistic crisp CCR DEA model (14) respectively.

$$\text{Maximize}\left[\Re\left(\tilde{E}_p^B\right) = \sum_{r=1}^s u_r^L y_{rp}^L + \sum_{r=1}^s u_r^M y_{rp}^M + \sum_{r=1}^s u_r^N y_{rp}^N + \sum_{r=1}^s u_r^U y_{rp}^U\right]$$

Subject to

$$\sum_{i=1}^m v_i^L x_{ip}^L + \sum_{i=1}^m v_i^M x_{ip}^M + \sum_{i=1}^m v_i^N x_{ip}^N + \sum_{i=1}^m v_i^U x_{ip}^U = 1$$

$$\left(\sum_{r=1}^s u_r^L y_{rj}^L + \sum_{r=1}^s u_r^M y_{rj}^M + \sum_{r=1}^s u_r^N y_{rj}^N + \sum_{r=1}^s u_r^U y_{rj}^U\right) \tag{13}$$

$$-\left(\sum_{i=1}^m v_i^L x_{ij}^L + \sum_{i=1}^m v_i^M x_{ij}^M + \sum_{i=1}^m v_i^N x_{ij}^N + \sum_{i=1}^m v_i^U x_{ij}^U\right) \le 0, \forall j$$

$$0 \le u_r^L \le u_r^M \le u_r^N \le u_r^U; 0 \le v_i^L \le v_i^M \le v_i^N \le v_i^U, i = 1, \ldots, m, r = 1, \ldots, s.$$

$$\text{Maximize}\left[\Re\left(\tilde{E}_p^B\right) = \sum_{r=1}^s u_r^L y_{rp}^L + \sum_{r=1}^s u_r^M y_{rp}^M + \sum_{r=1}^s u_r^N y_{rp}^N + \sum_{r=1}^s u_r^U y_{rp}^U\right]$$

Subject to

$$\sum_{i=1}^m v_i^L x_{ip}^L + \sum_{i=1}^m v_i^M x_{ip}^M + \sum_{i=1}^m v_i^N x_{ip}^N + \sum_{i=1}^m v_i^U x_{ip}^U = 1$$

$$\left(\sum_{r=1}^s u_r^L y_{rj}^L + \sum_{r=1}^s u_r^M y_{rj}^M + \sum_{r=1}^s u_r^N y_{rj}^N + \sum_{r=1}^s u_r^U y_{rj}^U\right) \tag{14}$$

$$-\left(\sum_{i=1}^m v_i^L x_{ij}^L + \sum_{i=1}^m v_i^M x_{ij}^M + \sum_{i=1}^m v_i^N x_{ij}^N + \sum_{i=1}^m v_i^U x_{ij}^U\right) \ge 0, \forall j$$

$$0 \le u_r^L \le u_r^M \le u_r^N \le u_r^U; 0 \le v_i^L \le v_i^M \le v_i^N \le v_i^U i = 1, \ldots, m, r = 1, \ldots, s.$$

Step 7: Find the optimal value $\Re\left(\tilde{E}_p^B\right) = E_p^B$, representing the best relative crisp efficiency of pth DMU, by solving optimistic crisp CCR DEA model (13).

Step 8: Find the optimal value $\Re\left(\tilde{E}_p^W\right) = E_p^W$, representing the worst relative crisp efficiency of pth DMU, by solving pessimistic crisp CCR DEA model (14).

Step 9: Find the relative geometric average crisp efficiency $E_P^{GEOMETRIC}$ of pth DMU by putting the values E_p^B and E_p^W, obtained in Step 4 and Step 5, in $E_P^{GEOMETRIC} = \sqrt{E_P^B \times E_p^W}$.

4 Flaws in the Existing Method

If $(a^L, a^M, a^N, a^U) \,\&\, (b^L, b^M, b^N, b^U).$ are two trapezoidal fuzzy numbers then

$$\Re\left(\frac{(a^L, a^M, a^N, a^U)}{(b^L, b^M, b^N, b^U)}\right) = \Re\left(\frac{a^L}{b^U}, \frac{a^M}{b^N}, \frac{a^N}{b^M}, \frac{a^U}{b^L}\right) = \frac{1}{4}\left(\frac{a^L}{b^U} + \frac{a^M}{b^N} + \frac{a^N}{b^M} + \frac{a^U}{b^L}\right) \text{ and}$$

$$\frac{\Re(a^L, a^M, a^N, a^U)}{\Re(b^L, b^M, b^N, b^U)} = \frac{a^L + a^M + a^N + a^U}{b^L + b^M + b^N + b^U}$$

It is obvious that

$$\Re\left(\frac{(a^L, a^M, a^N, a^U)}{(b^L, b^M, b^N, b^U)}\right) \neq \frac{\Re(a^L, a^M, a^N, a^U)}{\Re(b^L, b^M, b^N, b^U)}$$

However, Wang and Chin [25] have used the property

$$\Re\left(\frac{(a^L, a^M, a^N, a^U)}{(b^L, b^M, b^N, b^U)}\right) = \frac{\Re(a^L, a^M, a^N, a^U)}{\Re(b^L, b^M, b^N, b^U)}$$

in Step 4 of their proposed method. Therefore, the method, proposed by Wang and Chin [25], is not valid.

5 Proposed Fuzzy CCR DEA Approach

In this section, to resolve the flaws of the existing optimistic fuzzy CCR DEA model (1) and pessimistic fuzzy CCR DEA model (2), proposed by Wang and Chin [25], new optimistic fuzzy CCR DEA model (15) and pessimistic fuzzy CCR DEA model (16) are proposed.

$$\text{Maximize}\left[\tilde{E}_p^B \approx \left(E_{p1}^B, E_{p2}^B, E_{p3}^B, E_{p4}^B\right) \approx \frac{\sum_{r=1}^s \left(u_r^L, u_r^M, u_r^N, u_r^U\right)\left(y_{rp}^L, y_{rp}^M, y_{rp}^N, y_{rp}^U\right)}{\sum_{i=1}^m \left(v_i^L, v_i^M, v_i^N, v_i^U\right)\left(x_{ip}^L, x_{ip}^M, x_{ip}^N, x_{ip}^U\right)]}\right]$$

Subject to

$$\sum_{r=1}^s \left(u_r^L, u_r^M, u_r^N, u_r^U\right)\left(y_{rj}^L, y_{rj}^M, y_{rj}^N, y_{rj}^U\right) \preccurlyeq \sum_{i=1}^m \left(v_i^L, v_i^M, v_i^N, v_i^U\right)\left(x_{ij}^L, x_{ij}^M, x_{ij}^N, x_{ij}^U\right), \forall j \qquad (15)$$

$\left(u_r^L, u_r^M, u_r^N, u_r^U\right)$ and $\left(v_i^L, v_i^M, v_i^N, v_i^U\right), i = 1, \ldots, m, r = 1, \ldots, s,$ are non-negative trapezoidal fuzzy numbers.

$$\text{Minimize}\left[\tilde{E}_p^W \approx \left(E_{p1}^W, E_{p2}^W, E_{p3}^W, E_{p4}^W\right) \approx \frac{\sum_{r=1}^s \left(u_r^L, u_r^M, u_r^N, u_r^U\right)(y_{rp}^L, y_{rp}^M, y_{rp}^N, y_{rp}^U)}{\sum_{i=1}^m \left(v_i^L, v_i^M, v_i^N, v_i^U\right)(x_{ip}^L, x_{ip}^M, x_{ip}^N, x_{ip}^U)]}\right]$$

Subject to

$$\sum_{r=1}^s \left(u_r^L, u_r^M, u_r^N, u_r^U\right)\left(y_{rj}^L, y_{rj}^M, y_{rj}^N, y_{rj}^U\right) \succcurlyeq \sum_{i=1}^m \left(v_i^L, v_i^M, v_i^N, v_i^U\right)\left(x_{ij}^L, x_{ij}^M, x_{ij}^N, x_{ij}^U\right), \forall j \qquad (16)$$

$\left(u_r^L, u_r^M, u_r^N, u_r^U\right)$ and $\left(v_i^L, v_i^M, v_i^N, v_i^U\right), i = 1, \ldots, m, r = 1, \ldots, s,$ are non-negative trapezoidal fuzzy numbers.

To evaluate the best relative fuzzy efficiency as well as worst relative fuzzy efficiency and relative geometric crisp efficiency of DMUs considering the optimistic fuzzy CCR DEA model (15) and pessimistic fuzzy CCR DEA model (16) can be obtained by using the following steps:

Step 1: Using the product of trapezoidal fuzzy numbers, defined in Sect. 2.2, the optimistic fuzzy CCR DEA model (15) and pessimistic fuzzy CCR DEA model

(16) can be transformed into optimistic fuzzy CCR DEA model (17) and pessimistic fuzzy CCR DEA model (18) respectively.

$$\text{Maximize}\left[\tilde{E}_p^B \approx \left(E_{p1}^B, E_{p2}^B, E_{p3}^B, E_{p4}^B\right) \approx \frac{\sum_{r=1}^s \left(u_r^L y_{rp}^L, u_r^M y_{rp}^M, u_r^N y_{rp}^N, u_r^U y_{rp}^U\right)}{\sum_{i=1}^m \left(v_i^L x_{ip}^L, v_i^M x_{ip}^M, v_i^N x_{ip}^N, v_i^U x_{ip}^U\right)}\right] \quad (17)$$

Subject to
$$\sum_{r=1}^s \left(u_r^L y_{rj}^L, u_r^M y_{rj}^M, u_r^N y_{rj}^N, u_r^U y_{rj}^U\right) \preccurlyeq \sum_{i=1}^m \left(v_i^L x_{ij}^L, v_i^M x_{ij}^M, v_i^N x_{ij}^N, v_i^U x_{ij}^U\right), \forall j$$

$\left(u_r^L, u_r^M, u_r^N, u_r^U\right)$ and $\left(v_i^L, v_i^M, v_i^N, v_i^U\right), i = 1, \ldots, m, r = 1, \ldots, s,$ are non-negative trapezoidal fuzzy numbers.

$$\text{Minimize}\left[\tilde{E}_p^W \approx \left(E_{p1}^W, E_{p2}^W, E_{p3}^W, E_{p4}^W\right) \approx \frac{\sum_{r=1}^s \left(u_r^L y_{rp}^L, u_r^M y_{rp}^M, u_r^N y_{rp}^N, u_r^U y_{rp}^U\right)}{\sum_{i=1}^m \left(v_i^L x_{ip}^L, v_i^M x_{ip}^M, v_i^N x_{ip}^N, v_i^U x_{ip}^U\right)}\right] \quad (18)$$

Subject to
$$\sum_{r=1}^s \left(u_r^L y_{rj}^L, u_r^M y_{rj}^M, u_r^N y_{rj}^N, u_r^U y_{rj}^U\right) \succcurlyeq \sum_{i=1}^m \left(v_i^L x_{ij}^L, v_i^M x_{ij}^M, v_i^N x_{ij}^N, v_i^U x_{ij}^U\right), \forall j$$

$\left(u_r^L, u_r^M, u_r^N, u_r^U\right)$ and $\left(v_i^L, v_i^M, v_i^N, v_i^U\right), i = 1, \ldots, m, r = 1, \ldots, s,$ are non-negative trapezoidal fuzzy numbers.

Step 2: Using the division of trapezoidal fuzzy numbers, defined in Sect. 22, the optimistic fuzzy CCR DEA model (17) and pessimistic fuzzy CCR DEA model (18) can be transformed into optimistic fuzzy CCR DEA model (19) and pessimistic fuzzy CCR DEA model (20) respectively.

$$\text{Maximize}\left[\tilde{E}_p^B \approx \left(E_{p1}^B, E_{p2}^B, E_{p3}^B, E_{p4}^B\right) \approx \left(\frac{\sum_{r=1}^s u_r^L y_{rp}^L}{\sum_{i=1}^m v_i^U x_{ip}^U}, \frac{\sum_{r=1}^s u_r^M y_{rp}^M}{\sum_{i=1}^m v_i^N x_{ip}^N}, \frac{\sum_{r=1}^s u_r^N y_{rp}^N}{\sum_{i=1}^m v_i^M x_{ip}^M}, \frac{\sum_{r=1}^s u_r^U y_{rp}^U}{\sum_{i=1}^m v_i^L x_{ip}^L}\right)\right]$$

Subject to
$$\left(\sum_{r=1}^s u_r^L y_{rp}^L, \sum_{r=1}^s u_r^M y_{rp}^M, \sum_{r=1}^s u_r^N y_{rp}^N, \sum_{r=1}^s u_r^U y_{rp}^U\right)$$
$$\preccurlyeq \left(\sum_{i=1}^m v_i^L x_{ip}^L, \sum_{i=1}^m v_i^M x_{ip}^M, \sum_{i=1}^m v_i^N x_{ip}^N, \sum_{i=1}^m v_i^U x_{ip}^U\right), \forall j \quad (19)$$

$\left(u_r^L, u_r^M, u_r^N, u_r^U\right)$ and $\left(v_i^L, v_i^M, v_i^N, v_i^U\right), i = 1, \ldots, m, r = 1, \ldots, s,$ are non-negative trapezoidal fuzzy numbers.

$$\text{Minimize}\left[\tilde{E}_p^W \approx \left(E_{p1}^W, E_{p2}^W, E_{p3}^W, E_{p4}^W\right) \approx \left(\frac{\sum_{r=1}^s u_r^L y_{rp}^L}{\sum_{i=1}^m v_i^U x_{ip}^U}, \frac{\sum_{r=1}^s u_r^M y_{rp}^M}{\sum_{i=1}^m v_i^N x_{ip}^N}, \frac{\sum_{r=1}^s u_r^N y_{rp}^N}{\sum_{i=1}^m v_i^M x_{ip}^M}, \frac{\sum_{r=1}^s u_r^U y_{rp}^U}{\sum_{i=1}^m v_i^L x_{ip}^L}\right)\right]$$

Subject to
$$\left(\sum_{r=1}^s u_r^L y_{rp}^L, \sum_{r=1}^s u_r^M y_{rp}^M, \sum_{r=1}^s u_r^N y_{rp}^N, \sum_{r=1}^s u_r^U y_{rp}^U\right)$$
$$\succcurlyeq \left(\sum_{i=1}^m v_i^L x_{ip}^L, \sum_{i=1}^m v_i^M x_{ip}^M, \sum_{i=1}^m v_i^N x_{ip}^N, \sum_{i=1}^m v_i^U x_{ip}^U\right), \forall j \quad (20)$$

$\left(u_r^L, u_r^M, u_r^N, u_r^U\right)$ and $\left(v_i^L, v_i^M, v_i^N, v_i^U\right), i = 1, \ldots, m, r = 1, \ldots, s,$ are non-negative trapezoidal fuzzy numbers.

Step 3: Using the relation $(a^L, a^M, a^N, a^U) \leq (b^L, b^M, b^N, b^U)$, $a^L \leq b^L, a^M \leq b^M$, $a^N \leq b^N, a^U \leq b^U$, the optimistic fuzzy CCR DEA model (19) and pessimistic fuzzy CCR DEA model (20) can be transformed into optimistic fuzzy CCR DEA model (21) and pessimistic fuzzy CCR DEA model (22) respectively.

$$\text{Maximize}\left[\tilde{E}_p^B \approx \left(E_{p1}^B, E_{p2}^B, E_{p3}^B, E_{p4}^B\right) \approx \left(\frac{\sum_{r=1}^s u_r^L y_{rp}^L}{\sum_{i=1}^m v_i^U x_{ip}^U}, \frac{\sum_{r=1}^s u_r^M y_{rp}^M}{\sum_{i=1}^m v_i^N x_{ip}^N}, \frac{\sum_{r=1}^s u_r^N y_{rp}^N}{\sum_{i=1}^m v_i^M x_{ip}^M}, \frac{\sum_{r=1}^s u_r^U y_{rp}^U}{\sum_{i=1}^m v_i^L x_{ip}^L}\right)\right]$$

Subject to
$$\sum_{r=1}^s u_r^L y_{rj}^L \leq \sum_{i=1}^m v_i^L x_{ij}^L, \forall j$$
$$\sum_{r=1}^s u_r^M y_{rj}^M \leq \sum_{i=1}^m v_i^M x_{ij}^M, \forall j$$
$$\sum_{r=1}^s u_r^N y_{rp}^N \leq \sum_{i=1}^m v_i^N x_{ip}^N, \forall j \tag{21}$$
$$\sum_{r=1}^s u_r^U y_{rp}^U \leq \sum_{i=1}^m v_i^U x_{ip}^U, \forall j$$
$$0 \leq u_r^L \leq u_r^M \leq u_r^N \leq u_r^U; 0 \leq v_i^L \leq v_i^M \leq v_i^N \leq v_i^U, i=1,\ldots,m, r=1,\ldots,s.$$

$$\text{Minimize}\left[\tilde{E}_p^W \approx \left(E_{p1}^W, E_{p2}^W, E_{p3}^W, E_{p4}^W\right) \approx \left(\frac{\sum_{r=1}^s u_r^L y_{rp}^L}{\sum_{i=1}^m v_i^U x_{ip}^U}, \frac{\sum_{r=1}^s u_r^M y_{rp}^M}{\sum_{i=1}^m v_i^N x_{ip}^N}, \frac{\sum_{r=1}^s u_r^N y_{rp}^N}{\sum_{i=1}^m v_i^M x_{ip}^M}, \frac{\sum_{r=1}^s u_r^U y_{rp}^U}{\sum_{i=1}^m v_i^L x_{ip}^L}\right)\right]$$

Subject to
$$\sum_{r=1}^s u_r^L y_{rj}^L \geq \sum_{i=1}^m v_i^L x_{ij}^L, \forall j$$
$$\sum_{r=1}^s u_r^M y_{rj}^M \geq \sum_{i=1}^m v_i^M x_{ij}^M, \forall j$$
$$\sum_{r=1}^s u_r^N y_{rp}^N \geq \sum_{i=1}^m v_i^N x_{ip}^N, \forall j \tag{22}$$
$$\sum_{r=1}^s u_r^U y_{rp}^U \geq \sum_{i=1}^m v_i^U x_{ip}^U, \forall j$$
$$0 \leq u_r^L \leq u_r^M \leq u_r^N \leq u_r^U; 0 \leq v_i^L \leq v_i^M \leq v_i^N \leq v_i^U, i=1,\ldots,m, r=1,\ldots,s.$$

Step 4: The fuzzy optimal value $\tilde{E}_p^B \approx \left(E_{p1}^B, E_{p2}^B, E_{p3}^B, E_{p4}^B\right)$, representing the best relative fuzzy efficiency of pth DMU, as well as the fuzzy optimal value $\tilde{E}_p^W \approx \left(E_{p1}^W, E_{p2}^W, E_{p3}^W, E_{p4}^W\right)$, representing the worst relative fuzzy efficiency of pth DMU, can be obtained by solving the optimistic fuzzy CCR DEA model (21) and pessimistic fuzzy CCR DEA model (22) as follows:

Step 4(a): Find the optimal value $\left(E_{p1}^B\right)$ and $\left(E_{p1}^W\right)$ of the optimistic crisp CCR DEA model (23a) and pessimistic CCR DEA model (24a) by solving optimistic crisp CCR DEA model (23b) and pessimistic CCR DEA model (24b) equivalent to optimistic crisp CCR DEA model (23a) and pessimistic CCR DEA model (24a) respectively.

$$\text{Maximize}\left[E_{p1}^B = \frac{\sum_{r=1}^s u_r^L y_{rp}^L}{\sum_{i=1}^m v_i^U x_{ip}^U}\right]$$

Subject to
$$\frac{\sum_{r=1}^s u_r^L y_{rp}^L}{\sum_{i=1}^m v_i^U x_{ip}^U} \leq 1, \tag{23a}$$

& all the constraints of model 21.

$$\text{Maximize} \left[E_{p1}^B = \sum_{r=1}^s u_r^L y_{rp}^L \right]$$

Subject to

$$\sum_{i=1}^m v_i^U x_{ip}^U = 1, \tag{23b}$$

$$\sum_{r=1}^s u_r^L y_{rp}^L \leq \sum_{i=1}^m v_i^U x_{ip}^U,$$

& all the constraints of model 21.

$$\text{Minimize} \left[E_{p1}^W = \frac{\sum_{r=1}^s u_r^L y_{rp}^L}{\sum_{i=1}^m v_i^U x_{ip}^U} \right]$$

Subject to

$$\frac{\sum_{r=1}^s u_r^L y_{rp}^L}{\sum_{i=1}^m v_i^U x_{ip}^U} \geq 1, \tag{24a}$$

& all the constraints of model 22.

$$\text{Minimize} \left[E_{p1}^W = \sum_{r=1}^s u_r^L y_{rp}^L \right]$$

Subject to

$$\sum_{i=1}^m v_i^U x_{ip}^U = 1, \tag{24b}$$

$$\sum_{r=1}^s u_r^L y_{rp}^L \geq \sum_{i=1}^m v_i^U x_{ip}^U,$$

& all the constraints of model 22.

Step 4(b): Find the optimal value $\left(E_{p2}^B \right)$ and $\left(E_{p2}^W \right)$ of the optimistic crisp CCR DEA model (25a) and pessimistic CCR DEA model (26a) by solving optimistic crisp CCR DEA model (25b) and pessimistic CCR DEA model (26b) equivalent to optimistic crisp CCR DEA model (25a) and pessimistic CCR DEA model (26a) respectively.

$$\text{Maximize} \left[E_{p2}^B = \frac{\sum_{r=1}^s u_r^M y_{rp}^M}{\sum_{i=1}^m v_i^N x_{ip}^N} \right]$$

Subject to

$$\frac{\sum_{r=1}^s u_r^L y_{rp}^L}{\sum_{i=1}^m v_i^U x_{ip}^U} = E_{p1}^B, \tag{25a}$$

$$E_{p1}^B \leq \frac{\sum_{r=1}^s u_r^M y_{rp}^M}{\sum_{i=1}^m v_i^N x_{ip}^N} \leq 1,$$

& all the constraints of model (21).

$$\text{Maximize} \left[E_{p2}^B = \sum_{r=1}^s u_r^M y_{rp}^M \right]$$

Subject to

$$\sum_{i=1}^m v_i^N x_{ip}^N = 1,$$

$$\sum_{r=1}^s u_r^L y_{rp}^L = E_{p1}^B \left(\sum_{i=1}^m v_i^U x_{ip}^U \right), \tag{25b}$$

$$E_{p1}^B \left(\sum_{i=1}^m v_i^N x_{ip}^N \right) \leq \sum_{r=1}^s u_r^M y_{rp}^M \leq \sum_{i=1}^m v_i^N x_{ip}^N,$$

& all the constraints of model (21).

$$\text{Minimize}\left[E_{p2}^{W} = \frac{\sum_{r=1}^{s} u_r^M y_{rp}^M}{\sum_{i=1}^{m} v_i^N x_{ip}^N}\right]$$

Subject to

$$\frac{\sum_{r=1}^{s} u_r^L y_{rp}^L}{\sum_{i=1}^{m} v_i^U x_{ip}^U} = E_{p1}^{W}, \tag{26a}$$

$$E_{p1}^{W} \leq \frac{\sum_{r=1}^{s} u_r^M y_{rp}^M}{\sum_{i=1}^{m} v_i^N x_{ip}^N},$$

& all the constraints of model (22).

$$\text{Maximize}\left[E_{p2}^{W} = \sum_{r=1}^{s} u_r^M y_{rp}^M\right]$$

Subject to

$$\sum_{i=1}^{m} v_i^N x_{ip}^N = 1,$$

$$\sum_{r=1}^{s} u_r^L y_{rp}^L = E_{p1}^{W}\left(\sum_{i=1}^{m} v_i^U x_{ip}^U\right), \tag{26b}$$

$$E_{p1}^{W}\left(\sum_{i=1}^{m} v_i^N x_{ip}^N\right) \leq \sum_{r=1}^{s} u_r^M y_{rp}^M,$$

& all the constraints of model (22).

Step 4(c): Find the optimal value $\left(E_{p3}^{B}\right)$ and $\left(E_{p3}^{W}\right)$ of the optimistic crisp CCR DEA model (27a) and pessimistic CCR DEA model (28a) by solving optimistic crisp CCR DEA model (27b) and pessimistic CCR DEA model (28b) equivalent to optimistic crisp CCR DEA model (27a) and pessimistic CCR DEA model (28a) respectively.

$$\text{Maximize}\left[E_{p3}^{B} = \frac{\sum_{r=1}^{s} u_r^N y_{rp}^N}{\sum_{i=1}^{m} v_i^M x_{ip}^M}\right]$$

Subject to

$$\frac{\sum_{r=1}^{s} u_r^L y_{rp}^L}{\sum_{i=1}^{m} v_i^U x_{ip}^U} = E_{p1}^{B},$$

$$\frac{\sum_{r=1}^{s} u_r^M y_{rp}^M}{\sum_{i=1}^{m} v_i^N x_{ip}^N} = E_{p2}^{B}, \tag{27a}$$

$$E_{p2}^{B} \leq \frac{\sum_{r=1}^{s} u_r^N y_{rp}^N}{\sum_{i=1}^{m} v_i^M x_{ip}^M} \leq 1,$$

& all the constraints of model (21).

$$\text{Maximize}\left[E_{p3}^{B} = \sum_{r=1}^{s} u_r^N y_{rp}^N\right]$$

Subject to

$$\sum_{i=1}^{m} v_i^M x_{ip}^M = 1,$$

$$\sum_{r=1}^{s} u_r^L y_{rp}^L = E_{p1}^{B}\left(\sum_{i=1}^{m} v_i^U x_{ip}^U\right),$$

$$\sum_{r=1}^{s} u_r^M y_{rp}^M = E_{p2}^{B}\left(\sum_{i=1}^{m} v_i^N x_{ip}^N\right), \tag{27b}$$

$$E_{p2}^{B}\left(\sum_{i=1}^{m} v_i^M x_{ip}^M\right) \leq \sum_{r=1}^{s} u_r^N y_{rp}^N \leq \sum_{i=1}^{m} v_i^M x_{ip}^M,$$

& all the constraints of model (21).

$$\text{Minimize} \left[E_{p3}^{W} = \frac{\sum_{r=1}^{s} u_r^N y_{rp}^N}{\sum_{i=1}^{m} v_i^M x_{ip}^M} \right]$$

Subject to

$$\frac{\sum_{r=1}^{s} u_r^L y_{rp}^L}{\sum_{i=1}^{m} v_i^U x_{ip}^U} = E_{p1}^{W},$$

$$\frac{\sum_{r=1}^{s} u_r^M y_{rp}^M}{\sum_{i=1}^{m} v_i^N x_{ip}^N} = E_{p2}^{W}, \tag{28a}$$

$$E_{p2}^{W} \le \frac{\sum_{r=1}^{s} u_r^N y_{rp}^N}{\sum_{i=1}^{m} v_i^M x_{ip}^M},$$

& all the constraints of model (22).

$$\text{Minimize} \left[E_{p3}^{W} = \sum_{r=1}^{s} u_r^N y_{rp}^N \right]$$

Subject to

$$\sum_{i=1}^{m} v_i^M x_{ip}^M = 1,$$

$$\sum_{r=1}^{s} u_r^L y_{rp}^L = E_{p1}^{W} \left(\sum_{i=1}^{m} v_i^U x_{ip}^U \right),$$

$$\sum_{r=1}^{s} u_r^M y_{rp}^M = E_{p2}^{W} \left(\sum_{i=1}^{m} v_i^N x_{ip}^N \right), \tag{28b}$$

$$E_{p2}^{W} \left(\sum_{i=1}^{m} v_i^M x_{ip}^M \right) \le \sum_{r=1}^{s} u_r^N y_{rp}^N,$$

& all the constraints of model (22).

Step 4(d): Find the optimal value $\left(E_{p4}^{B} \right)$ and $\left(E_{p4}^{W} \right)$ of the optimistic crisp CCR DEA model (29a) and pessimistic CCR DEA model (30a) by solving optimistic crisp CCR DEA model (29b) and pessimistic CCR DEA model (30b) equivalent to optimistic crisp CCR DEA model (29a) and pessimistic CCR DEA model (30a) respectively.

$$\text{Maximize} \left[E_{p4}^{B} = \frac{\sum_{r=1}^{s} u_r^U y_{rp}^U}{\sum_{i=1}^{m} v_i^L x_{ip}^L} \right]$$

Subject to

$$\frac{\sum_{r=1}^{s} u_r^L y_{rp}^L}{\sum_{i=1}^{m} v_i^U x_{ip}^U} = E_{p1}^{B},$$

$$\frac{\sum_{r=1}^{s} u_r^M y_{rp}^M}{\sum_{i=1}^{m} v_i^N x_{ip}^N} = E_{p2}^{B}, \tag{29a}$$

$$\frac{\sum_{r=1}^{s} u_r^N y_{rp}^N}{\sum_{i=1}^{m} v_i^M x_{ip}^M} = E_{p3}^{B},$$

$$E_{p3}^{B} \le \frac{\sum_{r=1}^{s} u_r^U y_{rp}^U}{\sum_{i=1}^{m} v_i^L x_{ip}^L} \le 1,$$

& all the constraints of model (21).

$$\text{Maximize}\left[E_{p4}^B = \sum_{r=1}^s u_r^U y_{rp}^U\right]$$

Subject to

$$\sum_{i=1}^m v_i^L x_{ip}^L = 1,$$

$$\sum_{r=1}^s u_r^L y_{rp}^L = E_{p1}^B\left(\sum_{i=1}^m v_i^U x_{ip}^U\right),$$

$$\sum_{r=1}^s u_r^M y_{rp}^M = E_{p2}^B\left(\sum_{i=1}^m v_i^N x_{ip}^N\right),$$

$$\sum_{r=1}^s u_r^N y_{rp}^N = E_{p3}^B\left(\sum_{i=1}^m v_i^M x_{ip}^M\right),$$

$$E_{p3}^B\left(\sum_{i=1}^m v_i^L x_{ip}^L\right) \leq \sum_{r=1}^s u_r^U y_{rp}^U \leq \sum_{i=1}^m v_i^L x_{ip}^L,$$

& all the constraints of model (21).

(29b)

$$\text{Minimize}\left[E_{p4}^W = \frac{\sum_{r=1}^s u_r^U y_{rp}^U}{\sum_{i=1}^m v_i^L x_{ip}^L}\right]$$

Subject to

$$\frac{\sum_{r=1}^s u_r^L y_{rp}^L}{\left(\sum_{i=1}^m v_i^U x_{ip}^U\right)} = E_{p1}^W,$$

$$\frac{\sum_{r=1}^s u_r^M y_{rp}^M}{\left(\sum_{i=1}^m v_i^N x_{ip}^N\right)} = E_{p2}^W,$$

$$\frac{\sum_{r=1}^s u_r^N y_{rp}^N}{\sum_{i=1}^m v_i^M x_{ip}^M} = E_{p3}^W,$$

$$E_{p3}^W \leq \frac{\sum_{r=1}^s u_r^U y_{rp}^U}{\sum_{i=1}^m v_i^L x_{ip}^L},$$

& all the constraints of model (22).

(30a)

$$\text{Minimize}\left[E_{p4}^W = \sum_{r=1}^s u_r^U y_{rp}^U\right]$$

Subject to

$$\sum_{i=1}^m v_i^L x_{ip}^L = 1,$$

$$\sum_{r=1}^s u_r^U y_{rp}^U = E_{p1}^W\left(\sum_{i=1}^m v_i^U x_{ip}^U\right),$$

$$\sum_{r=1}^s u_r^M y_{rp}^M = E_{p2}^W\left(\sum_{i=1}^m v_i^N x_{ip}^N\right),$$

$$\sum_{r=1}^s u_r^N y_{rp}^N = E_{p3}^W\left(\sum_{i=1}^m v_i^M x_{ip}^M\right),$$

$$E_{p3}^W\left(\sum_{i=1}^m v_i^L x_{ip}^L\right) \leq \sum_{r=1}^s u_r^U y_{rp}^U,$$

& all the constraints of model (22).

(30b)

Step 5: Using the values of $E_{p1}^B, E_{p2}^B, E_{p3}^B, E_{p4}^B$ and $E_{p1}^W, E_{p2}^W, E_{p3}^W, E_{p4}^W$, obtained in Step (4a) to Step (4d), find the fuzzy optimal value $\tilde{E}_p^B = \left(E_{p1}^B, E_{p2}^B, E_{p3}^B, E_{p4}^B\right)$ of optimistic fuzzy DEA model (21), representing the best relative fuzzy efficiency of pth DMU, as well as pessimistic fuzzy optimal value $\tilde{E}_p^W = \left(E_{p1}^W, E_{p2}^W, E_{p3}^W, E_{p4}^W\right)$ of pessimistic fuzzy DEA model (22), representing the worst relative fuzzy efficiency of pth DMU.

Step 5: Find the crisp optimal value $E_p^B = \Re\left(\tilde{E}_p^B\right)$, representing the best relative crisp efficiency of pth DMU.

Step 6: Find the crisp optimal value $E_p^W = \Re\left(\tilde{E}_p^W\right)$, representing the worst relative crisp efficiency of pth DMU.

Step 7: Find the relative geometric crisp efficiency $E_P^{GEOMETRIC}$ of pth DMU by putting the values E_p^B and E_p^W, obtained in Step 5 and Step 6, in $E_P^{GEOMETRIC} = \sqrt{E_p^B \times E_p^W}$.

6 Application to Real Life Problem

Wang and Chin [25] evaluated the best relative intuitionistic fuzzy efficiency as well as worst relative intuitionistic fuzzy efficiency and hence, relative geometric crisp efficiency of by considering eight manufacturing enterprises (DMUs) of China with two inputs and two outputs shown in Table 1 and using the optimistic fuzzy CCR DEA model (1) as well as pessimistic fuzzy CCR DEA model (2). The eight manufacturing enterprises, all manufacture the same type of products but with different qualities. Both the gross output value (GOV) and product quality (PQ) are considered as outputs. Manufacturing cost (MC) and the number of employees (NOE) are considered as inputs. The data about the GOV and MC are uncertain due to the unavailability at the time of assessment and are therefore estimated as fuzzy numbers. The product quality is assessed by customers using fuzzy linguistic terms such as Excellent, Very Good, Average, Poor and Very Poor. The assessment results by customers are weighted and averaged.

However, as discussed in Sect. 4 that the optimistic fuzzy CCR DEA model (1) as well as pessimistic fuzzy CCR DEA model (2) are not valid. Therefore, the best relative fuzzy efficiency as well as worst relative fuzzy efficiency and hence, relative geometric crisp efficiency of 8 manufacturing enterprises, evaluated by Wang and Chin [25], is not exact. In this section, to illustrate the proposed method the exact best relative fuzzy efficiency as well as worst relative fuzzy efficiency and hence, relative geometric crisp efficiency of Enterprise A is evaluated by using the proposed method.

The best relative fuzzy efficiency and worst relative fuzzy efficiency of DMU_A can be obtained by solving the optimistic fuzzy CCR DEA models (31) and pessimistic CCR DEA models (32).

Table 1 Input and output data for eight manufacturing enterprises [25]

(DMUs)	Inputs (Two)		Outputs (two)	
	MC	NOE	GOV	PQ
A	(2120, 2170, 2210)	1870	(14500, 14790, 14860)	(3.1, 4.1, 4.9)
B	(1420, 1460, 1500)	1340	(12470, 12720, 12790)	(1.2, 2.1, 3.0)
C	(2510, 2570, 2610)	2360	(17900, 18260, 18400)	(3.3, 4.3, 5.0)
D	(2300, 2350, 2400)	2020	(14970, 15270, 15400)	(2.7, 3.7, 4.6)
E	(1480, 1520, 1560)	1550	(13980, 14260, 14330)	(1.0, 1.8, 2.7)
F	(1990, 2030, 2100)	1760	(14030, 14310, 14400)	(1.6, 2.6, 3.6)
G	(2200, 2260, 2300)	1980	(16540, 16870, 17000)	(2.4, 3.4, 4.4)
H	(2400, 2460, 2520)	2250	(17600, 17960, 18100)	(2.6, 3.6, 4.6)

$$\text{Maximize} \left[\tilde{E}_A^B = \left(E_{A1}^B, E_{A2}^B, E_{A3}^B \right) \right.$$

$$\approx \left[\frac{(14500, 14790, 14860)\left(u_1^L, u_1^M, u_1^U\right) + (3.1, 4.1, 4.9)\left(u_2^L, u_2^M, u_2^U\right)}{(2120, 2170, 2210)\left(v_1^L, v_1^M, v_1^U\right) + (1870, 1870, 1870)\left(v_2^L, v_2^M, v_2^U\right)} \right]$$

Subject to

$$(14500, 14790, 14860)\left(u_1^L, u_1^M, u_1^U\right) + (3.1, 4.1, 4.9)\left(u_2^L, u_2^M, u_2^U\right)$$
$$\leqslant (2120, 2170, 2210)\left(v_1^L, v_1^M, v_1^U\right) + (1870, 1870, 1870)\left(v_2^L, v_2^M, v_2^U\right)$$
$$(12470, 12720, 12790)\left(u_1^L, u_1^M, u_1^U\right) + (1.2, 2.1, 3.0)\left(u_2^L, u_2^M, u_2^U\right)$$
$$\leqslant (1420, 1460, 1500)\left(v_1^L, v_1^M, v_1^U\right) + (1340, 1340, 1340)\left(v_2^L, v_2^M, v_2^U\right)$$
$$(17900, 18260, 18400)\left(u_1^L, u_1^M, u_1^U\right) + (3.3, 4.3, 4.0)\left(u_2^L, u_2^M, u_2^U\right)$$
$$\leqslant (2510, 2570, 2610)\left(v_1^L, v_1^M, v_1^U\right) + (2360, 2360, 2360)\left(v_2^L, v_2^M, v_2^U\right) \qquad (31)$$
$$(14970, 15270, 15400)\left(u_1^L, u_1^M, u_1^U\right) + (2.7, 3.7, 4.6)\left(u_2^L, u_2^M, u_2^U\right)$$
$$\leqslant (2300, 2350, 2400)\left(v_1^L, v_1^M, v_1^U\right) + (2020, 2020, 2020)\left(v_2^L, v_2^M, v_2^U\right)$$
$$(13980, 14260, 14330)\left(u_1^L, u_1^M, u_1^U\right) + (1.0, 1.8, 2.7)\left(u_2^L, u_2^M, u_2^U\right)$$
$$\leqslant (1480, 1520, 1560)\left(v_1^L, v_1^M, v_1^U\right) + (1550, 1550, 1550)\left(v_2^L, v_2^M, v_2^U\right)$$
$$(14030, 14310, 14400)\left(u_1^L, u_1^M, u_1^U\right) + (1.6, 2.6, 3.6)\left(u_2^L, u_2^M, u_2^U\right)$$
$$\leqslant (1990, 2030, 2100)\left(v_1^L, v_1^M, v_1^U\right) + (1760, 1760, 1760)\left(v_2^L, v_2^M, v_2^U\right)$$
$$(16540, 16870, 17000)\left(u_1^L, u_1^M, u_1^U\right) + (2.4, 3.4, 4.4)\left(u_2^L, u_2^M, u_2^U\right)$$
$$\leqslant (2200, 2260, 2300)\left(v_1^L, v_1^M, v_1^U\right) + (1980, 1980, 1980)\left(v_2^L, v_2^M, v_2^U\right)$$
$$(17600, 17960, 18100)\left(u_1^L, u_1^M, u_1^U\right) + (2.6, 3.6, 4.6)\left(u_2^L, u_2^M, u_2^U\right)$$
$$\leqslant (2400, 2460, 2520)\left(v_1^L, v_1^M, v_1^U\right) + (2250, 2250, 2250)\left(v_2^L, v_2^M, v_2^U\right),$$

$\left(u_r^L, u_r^M, u_r^N\right)$ and $\left(v_i^L, v_i^M, v_i^U\right), i=1,2, r=1,2,$ are non-negative triangular fuzzy numbers.

$$\text{Minimize} \left[\tilde{E}_A^W = \left(E_{A1}^W, E_{A2}^W, E_{A3}^W \right) \right.$$

$$\left. \approx \left[\frac{(14500, 14790, 14860)\left(u_1^L, u_1^M, u_1^U\right) + (3.1, 4.1, 4.9)\left(u_2^L, u_2^M, u_2^U\right)}{(2120, 2170, 2210)\left(v_1^L, v_1^M, v_1^U\right) + (1870, 1870, 1870)\left(v_2^L, v_2^M, v_2^U\right)} \right] \right]$$

Subject to

$$(14500, 14790, 14860)\left(u_1^L, u_1^M, u_1^U\right) + (3.1, 4.1, 4.9)\left(u_2^L, u_2^M, u_2^U\right)$$
$$\geq (2120, 2170, 2210)\left(v_1^L, v_1^M, v_1^U\right) + (1870, 1870, 1870)\left(v_2^L, v_2^M, v_2^U\right)$$
$$(12470, 12720, 12790)\left(u_1^L, u_1^M, u_1^U\right) + (1.2, 2.1, 3.0)\left(u_2^L, u_2^M, u_2^U\right)$$
$$\geq (1420, 1460, 1500)\left(v_1^L, v_1^M, v_1^U\right) + (1340, 1340, 1340)\left(v_2^L, v_2^M, v_2^U\right)$$
$$(17900, 18260, 18400)\left(u_1^L, u_1^M, u_1^U\right) + (3.3, 4.3, 4.0)\left(u_2^L, u_2^M, u_2^U\right)$$
$$\geq (2510, 2570, 2610)\left(v_1^L, v_1^M, v_1^U\right) + (2360, 2360, 2360)\left(v_2^L, v_2^M, v_2^U\right)$$
$$(14970, 15270, 15400)\left(u_1^L, u_1^M, u_1^U\right) + (2.7, 3.7, 4.6)\left(u_2^L, u_2^M, u_2^U\right)$$
$$\geq (2300, 2350, 2400)\left(v_1^L, v_1^M, v_1^U\right) + (2020, 2020, 2020)\left(v_2^L, v_2^M, v_2^U\right)$$
$$(13980, 14260, 14330)\left(u_1^L, u_1^M, u_1^U\right) + (1.0, 1.8, 2.7)\left(u_2^L, u_2^M, u_2^U\right)$$
$$\geq (1480, 1520, 1560)\left(v_1^L, v_1^M, v_1^U\right) + (1550, 1550, 1550)\left(v_2^L, v_2^M, v_2^U\right)$$
$$(14030, 14310, 14400)\left(u_1^L, u_1^M, u_1^U\right) + (1.6, 2.6, 3.6)\left(u_2^L, u_2^M, u_2^U\right)$$
$$\geq (1990, 2030, 2100)\left(v_1^L, v_1^M, v_1^U\right) + (1760, 1760, 1760)\left(v_2^L, v_2^M, v_2^U\right)$$
$$(16540, 16870, 17000)\left(u_1^L, u_1^M, u_1^U\right) + (2.4, 3.4, 4.4)\left(u_2^L, u_2^M, u_2^U\right)$$
$$\geq (2200, 2260, 2300)\left(v_1^L, v_1^M, v_1^U\right) + (1980, 1980, 1980)\left(v_2^L, v_2^M, v_2^U\right)$$
$$(17600, 17960, 18100)\left(u_1^L, u_1^M, u_1^U\right) + (2.6, 3.6, 4.6)\left(u_2^L, u_2^M, u_2^U\right)$$
$$\geq (2400, 2460, 2520)\left(v_1^L, v_1^M, v_1^U\right) + (2250, 2250, 2250)\left(v_2^L, v_2^M, v_2^U\right),$$

(32)

$\left(u_r^L, u_r^M, u_r^N\right)$ and $\left(v_i^L, v_i^M, v_i^N\right), i = 1, 2, r = 1, 2$, are non-negative triangular fuzzy numbers.

Using the method, proposed in Sect. 5, the exact best relative fuzzy efficiency as well as worst relative fuzzy efficiency and hence relative geometric crisp efficiency of DMU$_A$ can be obtained as follows:

Step 1: Using the product of triangular fuzzy numbers, defined in Sect. 2.2, the optimistic fuzzy CCR DEA model (31) and pessimistic fuzzy CCR DEA model (32) can be transformed into optimistic fuzzy CCR DEA model (33) and pessimistic fuzzy CCR DEA model (34) respectively.

$$\text{Maximize}\left[\tilde{E}_A^B = \left(E_{A1}^B, E_{A2}^B, E_{A3}^B\right) \approx \left[\frac{\left(14500u_1^L + 3.1u_2^L, 14790u_1^M + 4.1u_2^M, 14860u_1^U + 4.9u_2^U\right)}{\left(2120v_1^L + 4.9v_2^L, 2170v_1^M + 1870v_2^M, 2210v_1^U + 1870v_2^U\right)}\right]\right]$$

Subject to

$$\left(14500u_1^L + 3.1u_2^L, 14790u_1^M + 4.1u_2^M, 14860u_1^U + 4.9u_2^U\right)$$
$$\leqslant \left(2120v_1^L + 1870v_2^L, 2170v_1^M + 1870v_2^M, 2210v_1^U + 1870v_2^U\right)$$
$$\left(12470u_1^L + 1.2u_2^L, 12720u_1^M + 2.1u_2^M, 12790u_1^U + 3.0u_2^U\right)$$
$$\leqslant \left(1420v_1^L + 1340v_2^L, 1460v_1^M + 1340v_2^M, 1500v_1^U + 1340v_2^U\right)$$
$$\left(17900u_1^L + 3.3u_2, 18260u_1^M + 4.3u_2^M, 18400u_1^U + 4.0u_2^U\right)$$
$$\leqslant \left(2510v_1^L + 2360v_2^L, 2570v_1^M + 2360v_2^M, 2610v_1^U + 2360v_2^U\right)$$
$$\left(14970u_1^L + 2.7u_2^L, 15270u_1^M + 3.7u_2^M, 15400u_1^U + 4.6u_2^U\right)$$
$$\leqslant \left(2300v_1^L + 2020v_2^L, 2350v_1^M + 2020v_2^M, 2400v_1^U + 2020v_2^U\right)$$
$$\left(13980u_1^L + 1.0u_2^L, 14260u_1^M + 1.8u_2^M, 14330u_1^U + 2.7u_2^U\right)$$
$$\leqslant \left(1480v_1^L + 1550v_2^L, 1520v_1^M + 1550v_2^M, 1560v_1^U + 1550v_2^U\right)$$
$$\left(14030u_1^L + 1.6u_2^L, 14310u_1^M + 2.6u_2^M, 14400u_1^U + 3.6u_2^U\right)$$
$$\leqslant \left(1990v_1^L + 1760v_2^L, 2030v_1^M + 1760v_2^M, 2100v_1^U + 1760v_2^U\right)$$
$$\left(16540u_1^L + 2.4u_2^L, 16870u_1^M + 3.4u_2^M, 17000u_1^U + 4.4u_2^U\right)$$
$$\leqslant \left(2200v_1^L + 1980v_2^L, 2260v_1^M + 1980v_2^M, 2300v_1^U + 1980v_2^U\right)$$
$$\left(17600u_1^L + 2.6u_2^L, 17960u_1^M + 3.6u_2^M, 18100u_1^U + 4.6u_2^U\right)$$
$$\leqslant \left(2400v_1^L + 2250v_2^L, 2460v_1^M + 2250v_2^M, 2520v_1^U + 2250v_2^U\right),$$
$$0 \leq u_r^L \leq u_r^M \leq u_r^N \leq u_r^U; 0 \leq v_i^L \leq v_i^M \leq v_i^N \leq v_i^U, i = 1, 2, r = 1, 2.$$

$$(33)$$

$$\text{Minimize}\left[\tilde{E}_A^W = \left(E_{A1}^W, E_{A2}^W, E_{A3}^W\right) \approx \left[\frac{\left(14500u_1^L + 3.1u_2^L, 14790u_1^M + 4.1u_2^M, 14860u_1^U + 4.9u_2^U\right)}{\left(2120v_1^L + 4.9v_2^L, 2170v_1^M + 1870v_2^M, 2210v_1^U + 1870v_2^U\right)}\right]\right]$$

Subject to

$$\left(14500u_1^L + 3.1u_2^L, 14790u_1^M + 4.1u_2^M, 14860u_1^U + 4.9u_2^U\right)$$
$$\geqslant \left(2120v_1^L + 1870v_2^L, 2170v_1^M + 1870v_2^M, 2210v_1^U + 1870v_2^U\right)$$
$$\left(12470u_1^L + 1.2u_2^L, 12720u_1^M + 2.1u_2^M, 12790u_1^U + 3.0u_2^U\right)$$
$$\geqslant \left(1420v_1^L + 1340v_2^L, 1460v_1^M + 1340v_2^M, 1500v_1^U + 1340v_2^U\right)$$
$$\left(17900u_1^L + 3.3u_2, 18260u_1^M + 4.3u_2^M, 18400u_1^U + 4.0u_2^U\right)$$
$$\geqslant \left(2510v_1^L + 2360v_2^L, 2570v_1^M + 2360v_2^M, 2610v_1^U + 2360v_2^U\right)$$
$$\left(14970u_1^L + 2.7u_2^L, 15270u_1^M + 3.7u_2^M, 15400u_1^U + 4.6u_2^U\right)$$
$$\geqslant \left(2300v_1^L + 2020v_2^L, 2350v_1^M + 2020v_2^M, 2400v_1^U + 2020v_2^U\right)$$
$$\left(13980u_1^L + 1.0u_2^L, 14260u_1^M + 1.8u_2^M, 14330u_1^U + 2.7u_2^U\right)$$
$$\geqslant \left(1480v_1^L + 1550v_2^L, 1520v_1^M + 1550v_2^M, 1560v_1^U + 1550v_2^U\right)$$
$$\left(14030u_1^L + 1.6u_2^L, 14310u_1^M + 2.6u_2^M, 14400u_1^U + 3.6u_2^U\right)$$
$$\geqslant \left(1990v_1^L + 1760v_2^L, 2030v_1^M + 1760v_2^M, 2100v_1^U + 1760v_2^U\right)$$
$$\left(16540u_1^L + 2.4u_2^L, 16870u_1^M + 3.4u_2^M, 17000u_1^U + 4.4u_2^U\right)$$
$$\geqslant \left(2200v_1^L + 1980v_2^L, 2260v_1^M + 1980v_2^M, 2300v_1^U + 1980v_2^U\right)$$
$$\left(17600u_1^L + 2.6u_2^L, 17960u_1^M + 3.6u_2^M, 18100u_1^U + 4.6u_2^U\right)$$
$$\geqslant \left(2400v_1^L + 2250v_2^L, 2460v_1^M + 2250v_2^M, 2520v_1^U + 2250v_2^U\right),$$
$$0 \leq u_r^L \leq u_r^M \leq u_r^N \leq u_r^U; 0 \leq v_i^L \leq v_i^M \leq v_i^N \leq v_i^U, i = 1, 2, r = 1, 2,$$

$$(34)$$

Step 2: The optimistic fuzzy CCR DEA model (33) and pessimistic fuzzy CCR DEA model (34) can be transformed into optimistic fuzzy CCR DEA model (35) and pessimistic fuzzy CCR DEA model (36) respectively.

$$\text{Maximize} \left[\tilde{E}_A^B = \left(E_{A1}^B, E_{A2}^B, E_{A3}^B \right) \approx \left[\frac{\left(14500u_1^L + 3.1u_2^L, 14790u_1^M + 4.1u_2^M, 14860u_1^U + 4.9u_2^U \right)}{\left(2120v_1^L + 4.9v_2^L, 2170v_1^M + 1870v_2^M, 2210v_1^U + 1870v_2^U \right)} \right] \right]$$

Subject to

$$14500u_1^L + 3.1u_2^L \le 2120v_1^L + 1870v_2^L, \qquad 14790u_1^M + 4.1u_2^M \le 2170v_1^M + 1870v_2^M,$$

$$14860u_1^U + 4.9u_2^U \le 2210v_1^U + 1870v_2^U, \qquad 12470u_1^L + 1.2u_2^L \le 1420v_1^L + 1340v_2^L,$$

$$12720u_1^M + 2.1u_2^M \le 1460v_1^M + 1340v_2^M, \qquad 12790u_1^U + 3.0u_2^U \le 1500v_1^U + 1340v_2^U,$$

$$17900u_1^L + 3.3u_2^L \le 2510v_1^L + 2360v_2^L, \qquad 18260u_1^M + 4.3u_2^M \le 2570v_1^M + 2360v_2^M,$$

$$18400u_1^U + 4.0u_2^U \le 2610v_1^U + 2360v_2^U, \qquad 14970u_1^L + 2.7u_2^L \le 2300v_1^L + 2020v_2^L,$$

$$15270u_1^M + 3.7u_2^M \le 2350v_1^M + 2020v_2^M, \qquad 15400u_1^U + 4.6u_2^U \le 2400v_1^U + 2020v_2^U,$$

$$13980u_1^L + 1.0u_2^L \le 1480v_1^L + 1550v_2^L, \qquad 14260u_1^M + 1.8u_2^M \le 1520v_1^M + 1550v_2^M,$$

$$14330u_1^U + 2.7u_2^U \le 1560v_1^U + 1550v_2^U, \qquad 14030u_1^L + 1.6u_2^L \le 1990v_1^L + 1760v_2^L,$$

$$14310u_1^M + 2.6u_2^M \le 2030v_1^M + 1760v_2^M, \qquad 14400u_1^U + 3.6u_2^U \le 2100v_1^U + 1760v_2^U,$$

$$16540u_1^L + 2.4u_2^L \le 2200v_1^L + 1980v_2^L, \qquad 16870u_1^M + 3.4u_2^M \le 2260v_1^M + 1980v_2^M,$$

$$17000u_1^U + 4.4u_2^U \le 2300v_1^U + 1980v_2^U, \qquad 17600u_1^L + 2.6u_2^L \le 2400v_1^L + 2250v_2^L,$$

$$17960u_1^M + 3.6u_2^M \le 2460v_1^M + 2250v_2^M, \qquad 18100u_1^U + 4.6u_2^U \le 2520v_1^U + 2250v_2^U.$$

$$0 \le u_r^L \le u_r^M \le u_r^N \le u_r^U; 0 \le v_i^L \le v_i^M \le v_i^N \le v_i^U, i = 1, 2, r = 1, 2.$$

(35)

$$\text{Minimize} \left[\tilde{E}_A^W = \left(E_{A1}^W, E_{A2}^W, E_{A3}^W \right) \approx \left[\frac{\left(14500u_1^L + 3.1u_2^L, 14790u_1^M + 4.1u_2^M, 14860u_1^U + 4.9u_2^U \right)}{\left(2120v_1^L + 4.9v_2^L, 2170v_1^M + 1870v_2^M, 2210v_1^U + 1870v_2^U \right)} \right] \right]$$

Subject to

$$14500u_1^L + 3.1u_2^L \ge 2120v_1^L + 1870v_2^L, \qquad 14790u_1^M + 4.1u_2^M \ge 2170v_1^M + 1870v_2^M,$$

$$14860u_1^U + 4.9u_2^U \ge 2210v_1^U + 1870v_2^U, \qquad 12470u_1^L + 1.2u_2^L \ge 1420v_1^L + 1340v_2^L,$$

$$12720u_1^M + 2.1u_2^M \ge 1460v_1^M + 1340v_2^M, \qquad 12790u_1^U + 3.0u_2^U \ge 1500v_1^U + 1340v_2^U,$$

$$17900u_1^L + 3.3u_2^L \ge 2510v_1^L + 2360v_2^L, \qquad 18260u_1^M + 4.3u_2^M \ge 2570v_1^M + 2360v_2^M,$$

$$18400u_1^U + 4.0u_2^U \ge 2610v_1^U + 2360v_2^U, \qquad 14970u_1^L + 2.7u_2^L \ge 2300v_1^L + 2020v_2^L,$$

$$15270u_1^M + 3.7u_2^M \ge 2350v_1^M + 2020v_2^M, \qquad 15400u_1^U + 4.6u_2^U \ge 2400v_1^U + 2020v_2^U,$$

$$13980u_1^L + 1.0u_2^L \ge 1480v_1^L + 1550v_2^L, \qquad 14260u_1^M + 1.8u_2^M \ge 1520v_1^M + 1550v_2^M,$$

$$14330u_1^U + 2.7u_2^U \ge 1560v_1^U + 1550v_2^U, \qquad 14030u_1^L + 1.6u_2^L \ge 1990v_1^L + 1760v_2^L,$$

$$14310u_1^M + 2.6u_2^M \ge 2030v_1^M + 1760v_2^M, \qquad 14400u_1^U + 3.6u_2^U \ge 2100v_1^U + 1760v_2^U,$$

$$16540u_1^L + 2.4u_2^L \ge 2200v_1^L + 1980v_2^L, \qquad 16870u_1^M + 3.4u_2^M \ge 2260v_1^M + 1980v_2^M,$$

$$17000u_1^U + 4.4u_2^U \ge 2300v_1^U + 1980v_2^U, \qquad 17600u_1^L + 2.6u_2^L \ge 2400v_1^L + 2250v_2^L,$$

$$17960u_1^M + 3.6u_2^M \ge 2460v_1^M + 2250v_2^M, \qquad 18100u_1^U + 4.6u_2^U \ge 2520v_1^U + 2250v_2^U.$$

$$0 \le u_r^L \le u_r^M \le u_r^N \le u_r^U; 0 \le v_i^L \le v_i^M \le v_i^N \le v_i^U, i = 1, 2, r = 1, 2,$$

(36)

Step 3: The optimistic fuzzy CCR DEA model (35) and pessimistic fuzzy CCR DEA model (36) can be transformed into optimistic fuzzy CCR DEA model (37) and pessimistic fuzzy CCR DEA model (38) respectively.

$$\text{Maximize}\left[\tilde{E}_A^B = \left(E_{A1}^B, E_{A2}^B, E_{A3}^B\right) \approx \left[\left(\frac{14500u_1^L + 3.1u_2^L}{2210v_1^U + 1870v_2^U}, \frac{14790u_1^M + 4.1u_2^M}{2170v_1^M + 1870v_2^M}, \frac{14860u_1^U + 4.9u_2^U}{2120v_1^L + 4.9v_2^L}\right)\right]\right]$$

Subject to

All the constraints of model (35).

(37)

$$\text{Minimize}\left[\tilde{E}_A^W = \left(E_{A1}^W, E_{A2}^W, E_{A3}^W\right) \approx \left[\left(\frac{14500u_1^L + 3.1u_2^L}{2210v_1^U + 1870v_2^U}, \frac{14790u_1^M + 4.1u_2^M}{2170v_1^M + 1870v_2^M}, \frac{14860u_1^U + 4.9u_2^U}{2120v_1^L + 4.9v_2^L}\right)\right]\right]$$

Subject to

All the constraints of model (36).

(38)

Step 4: The fuzzy optimal value $\tilde{E}_A^B \approx \left(E_{A1}^B, E_{A2}^B, E_{A3}^B\right)$, representing the best relative fuzzy efficiency of DMU_A, as well as the fuzzy optimal value $\tilde{E}_A^W \approx \left(E_{A1}^W, E_{A2}^W, E_{A3}^W\right)$, representing the worst relative fuzzy efficiency of DMU_A, can be obtained by solving the optimistic fuzzy CCR DEA model (37) and pessimistic fuzzy CCR DEA model (38) as follows:

Step 4(a): The optimal value $\left(E_{A1}^B\right)$ and $\left(E_{A1}^W\right)$ of the optimistic crisp CCR DEA model (39a) and pessimistic CCR DEA model (40a) by solving optimistic crisp CCR DEA model (39b) and pessimistic CCR DEA model (39b) equivalent to optimistic crisp CCR DEA model (39a) and pessimistic CCR DEA model (40a) are 0.812 and 1 respectively.

$$\text{Maximize}\left[E_{A1}^B = \frac{14500u_1^L + 3.1u_2^L}{2210v_1^U + 1870v_2^U}\right]$$

Subject to

$$\frac{14500u_1^L + 3.1u_2^L}{2210v_1^U + 1870v_2^U} \le 1,$$

& all the constraints of model (35).

(39a)

$$\text{Maximize}\left[E_{A1}^B = 14500u_1^L + 3.1u_2^L\right]$$

Subject to

$$2210v_1^U + 1870v_2^U = 1$$
$$14500u_1^L + 3.1u_2^L \le 2210v_1^U + 1870v_2^U,$$

& all the constraints of model (35).

(39b)

$$\text{Minimize}\left[E_{A1}^W = \frac{14500u_1^L + 3.1u_2^L}{2210v_1^U + 1870v_2^U}\right]$$

Subject to

$$\frac{14500u_1^L + 3.1u_2^L}{2210v_1^U + 1870v_2^U} \ge 1,$$

& all the constraints of model (36).

(40a)

$$\text{Minimize}\left[E_{A1}^W = 14500u_1^L + 3.1u_2^L\right]$$

Subject to

$$2210v_1^U + 1870v_2^U = 1$$
$$14500u_1^L + 3.1u_2^L \ge 2210v_1^U + 1870v_2^U,$$

& all the constraints of model (36).

(40b)

Step 4(b): The optimal value $\left(E_{A2}^{B}\right)$ and $\left(E_{A2}^{W}\right)$ of the optimistic crisp CCR DEA model (41a) and pessimistic CCR DEA model (42a) by solving optimistic crisp CCR DEA model (41b) and pessimistic CCR DEA model (42b) equivalent to optimistic crisp CCR DEA model (41a) and pessimistic CCR DEA model (42b) are 0.833 and 1.046 respectively.

$$\text{Maximize}\left[E_{A2}^{B} = \frac{14790u_1^M + 4.1u_2^M}{2170v_1^M + 1870v_2^M}\right]$$
$$\text{Subject to}$$
$$\frac{14500u_1^L + 3.1u_2^L}{2210v_1^U + 1870v_2^U} = 0.812 \tag{41a}$$
$$0.812 \le \frac{14790u_1^M + 4.1u_2^M}{2170v_1^M + 1870v_2^M} \le 1$$
$$\text{\& all the constraints of model (38).}$$

$$\text{Maximize}\left[E_{A2}^{B} = 14790u_1^M + 4.1u_2^M\right]$$
$$\text{Subject to}$$
$$2170v_1^M + 1870v_2^M = 1, \tag{41b}$$
$$14500u_1^L + 3.1u_2^L = (0.812)\left(2210v_1^U + 1870v_2^U\right),$$
$$(0.812)\left(2170v_1^M + 1870v_2^M\right) \le 14790u_1^M + 4.1u_2^M \le 2170v_1^M + 1870v_2^M$$
$$\text{\& all the constraints of model (35).}$$

$$\text{Minimize}\left[E_{A2}^{W} = \frac{14790u_1^M + 4.1u_2^M}{2170v_1^M + 1870v_2^M}\right]$$
$$\text{Subject to}$$
$$\frac{14500u_1^L + 3.1u_2^L}{2210v_1^U + 1870v_2^U} = 1, \tag{42a}$$
$$\frac{14790u_1^M + 4.1u_2^M}{2170v_1^M + 1870v_2^M} \ge 1$$
$$\text{\& all the constraints of model (36).}$$

$$\text{Minimize}\left[E_{A2}^{B} = 14790u_1^M + 4.1u_2^M\right]$$
$$\text{Subject to}$$
$$2170v_1^M + 1870v_2^M = 1,$$
$$14500u_1^L + 3.1u_2^L = 2210v_1^U + 1870v_2^U, \tag{42b}$$
$$2170v_1^M + 1870v_2^M \le 14790u_1^M + 4.1u_2^M$$
$$\text{\& all the constraints of model (36).}$$

Step 4(c): The optimal value $\left(E_{A2}^{B}\right)$ and $\left(E_{A2}^{W}\right)$ of the optimistic crisp CCR DEA model (43a) and pessimistic CCR DEA model (44a) by solving optimistic crisp CCR DEA model (43b) and pessimistic CCR DEA model (P29b) equivalent to optimistic crisp CCR DEA model (43a) and pessimistic CCR DEA model (44a) are 0.854 and 1.072 respectively.

$$\text{Maximize}\left[E_{A3}^B = \frac{14860u_1^U + 4.9u_2^U}{2120v_1^L + 4.9v_2^L}\right]$$

Subject to

$$\frac{14500u_1^L + 3.1u_2^L}{2210v_1^U + 1870v_2^U} = 0.812,$$

$$\frac{14790u_1^M + 4.1u_2^M}{2170v_1^M + 1870v_2^M} = 0.833,$$

$$0.833 \le \frac{14860u_1^U + 4.9u_2^U}{2120v_1^L + 4.9v_2^L} \le 1,$$

& all the constraints of model (37).

(43a)

$$\text{Maximize}\left[E_{A3}^B = 14860u_1^U + 4.9u_2^U\right]$$

Subject to

$$2120v_1^L + 4.9v_2^L = 1,$$
$$14500u_1^L + 3.1u_2^L = (0.812)\left(2210v_1^U + 1870v_2^U\right)$$
$$14790u_1^M + 4.1u_2^M = (0.833)\left(2170v_1^M + 1870v_2^M\right),$$
$$(0.833)\left(2120v_1^L + 4.9v_2^L\right) \le 14860u_1^U + 4.9u_2^U \le 2120v_1^L + 4.9v_2^L.$$

& all the constraints of model (37).

(43b)

$$\text{Minimize}\left[E_{A3}^W = \frac{14860u_1^U + 4.9u_2^U}{2120v_1^L + 4.9v_2^L}\right]$$

Subject to

$$\frac{14500u_1^L + 3.1u_2^L}{2210v_1^U + 1870v_2^U} = 1,$$

$$\frac{14790u_1^M + 4.1u_2^M}{2170v_1^M + 1870v_2^M} = 1.046,$$

$$\frac{14860u_1^U + 4.9u_2^U}{2120v_1^L + 4.9v_2^L} \ge 1.046,$$

& all the constraints of model (36).

(44a)

$$\text{Minimize}\left[E_{A3}^W = 14860u_1^U + 4.9u_2^U\right]$$

Subject to

$$2120v_1^L + 4.9v_2^L = 1,$$
$$14500u_1^L + 3.1u_2^L = 2210v_1^U + 1870v_2^U$$
$$14790u_1^M + 4.1u_2^M = (1.046)\left(2170v_1^M + 1870v_2^M\right),$$
$$(1.046)\left(2120v_1^L + 4.9v_2^L\right) \le 14860u_1^U + 4.9u_2^U.$$

& all the constraints of model (36).

(44b)

Step 5: Using the values of $E_{A1}^B, E_{A2}^B, E_{A3}^B$ and $E_{A1}^W, E_{A2}^W, E_{A3}^W$, obtained in Step (4a) to Step (4c), the fuzzy optimal value $\tilde{E}_A^B = \left(E_{A1}^B, E_{A2}^B, E_{A3}^B\right)$ of optimistic fuzzy DEA model-4.16, representing the best relative fuzzy efficiency of DMU$_A$, is $\tilde{E}_A^B = \left(E_{A1}^B, E_{A2}^B, E_{A3}^B\right) = (0.812, 0.833, 0.854)$, as well as pessimistic fuzzy optimal value $\tilde{E}_A^W = \left(E_{A1}^W, E_{A2}^W, E_{A3}^W\right)$ of pessimistic fuzzy DEA model (31), representing the worst relative fuzzy efficiency of DMU$_A$, is $\tilde{E}_A^W = \left(E_{A1}^W, E_{A2}^W, E_{A3}^W\right) = (1, 1.046, 1.072)$.

Step 6: The crisp optimal value $E_A^B = \Re\left(\tilde{E}_A^B\right) = \Re(0.812, 0.833, 0.854)$, representing the best relative crisp efficiency of DMU$_A$, is 0.833.

Step 7: The crisp optimal value $E_A^W = \Re\left(\tilde{E}_A^W\right) = \Re(1, 1.046, 1.072)$, representing the worst relative crisp efficiency of DMU$_A$, is 1.041.

Step 8: The geometric average crisp efficiency $E_A^{GEOMETRIC}$ of Ath DMU by putting the values E_A^B and E_A^W, obtained in Step 5 and Step 6, in $E_A^{GEOMETRIC} = \sqrt{E_A^B \times E_A^W}$ is 0.931.

7 Results

The exact best relative fuzzy efficiency, exact worst relative fuzzy efficiency and relative geometric crisp efficiency of all the DMUs, obtained by using the proposed method are shown in Table 2.

It is obvious from Table 2 that $\Re\left(\tilde{E}_B\right) > R\left(\tilde{E}_E\right) > R\left(\tilde{E}_G\right) = \Re\left(\tilde{E}_F\right) > R\left(\tilde{E}_H\right) > R\left(\tilde{E}_A\right) > R\left(\tilde{E}_C\right) > R\left(\tilde{E}_D\right)$. Therefore, $\tilde{E}_B \succ \tilde{E}_E \succ \tilde{E}_G \approx \tilde{E}_F \succ \tilde{E}_A \succ \tilde{E}_H \succ \tilde{E}_C \succ \tilde{E}_D$.

Table 2 Results obtained by using the proposed method

DMU$_j$	Best relative fuzzy efficiency	Worst relative fuzzy efficiency	Relative geometric crisp efficiency
A	(0.81238, 0.833217, 0.85407)	(1, 1.04625, 1.07227)	0.931
B	(0.97498, 1, 1)	(1, 1.12689, 1.28793)	1.062
C	(0.79661, 0.815201, 0.83732)	(1, 1.01738, 1.055204)	0.913
D	(0.77643, 0.79636, 0.81927)	(1, 1, 1.028724)	0.898
E	(0.97303, 1, 1)	(1, 1, 1)	0.996
F	(0.83517, 0.85653, 0.879205)	(1, 1, 1.104026)	0.938
G	(0.87519, 0.89757, 0.92263)	(1, 1.07384, 1.1585)	0.938
H	(0.819529, 0.84089, 0.86447)	(1, 1.009602, 1.08548)	0.929

8 Conclusions

On the basis the present study, it can be concluded that there are flaws in the existing method [25] and hence, the existing method [25] cannot be used for evaluating the best relative geometric crisp efficiency of DMUs. Also, to resolve the flaws of the existing method [25], a new approach is proposed to solve the proposed fuzzy CCR DEA models for evaluating the best relative geometric crisp efficiency of DMUs.

Acknowledgements Dr. Amit Kumar would like to acknowledge the adolescent inner blessings of Mehar (lovely daughter of his cousin sister Dr. Parmpreet Kaur). Dr. Amit Kumar believes that Mata Vaishno Devi has appeared on the earth in the form of Mehar and without her blessings it would not be possible to think the ideas presented in this chapter. The second author would like to acknowledge the financial support given by UGC under the UGC Dr. D.S. Kothari Postdoctoral Fellowship Scheme.

References

1. Banker RD, Charnes A, Cooper WW (1984) Some models for estimating technical and scale inefficiency in data envelopment analysis. Manag Sci 30:1078–1092
2. Charnes A, Cooper WW, Rhodes E (1978) Measuring the efficiency of decision making units. Eur J Oper Res 2:429–444
3. Dia M (2004) A model of fuzzy data envelopment analysis. INFOR: Inf Syst Oper Res 42:267–279
4. Garcia PAA, Schirru R, Melo PFFE (2005) A fuzzy data envelopment analysis approach for FMEA. Prog Nucl Energy 46:35–373
5. Guo P, Tanaka H (2001) Fuzzy DEA: a perceptual evaluation method. Fuzzy Sets Syst 119:149–160
6. Jahanshahloo GR, Soleimani-damaneh M, Nasrabadi E (2004) Measure of efficiency in DEA with fuzzy input–output levels: a methodology for assessing, ranking and imposing of weights restrictions. Appl Math Comput 156:175–187
7. Kao C, Liu ST (2000) Fuzzy efficiency measures in data envelopment analysis. Fuzzy Sets Syst 113:427–437
8. Kao C, Liu ST (2000) Data envelopment analysis with missing data: an application to University libraries in Taiwan. J Oper Res Soc 51:897–905
9. Kao C, Liu ST (2003) A mathematical programming approach to fuzzy efficiency ranking. Int J Prod Econ 86:45–154
10. Kao C, Liu ST (2005) Data envelopment analysis with imprecise data: an application of Taiwan machinery firms. Int J Uncertain Fuzziness Knowl-Based Syst 13:225–240
11. León T, Liern V, Ruiz JL, Sirvent I (2003) A fuzzy mathematical programming approach to the assessment of efficiency with DEA models. Fuzzy Sets Syst 139:407–419
12. Lertworasirikul S, Fang SC, Joines JA, Nuttle HLW (2003) Fuzzy data envelopment analysis (DEA): a possibility approach. Fuzzy Sets Syst 139:379–394
13. Lertworasirikul S, Fang SC, Joines JA, Nuttle HLW (2003) Fuzzy data envelopment analysis: a credibility approach. In: Verdegay JL (ed) Fuzzy sets based heuristics for optimization. Springer, Berlin, pp 141–158
14. Lertworasirikul S, Fang SC, Nuttle HLW, Joines JA (2003) Fuzzy BCC model for data envelopment analysis. Fuzzy Optim Decis Making 2:337–358

15. Liu ST (2008) A fuzzy DEA/AR approach to the selection of flexible manufacturing systems. Comput Ind Eng 54:66–76
16. Liu ST, Chuang M (2009) Fuzzy efficiency measures in fuzzy DEA/AR with application to university libraries. Expert Syst Appl 36:1105–1113
17. Moheb-Alizadeh H, Rasouli SM, Tavakkoli-Moghaddam R (2011) The use of multi-criteria data envelopment analysis (MCDEA) for location-allocation problems in a fuzzy environment. Expert Syst Appl 38:5687–5695
18. Puri J, Yadav SP (2013) A concept of fuzzy input mix-efficiency in fuzzy DEA and its application in banking sector. Expert Syst Appl 40:1437–1450
19. Puri J, Yadav SP (2015) Intuitionistic fuzzy data envelopment analysis: an application to the banking sector in India. Expert Syst Appl 42:4982–4998
20. Saati S, Memariani A (2005) Reducing weight flexibility in fuzzy DEA. Appl Math Comput 161:611–622
21. Saati S, Menariani A, Jahanshahloo GR (2002) Efficiency analysis and ranking of DMUs with fuzzy data. Fuzzy Optim Decis Making 1:255–267
22. Sengupta JK (1992) A fuzzy systems approach in data envelopment analysis. Comput Math Appl 24:259–266
23. Triantis K (2003) Fuzzy non-radial data envelopment analysis (DEA) measures of technical efficiency in support of an integrated performance measurement system. Int J Automot Technol Manag 3:328–353
24. Triantis K, Girod O (1998) A mathematical programming approach for measuring technical efficiency in a fuzzy environment. J Prod Anal 10:85–102
25. Wang YM, Chin KS (2011) Fuzzy data envelopment analysis: a fuzzy expected value approach. Expert Syst Appl 38:11678–11685
26. Wang YM, Luo Y, Liang L (2009) Fuzzy data envelopment analysis based upon fuzzy arithmetic with an application to performance assessment of manufacturing enterprises. Expert Syst Appl 36:5205–5211
27. Wu D, Yang Z, Liang L (2006) Efficiency analysis of cross-region bank branches using fuzzy data envelopment analysis. Appl Math Comput 181:271–281
28. Zerafat ALM, Emrouznejad A, Mustafa A (2012) Fuzzy data envelopment analysis: a discrete approach. Expert Syst Appl 39:2263–2269
29. Zimmermann HZ (1996) Fuzzy set theory and its applications, 3rd edn. Kluwer Nijhoff, Boston

A New Fuzzy CCR Data Envelopment Analysis Model and Its Application to Manufacturing Enterprises

Bindu Bhardwaj, Jagdeep Kaur and Amit Kumar

Abstract Wang et al. (Expert Syst Appl 36:5205–5211, 2009, [26]) pointed out that in the literature the solution of a fuzzy CCR data envelopment analysis (DEA) model which is a fuzzy fractional programming problem, is obtained by transforming it into a fuzzy linear programming problem. While, due to the property $\frac{\tilde{A}}{\tilde{A}} \neq \tilde{1}$, where $\tilde{A}$ is a fuzzy number, the fuzzy CCR DEA model cannot be transformed into a fuzzy linear programming problem. To resolve this flaw of the existing methods, Wang et al. proposed two new methods to solve the fuzzy CCR DEA model without transforming it into a fuzzy linear programming problem. In this chapter, flaws in the methods, proposed by Wang et al. as well as flaws in the fuzzy CCR DEA model, proposed by Wang et al., are pointed out. To resolve these flaws, a new fuzzy CCR DEA model as well as a new method to solve fuzzy CCR DEA model are proposed. To illustrate the proposed method, the real life problem, chosen by Wang et al., is solved.

Keywords Data envelopment analysis · Fuzzy input and fuzzy output data · Fuzzy efficiency

B. Bhardwaj
LMT School of Management, Thapar University, Patiala, Punjab, India
e-mail: bhardwajbindu1@gmail.com

J. Kaur (✉) · A. Kumar
School of Mathematics, Thapar University, Patiala, Punjab, India
e-mail: sidhu.deepi87@gmail.com

A. Kumar
e-mail: amitkdma@gmail.com

© Springer International Publishing AG 2018
M. Collan and J. Kacprzyk (eds.), *Soft Computing Applications for Group Decision-making and Consensus Modeling*, Studies in Fuzziness and Soft Computing 357, DOI 10.1007/978-3-319-60207-3_21

1 Introduction

The efficiency evaluation of every system is important to find its weakness so that subsequent improvements can be made. DEA, a non-parametric approach proposed by Charnes et al. [3], is an approach to measure the relative efficiency of homogenous units called decision making units (DMUs) which consume the same type of inputs and produce the same type of outputs. Since, the pioneering work Charnes et al. [3], DEA has been extensively used for evaluating the performance of many activities.

The conventional CCR and BCC DEA models [2, 3] require accurate measurement of both the inputs and outputs. However, inputs and outputs of DMUs in real world problems may be imprecise. Imprecise evaluations may be the result of unquantifiable, incomplete and non-obtainable information. In recent years, fuzzy set theory has been proven to be useful as a way to quantify imprecise and vague data in DEA models. The DEA model with fuzzy data, called "fuzzy DEA" models, can more realistically represent real world problems than the conventional DEA models. Several approaches have been developed and many new are coming for handling fuzzy input and output data in DEA.

Several authors [1, 4–25, 27, 28] have proposed methods to solve fuzzy DEA models and to solve fuzzy DEA models two approaches are used: (1) by transforming the fuzzy fractional programming $\sum_{r=1}^{s} u_r \tilde{y}_{rj} / \sum_{i=1}^{m} v_i \tilde{x}_{ij}$ into a fuzzy linear programming model $\sum_{r=1}^{s} u_r \tilde{y}_{rj}$ by setting $\sum_{i=1}^{m} v_i \tilde{x}_{ij} \approx \tilde{1}$ (2) by transforming fuzzy DEA models into two respective pessimistic and optimistic crisp DEA models using α-cut technique. Wang et al. [26] pointed out that the former ignores the fact that a fuzzy fractional programming cannot be transformed into a linear programming model as we do for a crisp fractional programming unless $\tilde{1}$ is assumed to be a crisp number; while the latter requires the solution of a series of linear programming models based on different α-level sets and therefore it requires much computational efforts to get the fuzzy efficiencies of DMUs.

To the best of our knowledge, the methods, proposed by Wang et al. [26], are the only existing methods to solve fuzzy DEA models without using the aforementioned transformations (1) and (2). In this chapter, the shortcomings of the existing methods [26] are pointed out and a new method is proposed for the same.

The rest of the chapter is organized as follows. In Sect. 2, some basic definitions and arithmetic operations on fuzzy numbers are presented. In Sect. 3, the existing methods [26] for solving fuzzy DEA problems are reviewed. In Sect. 4, the flaws of the existing methods and in Sect. 5, the flaws of the existing fuzzy CCR DEA model are pointed out. To overcome these flaws, a new fuzzy CCR DEA model is proposed in Sect. 6 and a new method is proposed in Sect. 7 for solving the same. The proposed method is illustrated with the help of a real world problem in Sect. 8 and the obtained results are discussed in Sect. 9. Finally, the conclusions are discussed in Sect. 10.

2 Basic Definitions and Arithmetic Operations

In this section, some basic definitions and arithmetic operations on fuzzy numbers are reviewed [29].

2.1 Basic Definitions

In this section, some basic definitions are reviewed [29].

Definition 1 A fuzzy number $\widetilde{A} = (a^L, a^M, a^R)$ is said to be a triangular fuzzy number if its membership function is given by

$$\mu_{\widetilde{A}}(x) = \begin{cases} \frac{(x-a^L)}{(a^M - a^L)}, & a^L \leq x < a^M \\ 1 & x = a^M \\ \frac{(x-a^R)}{(a^M - a^R)}, & a^M < x \leq a^R \\ 0, & \text{otherwise.} \end{cases}$$

Definition 2 A triangular fuzzy number $\widetilde{A} = (a^L, a^M, a^R)$ is said to be non-negative triangular fuzzy number if and only if $a^L \geq 0$.

Definition 3 A triangular fuzzy number $\widetilde{A} = (a^L, a^M, a^R)$ is said to be positive triangular fuzzy number if and only if $a^L > 0$.

2.2 Arithmetic Operations on Triangular Fuzzy Numbers

Let $\widetilde{A} = (a^L, a^M, a^R)$ and $\widetilde{B} = (b^L, b^M, b^R)$ be two arbitrary triangular fuzzy numbers. Then,

(i) $\widetilde{A} + \widetilde{B} = (a^L + b^L, a^M + b^M, a^R + b^R)$

(ii) $\widetilde{A} - \widetilde{B} = (a^L - b^R, a^M - b^M, a^R - b^L)$

(iii) $\widetilde{A}\widetilde{B} = (a^L b^L, a^M b^M, a^R b^R)$, where $\widetilde{A}$ and $\widetilde{B}$ are non-negative triangular fuzzy numbers.

(iv) $\frac{\widetilde{A}}{\widetilde{B}} = \left(\frac{a^L}{b^R}, \frac{a^M}{b^M}, \frac{a^R}{b^L} \right)$, where $\widetilde{A}$ is a non-negative triangular fuzzy number and $\widetilde{B}$ is a positive triangular fuzzy number.

3   An Overview of the Existing Fuzzy CCR DEA Methods

In this section, the existing methods [26] for solving fuzzy CCR DEA model are reviewed.

Suppose that there are n DMUs to be assessed where each $DMU_j(j = 1, 2, \ldots, n)$ consumes m inputs $\tilde{x}_{ij}(i = 1, 2, \ldots, m)$ to produce s outputs $\tilde{y}_{rj}(r = 1, 2, \ldots s)$. Wang et al. [26] proposed the following two methods for solving fuzzy CCR DEA model (1) by considering input and output data as triangular fuzzy numbers $\tilde{x}_{ij} = (x_{ij}^L, x_{ij}^M, x_{ij}^U)$ and $\tilde{y}_{rj} = (y_{rj}^L, y_{rj}^M, y_{rj}^U)$ respectively.

$$\text{Maximize} \left[\tilde{E}_p \approx \left(E_p^L, E_p^M, E_p^U \right) \approx \frac{\sum_{r=1}^s u_r \left(y_{rp}^L, y_{rp}^M, y_{rp}^U \right)}{\sum_{i=1}^m v_i \left(x_{ip}^L, x_{ip}^M, x_{ip}^U \right)} \right]$$

Subject to

$$\frac{\sum_{r=1}^s u_r \left(y_{rj}^L, y_{rj}^M, y_{rj}^U \right)}{\sum_{i=1}^m v_i \left(x_{ij}^L, x_{ij}^M, x_{ij}^U \right)} \preccurlyeq (1, 1, 1), \; \forall j$$

$$u_r, v_i \geq 0, \; \forall i, r.$$

(1)

3.1   First Method

The steps of the method, proposed by Wang et al. [26] for solving fuzzy CCR DEA model (1), are as follows:

Step 1:   Using the product of triangular fuzzy numbers, defined in Sect. 2.2, the fuzzy CCR DEA model (1) can be transformed into fuzzy CCR DEA model (2).

$$\text{Maximize} \left[\tilde{E}_p \approx \left(E_p^L, E_p^M, E_p^U \right) \approx \frac{\left(\sum_{r=1}^s u_r y_{rp}^L, \; \sum_{r=1}^s u_r y_{rp}^M, \; \sum_{r=1}^s u_r y_{rp}^U \right)}{\left(\sum_{i=1}^m v_i x_{ip}^L, \; \sum_{i=1}^m v_i x_{ip}^M \; \sum_{i=1}^m v_i x_{ip}^U \right)} \right]$$

Subject to

$$\frac{\left(\sum_{r=1}^s u_r y_{rj}^L, \; \sum_{r=1}^s u_r y_{rj}^M, \; \sum_{r=1}^s u_r y_{rj}^U \right)}{\left(\sum_{i=1}^m v_i x_{ij}^L, \; \sum_{i=1}^m v_i x_{ij}^M \; \sum_{i=1}^m v_i x_{ij}^U \right)} \preccurlyeq (1, 1, 1), \; \forall j$$

$$u_r, v_i \geq 0, \; \forall i, r.$$

(2)

Step 2: Using division of triangular fuzzy number, defined in Sect. 2.2, the fuzzy
CCR DEA model (2) can be transformed into fuzzy CCR DEA model (3).

$$\text{Maximize}\left[\widetilde{E}_p \approx \left(E_p^L, E_p^M, E_p^U\right) \approx \left(\frac{\sum_{r=1}^{s} u_r y_{rp}^L}{\sum_{i=1}^{m} v_i x_{ip}^U}, \frac{\sum_{r=1}^{s} u_r y_{rp}^M}{\sum_{i=1}^{m} v_i x_{ip}^M}, \frac{\sum_{r=1}^{s} u_r y_{rp}^U}{\sum_{i=1}^{m} v_i x_{ip}^L}\right)\right]$$

Subject to

$$\left(\frac{\sum_{r=1}^{s} u_r y_{rj}^L}{\sum_{i=1}^{m} v_i x_{ij}^U}, \frac{\sum_{r=1}^{s} u_r y_{rj}^M}{\sum_{i=1}^{m} v_i x_{ij}^M}, \frac{\sum_{r=1}^{s} u_r y_{rj}^U}{\sum_{i=1}^{m} v_i x_{ij}^L}\right) \preccurlyeq (1,1,1), \forall j \tag{3}$$

$$u_r, v_i \geq 0, \forall i, r.$$

Step 3: Using   the   relation $((a^L, a^M, a^U) \leq (b^L, b^M, b^U) \Rightarrow a^L \leq b^L, a^M \leq b^M$
$comma a^U \leq b^U$, the fuzzy CCR DEA model (3) can be transformed into
fuzzy CCR DEA model (4).

$$\text{Maximize}\left[\widetilde{E}_p \approx \left(E_p^L, E_p^M, E_p^U\right) \approx \left(\frac{\sum_{r=1}^{s} u_r y_{rp}^L}{\sum_{i=1}^{m} v_i x_{ip}^U}, \frac{\sum_{r=1}^{s} u_r y_{rp}^M}{\sum_{i=1}^{m} v_i x_{ip}^M}, \frac{\sum_{r=1}^{s} u_r y_{rp}^U}{\sum_{i=1}^{m} v_i x_{ip}^L}\right)\right]$$

Subject to

$$\frac{\sum_{r=1}^{s} u_r y_{rj}^L}{\sum_{i=1}^{m} v_i x_{ij}^U} \leq 1, \forall j$$

$$\frac{\sum_{r=1}^{s} u_r y_{rj}^M}{\sum_{i=1}^{m} v_i x_{ij}^M} \leq 1, \forall j \tag{4}$$

$$\frac{\sum_{r=1}^{s} u_r y_{rj}^U}{\sum_{i=1}^{m} v_i x_{ij}^L} \leq 1, \forall j$$

$$u_r, v_i \geq 0, \ \forall i, r.$$

Step 4: For a fuzzy number (a^L, a^M, a^R), if $a^R \leq 1$ then the condition $a^L \leq 1$ and
$a^M \leq 1$ will automatically be satisfied, so the fuzzy CCR DEA model (4)
can transformed into fuzzy CCR DEA model (5).

$$\text{Maximize}\left[\widetilde{E}_p \approx \left(E_p^L, E_p^M, E_p^U\right) \approx \left(\frac{\sum_{r=1}^{s} u_r y_{rp}^L}{\sum_{i=1}^{m} v_i x_{ip}^U}, \frac{\sum_{r=1}^{s} u_r y_{rp}^M}{\sum_{i=1}^{m} v_i x_{ip}^M}, \frac{\sum_{r=1}^{s} u_r y_{rp}^U}{\sum_{i=1}^{m} v_i x_{ip}^L}\right)\right]$$

Subject to

$$\frac{\sum_{r=1}^{s} u_r y_{rj}^U}{\sum_{i=1}^{m} v_i x_{ij}^L} \leq 1, \forall j \tag{5}$$

$$u_r, v_i \geq 0, \ \forall i, r.$$

Step 5: Transform the fuzzy CCR DEA model (5) into crisp CCR DEA models (6)–(8).

$$\text{Maximize}\left[E_p^L = \frac{\sum_{r=1}^{s} u_r y_{rp}^L}{\sum_{i=1}^{m} v_i x_{ip}^U}\right]$$

Subject to

$$\frac{\sum_{r=1}^{s} u_r y_{rj}^U}{\sum_{i=1}^{m} v_i x_{ij}^L} \leq 1, \forall j \tag{6}$$

$$u_r, v_i \geq 0, \ \forall i, r.$$

$$\text{Maximize}\left[E_p^M = \frac{\sum_{r=1}^{s} u_r y_{rp}^M}{\sum_{i=1}^{m} v_i x_{ip}^M}\right]$$

Subject to

$$\frac{\sum_{r=1}^{s} u_r y_{rj}^U}{\sum_{i=1}^{m} v_i x_{ij}^L} \leq 1, \forall j \tag{7}$$

$$u_r, v_i \geq 0, \ \forall i, r.$$

$$\text{Maximize}\left[E_p^U = \frac{\sum_{r=1}^{s} u_r y_{rp}^U}{\sum_{i=1}^{m} v_i x_{ip}^L}\right]$$

Subject to

$$\frac{\sum_{r=1}^{s} u_r y_{rj}^U}{\sum_{i=1}^{m} v_i x_{ij}^L} \leq 1, \forall j \tag{8}$$

$$u_r, v_i \geq 0, \ \forall i, r.$$

Step 6: The crisp CCR DEA models (6)–(8) can be transformed into crisp CCR DEA models (9) to (11) respectively.

$$\text{Maximize}\left[E_p^L = \sum_{r=1}^{s} u_r y_{rp}^L\right]$$

Subject to

$$\sum_{i=1}^{m} v_i x_{ip}^U = 1, \tag{9}$$

$$\sum_{r=1}^{s} u_r y_{rj}^U - \sum_{i=1}^{m} v_i x_{ij}^L \leq 0, \ \forall j$$

$$u_r, v_i \geq 0, \ \forall i, r.$$

$$\text{Maximize}\left[E_p^M = \sum_{r=1}^{s} u_r y_{rp}^M\right]$$

Subject to

$$\sum_{i=1}^{m} v_i x_{ip}^M = 1, \tag{10}$$

$$\sum_{r=1}^{s} u_r y_{rj}^U - \sum_{i=1}^{m} v_i x_{ij}^L \leq 0, \forall j$$

$$u_r, v_i \geq 0, \ \forall i, r.$$

$$\text{Maximize}\left[E_p^U = \sum_{r=1}^{s} u_r y_{rp}^U\right]$$

Subject to

$$\sum_{i=1}^{m} v_i x_{ip}^L = 1, \tag{11}$$

$$\sum_{r=1}^{s} u_r y_{rj}^U - \sum_{i=1}^{m} v_i x_{ij}^L \leq 0, \forall j$$

$$u_r, v_i \geq 0, \ \forall i, r.$$

Step 7: Find the optimal value E_p^L, E_p^M and E_p^U of crisp CCR DEA models (9) to (11) respectively.

Step 8: Using the optimal values of E_p^L, E_p^M and E_p^U, obtained in Step 6, the fuzzy optimal value $\left(E_p^L, E_p^M, E_p^U\right)$ (best relative fuzzy efficiency of DMU$_p$) of fuzzy CCR DEA model (1) can be obtained.

3.2 Second Method

The steps of the method, proposed by Wang et al. [26] for solving fuzzy CCR DEA model (1), are as follows:

Step 1: Using the product of triangular fuzzy numbers, defined in Sect. 2.2, the fuzzy CCR DEA model (1) can be transformed into fuzzy CCR DEA model (12).

$$\text{Maximize}\left[\tilde{E}_p \approx \left(E_p^L, E_p^M, E_p^U\right) \approx \frac{\left(\sum_{r=1}^{s} u_r y_{rp}^L, \ \sum_{r=1}^{s} u_r y_{rp}^M \ \sum_{r=1}^{s} \ \sum_{r=1}^{s} u_r y_{rp}^U,\right)}{\left(\sum_{i=1}^{m} v_i x_{ip}^L, \ \sum_{i=1}^{m} v_i x_{ip}^M, \ \sum_{i=1}^{m} v_i x_{ip}^U\right)}\right]$$

Subject to

$$\frac{\left(\sum_{r=1}^{s} u_r y_{rj}^L, \ \sum_{r=1}^{s} u_r y_{rj}^M, \ \sum_{r=1}^{s} u_r y_{rj}^U\right)}{\left(\sum_{i=1}^{m} v_i x_{ij}, \ \sum_{i=1}^{m} v_i x_{ij}^M, \ \sum_{i=1}^{m} v_i x_{ij}^U\right)} \leq (1, 1, 1), \forall j$$

$$u_r, v_i \geq 0, \ \forall i, r.$$

$$\tag{12}$$

Step 2: Using the division of triangular fuzzy numbers, defined in Sect. 2.2, the fuzzy CCR DEA model (12) can be transformed into fuzzy CCR DEA model (13).

$$\text{Maximize}\left[\tilde{E}_p \approx \left(E_p^L, E_p^M, E_p^U\right) \approx \left(\frac{\sum_{r=1}^s u_r y_{rp}^L}{\sum_{i=1}^m v_i x_{ip}^U}, \frac{\sum_{r=1}^s u_r y_{rp}^M}{\sum_{i=1}^m v_i x_{ip}^M}, \frac{\sum_{r=1}^s u_r y_{rp}^U}{\sum_{i=1}^m v_i x_{ip}^L}\right)\right]$$

Subject to

$$\left(\frac{\sum_{r=1}^s u_r y_{rj}^L}{\sum_{i=1}^m v_i x_{ij}^U}, \frac{\sum_{r=1}^s u_r y_{rj}^M}{\sum_{i=1}^m v_i x_{ij}^M}, \frac{\sum_{r=1}^s u_r y_{rj}^U}{\sum_{i=1}^m v_i x_{ij}^L}\right) \preccurlyeq ()1, 1, 1, \forall j$$

$$u_r, v_i \geq 0, \ \forall i, r.$$

$$(13)$$

Step 3: Using the relation $((a^L, a^M, a^U) \preccurlyeq (b^L, b^M, b^U) \Rightarrow a^L \leq b^L, a^M \leq b^M, a^U \leq b^U$, the fuzzy CCR DEA model (13) can be transformed into fuzzy CCR DEA model (14).

$$\text{Maximize}\left[\tilde{E}_p \approx \left(E_p^L, E_p^M, E_p^U\right) \approx \left(\frac{\sum_{r=1}^s u_r y_{rp}^L}{\sum_{i=1}^m v_i x_{ip}^U}, \frac{\sum_{r=1}^s u_r y_{rp}^M}{\sum_{i=1}^m v_i x_{ip}^M}, \frac{\sum_{r=1}^s u_r y_{rp}^U}{\sum_{i=1}^m v_i x_{ip}^L}\right)\right]$$

Subject to

$$\frac{\sum_{r=1}^s u_r y_{rj}^L}{\sum_{i=1}^m v_i x_{ij}^U} \leq 1, \forall j$$

$$\frac{\sum_{r=1}^s u_r y_{rj}^M}{\sum_{i=1}^m v_i x_{ij}^M} \leq 1, \forall j$$

$$\frac{\sum_{r=1}^s u_r y_{rj}^U}{\sum_{i=1}^m v_i x_{ij}^L} \leq 1, \forall j$$

$$u_r, v_i \geq 0, \ \forall i, r.$$

$$(14)$$

Step 4: As for fuzzy CCR DEA model (14), since there are no restrictions imposed subjectively on the support of the fuzzy numbers 1, its lower and upper bounds are therefore viewed as free. Based upon this point of view fuzzy CCR DEA model (14) can be transformed into fuzzy CCR DEA model (15).

$$\text{Maximize}\left[\widetilde{E}_p \approx \left(E_p^L, E_p^M, E_p^U\right) \approx \left(\frac{\sum_{r=1}^{s} u_r y_{rp}^L}{\sum_{i=1}^{m} v_i x_{ip}^U}, \frac{\sum_{r=1}^{s} u_r y_{rp}^M}{\sum_{i=1}^{m} v_i x_{ip}^M}, \frac{\sum_{r=1}^{s} u_r y_{rp}^U}{\sum_{i=1}^{m} v_i x_{ip}^L}\right)\right]$$

Subject to

$$\frac{\sum_{r=1}^{s} u_r y_{rj}^M}{\sum_{i=1}^{m} v_i x_{ij}^M} \le 1, \forall j$$

$$u_r, v_i \ge 0, \ \forall i, r.$$

$$(15)$$

Step 5: Transform the fuzzy CCR DEA model (15) into crisp CCR DEA models (16)–(18).

$$\text{Maximize}\left[E_p^L = \frac{\sum_{r=1}^{s} u_r y_{rp}^L}{\sum_{i=1}^{m} v_i x_{ip}^U}\right]$$

Subject to

$$\frac{\sum_{r=1}^{s} u_r y_{rj}^M}{\sum_{i=1}^{m} v_i x_{ij}^M} \le 1, \forall j$$

$$u_r, v_i \ge 0, \ \forall i, r.$$

$$(16)$$

$$\text{Maximize}\left[E_p^M = \frac{\sum_{r=1}^{s} u_r y_{rp}^M}{\sum_{i=1}^{m} v_i x_{ip}^M}\right]$$

Subject to

$$\frac{\sum_{r=1}^{s} u_r y_{rj}^M}{\sum_{i=1}^{m} v_i x_{ij}^M} \le 1, \forall j$$

$$u_r, v_i \ge 0, \ \forall i, r.$$

$$(17)$$

$$\text{Maximize}\left[E_p^U = \frac{\sum_{r=1}^{s} u_r y_{rp}^U}{\sum_{i=1}^{m} v_i x_{ip}^L}\right]$$

Subject to

$$\frac{\sum_{r=1}^{s} u_r y_{rj}^M}{\sum_{i=1}^{m} v_i x_{ij}^M} \le 1, \forall j$$

$$u_r, v_i \ge 0, \ \forall i, r.$$

$$(18)$$

Step 6: The crisp CCR DEA models (16)–(18) can be transformed into crisp CCR DEA models (19) to (21) respectively.

$$\text{Maximize} \left[E_p^L = \sum_{r=1}^s u_r y_{rp}^L \right]$$

Subject to

$$\sum_{i=1}^m v_i x_{ip}^U = 1, \tag{19}$$

$$\sum_{r=1}^s u_r y_{rj}^M - \sum_{i=1}^m v_i x_{ij}^M \leq 0, \forall j$$

$$u_r, v_i \geq 0, \ \forall i, r.$$

$$\text{Maximize} \left[E_p^M = \sum_{r=1}^s u_r y_{rp}^M \right]$$

Subject to

$$\sum_{i=1}^m v_i x_{ip}^M = 1, \tag{20}$$

$$\sum_{r=1}^s u_r y_{rj}^M, - \sum_{i=1}^m v_i x_{ij}^M \leq 0, \forall j$$

$$u_r, v_i \geq 0, \ \forall i, r.$$

$$\text{Maximize} \left[E_p^U = \sum_{r=1}^s u_r y_{rp}^U \right]$$

Subject to

$$\sum_{i=1}^m v_i x_{ip}^L = 1, \tag{21}$$

$$\sum_{r=1}^s u_r y_{rj}^M - \sum_{i=1}^m v_i x_{ij}^M \leq 0, \ \forall j$$

$$u_r, v_i \geq 0, \ \forall i, r.$$

Step 7: Find the optimal value E_p^L, E_p^M and E_p^U of crisp CCR DEA models (19) to (21) respectively.

Step 8: Using the optimal values of E_p^L, E_p^M and E_p^U, obtained in Step 6, the fuzzy optimal value $\left(E_p^L, E_p^M, E_p^U \right)$ (best relative fuzzy efficiency of DMU$_p$) of fuzzy CCR DEA model (1) can be obtained.

4 Flaws in the Existing Methods

In this section, the flaws in the methods, proposed by Wang et al. [26], are pointed out.

(i) In the fuzzy CCR DEA model (1), u_r, v_i are crisp numbers. Therefore, if these numbers are represented as triangular fuzzy numbers $u_r = (u_r^L, u_r^M, u_r^U)$ and $v_i = (v_i^L, v_i^M, v_i^U)$ then the condition $u_r^L = u_r^M = u_r^U, v_i^L = v_i^M = v_i^U$ should always be satisfied. However, in the methods, proposed by Wang et al. [26] the values of E_p^L, E_p^M, E_p^U are obtained by solving the crisp CCR DEA models (9)–(11) independently. Since, all these models are solved independently so the optimal values of u_r and v_i, obtained on solving these models, will not necessarily be same i.e., if $u_r^L, v_i^L, u_r^M, v_i^M$ and u_r^U, v_i^U represents the optimal values of u_r and v_i, obtained on solving crisp CCR DEA models (9) to (11) respectively, then the condition $u_r^L = u_r^M = u_r^U$ and $v_i^L = v_i^M = v_i^U$ will not necessarily be satisfied. Hence, using the methods, proposed by Wang et al. [26] the obtained optimal solution $\{u_r^L, v_i^L, u_r^M, v_i^M, u_r^U, v_i^U\}$ is not an optimal solution of fuzzy CCR DEA model (1) i.e., using the method, proposed by Wang et al. [26], it is not possible to find a crisp optimal solution $\{u_r, v_i\}$ of fuzzy CCR DEA model (1).

(ii) It is well known that for a triangular fuzzy number (a, b, c), the property $a \leq b \leq c$ should always be satisfied. However, it is obvious from Step 6 of the existing methods, presented in Sects. 1 and 2, that the values of E_p^L, E_p^M, E_p^U are obtained by solving three independent crisp CCR DEA models so the restriction $E_p^L \leq E_p^M \leq E_p^U$ may or may not be satisfied.

5 Flaws in the Existing Fuzzy DEA Model

It is obvious from Step 2 and Step 3 of the existing methods [26], presented in Sect. 3, that Wang et al. [26] have transformed the fuzzy constraints

$$\left(\frac{\sum_{r=1}^{s} u_r y_{rj}^M}{\sum_{i=1}^{m} v_i x_{ij}^U}, \frac{\sum_{r=1}^{s} u_r y_{rj}^M}{\sum_{i=1}^{m} v_i x_{ij}^M}, \frac{\sum_{r=1}^{s} u_r y_{rj}^U}{\sum_{i=1}^{m} v_i x_{ij}^L} \right) \preccurlyeq (1, 1, 1)$$

into the crisp constraints $\frac{\sum_{r=1}^{s} u_r y_{rj}^L}{\sum_{i=1}^{m} v_i x_{ij}^U} \leq 1 \forall j$, $\frac{\sum_{r=1}^{s} u_r y_{rj}^M}{\sum_{i=1}^{m} v_i x_{ij}^M} \leq 1 \forall j$ and $\frac{\sum_{r=1}^{s} u_r y_{rj}^U}{\sum_{i=1}^{m} v_i x_{ij}^L} \leq 1, \forall j$.

Since, $\left(\frac{\sum_{r=1}^{s} u_r y_{rj}^L}{\sum_{i=1}^{m} v_i x_{ij}^U}, \frac{\sum_{r=1}^{s} u_r y_{rj}^M}{\sum_{i=1}^{m} v_i x_{ij}^M}, \frac{\sum_{r=1}^{s} u_r y_{rj}^U}{\sum_{i=1}^{m} v_i x_{ij}^L} \right) \preccurlyeq (1, 1, 1) \forall j$ so there should exist a non-negative triangular fuzzy number (S^L, S^M, S^U) such that

$$\left(\frac{\sum_{r=1}^{s} u_r y_{rj}^L}{\sum_{i=1}^{m} v_i x_{ij}^U}, \frac{\sum_{r=1}^{s} u_r y_{rj}^M}{\sum_{i=1}^{m} v_i x_{ij}^M}, \frac{\sum_{r=1}^{s} u_r y_{rj}^U}{\sum_{i=1}^{m} v_i x_{ij}^L} \right) + \left(S^L, S^M, S^U \right) = (1, 1, 1), \forall j.$$

However, the following clearly indicates that there will never exist a non-negative triangular fuzzy number (S^L, S^M, S^U), where $S^L \leq S^M \leq S^U$, and hence the constraint

$$\left(\frac{\sum_{r=1}^{s} u_r y_{rj}^L}{\sum_{i=1}^{m} v_i x_{ij}^U}, \frac{\sum_{r=1}^{s} u_r y_{rj}^M}{\sum_{i=1}^{m} v_i x_{ij}^M}, \frac{\sum_{r=1}^{s} u_r y_{rj}^U}{\sum_{i=1}^{m} v_i x_{ij}^L} \right) \preccurlyeq (1, 1, 1), \forall j$$

cannot be transformed into the constraints $\frac{\sum_{r=1}^{s} u_r y_{rj}^L}{\sum_{i=1}^{m} v_i x_{ij}^U} \leq 1 \forall j$, $\frac{\sum_{r=1}^{s} u_r y_{rj}^M}{\sum_{i=1}^{m} v_i x_{ij}^M} \leq 1 \forall j$ and

$\frac{\sum_{r=1}^{s} u_r y_{rj}^U}{\sum_{i=1}^{m} v_i x_{ij}^L} \leq 1, \forall j.$

$$\left(\frac{\sum_{r=1}^{s} u_r y_{rj}^L}{\sum_{i=1}^{m} v_i x_{ij}^U}, \frac{\sum_{r=1}^{s} u_r y_{rj}^M}{\sum_{i=1}^{m} v_i x_{ij}^M}, \frac{\sum_{r=1}^{s} u_r y_{rj}^U}{\sum_{i=1}^{m} v_i x_{ij}^L} \right) \preccurlyeq (1,1,1)$$

$$\Rightarrow \left(\frac{\sum_{r=1}^{s} u_r y_{rj}^L}{\sum_{i=1}^{m} v_i x_{ij}^U}, \frac{\sum_{r=1}^{s} u_r y_{rj}^M}{\sum_{i=1}^{m} v_i x_{ij}^M}, \frac{\sum_{r=1}^{s} u_r y_{rj}^U}{\sum_{i=1}^{m} v_i x_{ij}^L} \right) + \left(S^L, S^M, S^U \right) = (1,1,1)$$

$$\Rightarrow \frac{\sum_{r=1}^{s} u_r y_{rj}^L}{\sum_{i=1}^{m} v_i x_{ij}^U} + S^L = 1, \quad \frac{\sum_{r=1}^{s} u_r y_{rj}^M}{\sum_{i=1}^{m} v_i x_{ij}^M} + S^M = 1, \quad \frac{\sum_{r=1}^{s} u_r y_{rj}^U}{\sum_{i=1}^{m} v_i x_{ij}^L} + S^U = 1$$

$$\Rightarrow \frac{\sum_{r=1}^{s} u_r y_{rj}^L}{\sum_{i=1}^{m} v_i x_{ij}^U} = 1 - S^L, \quad \frac{\sum_{r=1}^{s} u_r y_{rj}^M}{\sum_{i=1}^{m} v_i x_{ij}^M} = 1 - S^M, \quad \frac{\sum_{r=1}^{s} u_r y_{rj}^U}{\sum_{i=1}^{m} v_i x_{ij}^L} = 1 - S^U.$$

Now,

$$\frac{\sum_{r=1}^{s} u_r y_{rj}^L}{\sum_{i=1}^{m} v_i x_{ij}^U} \leq \frac{\sum_{r=1}^{s} u_r y_{rj}^M}{\sum_{i=1}^{m} v_i x_{ij}^M} \leq \frac{\sum_{r=1}^{s} u_r y_{rj}^U}{\sum_{i=1}^{m} v_i x_{ij}^L}$$

$$\Rightarrow 1 - S^L \leq 1 - S^M \leq 1 - S^U$$

$$\Rightarrow -S^L \leq -S^M \leq -S^U$$

$$\Rightarrow S^L \geq S^M \geq S^U$$

$$\Rightarrow \left(S^L, S^M, S^U \right) \text{ is not a triangular fuzzy number.}$$

6 Proposed Fuzzy CCR DEA Model

In this section, to resolve the flaws of the existing fuzzy CCR DEA model, pointed out in Sect. 3, a modified fuzzy CCR DEA model is proposed.

If there are n DMUs then the best relative crisp efficiency (E_p) of pth DMU can be obtained by solving the CCR DEA model (22).

$$\text{Maximize} \left[E_p = \frac{\text{Virtual output of } p\text{th DMU}}{\text{Virtual input of } p\text{th MU}} \right]$$

Subject to

Virtual output of jth DMU $\leq$ Virtual input of jth DMU, $\forall j$.

(22)

If each DMU uses m inputs (x_{ij}; $i = 1, 2, \ldots, m$; $j = 1, 2, \ldots, n$) to produce s outputs (y_{rj}; $r = 1, 2, \ldots s$; $j = 1, 2, \ldots, n$) then CCR DEA model (22) is transformed into crisp CCR DEA model (23).

$$\text{Maximize}\left[E_p = \frac{\sum_{r=1}^{s} u_r y_{rp}}{\sum_{i=1}^{m} v_i x_{ip}}\right]$$

Subject to   (23)

$$\sum_{r=1}^{s} u_r y_{rj} \leq \sum_{i=1}^{m} v_i x_{ij}, \ \forall j$$

$$u_r, v_i \geq 0, \ \forall i, r.$$

where $u_r(r = 1, \ldots .s)$ and $v_i(i = 1, \ldots, m)$ are the weights assigned to the rth output and ith input, respectively.

Replacing the crisp input data (x_{ij}) and crisp output data (y_{rj}) by triangular fuzzy numbers $\tilde{x}_{ij} = (x_{ij}^L, x_{ij}^M, x_{ij}^U)$ and $\tilde{y}_{rj} = (y_{rj}^L, y_{rj}^M, y_{rj}^U)$ respectively, the CCR DEA model (23) can be transformed into fuzzy CCR DEA model (24).

$$\text{Maximize}\left[\tilde{E}_j \approx \left(E_p^L, E_p^M, E_p^U\right) \approx \frac{\sum_{r=1}^{s} u_r \left(y_{rp}^L, y_{rp}^M, y_{rp}^U\right)}{\sum_{i=1}^{m} v_i \left(x_{ip}^L, x_{ip}^M, x_{ip}^U\right)]}\right]$$

Subject to   (24)

$$\sum_{r=1}^{s} u_r \left(y_{rj}^L, y_{rj}^M, y_{rj}^U\right) \preccurlyeq \sum_{i=1}^{m} v_i \left(x_{ij}^L, x_{ij}^M, x_{ij}^U\right), \forall j$$

$$u_r, v_i \geq 0, \forall i, r.$$

7   Proposed Method

In Sect. 6, it is shown that the fuzzy CCR DEA model (5), proposed by Wang et al. [26], are not valid and hence cannot be used for evaluating the best relative fuzzy efficiency of DMUs. In this section, a new method is proposed to evaluate the best relative fuzzy efficiency of DMUs by using the fuzzy CCR DEA model (24).

Step 1:   Using the product of triangular fuzzy numbers, defined in Sect. 2.2, the fuzzy CCR DEA model (24) can be transformed into fuzzy CCR DEA model (25).

$$\text{Maximize}\left[\tilde{E}_j \approx \left(E_p^L, E_p^M, E_p^U\right) \approx \frac{\sum_{r=1}^{s} \left(u_r y_{rp}^L, u_r y_{rp}^M, u_r y_{rp}^U\right)}{\sum_{i=1}^{m} \left(v_i x_{ip}^L, v_i x_{ip}^M, v_i x_{ip}^U\right)}\right]$$

Subject to   (25)

$$\sum_{r=1}^{s} \left(u_r y_{rj}^L, u_r y_{rj}^M, u_r y_{rj}^U\right) \preccurlyeq \sum_{i=1}^{m} \left(v_i x_{ij}^L, v_i x_{ij}^M, v_i x_{ij}^U\right), \forall j$$

$$u_r, v_i \geq 0, \ \forall i, r.$$

Step 2: Using the division of triangular fuzzy numbers defined in Sect. 2.2, the
fuzzy CCR DEA model (25) can be transformed into fuzzy CCR DEA
model (26).

$$
\text{Maximize} \left[\widetilde{E}_p \approx \left(E_p^L, E_p^M, E_p^U \right) \approx \left(\frac{\sum_{r=1}^{s} u_r y_{rp}^L}{\sum_{i=1}^{m} v_i x_{ip}^U}, \frac{\sum_{r=1}^{s} u_r y_{rp}^M}{\sum_{i=1}^{m} v_i x_{ip}^M}, \frac{\sum_{r=1}^{s} u_r y_{rp}^U}{\sum_{i=1}^{m} v_i x_{ip}^L} \right) \right]
$$

Subject to

$$
\sum_{r=1}^{s} \left(u_r y_{rj}^L, u_r y_{rj}^M, u_r y_{rj}^U \right) \preccurlyeq \sum_{i=1}^{m} \left(v_i x_{ij}^L, v_i x_{ij}^M, v_i x_{ij}^U \right), \forall j
$$

$$
u_r, v_i \geq 0, \ \forall i, r.
$$

$$
(26)
$$

Step 3: Using   the   relation $\left((a^L, a^M, a^U) \preccurlyeq (b^L, b^M, b^U) \Rightarrow a^L \leq b^L, a^M \leq b^M, a^U \leq b^U \right)$, the fuzzy CCR DEA model (26) can be transformed into fuzzy
CCR DEA model (27).

$$
\text{Maximize} \left[\widetilde{E}_p \approx \left(E_p^L, E_p^M, E_p^U \right) \approx \left(\frac{\sum_{r=1}^{s} u_r y_{rp}^L}{\sum_{i=1}^{m} v_i x_{ip}^U}, \frac{\sum_{r=1}^{s} u_r y_{rp}^M}{\sum_{i=1}^{m} v_i x_{ip}^M}, \frac{\sum_{r=1}^{s} u_r y_{rp}^U}{\sum_{i=1}^{m} v_i x_{ip}^L} \right) \right]
$$

Subject to

$$
\sum_{r=1}^{s} u_r y_{rj}^L \leq \sum_{i=1}^{m} v_i x_{ij}^L, \forall j
$$

$$
\sum_{r=1}^{s} u_r y_{rj}^M \leq \sum_{i=1}^{m} v_i x_{ij}^M, \forall j
$$

$$
\sum_{r=1}^{s} u_r y_{rj}^U \leq \sum_{i=1}^{m} v_i x_{ij}^U, \forall j
$$

$$
u_r, v_i \geq 0, \ \forall i, r.
$$

$$
(27)
$$

Step 4: The fuzzy optimal value $\widetilde{E}_p \approx \left(E_p^L, E_p^M, E_p^U \right)$, representing the best relative
fuzzy efficiency of pth DMU, can be obtained by solving the fuzzy
CCR DEA model (27) as follows:

Step 4(a): Find the optimal value $\left(E_p^L \right)$ of the crisp CCR DEA model
(28a) by solving crisp CCR DEA model (28b) equivalent to
crisp CCR DEA model (28a).

$$\text{Maximize}\left[E_p^L = \frac{\sum_{r=1}^{s} u_r y_{rp}^L}{\sum_{i=1}^{m} v_i x_{ip}^U}\right]$$

Subject to

$$\frac{\sum_{r=1}^{s} u_r y_{rp}^L}{\sum_{i=1}^{m} v_i x_{ip}^U} \leq 1,$$

(28a)

& all the constraints of model M − 27.

$$\text{Maximize}\left[E_p^L = \sum_{r=1}^{s} u_r y_{rp}^L\right]$$

Subject to

$$\sum_{i=1}^{m} v_i x_{ip}^U = 1,$$

$$\sum_{r=1}^{s} u_r y_{rp}^L \leq \sum_{i=1}^{m} v_i x_{ip}^U,$$

(28b)

& all the constraints of model M − 27.

Step 4(b): Find the optimal value $\left(E_p^M\right)$ of the crisp CCR DEA model (29a) by solving crisp CCR DEA model (29b) equivalent to crisp CCR DEA model (29a).

$$\text{Maximize}\left[E_p^M = \frac{\sum_{r=1}^{s} u_r y_{rp}^M}{\sum_{i=1}^{m} v_i x_{ip}^M}\right]$$

Subject to

$$\frac{\sum_{r=1}^{s} u_r y_{rp}^L}{\sum_{i=1}^{m} v_i x_{ip}^U} = E_p^L,$$

$$E_p^L \leq \frac{\sum_{r=1}^{s} u_r y_{rp}^M}{\sum_{i=1}^{m} v_i x_{ip}^M} \leq 1,$$

(29a)

& all the constraints of model M − 27.

$$\text{Maximize}\left[E_p^M = \sum_{r=1}^{s} u_r y_{rp}^M\right]$$

Subject to

$$\sum_{r=1}^{s} v_i x_{ip}^M = 1,$$

$$\sum_{r=1}^{s} u_r y_{rp}^L = E_p^L\left(\sum_{i=1}^{m} v_i x_{ip}^U\right),$$

$$E_p^L\left(\sum_{i=1}^{m} v_i x_{ip}^M\right) \leq \sum_{r=1}^{s} u_r y_{rp}^M$$

(29b)

& all the constraints of model M − 27.

Step 4(c): Find the optimal value $\left(E_p^U\right)$ of the crisp CCR DEA model (30a) by solving crisp CCR DEA model (30b) equivalent to crisp CCR DEA model (30a).

$$\text{Maximize}\left[E_p^U = \frac{\sum_{r=1}^{s} u_r y_{rp}^U}{\sum_{i=1}^{m} v_i x_{ip}^L}\right]$$

Subject to

$$\frac{\sum_{r=1}^{s} u_r y_{rp}^L}{\sum_{i=1}^{m} v_i x_{ip}^U} = E_p^L$$

$$\frac{\sum_{r=1}^{s} u_r y_{rp}^M}{\sum_{i=1}^{m} v_i x_{ip}^M} = E_p^M,$$

$$(30a)$$

$$E_p^M \leq \frac{\sum_{r=1}^{s} u_r y_{rp}^U}{\sum_{i=1}^{m} v_i x_{ip}^L} \leq 1,$$

& all the constraints of model M − 27.

$$\text{Maximize}\left[E_p^U = \sum_{r=1}^{s} u_r y_{rp}^U\right]$$

Subject to

$$\sum_{i=1}^{m} v_i x_{ip}^L = 1,$$

$$\sum_{r=1}^{s} u_r y_{rp}^L = E_p^L\left(\sum_{i=1}^{m} v_i x_{ip}^U\right),$$

$$\sum_{r=1}^{s} u_r y_{rp}^M = E_p^M\left(\sum_{i=1}^{m} v_i x_{ip}^M\right),$$

$$(30b)$$

$$E_p^M\left(\sum_{i=1}^{m} v_i x_{ip}^L\right) \leq \sum_{r=1}^{s} u_r y_{rp}^U$$

$$\sum_{r=1}^{s} u_r y_{rp}^U \leq \sum_{i=1}^{m} v_i x_{ip}^L$$

& all the constraints of model M − 27.

Step 5: Using the optimal values of E_p^L, E_p^M and E_p^U, obtained in Step 4, the fuzzy optimal value (best relative fuzzy efficiency of DMU$_P$) is $\left(E_p^L, E_p^M, E_p^U\right)$.

8 Exact Fuzzy Efficiency of Real Life Problem

Wang et al. [26] solved a problem to illustrate his proposed approach. However, as discussed in Sect. 4 that there are flaws in the existing methods [26]. So, the results of this problem, obtained by using Wang et al.'s approach [26], are not exact. In this section, the exact results of the same problem are obtained by using the proposed method.

8.1 Problem Description

Consider a performance assessment problem in China where eight manufacturing enterprises (DMUs) are to be evaluated in terms of two inputs and two outputs. The eight manufacturing enterprises all manufacture the same type of products but with different qualities. Both the gross output value (GOV) and product quality (PQ) are considered as outputs. Manufacturing cost (MC) and the number of employees (NOE) are considered as inputs. The data about the GOV and MC are uncertain due to the unavailability at the time of assessment and are therefore estimated as fuzzy numbers. The product quality is assessed by customers using fuzzy linguistic terms such as Excellent, Very Good, Average, Poor and Very Poor. The assessment results by customers are weighted and averaged. The input and output data for the eight manufacturing enterprises is summarized in Table 1.

The fuzzy CCR DEA model to find the best relative fuzzy efficiency of DMU_A can be formulated as model (31).

$$\text{Maximize}\left[\widetilde{E}_A \approx \left(E_A^L, E_A^M, E_A^U\right)\right.$$
$$\left.\approx \left[\frac{(14500, 14790, 14860)u_1 + (3.1, 4.1, 4.9)u_2}{(2120, 2170, 2210)v_1 + (1870, 1870, 1870)v_2}\right]\right]$$

Subject to

$$(14500u_1 + 3.1u_2, 14790u_1 + 4.1u_2, 14860u_1 + 4.9u_2)$$
$$\leqslant (2120v_1 + 1870v_2, 2170v_1 + 1870v_2, 2210v_1 + 1870v_2),$$
$$(12470u_1 + 1.2u_2, 12720u_1 + 2.1u_2, 12790u_1 + 3.0u_2)$$
$$\leqslant (1420v_1 + 1340v_2, 1460v_1 + 1340v_2, 1500v_1 + 1340v_2),$$
$$(17900u_1 + 3.3u_2, 18260u_1 + 4.3u_2, 18400u_1 + 5.0u_2)$$
$$\leqslant (2510v_1 + 2360v_2, 2570v_1 + 2360v_2, 2610v_1 + 2360v_2),$$
$$(14970u_1 + 2.7u_2, 15270u_1 + 3.7u_2, 15400u_1 + 4.6u_2) \qquad (31)$$
$$\leqslant (2300v_1 + 2020v_2, 2350v_1 + 2020v_2, 2400v_1 + 2020v_2),$$
$$(13980u_1 + 1.0u_2, 14260u_1 + 1.8u_2, 14330u_1 + 2.7u_2)$$
$$\leqslant (1480v_1 + 1550v_2, 1520v_1 + 1550v_2, 1560v_1 + 1550v_2),$$
$$(14030u_1 + 1.6u_2, 14310u_1 + 2.6u_2, 14400u_1 + 3.6u_2)$$
$$\leqslant (1990v_1 + 1760v_2, 2030v_1 + 1760v_2, 2100v_1 + 1760v_2),$$
$$(16540u_1 + 2.4u_2, 16870u_1 + 3.4u_2, 17000u_1 + 4.4u_2)$$
$$\leqslant (2200v_1 + 1980v_2, 2260v_1 + 1980v_2, 2300v_1 + 1980v_2),$$
$$(17600u_1 + 2.6u_2, 17960u_1 + 3.6u_2, 18100u_1 + 4.6u_2)$$
$$\leqslant (2400v_1 + 2250v_2, 2460v_1 + 2250v_2, 2520v_1 + 2250v_2),$$
$$u_1, u_2, v_1, v_2 \geq 0.$$

Table 1 Input and output data for eight manufacturing enterprises [26]

Enterprises (DMUs)	Inputs		Outputs	
	MC	NOE	GOV	PQ
A	(2120, 2170, 2210)	1870	(14500, 14790, 14860)	(3.1, 4.1, 4.9)
B	(1420, 1460, 1500)	1340	(12470, 12720, 12790)	(1.2, 2.1, 3.0)
C	(2510, 2570, 2610)	2360	(17900, 18260, 18400)	(3.3, 4.3, 5.0)
D	(2300, 2350, 2400)	2020	(14970, 15270, 15400)	(2.7, 3.7, 4.6)
E	(1480, 1520, 1560)	1550	(13980, 14260, 14330)	(1.0, 1.8, 2.7)
F	(1990, 2030, 2100)	1760	(14030, 14310, 14400)	(1.6, 2.6, 3.6)
G	(2200, 2260, 2300)	1980	(16540, 16870, 17000)	(2.4, 3.4, 4.4)
H	(2400, 2460, 2520)	2250	(17600, 17960, 18100)	(2.6, 3.6, 4.6)

Using the method, proposed in Sect. 7, the exact best relative fuzzy efficiency of DMU_A can be obtained by using the following steps:

Step 1: Using the arithmetic operations of triangular fuzzy numbers, defined in Sect. 2.2, the fuzzy CCR DEA model (31) can be transformed into fuzzy CCR DEA model (32).

$$\text{Maximize}\left[\widetilde{E}_A \approx \left(E_A^L, E_A^M, E_A^U\right)\right.$$
$$\approx \left[\frac{(14500u_1 + 3.1u_2, 14790u_1 + 4.1u_2, 14860u_1 + 4.9u_2)}{(2120v_1 + 4.9u_2, 2170v_1 + 1870v_2, 2210v_1 + 1870v_2)}\right]\right]$$

Subject to
$$(14500u_1 + 3.1u_2, 14790u_1 + 4.1u_2, 14860u_1 + 4.9u_2)$$
$$\leqslant (2120v_1 + 1870v_2, 2170v_1 + 1870v_2, 2210v_1 + 1870v_2),$$
$$(12470u_1 + 1.2u_2, 12720u_1 + 2.1u_2, 12790u_1 + 3.0u_2)$$
$$\leqslant (1420v_1 + 1340v_2, 1460v_1 + 1340v_2, 1500v_1 + 1340v_2),$$
$$(17900u_1 + 3.3u_2, 18260u_1 + 4.3u_2, 18400u_1 + 5.0u_2)$$
$$\leqslant (2510v_1 + 2360v_2, 2570v_1 + 2360v_2, 2610v_1 + 2360v_2),$$
$$(14970u_1 + 2.7u_2, 15270u_1 + 3.7u_2, 15400u_1 + 4.6u_2) \tag{32}$$
$$\leqslant (2300v_1 + 2020v_2, 2350v_1 + 2020v_2, 2400v_1 + 2020v_2),$$
$$(13980u_1 + 1.0u_2, 14260u_1 + 1.8u_2, 14330u_1 + 2.7u_2)$$
$$\leqslant (1480v_1 + 1550v_2, 1520v_1 + 1550v_2, 1560v_1 + 1550v_2),$$
$$(14030u_1 + 1.6u_2, 14310u_1 + 2.6u_2, 14400u_1 + 3.6u_2)$$
$$\leqslant (1990v_1 + 1760v_2, 2030v_1 + 1760v_2, 2100v_1 + 1760v_2),$$
$$(16540u_1 + 2.4u_2, 16870u_1 + 3.4u_2, 17000u_1 + 4.4u_2)$$
$$\leqslant (2200v_1 + 1980v_2, 2260v_1 + 1980v_2, 2300v_1 + 1980v_2),$$
$$(17600u_1 + 2.6u_2, 17960u_1 + 3.6u_2, 18100u_1 + 4.6u_2)$$
$$\leqslant (2400v_1 + 2250v_2, 2460v_1 + 2250v_2, 2520v_1 + 2250v_2),$$
$$u_1, u_2, v_1, v_2 \geq 0.$$

Step 2: The fuzzy CCR DEA model (32) can be transformed into fuzzy CCR DEA model (33).

$$\text{Maximize}\left[\widetilde{E}_A \approx \left(E_A^L, E_A^M, E_A^U\right)\right]$$

$$\approx \left[\frac{\left(14500u_1 + 3.1u_2,\, 14790u_1 + 4.1u_2,\, 14860u_1 + 4.9u_2\right)}{\left(2120v_1 + 4.9u_2,\, 2170v_1 + 1870v_2,\, 2210v_1 + 1870v_2\right)}\right]$$

Subject to

$14500u_1 + 3.1u_2 \le 2120v_1 + 1870v_2, \qquad 14790u_1 + 4.1u_2 \le 2170v_1 + 1870v_2,$

$14860u_1 + 4.9u_2 \le 2210v_1 + 1870v_2, \qquad 12470u_1 + 1.2u_2 \le 1420v_1 + 1340v_2,$

$12720u_1 + 2.1u_2 \le 1460v_1 + 1340v_2, \qquad 12790u_1 + 3.0u_2 \le 1500v_1 + 1340v_2,$

$17900u_1 + 3.3u_2 \le 2510v_1 + 2360v_2, \qquad 18260u_1 + 4.3u_2 \le 2570v_1 + 2360v_2,$

$18400u_1 + 5.0u_2 \le 2610v_1 + 2360v_2, \qquad 14970u_1 + 2.7u_2 \le 300v_1 + 2020v_2,$

$15270u_1 + 3.7u_2 \le 2350v_1 + 2020v_2, \qquad 15400u_1 + 4.6u_2 \le 2400v_1 + 2020v_2,$

$3980u_1 + 1.0u_2 \le 1480v_1 + 1550v_2, \qquad 14260u_1 + 1.8u_2 \le 1520v_1 + 1550v_2,$

$14330u_1 + 2.7u_2 \le 1560v_1 + 1550v_2, \qquad 14030u_1 + 1.6u_2 \le 1990v_1 + 1760v_2,$

$14310u_1 + 2.6u_2 \le 2030v_1 + 1760v_2, \qquad 14400u_1 + 3.6u_2 \le 2100v_1 + 1760v_2,$

$16540u_1 + 2.4u_2 \le 2200v_1 + 1980v_2, \qquad 16870u_1 + 3.4u_2 \le 2260v_1 + 1980v_2,$

$17000u_1 + 4.4u_2 \le 2300v_1 + 1980v_2, \qquad 17600u_1 + 2.6u_2 \le 2400v_1 + 2250v_2,$

$17960u_1 + 3.6u_2 \le 2460v_1 + 2250v_2, \qquad 18100u_1 + 4.6u_2 \le 2520v_1 + 2250v_2.$

$$u_1,\ u_2, v_1, v_2 \ge 0.$$

$$(33)$$

Step 3: The fuzzy CCR DEA model (33) can be transformed into fuzzy CCR DEA model (34).

$$\text{Maximize}\left[\widetilde{E}_A \approx \left(E_A^L, E_A^M, E_A^U\right) \approx \left[\left(\frac{14500u_1 + 3.1u_2}{2210v_1 + 1870v_2},\, \frac{14790u_1 + 4.1u_2}{2170v_1 + 1870v_2},\, \frac{14860u_1 + 4.9u_2}{2120v_1 + 4.9u_2}\right)\right]\right]$$

Subject to

All the constraints of model M − 33.

$$(34)$$

Step 4: The fuzzy optimal value $\widetilde{E}_A \approx \left(E_A^L, E_A^M, E_A^U\right)$, representing the best relative fuzzy efficiency of DMU_A, can be obtained by solving the CCR fuzzy DEA model (34) as follows:

   Step 4(a): The left optimal value $\left(E_p^L\right)$ of fuzzy efficiency of the fuzzy CCR DEA model (34) can be obtained by solving the following models (35) and (36).

$$\text{Maximize}\left[E_A^L = \frac{14500u_1 + 3.1u_2}{2210v_1 + 1870v_2}\right]$$

Subject to

$$\frac{14500u_1 + 3.1u_2}{2210v_1 + 1870v_2} \le 1, \tag{35}$$

& all the constraints of model M − 33.

The model (35) can be transformed into the following linear model (36).

$$\text{Maximize}\left[E_A^L = 14500u_1 + 3.1u_2\right]$$

Subject to

$$2210v_1 + 1870v_2 = 1, \tag{36}$$

$$14500u_1 + 3.1u_2 \le 2210v_1 + 1870v_2,$$

& all the constraints of model M − 33.

On solving the linear model (36), the obtained left optimal value $\left(E_p^L\right)$ of the fuzzy efficiency of CCR DEA model (34) is 0.812.

Step 4(b): The middle optimal value $\left(E_p^M\right)$ of fuzzy efficiency of the fuzzy CCR DEA model (34) can be obtained by solving the following models (37) and (38).

$$\text{Maximize}\left[E_A^M = \frac{14790u_1 + 4.1u_2}{2170v_1 + 1870v_2}\right]$$

Subject to

$$\frac{14500u_1 + 3.1u_2}{2210v_1 + 1870v_2} = 0.812, \tag{37}$$

$$0.812 \le \frac{14790u_1 + 4.1u_2}{2170v_1 + 1870v_2} \le 1,$$

& all the constraints of model M − 33.

The model (37) can be transformed into the following linear model (38).

$$\text{Maximize}\left[E_A^M = 14790u_1 + 4.1u_2\right]$$

Subject to

$$2170v_1 + 1870v_2 = 1,$$
$$14500u_1 + 3.1u_2 = 0.812(2210v_1 + 1870v_2),$$
$$0.812(2170v_1 + 1870v_2) \le 14790u_1 + 4.1u_2 \le 2170v_1 + 1870v_2,$$

& all the constraints of model M − 33.

$$(38)$$

On solving the linear model (38), the obtained middle optimal value $\left(E_p^M\right)$ of the fuzzy efficiency of CCR DEA model (34) is 0.829.

Step 4(c): The right optimal value $\left(E_p^R\right)$ of fuzzy efficiency of the fuzzy CCR DEA model (34) can be obtained by solving the following models (39) and (40).

$$\text{Maximize}\left[E_A^U = \frac{14860u_1 + 4.9u_2}{2120v_1 + 4.9u_2}\right]$$

Subject to

$$\frac{14500u_1 + 3.1u_2}{2210v_1 + 1870v_2} = 0.812,$$
$$\frac{14790u_1 + 4.1u_2}{2170v_1 + 1870v_2} = 0.829,$$
$$0.829 \le \frac{14860u_1 + 4.9u_2}{2120v_1 + 4.9u_2} \le 1,$$

$$(39)$$

& all the constraints of model M − 33.

The model (39) an be transformed into the following linear model (40).

$$\text{Maximize}\left[E_A^U = 14860u_1 + 4.9u_2\right]$$

Subject to

$$2120v_1 + 4.9u_2 = 1,$$
$$14500u_1 + 3.1u_2 = 0.812(2210v_1 + 1870v_2),$$
$$14790u_1 + 4.1u_2 = 0.829(2170v_1 + 1870v_2),$$
$$0.829(2120v_1 + 4.9u_2) \le 14860u_1 + 4.9u_2 \le 2120v_1 + 4.9u_2,$$

& all the constraints of model M − 33.

$$(40)$$

On solving the linear model (40), the obtained middle optimal value $\left(E_p^R\right)$ of the fuzzy efficiency of CCR DEA model (34) is 0.833.

Step 5:   Using the optimal values of E_A^L, E_A^M and E_A^U, obtained in Step 4, the fuzzy optimal value (best relative fuzzy efficiency of DMU$_A$ of fuzzy CCR DEA model (34) is $\left(E_A^L, E_A^M, E_A^U\right) = (0.812, 0.829, 0.833)$.

Similarly, the best relative fuzzy efficiency of the remaining DMUs can also be obtained.

9   Results

The exact best relative fuzzy efficiency of all the DMUs obtained by using the proposed method are shown in Table 2.

The DMUs are ranked in Table 3 according to their degree of preferences by using the Wang et al.'s ranking approach [26]. According to the existing ranking approach [26] it can be concluded that $B >^{51.92\%} E >^{100\%} G >^{100\%} F >^{94.07\%} H >^{80.80\%} A >^{87.23\%} C >^{99.69\%} D$.

Table 2 Exact best relative fuzzy efficiencies of DMUs

DMUs	Exact best relative fuzzy efficiency of jth DMU $(\widetilde{E}_j)$
A	(0.812, 0.829, 0.833)
B	(0.975, 1, 1)
C	(0.797, 0.815, 0.825)
D	(0.776, 0.792, 0.799)
E	(0.973, 1, 1)
F	(0.835, 0.852, 0.857)
G	(0.875, 0.893, 0.900)
H	(0.820, 0.836, 0.843)

Table 3 Degree of preference for fuzzy efficiencies and their ranking

Enterprises	A	B	C	D	E	F	G	H	Rank
A	–	0	**0.8723**	1	0	0	0	0.1920	6
B	1	–	1	1	**0.5192**	1	1	1	1
C	0.1277	0	–	**0.9969**	0	0	0	0.0188	7
D	0	0	0.0031	–	0	0	0	0	8
E	1	0.4808	1	1	–	1	**1**	1	2
F	1	0	1	1	0	–	0	**0.9407**	4
G	1	0	1	1	0	**1**	–	1	3
H	**0.8080**	0	0.9812	1	0	0.0593	0	–	5

10 Conclusions

On the basis the present study, it can be concluded that there are flaws in the fuzzy CCR DEA model as well as in the method, proposed by Wang et al. [26], and hence neither the fuzzy CCR DEA model nor the method, proposed by Wang et al. [26], should be used for evaluating the best relative fuzzy efficiency of DMUs. Also, to resolve the flaws of the fuzzy CCR DEA model, a new fuzzy CCR model is proposed. Further, a new method is proposed to solve the proposed fuzzy CCR DEA model for evaluating the best relative fuzzy efficiency of DMUs and the exact best relative fuzzy efficiency of the DMUs, considered by Wang et al. [26], is evaluated by using the proposed method.

Acknowledgements Dr. Amit Kumar would like to acknowledge the adolescent inner blessings of Mehar (lovely daughter of his cousin sister Dr. Parmpreet Kaur). Dr. Amit Kumar believes that Mata Vaishno Devi has appeared on the earth in the form of Mehar and without her blessings it would not be possible to think the ideas presented in this chapter. The second author would like to acknowledge the financial support given by UGC under the UGC Dr. D.S. Kothari Postdoctoral Fellowship Scheme.

References

1. Angiz L, Emrouznejad MZ, Mustafa A (2012) Fuzzy data envelopment analysis: a discrete approach. Expert Syst Appl 39:2263–2269
2. Banker RD, Charnes A, Cooper WW (1984) Some models for estimating technical and scale inefficiency in data envelopment analysis. Manage Sci 30:1078–1092
3. Charnes A, Cooper WW, Rhodes E (1978) Measuring the efficiency of decision making units. Eur J Oper Res 2:429–444
4. Dia, M. (2004). A model of fuzzy data envelopment analysis. INFOR: Inf Syst Oper Res 42:267–279
5. Garcia PAA, Schirru R, Melo PFFE (2005) A fuzzy data envelopment analysis approach for FMEA. Prog Nucl Energy 46:35–373
6. Guo P, Tanaka H (2001) Fuzzy DEA: A perceptual evaluation method. Fuzzy Sets Syst 119:149–160
7. Jahanshahloo GR, Soleimani-damaneh M, Nasrabadi E (2004) Measure of efficiency in DEA with fuzzy input–output levels: A methodology for assessing, ranking and imposing of weights restrictions. Appl Math Comput 156:175–187
8. Kao C, Liu ST (2000) Fuzzy efficiency measures in data envelopment analysis. Fuzzy Sets Syst 113:427–437
9. Kao C, Liu ST (2000) Data envelopment analysis with missing data: an application to University libraries in Taiwan. J Oper Res Soc 51:897–905
10. Kao C, Liu ST (2003) A mathematical programming approach to fuzzy efficiency ranking. Int J Prod Econ 86:45–154
11. Kao C, Liu ST (2005) Data envelopment analysis with imprecise data: an application of Taiwan machinery firms. Int J Uncertain Fuzziness Knowl-Based Syst 13:225–240
12. León T, Liern V, Ruiz JL, Sirvent I (2003) A fuzzy mathematical programming approach to the assessment of efficiency with DEA models. Fuzzy Sets Syst 139:407–419
13. Lertworasirikul S, Fang SC, Joines JA, Nuttle HLW (2003) Fuzzy data envelopment analysis (DEA): A possibility approach. Fuzzy Sets Syst 139:379–394

14. Lertworasirikul S, Fang SC, Joines JA, Nuttle HLW (2003) Fuzzy data envelopment analysis: a credibility approach. In: Verdegay JL (ed) Fuzzy sets based heuristics for optimization. Springer, Berlin, pp 141–158
15. Lertworasirikul S, Fang SC, Nuttle HLW, Joines JA (2003) Fuzzy BCC model for data envelopment analysis. Fuzzy Optim Decis Making 2:337–358
16. Liu ST (2008) A fuzzy DEA/AR approach to the selection of flexible manufacturing systems. Comput Ind Eng 54:66–76
17. Liu ST, Chuang M (2009) Fuzzy efficiency measures in fuzzy DEA/AR with application to university libraries. Expert Syst Appl 36:1105–1113
18. Moheb-Alizadeh H, Rasouli SM, Tavakkoli-Moghaddam R (2011) The use of multi-criteria data envelopment analysis (MCDEA) for location-allocation problems in a fuzzy environment. Expert Syst Appl 38:5687–5695
19. Puri J, Yadav SP (2013) A concept of fuzzy input mix-efficiency in fuzzy DEA and its application in banking sector. Expert Syst Appl 40:1437–1450
20. Puri J, Yadav SP (2015) Intuitionistic fuzzy data envelopment analysis: an application to the banking sector in India. Expert Syst Appl 42:4982–4998
21. Saati S, Memariani A (2005) Reducing weight flexibility in fuzzy DEA. Appl Math Comput 161:611–622
22. Saati S, Menariani A, Jahanshahloo GR (2002) Efficiency analysis and ranking of DMUs with fuzzy data. Fuzzy Optim Decis Making 1:255–267
23. Sengupta JK (1992) A fuzzy systems approach in data envelopment analysis. Comput Math Appl 24:259–266
24. Triantis K (2003) Fuzzy non-radial data envelopment analysis (DEA) measures of technical efficiency in support of an integrated performance measurement system. Int J Automot Technol Manag 3:328–353
25. Triantis K, Girod O (1998) A mathematical programming approach for measuring technical efficiency in a fuzzy environment. J Prod Anal 10:85–102
26. Wang YM, Luo Y, Liang L (2009) Fuzzy data envelopment analysis based upon fuzzy arithmetic with an application to performance assessment of manufacturing enterprises. Expert Syst Appl 36:5205–5211
27. Wu D, Yang Z, Liang L (2006) Efficiency analysis of cross-region bank branches using fuzzy data envelopment analysis. Appl Math Comput 181:271–281
28. Zerafat Angiz LM, Emrouznejad A, Mustafa A (2012) Fuzzy data envelopment analysis: a discrete approach. Expert Syst Appl 39:2263–2269
29. Zimmermann HZ (1996) Fuzzy set theory and its applications, 3rd edn. Kluwer Nijhoff, Boston

Part V
Multiperson Decision Making and Consensus Reaching

A Feedback Mechanism Based on Granular Computing to Improve Consensus in GDM

Francisco Javier Cabrerizo, Francisco Chiclana, Ignacio Javier Pérez, Francisco Mata, Sergio Alonso and Enrique Herrera-Viedma

Abstract Group decision making is an important task in real world activities. It consists in obtaining the best solution to a particular problem according to the opinions given by a set of decision makers. In such a situation, an important issue is the level of consensus achieved among the decision makers before making a decision. For this reason, different feedback mechanisms, which help decision makers for reaching the highest degree of consensus possible, have been proposed in the literature. In this contribution, we present a new feedback mechanism based on granular computing to improve consensus in group decision making problems. Granular computing is a framework of designing, processing, and interpretation of information granules, which can be used to obtain a required flexibility to improve the level of consensus within the group of decision makers.

Keywords Group decision making · Consensus · Feedback mechanism · Granular computing

F.J. Cabrerizo (✉) · E. Herrera-Viedma
Department of Computer Science and Artificial Intelligence, University of Granada,
C/ Periodista Daniel Saucedo Aranda s/n, 18071 Granada, Spain
e-mail: cabrerizo@decsai.ugr.es

E. Herrera-Viedma
e-mail: viedma@decsai.ugr.es

F. Chiclana
Faculty of Technology, De Montfort University, Leicester, England
e-mail: chiclana@dmu.ac.uk

I.J. Pérez
Department of Computer Sciences and Engineering, University of Cádiz, Cádiz, Spain
e-mail: ignaciojavier.perez@uca.es

F. Mata
Department of Computer Science, University of Jaén, Jaén, Spain
e-mail: fmata@ujaen.es

S. Alonso
Department of Software Engineering, University of Granada, Granada, Spain
e-mail: zerjioi@ugr.es

© Springer International Publishing AG 2018   371
M. Collan and J. Kacprzyk (eds.), *Soft Computing Applications for Group Decision-making and Consensus Modeling*, Studies in Fuzziness and Soft Computing 357, DOI 10.1007/978-3-319-60207-3_22

1 Introduction

Group decision making (GDM) is utilized to get the best solution or solutions for a given problem using the preferences or opinions expressed by a group of decision makers [11, 18, 44]. In such a situation, each decision maker usually approaches the decision process from a different point of view. However, the decision makers have a common interest in obtaining a consensus or agreement before making the decision. In particular, in a GDM situation, there is a set of different alternatives to solve the problem and a group of decision makers that are usually required to express their opinions about the alternatives by means of a particular preference structure [13, 16].

An important issue in a GDM situation is the level of consensus achieved among the decision makers before making the decision. Usually, when decisions are made by a group of individuals, it is advisable that the decision makers are involved in a discussion process in which they talk about their reasons for making decisions with the aim of arriving at a sufficient level of consensus acceptable to all [6, 27]. If this discussion process is not carried out, solutions which are not well accepted by some decision makers could be obtained [6, 45], and therefore the decision makers might reject them. Due to it, a consensus process is usually carried out before obtaining a final solution in a GDM situation [1, 10, 15, 17, 27, 35, 51].

In a consensus process, an important step is the recommendations provided to the decision makers to improve the level of consensus. From this point of view, the first consensus approaches presented by the researchers of the GDM field can be considered as basic approaches because they are based on a moderator who gives the advice to the decision makers [5, 19, 20, 28–30]. The objective of the moderator in each discussion round is to address the consensus process towards success by achieving the highest consensus degree and reducing the number of decision makers outside of the agreement. However, a drawback of these approaches is that the moderator can introduce some subjectivity in the discussion process. To overcome it, new consensus approaches have been presented by providing to the moderator with better analysis tools or substituting the moderator figure. It makes more effective and efficient the discussion process.

In consensus approaches incorporating a feedback mechanism, which substitutes the moderator's actions, proximity measures are computed to evaluate the proximity between individual decision makers' opinions and the collective one [7, 22, 24, 25, 49]. These proximity measures are utilized to identify the opinions given by the decision makers which are contributing less to reach a high consensus level. The goal of the feedback mechanism is to give advice to those decision makers to find out the modifications they need to make in their preferences to achieve a solution with better consensus.

On the other hand, a novel data mining tool [31], the so called action rules [37], has been incorporated in consensus approaches to support and stimulate the discussion in the group. The aim of an action rule is to show how a subset of flexible attributes should be modified to achieve an expected change of the decision attribute for a subset of objects characterized by some values of the subset of stable attributes. In such a way, these action rules are utilized to suggest and indicate to the moderator with which decision makers and with respect to which preferences it may be expedient to deal.

In any case, decisions makers have to allow a certain degree of flexibility and be ready to make changes on their first opinions to obtain a higher level of consensus. In such a situation, information granularity [38, 40, 41, 50] may become relevant because it gives to the decision makers a level of flexibility using some first opinions that can be adjusted in order to improve the consensus level among the decision makers.

The objective of this contribution is to develop a new feedback mechanism based on granular computing to improve the consensus achieved among the decision makers in a GDM situation. Granular computing is a paradigm that represents and processes information in form of information granules [2, 38], that are complex information entities arising in the process of abstraction of data and derivation of knowledge from information [4]. In particular, an allocation of information granularity is used in the feedback mechanism as a key component to suggest advice to the decision makers in order to improve the consensus.

This contribution is organized as follows. In Sect. 2, we introduce the description of a GDM situation and describe the process carried out to solve it. Section 3 presents the feedback mechanism based on granular computing proposed here to improve the consensus achieve among the decision makers involved in a GDM situation. An example of application of the feedback mechanism is illustrated in Sect. 4. Finally, some conclusions and future work are pointed out in Sect. 5.

2　GDM Process

A GDM process is defined as a situation in which a group of two or more decision makers, $E = \{e_1, e_2, \ldots, e_m\}$ ($m \geq 2$), provide their opinions or preferences about a solution set of possible alternatives, $X = \{x_1, x_2, \ldots, x_n\}$ ($n \geq 2$), to achieve a common solution [11, 18, 27]. In particular, if the decision process is defined in a fuzzy context, the goal is to rank the alternatives from best to worst, associating with the alternatives some degrees of preferences given in the unit interval.

In the literature we can find different representation structures in which the decision makers can convey their judgments [13, 14]. Among them, the fuzzy preference relation [34, 47, 52] has been widely utilized by the researchers because this representation structure offers a very expressive representation and, in addition, it presents good properties allowing to operate with it easily [13, 23].

Definition 1 A fuzzy preference relation PR on a set of alternatives X is a fuzzy set on the Cartesian product $X \times X$, i.e., it is characterized by a membership function $\mu_{PR} : X \times X \rightarrow [0, 1]$.

A fuzzy preference relation PR is usually represented by the $n \times n$ matrix $PR = (pr_{ij})$, being $pr_{ij} = \mu_{PR}(x_i, x_j)$ $(\forall i, j \in \{1, \ldots, n\})$ interpreted as the preference degree or intensity of the alternative x_i over x_j: $pr_{ij} = 0.5$ indicates indifference between x_i and x_j $(x_i \sim x_j)$, $pr_{ij} = 1$ indicates that x_i is absolutely preferred to x_j, and $pr_{ij} > 0.5$ indicates that x_i is preferred to x_j $(x_i > x_j)$. Based on this interpretation, we have that $pr_{ii} = 0.5$ $\forall i \in \{1, \ldots, n\}$ $(x_i \sim x_i)$. Since pr_{ii}'s (as well as the corresponding elements on the main diagonal in some other matrices) do not matter, it will be written as '–' instead of 0.5 [25, 28].

GDM processes are usually faced by carrying out two processes before a final solution can be provided [1, 30]:

- A consensus process referring to how to get the highest degree of agreement among the decision makers.
- A selection process obtaining the final solution using the opinions expressed by the group of decision makers.

In the following subsections, both the consensus process and the selection process are described in detail.

2.1 Consensus Process

A consensus process is an iterative and a dynamic discussion process carried out among the members of a group, coordinated by a moderator who helps them bring their preferences closer. On the one hand, if the agreement among the decision makers is lower than a threshold, the moderator would urge them to discuss their preferences further in an effort to bring them closer. On the other hand, if the consensus level is higher than the threshold, the moderator would apply the selection process with the aim of obtaining the final consensus solution to the problem [27, 36].

An important step of a consensus process is the assessment of the agreement achieved among the group of decision makers. To obtain it, coincidence existing among the decision makers is computed [8, 21]. Consensus approaches usually obtain consensus degrees, utilized to evaluate the current level of agreement among the decision makers' preferences, given at three different levels of a fuzzy preference relation [8, 19]: pairs of alternatives, alternatives, and relation. According to it, the computation of the consensus degrees is performed as follows once the fuzzy preference relations have been provided by all the decision makers within the group [8, 25, 49]:

1. For each pair of decision makers (e_k, e_l) $(k = 1, \ldots, m - 1, \ l = k + 1, \ldots, m)$ a similarity matrix, $SM^{kl} = (sm_{ij}^{kl})$, is defined as:

$$sm_{ij}^{kl} = 1 - |pr_{ij}^k - pr_{ij}^l| \tag{1}$$

2. Then, a consensus matrix, $CM = (cm_{ij})$, is calculated by aggregating all the $(m-1) \times (m-2)$ similarity matrices using an aggregation function, ϕ:

$$cm_{ij} = \phi(sm_{ij}^{kl}), \; k = 1, \ldots, m-1, \; l = k+1, \ldots, m \qquad (2)$$

Here, the arithmetic mean is utilized as aggregation function. However, different aggregation operators could be utilized according to the particular properties that we want to implement.

3. Once the consensus matrix has been calculated, the consensus degrees are obtained at the three different levels of a fuzzy preference relation:

   a. *Consensus degree on the pairs of alternatives.* The consensus degree on a pair of alternatives (x_i, x_j), called cp_{ij}, is defined to measure the consensus degree among all the decision makers on that pair of alternatives. In this case, this is expressed by the element of the collective similarity matrix CM:

   $$cp_{ij} = cm_{ij} \qquad (3)$$

   The closer cp_{ij} to 1, the greater the agreement among all the decision makers on the pair of alternatives (x_i, x_j).

   b. *Consensus degree on the alternatives.* The consensus degree on the alternative x_i, called ca_i, is defined to measure the consensus degree among all the decision makers on that alternative:

   $$ca_i = \frac{\sum_{j=1;j\neq i}^{n}(cp_{ij} + cp_{ji})}{2(n-1)} \qquad (4)$$

   c. *Consensus degree on the relation.* The consensus degree on the relation, called cr, expresses the global consensus degree among all the decision makers' opinions. It is computed as the average of all the consensus degree for the alternatives:

   $$cr = \frac{\sum_{i=1}^{n} ca_i}{n} \qquad (5)$$

The consensus degree of the relation, cr, is the value used to control the consensus state. The closer cr is to 1, the greater the agreement among all the decision makers' preferences.

2.2 Selection Process

Once the consensus level is higher than a specified threshold, the selection process is carried out in two sequential steps:

1. Aggregation step defining a collective fuzzy preference relation that indicates the global preference between every pair of alternatives.
2. Exploitation step transforming the global information about the alternatives into a global ranking of them, from which a set of alternatives is derived.

In what follows, we present in more detail both the aggregation step and the exploitation step of a selection process.

2.2.1 Aggregation Step

The aim of this step is to obtain a collective fuzzy preference relation, $PR^c = (pr_{ij}^c)$, by aggregating all individual fuzzy preference relations, $\{PR^1, \ldots, PR^m\}$, given by the decision makers involved in the problem. Each value pr_{ij}^c represents the preference of the alternative x_i over the alternative x_j according to the majority of the decision makers' assessments. To do so, an OWA operator is used [53].

Definition 2 An OWA operator of dimension n is a function $\phi : [0, 1]^n \longrightarrow [0, 1]$, that has a weighting vector associated with it, $W = (w_1, \ldots, w_n)$, with $w_i \in [0, 1]$, $\sum_{i=1}^{n} w_i = 1$, and it is defined according to the following expression:

$$\phi_W(a_1, \ldots, a_n) = W \cdot B^T = \sum_{i=1}^{m} w_i \cdot a_{\sigma(i)} \tag{6}$$

being $\sigma : \{1, \ldots, n\} \longrightarrow \{1, \ldots, n\}$ a permutation such that $p_{\sigma(i)} \geq a_{\sigma(i+1)}$, $\forall i = 1, \ldots, n-1$, i.e., $a_{\sigma(i)}$ is the i-highest value in the set $\{a_1, \ldots, a_n\}$.

OWA operators fill the gap between the operators Min and Max. It can be immediately verified that OWA operators are commutative, increasing monotonous and idempotent, but in general not associative.

In order to classify OWA aggregation operators with regards to their localization between "or" and "and", Yager [53] introduced the measure of *orness* associated with any vector W expressed as:

$$\text{orness}(W) = \frac{1}{n-1} \sum_{i=1}^{n} (n-i)w_i \tag{7}$$

This measure, which lies in the unit interval, characterizes the degree to which the aggregation is like an "or" (Max) operation. Note that the nearer W is to an "or", the closer its measure is to one; while the nearer it is to an "and", the closer is to zero. As we move weight up the vector we increase the orness(W), while moving weight down causes us to decrease orness(W). Therefore, an OWA operator with much of nonzero weights near the top will be an "orlike" operator (orness(W) ≥ 0.5), and when much of the weights are nonzero near the bottom, the OWA operator will be "andlike" (orness(W) < 0.5).

A natural question in the definition of the OWA operator is how to obtain the associated weighting vector. In [53], it was defined an expression to obtain W that allows to represent the concept of fuzzy majority [28] by means of a fuzzy linguistic non-decreasing quantifier Q [58]:

$$w_i = Q\left(\frac{i}{n}\right) - Q\left(\frac{i-1}{n}\right), \quad i = 1, \ldots, n \tag{8}$$

The membership function of Q is given by Eq. (9), with $a, b, r \in [0, 1]$. Some examples of non-decreasing proportional fuzzy linguistic quantifiers are: "most" (0.3, 0.8), "at least half" (0, 0.5), and "as many as possible" (0.5, 1).

$$Q(r) = \begin{cases} 0 & \text{if } r < a \\ \frac{r-a}{b-a} & \text{if } a \leq r \leq b \\ 1 & \text{if } r > a \end{cases} \tag{9}$$

When a fuzzy quantifier Q is used to compute the weights of the OWA operator ϕ, it is symbolized by ϕ_Q.

2.2.2 Exploitation Step

The aim of this step is to obtain a rank of the alternatives. To do so, the concept of fuzzy majority (of alternatives) and the OWA operator are used to compute two choice degrees of alternatives: the quantifier-guided dominance degree ($QGDD$) and the quantifier-guided non-dominance degree ($QGNDD$) [9, 26]. They will act over the collective preference relation resulting in a global ranking of the alternatives, from which the solution will be obtained.

- $QGDD_i$: It quantifies the dominance that one alternative has over all the others in a fuzzy majority sense. It is obtained as follows:

$$QGDD_i = \phi_Q(pr^c_{i1}, pr^c_{i2}, \ldots, pr^c_{i(i-1)}, pr^c_{i(i+1)}, \ldots, pr^c_{in}) \tag{10}$$

- $QGNDD_i$: It gives the degree in which each alternative is not dominated by a fuzzy majority of the remaining alternatives. It is obtained as follows:

$$QGNDD_i = \phi_Q(1 - p^s_{1i}, 1 - p^s_{2i}, \ldots, 1 - p^s_{(i-1)i}, 1 - p^s_{(i+1)i}, \ldots, 1 - p^s_{ni}) \tag{11}$$

where $p^s_{ji} = max\{pr^c_{ji} - pr^c_{ij}, 0\}$ represents the degree in which x_i is strictly dominated by x_j. When the fuzzy quantifier represents the statement "all", whose algebraic aggregation corresponds to the conjunction operator Min, this non-dominance degree coincides with Orlovski's non-dominated alternative concept [34].

Two different policies can be used to carry out the application of both choice degrees: a sequential policy or a conjunctive policy [12, 26]. In the sequential policy, one of the choice degrees is selected and applied to the set of alternatives according to the opinions given by the decision makers, obtaining a selection set of alternatives. If there is more than one alternative in this selection set, then, the other choice degree is applied to select the alternative of this set with the best second choice degree. In the conjunctive policy, both choice degrees are applied to the set of alternatives, obtaining two selection sets of alternatives. The final selection set of alternatives is obtained as the intersection of these two selection sets of alternatives. As it is possible to get and empty selection set, the latter conjunction selection process is more restrictive than the former sequential selection process.

3   A Feedback Mechanism Based on Granular Computing

In the discussion process, if the consensus achieved among the decision makers is lower than a consensus threshold, the decision makers must discuss and modify their opinions. It is done by a feedback mechanism, which gives advice to the decision makers on how to change their preferences in order to increase the consensus. In addition, the feedback mechanism usually substitutes the moderators' actions with the aim of avoiding the subjectivity that the moderator can introduce in the discussion process.

In order to improve the consensus, the decision makers have to accept some modifications in their initial preferences by allowing a certain flexibility. If fuzzy preference relations are used to represent the assessments provided by the decision makers, this flexibility could be brought by allowing the fuzzy preference relations to be granular rather than numeric. That is, the feedback mechanism proposed here assumes that the entries of a fuzzy preference relation are information granules instead of plain numbers. In such a way, the feedback mechanism elevates the fuzzy preference relations to their granular format.

To emphasize that the feedback mechanism uses granular fuzzy preference relations, the notation $G(PR)$ is employed. Here, $G(.)$ represents a specific granular formalism being utilized. For example, as information granules we could use fuzzy sets [54–57], rough sets [46], probability density functions [59], intervals [3], and others. In particular, information granularity is used here by the feedback mechanism as an important computational and conceptual resource being exploited as a means to give advice to the decision makers in order to improve the consensus among them. That is, granularity is used as synonymous of flexibility. It facilitates the increase of the agreement achieved among the group of decision makers.

In this contribution, the feedback mechanism uses intervals to articulate the granularity of information. Therefore, the length of the intervals can be sought as a level of granularity α. In addition, because interval-valued fuzzy preference relations are used, $G(PR) = P(PR)$, where $P(.)$ denotes a family of intervals.

The concept of interval-valued fuzzy preference relations is employed by the feedback mechanism to generate recommendations to the decision makers in order to improve the consensus among them. Specifically, the level of consensus achieved among the decision makers is used as a performance index.

In what follows, we give the details both the performance index to be optimized and its optimization, which, given the nature of the required task, is carried out by the Particle Swarm Optimization (PSO) framework [32].

3.1 The Performance Index

The level of granularity is used by the feedback mechanism to improve the agreement achieved among the decision makers by generating recommendations in order to bring all preferences close to each other. Decision makers should feel comfortable when accepting the modifications provided by the feedback mechanism located within the bounds established by the fixed level of granularity α.

Advice is generated by the feedback mechanism by maximizing the global consensus degree among the decision makers. It is calculated in term of the consensus degree on the relation (see Sect. 2.1):

$$O = cr \tag{12}$$

The optimization problem reads as follows:

$$\text{Max}_{PR^1, PR^2, \dots, PR^m \in P(PR)} O \tag{13}$$

This maximization problem is performed by the feedback mechanism for all interval-valued fuzzy preference relations that are possible according to the fixed level of information granularity α. This truth is emphasized by incorporating the granular form of the fuzzy preference relations, that is, $PR^1, \dots, PR^m$, are elements of the family of interval-valued fuzzy preference relations, $P(PR)$.

Due to the nature of the not straight relationship between the optimized fuzzy preference relations, this optimization problem is not an easy task. The optimized fuzzy preference relations are chosen from a quite large search space formed by $P(PR)$ and, therefore, it requires the use of an advanced technique of global optimization.

Among the different techniques of global optimization, the PSO framework [32] is used in this contribution because it does not come with a prohibitively high level of computational overhead as this is the case of other global optimization techniques and it offers a substantial level of optimization flexibility, being a viable alternative for this problem. However, it should be noted that other techniques as genetic algorithms, evolutionary optimization, simulated annealing, and so on, could be also used.

3.2   PSO Framework

As aforementioned, the PSO environment is employed in the feedback mechanism to optimize the fuzzy preference relations coming from the space of interval-valued fuzzy preference relations because this technique is a viable alternative for the problem at hand.

PSO is an evolutionary computational method based on the social behavior metaphor, which was developed by Kennedy and Eberhart [32, 33]. In this technique, a population of random candidate solutions, called particles, is initialized. Then, a randomized velocity is assigned to each particle, which is iteratively moved though the search-space according to simple mathematical formulae over the particle's velocity and position. The movement of each particle is attracted towards the position of the best fitness achieved so far by the particle itself (z_p) and by the position of the best fitness achieved so far across the whole population (z_g) [32, 48] (see Fig. 1).

An important issue in the PSO framework is how to find a suitable mapping between the representation of the particle and the problem solution. In a GDM context, each particle represents a vector in which the elements are located in the unit interval. That is, if the GDM problem is set up with a group of m decision makers and a set of n alternatives, the number of elements of the particle will be $m \cdot n(n-1)$.

Let us consider an element pr_{ij} and assume a level of granularity α located in the $[0, 1]$ interval. If we use an initial fuzzy preference relation expressed by a decision maker, the interval of admissible values of this element of $P(PR)$ is equal to:

$$[a, b] = [\max(0, pr_{ij} - \alpha/2), \min(1, pr_{ij} + \alpha/2)] \tag{14}$$

As an example, if we have $pr_{ij} = 0.8$, being the level of granularity α equal to 0.2 and the corresponding element of the particle x equal to 0.3, then, the corresponding interval of the interval-valued fuzzy preference relation calculated using Eq. (14) is $[a, b] = [0.70, 0.90]$. Using the expression $z = a + (b - a)x$, the modified value of pr_{ij} becomes equal to 0.76.

Another important question in the PSO framework is how to assess the performance of each particle during its movement. To do so, a performance index or fitness function is used. In the GDM context considered in this contribution, the PSO aims to maximize the level of agreement achieved among the decision makers involved in the problem. Hence, the following fitness function f will be used:

$$f = O \tag{15}$$

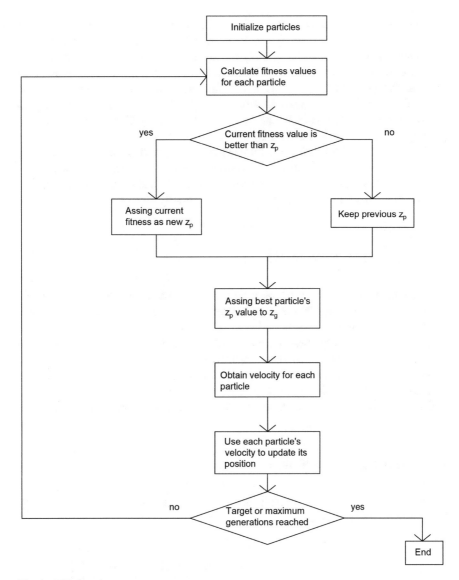

Fig. 1 PSO flowchart

where O is the optimization criterion presented previously. Here, the higher the value of f is, the better the particle is.

It should be pointed out that the generic form of the PSO framework is employed in this contribution. Therefore, the updates of the velocity of a particle are performed in the form $\mathbf{v}(t + 1) = w \times \mathbf{v}(t) + c_1\mathbf{a} \cdot (\mathbf{z}_p - \mathbf{z}) + c_2\mathbf{b} \cdot (\mathbf{z}_g - \mathbf{z})$. Here, $\cdot$ means a vector multiplication carried out coordinate-wise, "t" is an index of the generation, $\mathbf{z}_g$

denotes the best position overall and developed so far across the swarm, and z_p is the best position obtained so far for the particle under study. The inertia component, called w, scales the actual velocity $v(t)$ and stresses some effect of resistance to modify the actual velocity. Its value is usually 0.2 and it is kept constant through the process [39]. On the other hand, a and b represents vectors of random numbers that are drawn from the uniform distribution over the unit interval. These vectors help from a proper mix of the components of the velocity. Finally, in iteration "t + 1", the particle's position is calculated as: $z(t + 1) = z(t) + v(t + 1)$.

Once the PSO algorithm has optimized the fuzzy preference relations coming from the space of interval-valued fuzzy preference relation, the feedback mechanism advise the decision makers the modifications that they should put into practice in their opinions in order to improve the consensus among them.

4   An Illustrative Example

An example of application of the proposed feedback mechanism is presented in this section. It helps quantifying the improvement of the consensus when the feedback mechanism is applied.

Let us suppose that a patient presents some symptoms, being all of them common to several diseases, and some doctors, who are specialist in different diagnosis, have to jointly diagnose the disease that the patient has contracted. This situation can be defined as a GDM problem in which there are a set of four possible diseases (alternatives), $\{x_1, x_2, x_3, x_4\}$, and a set of four doctors (decision makers), $\{e_1, e_2, e_3, e_4\}$.

4.1   First Consensus Round

At the first stage of the discussion process, the four doctors express the following fuzzy preference relations:

$$PR^1 = \begin{pmatrix} - & 0.30 & 0.70 & 0.50 \\ 0.70 & - & 0.70 & 0.60 \\ 0.40 & 0.20 & - & 0.30 \\ 0.70 & 0.30 & 0.80 & - \end{pmatrix} \quad PR^2 = \begin{pmatrix} - & 0.30 & 0.60 & 0.70 \\ 0.80 & - & 0.70 & 0.20 \\ 0.20 & 0.40 & - & 0.50 \\ 0.20 & 0.60 & 0.50 & - \end{pmatrix}$$

$$PR^3 = \begin{pmatrix} - & 0.80 & 0.50 & 0.20 \\ 0.20 & - & 0.60 & 0.90 \\ 0.50 & 0.30 & - & 0.70 \\ 0.60 & 0.20 & 0.20 & - \end{pmatrix} \quad PR^4 = \begin{pmatrix} - & 0.90 & 0.20 & 0.70 \\ 0.30 & - & 0.60 & 0.30 \\ 0.90 & 0.40 & - & 0.50 \\ 0.40 & 0.90 & 0.50 & - \end{pmatrix}$$

4.1.1 Consensus Measures

Once the doctors have provided their opinions, the consensus measures are calculated as described in Sect. 2.1.

The consensus matrix is equal to:

$$CM = \begin{pmatrix} - & 0.62 & 0.73 & 0.72 \\ 0.63 & - & 0.93 & 0.60 \\ 0.63 & 0.88 & - & 0.80 \\ 0.72 & 0.60 & 0.70 & - \end{pmatrix}$$

The element (i, j) of the consensus matrix represents the consensus degrees on the pair of alternatives (x_i, x_j).

The consensus degrees on the alternatives are:

$$ca_1 = 0.67$$
$$ca_2 = 0.71$$
$$ca_3 = 0.78$$
$$ca_4 = 0.69$$

And the consensus on the relation is:

$$cr = 0.71$$

Assuming a minimum consensus threshold equal to 0.75, the selection process cannot be applied because the consensus achieved among the doctors is lower than the minimum consensus threshold. Therefore, the feedback mechanism has to be applied in order to improve the agreement.

4.1.2 Feedback Mechanism

The aim of the feedback mechanism is to support the doctors' changes in their fuzzy preference relations in order to increase the consensus.

First, it should be pointed out that, as a result of an intensive experimentation, the following values of the parameters were selected in the PSO algorithm:

- 50 particles formed the swarm. This value was found to obtain stable results. That is, identical or similar results were obtained in successive runs of the PSO algorithm.
- 200 generations or iterations were carried out as it was observed that were no further modifications of the values of the fitness functions after this number of iterations.
- c_1 and c_2 were set as 2 because these values are commonly found in the existing literature.

Considering a given level of granularity $\alpha = 0.4$, the recommended fuzzy preference relations generated by the feedback mechanism are as follows:

$$PR^1 = \begin{pmatrix} - & 0.31 & 0.60 & 0.57 \\ 0.63 & - & 0.58 & 0.63 \\ 0.35 & 0.22 & - & 0.31 \\ 0.73 & 0.40 & 0.80 & - \end{pmatrix} \quad PR^2 = \begin{pmatrix} - & 0.33 & 0.61 & 0.69 \\ 0.85 & - & 0.60 & 0.27 \\ 0.26 & 0.39 & - & 0.50 \\ 0.15 & 0.61 & 0.50 & - \end{pmatrix}$$

$$PR^3 = \begin{pmatrix} - & 0.78 & 0.45 & 0.13 \\ 0.19 & - & 0.49 & 0.78 \\ 0.46 & 0.35 & - & 0.65 \\ 0.58 & 0.13 & 0.30 & - \end{pmatrix} \quad PR^4 = \begin{pmatrix} - & 0.88 & 0.20 & 0.66 \\ 0.25 & - & 0.65 & 0.36 \\ 0.81 & 0.53 & - & 0.46 \\ 0.48 & 0.86 & 0.50 & - \end{pmatrix}$$

4.2 Second Consensus Round

In the second consensus round, we assume that the doctors agree the advice generated by the feedback mechanism. Then, the consensus measures are computed again.

4.2.1 Consensus Measures

The consensus matrix is equal to:

$$CM = \begin{pmatrix} - & 0.64 & 0.77 & 0.71 \\ 0.60 & - & 0.92 & 0.70 \\ 0.71 & 0.84 & - & 0.82 \\ 0.69 & 0.60 & 0.75 & - \end{pmatrix}$$

The element (i, j) of the consensus matrix represents the consensus degrees on the pair of alternatives (x_i, x_j).

The consensus degrees on the alternatives are:

$$ca_1 = 0.68$$
$$ca_2 = 0.72$$
$$ca_3 = 0.80$$
$$ca_4 = 0.71$$

And the consensus on the relation is:

$$cr = 0.73$$

Because the consensus achieved among the doctors is lower than the minimum consensus threshold, the feedback mechanism has to be applied again in order to increase the consensus.

4.2.2 Feedback Mechanism

Considering the same values of the parameters in the PSO algorithm as in the first round, the new recommended fuzzy preference relations provided by the feedback mechanism are the following:

$$PR^1 = \begin{pmatrix} - & 0.31 & 0.60 & 0.57 \\ 0.63 & - & 0.58 & 0.63 \\ 0.35 & 0.22 & - & 0.31 \\ 0.73 & 0.40 & 0.80 & - \end{pmatrix} \quad PR^2 = \begin{pmatrix} - & 0.33 & 0.61 & 0.69 \\ 0.85 & - & 0.60 & 0.27 \\ 0.26 & 0.39 & - & 0.50 \\ 0.15 & 0.61 & 0.50 & - \end{pmatrix}$$

$$PR^3 = \begin{pmatrix} - & 0.78 & 0.45 & 0.13 \\ 0.19 & - & 0.49 & 0.78 \\ 0.46 & 0.35 & - & 0.65 \\ 0.58 & 0.13 & 0.30 & - \end{pmatrix} \quad PR^4 = \begin{pmatrix} - & 0.88 & 0.20 & 0.66 \\ 0.25 & - & 0.65 & 0.36 \\ 0.81 & 0.53 & - & 0.46 \\ 0.48 & 0.86 & 0.50 & - \end{pmatrix}$$

4.3 Third Consensus Round

As in the above round, it is assumed that the doctors accept the preferences generated by the feedback mechanism and, therefore, the consensus measures are obtained again.

4.3.1 Consensus Measures

The consensus matrix is equal to:

$$CM = \begin{pmatrix} - & 0.65 & 0.78 & 0.71 \\ 0.60 & - & 0.90 & 0.82 \\ 0.68 & 0.84 & - & 0.82 \\ 0.69 & 0.64 & 0.78 & - \end{pmatrix}$$

The element (i, j) of the consensus matrix represents the consensus degrees on the pair of alternatives (x_i, x_j).

The consensus degrees on the alternatives are:

$$ca_1 = 0.69$$
$$ca_2 = 0.75$$
$$ca_3 = 0.81$$
$$ca_4 = 0.75$$

And the consensus on the relation is:

$$cr = 0.75$$

In this round, the consensus is equal to the minimum consensus threshold and, therefore, the selection process can be applied in order to rank the alternatives.

4.4 Selection Process

The goal of the selection process is to obtain a ranking of the alternatives from best to worst according to the preferences given by the doctors. To do so, an aggregation step and an exploitation step are carried out.

4.4.1 Aggregation

The OWA operator is used to aggregation the fuzzy preference relations given by the doctors. We make use of the linguistic quantifier "most", defined in Sect. 2.2.1, which, applying Eq. (8), generates a weighting vector of four values to obtain each collective preference value pr_{ij}^c. As example, the collective preference value pr_{12}^c is computed as follows:

$$w_1 = Q(1/4) - Q(0) = 0 - 0 = 0$$
$$w_2 = Q(2/4) - Q(1/4) = 0.4 - 0 = 0.4$$
$$w_3 = Q(3/4) - Q(2/4) = 0.9 - 0.4 = 0.5$$
$$w_4 = Q(1) - Q(3/4) = 1 - 0.9 = 0.1$$
$$pr_{12}^c = w_1 \cdot pr_{12}^4 + w_2 \cdot pr_{12}^3 + w_3 \cdot pr_{12}^2 + w_4 \cdot pr_{12}^1 = 0.51$$

Then, the collective fuzzy preference relation is:

$$PR^c = \begin{pmatrix} - & 0.51 & 0.48 & 0.56 \\ 0.40 & - & 0.58 & 0.46 \\ 0.38 & 0.35 & - & 0.46 \\ 0.49 & 0.46 & 0.48 & - \end{pmatrix}$$

4.4.2 Exploitation

Using again the same linguistic quantifier "most" and Eq. (8), we obtain the following weighting vector $W = (w_1, w_2, w_3)$:

$$w_1 = Q(1/3) - Q(0) = 0.07 - 0 = 0.07$$
$$w_2 = Q(2/3) - Q(1/3) = 0.73 - 0.07 = 0.66$$
$$w_3 = Q(1) - Q(2/3) = 1 - 0.73 = 0.27$$

Using, for example, the quantifier guided dominance degree, $QGDD_i$, we obtain the following values:

$$QGDD_1 = 0.51$$
$$QGDD_2 = 0.45$$
$$QGDD_3 = 0.38$$
$$QGDD_4 = 0.47$$

Finally, applying the sequential policy with the quantifier guided dominance degree, the following ranking of alternatives is obtained:

$$x_1 \succ x_4 \succ x_2 \succ x_3$$

Therefore, according to the doctors' judgments, the patient's symptoms correspond to the first disease.

Finally, it should be pointed out that here a granularity level of 0.4 has been used. However, the higher the level of granularity is, the higher the level of flexibility is and, hence, the possibility of obtaining a higher consensus. Anyway, if the level of granularity is very high, the fuzzy preference relations generated by the feedback mechanism could be very different in comparison with those provided by the decision makers and, in such a way, they could reject them.

5 Conclusions and Future Work

In this contribution, we have presented a feedback mechanism based on granular computing to improve the consensus achieved among the decision makers in a GDM situation. The feedback mechanism assumes the concept of granular fuzzy preference relation and accentuates the role of information granularity, which is regarded as an important resource to be exploited as a means to improve the consensus achieved among the decision makers involved in the problem. In particular, the granularity level has been treated as synonymous of flexibility, which has been used to optimize a certain optimization criterion capturing the essence of reconciliation of the individual preferences. It has also been shown that the PSO environment is a suitable optimization framework for this purpose. However, it should be noted that the

PSO optimizes the fitness function but there is no guarantee that the result is optimal rather that the solution is the best one being formed by the PSO environment.

In the future, it is worth continuing this research in several directions:

- In this contribution, intervals have been used as information granules in the granular representation of the fuzzy preference relations. However, other formalism as, for instance, rough sets or fuzzy sets, could be utilized in the granular representation of the preferences.
- The feedback mechanism has been proposed in a fixed framework, that is, in a situation in which the decision makers and the alternatives do not change during the decision making process. However, with the aim of making the process more realistic, the approach should be able to deal with changeable elements. In such a way, the feedback mechanism should be able to deal with a dynamic environment [42, 43].

Acknowledgements The authors would like to acknowledge FEDER financial support from the Projects TIN2013-40658-P and TIN2016-75850-P.

References

1. Alonso S, Pérez IJ, Cabrerizo FJ, Herrera-Viedma E (2013) A linguistic consensus model for web 2.0 communities. Appl Soft Comput 13(1):149–157
2. Balamash A, Pedrycz W, Al-Hmouz R, Morfeq A (2016) An expansion of fuzzy information granules through successive refinements of their information content and their use to system modeling. Expert Syst Appl 43(6):2985–2997
3. Bargiela A (2001)Interval and ellipsoidal uncertainty models. In: Pedrycz W (ed) Granular computing: an emerging paradigm. Physica-Verlag, pp 23–57
4. Bargiela A, Pedrycz W (2003) Granular computing: an introduction. Kluwer Academic Publishers, Dordrecht
5. Bordogna G, Fedrizzi M, Pasi A (1997) A linguistic modeling of consensus in group decision making based on OWA operators. IEEE Trans Syst Man Cybern—Part A: Syst Humans 27(1):126–133
6. Butler CT, Rothstein A (2006) On conflict and consensus: a handbook on formal consensus decision making. Tahoma Park
7. Cabrerizo FJ, Pérez IJ, Herrera-Viedma E (2010) Managing the consensus in group decision making in an unbalanced fuzzy linguistic context with incomplete information. Knowl-Based Syst 23(2):169–181
8. Cabrerizo FJ, Moreno JM, Pérez IJ, Herrera-Viedma E (2010) Analyzing consensus approaches in fuzzy group decision making: advantages and drawbacks. Soft Comput 14(5):451–463
9. Cabrerizo FJ, Heradio R, Pérez IJ, Herrera-Viedma E (2010) A selection process based on additive consistency to deal with incomplete fuzzy linguistic information. J Univers Comput Sci 16(1):62–81
10. Cabrerizo FJ, Chiclana F, Al-Hmouz R, Morfeq A, Balamash AS, Herrera-Viedma E (2015) Fuzzy decision making and consensus: challenges. J Intell Fuzzy Syst 29(3):1109–1118
11. Chen SJ, Hwang CL (1992) Fuzzy multiple attributive decision making: theory and its applications. Springer, Berlin
12. Chiclana F, Herrera F, Herrera-Viedma E, Poyatos MC (1996) A classification method of alternatives of multiple preference ordering criteria based on fuzzy majority. J Fuzzy Math 4(4):801–813

13. Chiclana F, Herrera F, Herrera-Viedma E (1998) Integrating three representation models in fuzzy multipurpose decision making based on fuzzy preference relations. Fuzzy Sets Syst 97(1):33–48
14. Chiclana F, Herrera F, Herrera-Viedma E (2002) A note on the internal consistency of various preference representations. Fuzzy Sets Syst 131(1):75–78
15. Chu J, Liu X, Wang Y, Chin K-S (2016) A group decision making model considering both the additive consistency and group consensus of intuitionistic fuzzy preference relations. Comput Ind Eng 101:227–242
16. Dong Y, Zhang H (2014) Multiperson decision making with different preference representation structures: a direct consensus framework and its properties. Knowl-Based Syst 58:45–57
17. Dong Y, Xiao J, Zhang H, Wang T (2016) Managing consensus and weights in iterative multiple-attribute group decision making. Appl Soft Comput 48:80–90
18. Fodor J, Roubens M (1994) Fuzzy preference modelling and multicriteria decision support. Kluwer, Dordrecht
19. Herrera F, Herrera-Viedma E, Verdegay JL (1996) A model of consensus in group decision making under linguistic assessments. Fuzzy Sets Syst 78(1):73–87
20. Herrera F, Herrera-Viedma E, Verdegay JL (1997) A rational consensus model in group decision making using linguistic assessments. Fuzzy Sets Syst 88(1):31–49
21. Herrera F, Herrera-Viedma E, Verdegay JL (1997) Linguistic measures based on fuzzy coincidence for reaching consensus in group decision making. Int J Approx Reason 16(3–4):309–334
22. Herrera-Viedma E, Herrera F, Chiclana F (2002) A consensus model for multiperson decision making with different preference structures. IEEE Trans Syst Man Cybern—Part A: Syst Humans 32(3):394–402
23. Herrera-Viedma E, Herrera F, Chiclana F, Luque M (2004) Some issues on consistency of fuzzy preference relations. Eur J Oper Res 154(1):98–109
24. Herrera-Viedma E, Martínez L, Mata F, Chiclana F (2005) A consensus support system model for group decision-making problems with multigranular linguistic preference relations. IEEE Trans Fuzzy Syst 13(5):644–658
25. Herrera-Viedma E, Alonso S, Chiclana F, Herrera F (2007) A consensus model for group decision making with incomplete fuzzy preference relations. IEEE Trans Fuzzy Syst 15(5):863–877
26. Herrera-Viedma E, Herrera F, Alonso S (2007) Group decision-making model with incomplete fuzzy preference relations based on additive consistency. IEEE Trans Syst Man Cybern—Part B: Cybern 37(1):176–189
27. Herrera-Viedma E, Cabrerizo FJ, Kacprzyk J, Pedrycz W (2014) A review of soft consensus models in a fuzzy environment. Inf Fusion 17:4–13
28. Kacprzyk J, Fedrizzi M (1986) 'Soft' consensus measures for monitoring real consensus reaching processes under fuzzy preferences. Control Cybern 15(3–4):309–323
29. Kacprzyk J, Fedrizzi M (1988) A 'soft' measure of consensus in the setting of partial (fuzzy) preferences. Eur J Oper Res 34(3):316–325
30. Kacprzyk J, Fedrizzi M, Nurmi H (1992) Group decision making and consensus under fuzzy preferences and fuzzy majority. Fuzzy Sets Syst 49(1):21–31
31. Kacprzyk K, Zadrozny S, Ras ZW (2010) How to support consensus reaching using action rules: a novel approach. Int J Uncertain Fuzziness Knowl-Based Syst 18(4):451–470
32. Kennedy J, Eberhart RC (1995) Particle swarm optimization. In: Proceedings of the IEEE international conference on neural networks, pp 1942–1948
33. Kennedy J, Eberhart RC, Shi Y (2001) Swarm intelligence. Morgan Kaufmann Publishers, San Francisco
34. Orlovski SA (1978) Decision-making with a fuzzy preference relation. Fuzzy Sets Syst 1(3):155–167
35. Ma L-C (2016) A new group ranking approach for ordinal preferences based on group maximum consensus sequences. Eur J Oper Res 251(1):171–181
36. Palomares I, Estrella FJ, Martínez L, Herrera F (2014) Consensus under a fuzzy context: taxonomy, analysis framework AFRYCA and experimental case of study. Inf Fusion 20:252–271
37. Pawlak Z (1981) Information systems theoretical foundations. Inf Syst 6(3):205–218

38. Pedrycz W (2011) The principle of justifiable granularity and an optimization of information granularity allocation as fundamentals of granular computing. J Inf Process Syst 7(3):397–412

39. Pedrycz A, Hirota K, Pedrycz W, Dong F (2012) Granular representation and granular computing with fuzzy sets. Fuzzy Sets Syst 203:17–32

40. Pedrycz W (2013) Granular computing: analysis and design of intelligent systems. CRC Press/Francis Taylor, Boca Raton

41. Pedrycz W (2013) Knowledge management and semantic modeling: a role of information granularity. Int J Softw Eng Knowl 23(1):5–12

42. Pérez IJ, Cabrerizo FJ, Herrera-Viedma E (2010) A mobile decision support system for dynamic group decision making problems. IEEE Trans Syst Man Cybern—Part A: Syst Humans 40(6):1244–1256

43. Pérez IJ, Cabrerizo FJ, Herrera-Viedma E (2011) Group decision making problems in a linguistic and dynamic context. Expert Syst Appl 38(3):1675–1688

44. Pérez IJ, Cabrerizo FJ, Alonso S, Herrera-Viedma E (2014) A new consensus model for group decision making problems with non homogeneous experts. IEEE Trans Syst Man Cybern: Syst 44(4):494–498

45. Saint S, Lawson JR (1994) Rules for reaching consensus: a modern approach to decision making. Jossey-Bass

46. Slowinski R, Greco S, Matarazzo B (2002) Rough set analysis of preference-ordered data. In: Alpigini JJ, Peters JF, Skowron A, Zhong N (eds) Rough sets and current trends in computing, pp 44–59. Springer

47. Tanino T (1984) Fuzzy preference orderings in group decision making. Fuzzy Sets Syst 12(2):117–131

48. Trelea IC (2003) The particle swarm optimization algorithm: convergence analysis and parameter selection. Inf Process Lett 85:317–325

49. Ureña MR, Cabrerizo FJ, Morente-Molinera JA, Herrera-Viedma E (2016) GDM-R: a new framework in R to support fuzzy group decision making processes. Inf Sci 357:161–181

50. Wang X, Pedrycz W, Gacek A, Liu X (2016) From numeric data to information granules: a design through clustering and the principle of justifiable granularity. Knowl-Based Syst 101:100–113

51. Wu Z, Xu J (2016) Managing consistency and consensus in group decision making with hesitant fuzzy linguistic preference relations. Omega 65:28–40

52. Xu Y, Patnayakuni R, Wang H (2013) The ordinal consistency of a fuzzy preference relation. Inf Sci 224:152–164

53. Yager RR (1988) On ordered weighted averaging aggregation operators in multicriteria decision making. IEEE Trans Syst Man Cybern 18(1):183–190

54. Zadeh LA (1965) Fuzzy sets. Inf Control 8(3):338–353

55. Zadeh LA (1975) The concept of a linguistic variable and its applications to approximate reasoning. Part I. Inf Sci 8(3):199–243

56. Zadeh LA (1975) The concept of a linguistic variable and its applications to approximate reasoning. Part II. Inf Sci 8(4):301–357

57. Zadeh LA (1975) The concept of a linguistic variable and its applications to approximate reasoning. Part III. Inf Sci 9(1):43–80

58. Zadeh LA (1983) A computational approach to fuzzy quantifiers in natural languages. Comput Math Appl 9(1):149–184

59. Zadeh LA (2002) Toward a perception-based theory of probabilistic reasoning with imprecise probabilities. J Stat Plan Inference 105(1):233–264

A Method for the Team Selection Problem Between Two Decision-Makers Using the Ant Colony Optimization

Marilyn Bello, Rafael Bello, Ann Nowé and María M. García-Lorenzo

Abstract The team selection issue is important in the management of human resources, in which the purpose is to conduct a personnel selection process to form teams according to certain preferences. This selection problem is usually solved by ranking the candidates based on the preferences of decision-makers and allowing the decision-makers to select a candidate on its turn. While this solution method is simple and might seem fair it usually results in an unfair allocation of candidates to the different teams, i.e. the quality of the teams might be quite different according to the rankings articulated by the decision-makers. In this paper we propose a new approach to the team selection problem in which two employers should form their teams selecting personnel from a set of candidates that is common to both; each decision-maker has a personal ranking of those candidates. The objective it to make teams of high quality according to the valuation of each of the decision-makers; this results in a method for the team selection problem which not only result in high quality teams, but also focuses on a fair composition of the teams. Our approach is based on the Ant Colony Optimization metaheuristic, and allows to solve large instances of the problem as shown in the experimental section of this paper.

1 Introduction

For a company to be successful in the environment in which it is immersed, managers must make good decisions about the personnel selection, since employees can be an important source of long-term competitive advantage. Personnel selection is the process by which one or more people are chosen depending on how suitable are their characteristics for a job. It is one of the main processes of any company or

M. Bello · R. Bello (✉) · M.M. García-Lorenzo
Department of Computer Science, Universidad Central "Marta Abreu" de Las Villas, Santa Clara, Cuba
e-mail: rbellop@uclv.edu.cu

A. Nowé
Artificial Intelligence Lab, Vrije Universiteit Brussel, Brussels, Belgium

© Springer International Publishing AG 2018
M. Collan and J. Kacprzyk (eds.), *Soft Computing Applications for Group Decision-making and Consensus Modeling*, Studies in Fuzziness and Soft Computing 357, DOI 10.1007/978-3-319-60207-3_23

organization, since it provides the right workers to perform the duties of each of the positions [1–3]. The process of personnel selection determines the quality of the joining staff, therefore it plays an important role in the human resource management, and the future of the company depends largely on the contribution of its staff in order to maintain a place in the market [4–6]. Qualities such as capability, knowledge, abilities and other personal characteristics play an important role in the success of the organization, so that an important objective is to achieve a ranking of the personnel in terms of those skills.

The personnel selection is a complex process that is expected to be able to place the right employee in the right job at the right time [7]. Today, many tools and techniques are used in this specific decision-making problem [8–10]. Among the first works where the problem was presented from the perspective of the intelligent systems was reported in [11]. Extensions based on Multi-Criteria Decision-Making (MCDM) have also been proposed [12] which are relevant is the decision-maker seeks to optimize a combination of criteria associated to the candidates [13].

The MCDM approaches focus on determining overall preferences of possible alternatives. According to this goal, they can be used to rank alternatives (to build a ranking) [14–17]. To obtain a ranking of candidates is especially interesting when the management of human resources is directed to organize, manage and lead a team instead of selecting an employee for a simple vacant; this contributes to the success of the project and creates a competitive advantage for the organization [18, 19].

The MCDM methods have been applied in many studies related to personnel selection, including [16, 20–22], in which they were used to evaluate candidates from the degree to which they meet the requirements or evaluation criteria; these methods provide a model of aggregation of this information. In [23] is proposed a mathematical model for the problem of personnel selection, which generates a ranking of candidates. Well-known methods to help decision-making as TOPSIS [24], ELECTRE [15], PROMETHEE [25] and AHP [26] have been used with this problem.

In [27] is proposed a method for personnel selection based on multi-criteria decision-making, including AHP and Stochastic Dynamic Programming methods. Another solution in which AHP and fuzzy sets are combined is proposed in [26]; another example of the use of a multi-criteria decision-making method based on fuzzy sets for personnel selection is presented in [28]. An application of the method VIKOR [29] to the personnel selection problem is presented in [17]. In [30] an extension of TOPSIS for the problem of personnel selection is proposed; other solutions based on TOPSIS are proposed in [16, 24, 31]. In [15], one solution using ELECTRE is proposed. The use of PROMETHEE for preparing the ranking of candidates is shown in [14].

The personnel selection problem can also be seen as the problem of forming a team, called team selection; in this case, the problem is not to select the most suitable for a job, but to select a set of people who must act as a team. There are many factors to consider in the team selection [32]. A variety of approaches are proposed for the selection of the members of a team, most of them aimed at forming teams in the field of business, industry and sport.

Many researches in this field focus on the selection of team members, given that they have to solve common tasks collaboratively to achieve a certain objective (the so-called teamwork) [33]. The employer needs to maximize profits selecting team members from the available candidates [34]. Given the importance and complexity of the creation of a suitable team, many researches are still proposing improvements, but it remains an open problem.

In [35], the problem of team selection is treated as a multi-objective decision-making problem in the event Cricket team selection. In [36] is considered the problem of the team formation for large-scale multi-agent system. In [33] is proposed the use of fuzzy inference systems to treat the selection of players and the formation of teams as a complex multi-criteria decision-making problem with conflicting objectives. In [37], the problem of team formation is developed by building rankings through a multi-criteria decision process that uses the TOPSIS method. Heuristic search techniques, as genetic algorithm and simulated annealing, have been proposed to optimize the process of selecting members of the team [38–40].

Most researches and publications on the subject analyze the factors or indicators to be considered for the selection of the team members, and how to use the decision-making methods to form the team considering these factors so as to achieve the teamwork; in many cases the result is to generate a ranking of candidates, which works as basis for selection In this research, this problem is tried in a framework different to the classic, due to the process of selection is realized in a competitive environment, that is, when two or more decision-makers should to form their teams by choosing the personnel from the same set of candidates; and at the same time, it is necessary to form two teams as similar as possible to the preferences defined in the rankings. More precisely, the purpose is to minimize the difference between the teams formed and the set of candidates placed in the top positions of the ranking given by the decision-makers. For example; suppose a company dedicated to develop projects, which have personnel for different roles. This company has to form the work teams for two projects that will be developed, so they have to choose the personnel for each team starting from their human resources. The purpose is to achieve that both teams will be enough efficient, since the company wants develop both project successfully.

This new problem can be seen as an optimization process. In this paper two method based on the Ant Colony Optimization metaheuristic [41–44] are proposed to solve it.

2 Formulation of the Personnel Selection Problem in a Competitive Environment

In order to form the work teams two decision-makers draw up a ranking of the N candidates according to their interests. An intuitive approach is to allow each decision-maker select a candidate alternately to form a team that consists of N/2 members. To do this, she/he takes into account the order of preference and selects the next best candidate who has not been selected yet. The purpose of each decision-maker is to get a set of candidates that is as similar as possible to those in the first N/2 ranking positions defined by her/him. However, given that both decision-makers select from the same set of candidates, there will be probably a conflict of interest.

Formally, we can state the problem as follows:

Given a set of N candidates $C = \{c1, c2, ..., cN\}$, every decision-maker has to select from this set of candidates to form a team with N/2 members (without loss of generality it can be considered that N is an even number).

Two rankings $R = \{R1, R2\}$ are formed from C, taking into account different preferences to evaluate candidates from the perspective of each decision-maker. A decision-maker D1 defines a ranking of candidates $R1 = \{r_{11}, r_{12}, ..., r_{1N}\}$ and another decision-maker D2 defines another ranking $R2 = \{r_{21}, r_{22}, ..., r_{2N}\}$ ordered by a decreasing order of preference, or let be $r_{11} \geq r_{12} \geq \cdots \geq r_{1N}$ and $r_{21} \geq r_{22} \geq \cdots \geq r_{2N}$.

A first strategy is to make the selection according to the order established in the rankings R1 and R2, as shown in Example 1. In case 1, there is no conflict of interest according to the rankings R1 and R2, so that if D1 and D2 alternately choose from C according to the order of preference, the resulting sets R1* and R2* will fully satisfy the preferences of both decision-makers; while in the case 2 there is a conflict of interest and D2 cannot fully meet their preferences.

Example 1: Let be $N = 4$ candidates; $C = \{1, 2, 3, 4\}$;

Case 1: Given $R1 = \{1, 2, 3, 4\}$ and $R2 = \{4, 3, 2, 1\}$, results in $R1* = \{1, 2\}$ and $R2* = \{4, 3\}$.

Case 2: Given $R1 = \{1, 2, 4, 3\}$ and $R2 = \{3, 2, 4, 1\}$, results in $R1* = \{1, 2\}$ and $R2* = \{3, 4\}$.

However, let us suppose that D2 had not strictly followed the order established by its ranking for Case 2, i.e. D1 will select candidate 1, but D2 will select candidate 2 instead of 3; now D1 will have to select candidate 4 and D2 will select candidate 3. See example 2 below.

Example 2: Let be $N = 6$ candidates; $C = \{1, 2, 3, 4, 5, 6\}$; given the rankings $R1 = \{1, 2, 4, 3, 5, 6\}$ and $R2 = \{3, 2, 6, 1, 5, 4\}$, the teams resulting from the process selection will be: $R1* = \{1, 4, 5\}$ and $R2* = \{2, 3, 6\}$.

In this example, the decision-maker D2 not strictly followed the established order in his ranking and gets a solution of higher quality than the decision-maker D1.

These examples show that there are alternatives to improve decision-makers options. The idea is to consider the ranking of each decision-maker as a reference of preferences, but not necessarily as a strict order to follow in the selection. The ranking of each decision-maker can be seen as a heuristic information for the selection process. Hence, the problem is how to develop a heuristic search to guide the selection process, for trying to meet the preferences of both decision-makers. In this work, the Ant Colony Optimization (ACO) metaheuristic is used to direct the process of team selection.

3 Ant Colony Optimization

The ACO is a relatively recent metaheuristic, which is inspired by the behavior governing the ants to find the shortest paths between the food sources and the anthill [42]. ACO algorithms are directly inspired by the behavior of real ant colonies to solve combinatorial optimization problems. They are based on a colony of artificial ants, i.e., simple computational agents that work cooperatively and communicate through artificial pheromone trails [43].

These algorithms are essentially constructive: at each iteration of the algorithm, each ant builds a solution for the problem, going through a construction graph. Each edge of the graph, which represents the possible steps that the ant can take, is associated with two types of information that guide the movement of the ant:

- Heuristic information, which measures the heuristic preference to move from node r to node s, that is, to cover the edge a_{rs}. It is denoted by η_{rs}. Ants do not modify this information during the execution of the algorithm.
- Information of the artificial pheromone trails, which measures the "learned desirability" of the movement from r to s. It imitates the real pheromone that natural ants deposit. This information is modified while running the algorithm depending on the solutions found by the ants. It is denoted by τ_{rs}.

The basic mode of operation of an ACO algorithm is as follows: the m artificial ants of the colony move concurrently and asynchronously through the adjacent states of the problem (which can be represented as a graph with weights). This movement is performed following a transition rule which is based on local heuristic and rote information available in the components (nodes). This local information includes the heuristic and the pheromone trails to guide the search information. When moving through the construction graph, ants incrementally build solutions. Once each ant has generated a solution, it is evaluated and it can deposit a quantity of pheromone according to the quality of the solution. Optionally, ants can lay pheromone every time they cross an arch (connection) while building the solution (online step by step pheromone trail). This information will guide the future search for the other ants in the colony. In addition, the operation mode of a generic ACO algorithm includes evaporation of the pheromone trails and it is used as a mechanism to avoid the deadlock in the search and it allows ants to seek and explore new regions of space. Optional actions that do not have a natural counterpoint can be also run and they are used to implement tasks from a global perspective that the ants cannot carry out because of the local perspective they offer. Examples of these are to observe the quality of all solutions generated and to deposit a new amount of additional pheromone just in the links associated with some solutions, or to apply a local search procedure to the solutions generated by the ants before upgrading the pheromone trails.

3.1 Ant System

Among the ACO algorithms, there is the Ant System (AS) [41]. The AS was the first ACO algorithm. Initially, three variants were presented: AS-density, AS-quantity and AS-cycle that differed in the way the pheromone trails were updated. In the first two, ants deposited pheromones while building solutions (that is, they applied an online update step to step of pheromone), except that the amount of pheromone deposited on the AS-density is constant, while the amount deposited in AS-quantity depended directly on the heuristic desirability of the transition η_{rs}. Finally, in the AS-cycle, the pheromone deposition takes place once the solution is complete (update online a posteriori pheromone). This last option was the one which got the best results, it is therefore known as the AS in literature and it is the one used in this research.

AS is characterized by the fact that the pheromone update is performed once all ants have completed their solutions, and it is performed as follows: first, all pheromone trails are reduced by a constant factor, implementing in this way the pheromone evaporation; next, every ant colony deposits an amount of pheromone which is in function of the solution's quality.

The solutions in AS are constructed as follows. At each step of construction, one ant k chooses to go to the next node with a probability that is calculated according to expression (1):

$$P_{rs}^{k}(t+1) = \begin{cases} \dfrac{[\tau_{rs}(t)]^{\alpha}*[\eta_{rs}]^{\beta}}{\sum_{s \in \mathcal{N}_{r}^{k}}[\tau_{rs}(t)]^{\alpha}*[\eta_{rs}]^{\beta}}, & \text{if } s \in \mathcal{N}_{r}^{k} \\ 0 & \text{if } s \notin \mathcal{N}_{r}^{k} \end{cases} \tag{1}$$

Where N_{r}^{k} is the feasible neighborhood by ant k when it is located at node r, and α, β are two parameters that weight the relative importance of the pheromone trails and the heuristic information. Each ant k stores the sequence it has followed so far and its memory is used to determine its neighborhood at each construction step.

As it has been exposed, the pheromone deposit is made once all ants have finished building their solutions. First, the pheromone trails associated with each arc are evaporated by reducing all pheromones by a constant factor $\rho \in (0, 1]$ that is the evaporation rate; see expression (2):

$$\tau_{rs}(t+1) = (1-\rho)*\tau_{rs}(t) \tag{2}$$

The next step of each ant is to walk back the way it has followed, that is, the way it is stored in its local memory and to deposit an amount of pheromone in each connection for which has traveled, as shown in Eq. (3):

$$\tau_{rs}(t+1) = \tau_{rs}(t) + \Delta\tau_{rs}^{k}, \quad \forall a_{rs} \epsilon S_{k} \tag{3}$$

Where $\Delta \tau_{rs}^k = f(C(S_k))$, that is, the amount of pheromone released depends on the quality $C(S_k)$ of the solution S_k built by ant k.

3.2 MAX-MIN Ant System

The MAX-MIN Ant System (MMAS) [45] was specifically developed to achieve stronger exploitation of solutions, avoiding stagnation states. In a nutshell, we could define a stagnation state as the situation where ants construct the same solution over and over again and the exploration stops. This model has the following features:

1. At the end of each iteration only an ant adds pheromone to its found way. This ant can be the best solution or the best solution global.
2. Second, all pheromone trails are limited in the range $[\tau_{max}, \tau_{min}]$, so that no pheromone trail is less than τ_{min} or greater than τ_{max}.
3. As a final point, pheromone trails are initialized with τ_{max} to ensure further exploration of the search space.

4 A Method for Solving the Personnel Selection Problem in a Competitive Environment Using Ant Colony Optimization

The method proposed in this section uses the multi-type ants' model [46]. In this model two types of ants and two types of pheromones are used: ants are attracted by the pheromones of the same type (cooperation) and they are repelled by the pheromones of the other type (competition).

The graph where the ants operate consists of N nodes, each node represents a candidate. It is a complete graph, meaning that there are links among all nodes. Two pheromone values are associated to each node, one for each type of ant, which correspond to the two decision-makers who make the selection of candidates. Two sets of ants (with equal number of ants) operate on the graph, one per decision-maker. Ants work in pairs, each pair has an ant of each type; at the end of each cycle every pair of ants has built a team for each decision-maker. When an ant visits a node, this is already disabled for its partner; that is, the ants in a pair cannot visit the same nodes. In each search cycle, ants visit N/2 nodes.

Another element of the proposed method is the heuristic information used by the ants. The objective of each ant is to select candidates according to the values of preference defined in the respective rankings for each type of ant. The heuristic value of each node for a given ant type is determined by its position in the corresponding ranking; the node corresponding to the first candidate of the ranking has the greatest heuristic value, while the node corresponding to the last element has the lowest heuristic value.

The initial value of the pheromone associated to the link between nodes i and j for each type of ant is calculated taking into account the distance of those candidates in the rankings; the greater the distance between the two candidates, the lower the value of the pheromone. Updating the pheromone value depends on the quality of the solution found by each ant. The quality of the solution depends on two aspects: (i) the similarity of the solution built by each ant in the part respect to the ranking of preference; (ii) the difference of the value of both solutions (trying to the ants in the pair reach solutions with similar quality).

This approach for assessing the quality of the solutions reached by each pair of ants is specified in the expressions (10) and (11); expression (10) has two terms, the first term takes into account to what extent each ant finds a solution similar to the ranking given by the decision-makers and the second term compares the quality of the solutions found by each pair of ants; in other words, expression (10) is looking for the solutions found by the pair of ants to be as close as possible to the ranking of preference for each decision-maker as well as the level of satisfaction of both decision-makers to be similar. As the first term of the expression (10) is an average, if the second is not considered, solutions could be preferred, in which an ant of the pair reaches a very good solution but the other ant obtains a very bad one.

In the expression (11) the quality of a solution according to a given ranking is calculated. If the candidate on the first position is included in the solution a 1 is taken into account in the sum, for the candidate on the second position a 2 is taken into account in the sum, and so on. The lower this sum, the more the solution satisfies a particular decision-makers preference.

Components of the Ant System to Team Selection method (AS-TS):

1. Build a complete graph with N nodes, where each node represents a candidate.
2. Generate two types of ants: type 1 (h^1) and type 2 (h^2), where each type of ant represents the decision-maker D1 and D2 respectively.
3. m ants of each type are used, which work in pairs (h^{1k}, h^{2k}) where k represents the kth ant of type 1 and 2 respectively, with k = 1, 2, ..., m. Hence, in each cycle there are working m pairs $\{(h^{11}, h^{21}), (h^{12}, h^{22})... (h^{1m}, h^{2m})\}$.
4. At the starting point of each iteration, ants are distributed randomly, but the ants of the same pair must be in different nodes; i.e., an ant of type 2 can only be placed in a node if the ant type 1 of that pair is not located in that node.
5. Two types of pheromones τ_{ij}^1 (amount of pheromone for ants of type 1 associated with the link between node i and node j) and τ_{ij}^2 (amount of pheromone for ants of type 2 between nodes i and j) are generated, where ants of type 1 are attracted by pheromones of type 1 (τ_{ij}^1) and repelled by the type 2 (τ_{ij}^2), and vice versa. Pheromone values are calculated by expressions (4) and (5) for ants of type 1 and 2 respectively:

$$\tau_{ij}^1(0) = 1/(1 + \delta(i,j)) \tag{4}$$

$$\tau_{ij}^2(0) = 1/(1 + \delta(i,j)) \tag{5}$$

Where $\delta(i,j)$ in τ_{ij}^1 and τ_{ij}^2 is the number of positions between candidates i and j in the rankings R1 and R2 respectively. This results in pheromone values in the range $[1/(N-2), 1]$ reaching the value 1 when the nodes i and j are adjacent in the ranking.

6. The heuristic value used to assess the quality of each possible successor node is denoted for ants of type 1 as η_{ij}^1 and for ants of type 2 as η_{ij}^2; it is calculated by the expressions (6) and (7) for ants of type 1 and 2 respectively:

$$\eta_{ij}^1 = 1/O(j) \tag{6}$$

$$\eta_{ij}^2 = 1/O(j) \tag{7}$$

Where $O(j)$ is the position of the candidate j in the rankings R1 and R2 for ants of type 1 and type 2 respectively.

7. The neighborhood at node i of the kth ant of type 1 (or type 2), denoted by V_i^{1k} (or V_i^{2k}), has all the nodes that have not been selected yet by the kth ant nor by its pair's ant.
8. The probabilistic rule, which decides the new node to visit, is defined by the expressions (8) and (9) for ants of type 1 and 2 respectively:

$$p_{ij}^{1k} = \begin{cases} 0 & \text{if } j \notin V_i^{1k} \\ \dfrac{[\tau_{ij}^1]^\alpha * [\eta_{ij}^1]^\beta / [\tau_{ij}^2]^\theta}{\sum_{j \in V_i^{1k}} [\tau_{ij}^1]^\alpha * [\eta_{ij}^1]^\beta / [\tau_{ij}^2]^\theta} & \text{if } j \in V_i^{1k} \end{cases} \tag{8}$$

$$p_{ij}^{2k} = \begin{cases} 0 & \text{if } j \notin V_i^{2k} \\ \dfrac{[\tau_{ij}^2]^\alpha * [\eta_{ij}^2]^\beta / [\tau_{ij}^1]^\theta}{\sum_{j \in V_i^{2k}} [\tau_{ij}^2]^\alpha * [\eta_{ij}^2]^\beta / [\tau_{ij}^1]^\theta} & \text{if } j \in V_i^{2k} \end{cases} \tag{9}$$

9. Each ant ends in a cycle when it has visited N/2 nodes. At the end of each cycle, each pair (h^{1k}, h^{2k}) will generate a solution where the ant h1 k has the solution for D1 (R1*) and the ant h^{2k} has the solution for D2 (R2*).
10. The solutions found for each pair of ants is evaluated at the end of each cycle by the expression (10):

$$\text{Eval}^k = [(\text{eval}(s(h^{1k}), R1) + \text{eval}(s(h^{2k}), R2))/2]$$
$$+ |\text{eval}(s(h^{1k}), R1) - \text{eval}(s(h^{2k}), R2)| \tag{10}$$

Where $s(h^{1k})$ and $s(h^{2k})$ are the solutions found by the ants h^{1k} and h^{2k} respectively; eval $(s(h^{1k}, R1)$ and eval $(s(h^{2k}), R2)$ values are obtained according to the expression (11):

$$\text{eval}(Ri^*, Ri) = \sum_{\forall c \in Ri^*} \pi(c) \tag{11}$$

Where $\pi(c)$ is the value of the candidate $c \in Ri^*$ according to its position in the ranking Ri; the function π assigns the value of 1 to the first place in the ranking, 2 to the second and so on, until the last place in the ranking is assigned to the value N. See example 3 below.

11. When a cycle is completed, that is, when all the ants have covered the N/2 nodes in the cycle, the evaporation of all pheromones is performed and then new pheromone is deposited. As the proposed model is based on the Ant System, the pheromone is deposited in all ij arcs, appearing in the solutions found by each ant in the cycle. All pheromone's values are decreased using the expression (12) where ρ is a value between 0 and 1.

$$\tau_{ij}(t+1) = \rho * \tau_{ij}(t) \tag{12}$$

The pheromone's deposit is calculated using Eqs. (13) y (14):

$$\tau_{ij}^1(t+1) = \tau_{ij}^1(t) + (1/\text{eval}(s(h^{1k}), R1)) \tag{13}$$

$$\tau_{ij}^2(t+1) = \tau_{ij}^2(t) + (1/\text{eval}(s(h^{2k}), R2)) \tag{14}$$

12. On the completion of the search, the solutions $s(h^{1k})$ and $s(h^{2k})$ associated with the pair (h^{1k}, h^{2k}) with the lowest value of Evalk will be the sets R1* and R2 * resulting for the decision-makers D1 and D2 respectively.

Example 3: Given the set of N = 6 candidates C = {0, 1, 2, 3, 4, 5}; the rankings R1 = {2, 0, 1, 3, 5, 4} and R2 = {3, 2, 1, 5, 0, 4}; and the resulting subsets R1* = {0, 2, 4} and R2* = {3, 5, 1}. Applying the expression (11) we have:

eval (R1*, R1) = 2 + 1 + 6 = 9
eval (R2*, R2) = 1 + 4 + 3 = 8

The algorithm Ant System to Team Selection (AS-TS) is defined using these components.

Algorithm **AS-TS**

Input: N (number of candidates); R1 and R2 (rankings of preferences of decision-makers D1 and D2 respectively); NmaxC (maximum number of cycles to run); m (number of ants of each type); ρ (parameter to decrease the pheromone); α, β, θ (exponents of the probabilistic rule's terms).
Output: the teams R1* and R2* with N/2 candidates.

P0:
To calculate the initial values of the pheromone $\tau_{ij}{}^1(0)$ and $\tau_{ij}{}^2(0)$ according to (4) and (5).
To calculate the heuristic values $\eta_{ij}{}^1$ and $\eta_{ij}{}^2$ according to (6) and (7).
Iteration← 0
P1: Repeat
    P1.1: Ants are distributed on the graph according to point 4.
    P1.2: For i=1 until N/2 to do
        For k=1 until m to do
            For ant h^{1k} to select the next node according to (8).
            For ant h^{2k} to select the next node according to (9).
        End
    End
    P1.3: At the end of P12 each pair (h^{1k}, h^{2k}) has reached a solution where the ant h^{1k} has the solution for D1, $s(h^{1k})$ and the ant , h^{2k} has the solution for D2, $s(h^{2k})$.
    P1.4: Evaporate the pheromones according to (12).
    P1.5: Increase the pheromone according to (13) and (14) in all ij arcs that appear in the solutions found by each ant.
    P1.6: Evaluate the solutions obtained for each pair of ants as (10) and to update sets R1* and R2* with the best solution obtained so far.
    P1.7: Iteration←Iteration+1
Until Iteration = = NmaxC
P2: To return sets R1* and R2*.

5 Another Proposal for Solving the Personnel Selection Problem in a Competitive Environment Using Ant Colony Optimization

In this research the algorithms Max-Min Ant System to Team Selection (MMAS-TS) is also proposed. This proposal is based on the MAX-MIN Ant System model and the components AS-TS algorithm. The components of the algorithm MMAS-TS are similar to those described for AS-TS algorithm except the three modifications enunciated below.

M1: The initial pheromone values associated with the link between nodes i and j for each type of ant take the value $\tau_{ij}^1(0) = \tau_{ij}^2(0) = \tau_{max}$.

M2: The pheromone deposit is made in each arch ij that appears in the solutions found by the best couple of ants in the iteration and by the pair of ants that has obtained the best solution from the execution start until now.

M3: Pheromone levels are delimited each time a cycle is completed; the values are limited to τ_{max} and τ_{min}. These values are calculated from the expressions (15) and (16) respectively.

$$\tau_{max} = (1/(1-\rho)) * (1/Eval^{mejor\_par\_global}) \qquad (15)$$

$$\tau_{min} = \tau_{max}/(10*N) \qquad (16)$$

Where N represents the number of candidates, ρ is the value of the constant evaporation and $Eval^{mejor\_par\_global}$ represents the quality of the best global solution found by the ant colony throughout the search process, and is calculated from the expression (10).

Algorithm MMAS-TS

Input: N (number of candidates); R1 and R2 (rankings of preferences of decision-makers D1 and D2 respectively); NmaxC (maximum number of cycles to run); m (number of ants of each type); ρ (parameter to decrease the pheromone); α, β, θ (exponents of the probabilistic rule's terms); τ_{max} (maximum value of pheromone). Output: the teams R1* and R2* with N/2 candidates.

P0:
To calculate the initial values of the pheromone $\tau_{ij}^{1}(0)$ and $\tau_{ij}^{2}(0)$ according to (M1).
To calculate the heuristic values η_{ij}^{1} and η_{ij}^{2} according to (6) and (7).
Iteration$\leftarrow$ 0
P1: Repeat
    P1.1: Ants are distributed on the graph according to point 4.
    P1.2: For i=1 until N/2 to do
        For k=1 until m to do
            For ant h^{1k} to select the next node according to (8).
            For ant h^{2k} to select the next node according to (9).
        End
        End
    P1.3: At the end of P12 each pair (h^{1k}, h^{2k}) has reached a solution where the ant h^{1k} has the solution for   D1, s(h^{1k}) and the ant h^{2k} has the solution for D2, s(h^{2k}).
    P1.4: Evaporate the pheromones according to (12).
    P1.5: Increase the pheromone according to (M2).
    P1.6: Evaluate the solutions obtained for each pair of ants as (10) and to update sets R1* and R2* with the best solution obtained so far.
    P1.7: To delimit the levels of pheromone according to (M3).
    P1.8: Iteration$\leftarrow$Iteration+1
Until Iteration $==$ NmaxC
P2: To return sets R1* and R2*.

In the following section the performance of the proposed methods is reported by an experimental study.

6   Experimental Study

The purpose of this empirical study is to illustrate the effectiveness of the algorithms AS-TS and MMAS-TS using some examples. For doing that, a simulation of the candidates' selection process is performed, in which two rankings R1 and R2 of N elements are generated randomly for decision-makers D1 and D2 respectively. In the study we considered values of 4, 6, 8, 10, 12, 14, 16, 18 and 20 for N. Different values were evaluated for the model's parameters. The results shown in the following tables were obtained with the following values of input's parameters for the algorithm: $\alpha = \beta = \theta = 1$, $\rho = 0.75$, NmaxC = 10 and m = N.

In Tables 1, 2, 3, 4, 5, 6, 7, 8 and 9 we give the results achieved using the proposed algorithms. Each table contains three rows corresponding to different examples generated for that number of candidates. The first column shows the rankings R1 and R2, which were randomly generated; the second column shows the results achieved by each of the decision-makers, if the selection of candidates is carried out in the order established of the rankings; in the third column the results obtained after applying the algorithms AS-TS and MMAS-TS are given.

In addition to the teams achieved by each method, we also report the distances between the teams R1* and R2* to the rankings R1 and R2 for every example, and the evaluation according to the expression (10). The best results in each case are indicated in bold.

Table 1 4 candidates

Rankings	Results according to the order established in the rankings	Results applying the algorithm AS-TS
		Results applying the algorithm MMAS-TS
R1 = {3, 1, 2, 0} R2 = {0, 3, 1, 2}	R1* = {3, 1} R2* = {0, 2} (3, 5) = 4 + 2 = 6.0	R1* = {2, 3} R2* = {1, 0} **(4, 4) = 4 + 0 = 4.0**
		R1* = {2, 3} R2* = {1, 0} **(4, 4) = 4 + 0 = 4.0**
R1 = {0, 2, 1, 3} R2 = {3, 1, 0, 2}	R1* = {0, 2} R2* = {3, 1} **(3, 3) = 3 + 0 = 3.0**	R1* = {2, 0} R2* = {1, 3} **(3, 3) = 3 + 0 = 3.0**
		R1* = {0, 2} R2* = {3, 1} **(3, 3) = 3 + 0 = 3.0**
R1 = {0, 2, 3, 1} R2 = {0, 1, 2, 3}	R1* = {0, 2} R2* = {1, 3} (3, 6) = 4.5 + 3 = 7.5	R1* = {0, 3} R2* = {2, 1} **(4, 5) = 4.5 + 1 = 5.5**
		R1* = {3, 0} R2* = {1, 2} **(4, 5) = 4.5 + 1 = 5.5**

Table 2 6 candidates

Rankings	Results according to the order established in the rankings	Results applying the algorithm AS-TS
		Results applying the algorithm MMAS-TS
R1 = {2, 0, 1, 3, 5, 4} R2 = {3, 2, 1, 5, 0, 4}	R1* = {2, 0, 5} R2* = {3, 1, 4} (8, 10) = 9 + 2 = 11.0	R1* = {0, 2, 4} R2* = {3, 5, 1} **(9, 8) = 8.5 + 1 = 9.5**
		R1* = {0, 2, 4} R2* = {1, 5, 3} **(9, 8) = 8.5 + 1 = 9.5**
R1 = {5, 3, 1, 4, 0, 2} R2 = {0, 3, 2, 5, 4, 1}	R1* = {5, 3, 1} R2* = {0, 2, 4} (6, 9) = 7.5 + 3 = 10.5	R1* = {1, 5, 4} R2* = {0, 3, 2} **(8, 6) = 7 + 2 = 9.0**
		R1* = {4, 5, 1} R2* = {2, 3, 0} **(8, 6) = 7 + 2 = 9.0**
R1 = {2, 1, 0, 4, 3, 5} R2 = {2, 0, 3, 4, 1, 5}	R1* = {2, 1, 4} R2* = {0, 3, 5} (7, 11) = 9 + 4 = 13.0	R1* = {1, 2, 5} R2* = {0, 3, 4} (9, 9) = 9 + 0 = 9.0
		R1* = {0, 3, 1} R2* = {5, 4, 2} **(7, 8) = 7.5 + 1 = 8.5**

Table 3 8 candidates

Rankings	Results according to the order established in the rankings	Results applying the algorithm AS-TS
		Results applying the algorithm MMAS-TS
R1 = {0, 1, 3, 7, 5, 6, 4, 2} R2 = {0, 1, 7, 5, 4, 6, 3, 2}	R1* = {0, 3, 5, 6} R2* = {1, 7, 4, 2} (15, 18) = 16.5 + 3 = 19.5	R1* = {4, 3, 0, 6} R2* = {2, 1, 5, 7} **(17, 17) = 17 + 0 = 17.0**
		R1* = {4, 0, 3, 6} R2* = {5, 7, 1, 2} **(17, 17) = 17 + 0 = 17.0**
R1 = {0, 2, 7, 5, 6, 3, 1, 4} R2 = {0, 2, 7, 5, 6, 3, 1, 4}	R1* = {0, 7,6, 1} R2* = {2, 5, 3, 4} (16, 20) = 18 + 4 = 22.0	R1* = {2, 7, 6, 4} R2* = {0, 3, 5, 1} **(18, 18) = 18 + 0 = 18.0**
		R1* = {0, 2, 1, 4} R2* = {6, 5, 7, 3} **(18,18) = 18 + 0 = 18.0**
R1 = {7, 6, 0, 1, 2, 5, 4, 3} R2 = {7, 6, 2, 1, 3, 5, 4, 0}	R1* = {7, 0, 1, 5} R2* = {6, 2, 3, 4} (14, 17) = 15.5 + 3 = 18.5	R1* = {1, 7, 0, 4} R2* = {6, 3, 2, 5} (15, 16) = 15.5 + 1 = 16.5
		R1* = {0, 7, 3, 2} R2* = {5, 4, 1, 6} **(10, 10) = 10 + 0 = 10.0**

Table 4 10 candidates

Rankings	Results according to the order established in the rankings	Results applying the algorithm AS-TS
		Results applying the algorithm MMAS-TS
R1 = {8, 2, 7, 5, 4, 6, 9, 0, 1, 3} R2 = {9, 8, 2, 4, 6, 3, 1, 0, 5, 7}	R1* = {8, 2, 7, 5, 0} R2* = {9, 4, 6, 3, 1} (18, 23) = 20.5 + 5 = 25.5	R1* = {5, 8, 7, 4, 0} R2* = {9, 2, 6, 3, 1} **(21, 22) = 21.5 + 1 = 22.5**
		R1* = {0, 4, 7, 5, 8} R2* = {6, 9, 2, 3, 1} **(21,22) = 21.5 + 1 = 22.5**
R1 = {8, 1, 4, 6, 7, 0, 9, 3, 2, 5} R2 = {6, 5, 8, 9, 4, 1, 3, 7, 0, 2}	R1* = {8, 1, 4, 7, 0} R2* = {6, 5, 9, 3, 2} (17, 24) = 20.5 + 7 = 27.5	R1* = {8, 0, 1, 7, 3} R2* = {9, 6, 5, 4, 2} **(22, 22) = 22 + 0 = 22.0**
		R1* = {7, 1, 8, 0, 3} R2* = {4, 6, 2, 9, 5} **(22, 22) = 22 + 0 = 22.0**
R1 = {9, 7, 5, 4, 3, 1, 0, 2, 6, 8} R2 = {9, 7, 5, 4, 3, 1, 0, 2, 6, 8}	R1* = {9, 5, 3, 0, 6} R2* = {7, 4, 1, 2, 8} (25, 30) = 27.5 + 5 = 32.5	R1* = {4, 7, 1, 2, 0} R2* = {9, 8, 6, 5, 3} **(27, 28) = 27.5 + 1 = 28.5**
		R1* = {7, 5, 2, 4, 8} R2* = {0, 9, 1, 3, 6} **(27, 28) = 27.5 + 1 = 28.5**

Table 5 12 candidates

Rankings	Results according to the order established in the rankings	Results applying the algorithm AS-TS
		Results applying the algorithm MMAS-TS
R1 = {11, 10, 5, 4, 3, 2, 9, 8, 1, 6, 7, 0} R2 = {11, 10, 4, 5, 2, 3, 9, 8, 6, 1, 7, 0}	R1* = {11, 5, 3, 9, 1, 7} R2* = {10, 4, 2, 8, 6, 0} (36, 39) = 37.5 + 3 = 40.5	R1* = {10, 5, 11, 0, 1, 7} R2* = {4, 8, 3, 9, 6, 2} **(38, 38) = 38 + 0 = 38.0**
		R1* = {2, 3, 5, 8, 9, 1} R2* = {11, 10, 4, 6, 7, 0} **(38, 38) = 38 + 0 = 38.0**
R1 = {9, 5, 6, 3, 0, 1, 10, 11, 7, 4, 2, 8} R2 = {9, 5, 6, 3, 11, 10, 1, 0, 7, 4, 2, 8}	R1* = {9, 6, 0, 1, 7, 2} R2* = {5, 3, 11, 10, 4, 8} (35, 39) = 37 + 4 = 41.0	R1* = {0, 3, 5, 7, 2, 1} R2* = {4, 6, 9, 11, 10, 8} **(37, 37) = 37 + 0 = 37.0**
		R1* = {6, 9, 1, 4, 0, 8} R2* = {7, 5, 3, 10, 11, 2} **(37, 37) = 37 + 0 = 37.0**
R1 = {0, 1, 4, 5, 10, 11, 7, 8, 3, 2, 9, 6} R2 = {0, 6, 1, 9, 4, 2, 5, 3, 10, 8, 11, 7}	R1* = {0, 1, 4, 5, 10, 11} R2* = {6, 9, 2, 3, 8, 7} (21, 42) = 31.5 + 21 = 52.5	R1* = {7, 4, 11, 1, 5, 8} R2* = {6, 0, 9, 10, 3, 2} **(30, 30) = 30 + 0 = 30.0**
		R1* = {11, 0, 4, 10, 7, 8} R2* = {9, 1, 5, 3, 6, 2} **(30, 30) = 30 + 0 = 30.0**

Table 6 14 candidates

Rankings	Results according to the order established in the rankings	Results applying the algorithm AS-TS
		Results applying the algorithm MMAS-TS
R1 = {4, 13, 7, 3, 10, 1, 0, 12, 6, 2, 11, 8, 5, 9} R2 = {8, 2, 11, 3, 7, 0, 1, 5, 13, 6, 12, 4, 10, 9}	R1* = {4, 13, 7, 3, 10, 12, 6} R2* = {8, 2, 11, 0, 1, 5, 9} (32, 41) = 36.5 + 9 = 45.5	R1* = {4, 3, 10, 7, 13, 9, 6} R2* = {8, 2, 1, 11, 0, 5, 12} (38, 38) = 38 + 0 = 38.0
		R1* = {3, 13, 4, 12,9, 10, 7} R2* = {0, 1, 5, 2,8, 6, 11} **(37, 37) = 37 + 0 = 37.0**
R1 = {12, 9, 1, 0, 3, 10, 2, 11, 7, 6, 4, 5, 13, 8} R2 = {6, 9, 0, 8, 12, 5, 10, 1, 2, 4, 3, 7, 13, 11}	R1* = {12, 9, 1, 3, 10, 11, 7} R2* = {6, 0, 8, 5, 2, 4, 13} (34, 46) = 40 + 12 = 52.0	R1* = {2, 12, 1, 3,10, 11, 7} R2* = {5, 0, 6, 9, 8, 4, 13} **(39, 39) = 39 + 0 = 39.0**
		R1* = {10, 12, 11, 1, 3, 7, 2} R2* = {13, 4, 8, 9, 0, 6, 5} **(39, 39) = 39 + 0 = 39.0**
R1 = {1, 5, 13, 0, 12, 10, 2, 8, 6, 9, 7, 4, 11, 3} R2 = {8, 6, 11, 4, 12, 7, 1, 13, 0, 2, 10, 3, 5, 9}	R1* = {1, 5, 13, 0, 12, 10, 9} R2* = {8, 6, 11, 4, 7, 2, 3} (31, 38) = 34.5 + 7 = 41.5	R1* = {9, 13, 5, 1, 10, 0, 2} R2* = {12, 4, 6, 11, 8, 7, 3} **(33, 33) = 33 + 0 = 33.0**
		R1* = {5, 1, 13, 9, 10, 0, 2} R2* = {11, 8, 6, 3, 7, 12, 4} **(33, 33) = 33 + 0 = 33.0**

Table 7 16 candidates

Rankings	Results according to the order established in the rankings	Results applying the algorithm AS-TS
		Results applying the algorithm MMAS-TS
R1 = {3, 4, 5, 6, 10, 11, 12, 13, 15, 2, 8, 0, 1, 9, 7, 14} R2 = {3, 4, 7, 6, 11, 10, 12, 14, 0, 1, 8, 15, 2, 9, 5, 13}	R1* = {3, 5, 6, 10, 13, 15, 2, 8} R2* = {4, 7, 11, 12, 14, 0, 1, 9} (51, 58) = 54.5 + 7 = 61.5	R1* = {12, 15, 3, 5,10, 13, 8, 2} R2* = {6, 4, 0, 11, 7, 1, 14, 9} **(54, 55) = 54.5 + 1 = 55.5**
		R1* = {3, 2, 10, 13, 5, 15, 6, 9} R2* = {7, 12, 0, 4, 11, 14, 8, 1} **(54, 55) = 54.5 + 1 = 55.5**
R1 = {10, 11, 0, 1, 7, 3, 4, 6, 12, 9, 5, 8, 14, 2, 15, 13} R2 = {10, 11, 1, 0, 2, 7, 3, 15, 4, 6, 13, 12, 14, 9, 5, 8}	R1* = {10, 0, 7, 3, 4, 12, 9, 5} R2* = {11, 1, 2, 15, 6, 13, 14, 8} (52, 68) = 60 + 16 = 76.0	R1* = {8, 10, 11, 3, 12, 9, 6, 5} R2* = {14, 13, 0, 1, 2, 7, 4, 15} **(59, 59) = 59 + 0 = 59.0**
		R1* = {12, 0, 6, 5, 7, 4, 3, 9} R2* = {8, 10, 1, 15, 11, 13, 14, 2} **(59, 59) = 59 + 0 = 59.0**
R1 = {15, 12, 8, 9, 0, 1, 3, 7, 14, 13, 2, 5, 10, 6, 11, 4} R2 = {4, 11, 8, 9, 0, 1, 15, 12, 2, 5, 10, 6, 3, 14, 7, 13}	R1* = {15, 12, 8, 0, 3, 7, 14, 13} R2* = {4, 11, 9, 1, 2, 5, 10, 6} (45, 55) = 50 + 10 = 60.0	R1* = {15, 12, 8, 14, 2, 7, 3, 13} R2* = {4, 9, 11, 0, 1, 6, 5, 10} **(51, 51) = 51 + 0 = 51.0**
		R1* = {3, 8, 15, 12, 2, 13, 14, 7} R2* = {0, 4, 11, 6, 5, 9, 10, 1} **(51, 51) = 51 + 0 = 51.0**

Table 8 18 candidates

Rankings	Results according to the order established in the rankings	Results applying the algorithm AS-TS
		Results applying the algorithm MMAS-TS
R1 = {17, 12, 1, 15, 16, 13, 9, 7, 8, 2, 6, 14, 3, 5, 4, 10, 11, 0} R2 = {0, 12, 17, 15, 16, 10, 1, 6, 8, 2, 14, 3, 5, 4, 9, 7, 13, 11}	R1* = {17, 12, 1, 13, 9, 7, 2, 3, 4} R2* = {0, 15, 16, 10, 6, 8, 14, 5, 11} (65, 75) = 70 + 10 = 80.0	R1* = {8, 17, 7, 1, 9, 16, 13, 11, 3} R2* = {10, 15, 0, 12, 5, 4, 2, 6, 14} (**69, 69**) = **69 + 0 = 69.0**
		R1* = {17, 7, 1, 12, 9, 13, 2, 4, 11} R2* = {15, 6, 0, 10, 5, 8, 16, 3, 14} (**69, 69**) = **69 + 0 = 69.0**
R1 = {11, 4, 9, 5, 16, 7, 13, 0, 1, 14, 2, 12, 3, 15, 8, 6, 17, 10} R2 = {1, 16, 4, 0, 6, 3, 2, 13, 7, 15, 8, 12, 9, 11, 5, 17, 10, 14}	R1* = {11, 4, 9, 5, 7, 13, 14, 2, 12, 3} R2* = {1, 16, 0, 6, 3, 2, 15, 8, 17, 10} (69, 79) = 74 + 10 = 84.0	R1* = {12, 4, 9, 11, 13, 5, 7, 14, 17} R2* = {1, 6, 0, 16, 3, 2, 8, 15, 10} (**62, 63**) = **62.5 + 1 = 63.5**
		R1* = {13, 9, 11, 7, 4, 5, 14, 17, 12} R2* = {2, 16, 1, 0, 8, 3, 15, 10, 6} (**62, 63**) = **62.5 + 1 = 63.5**
R1 = {12, 11, 16, 2, 0, 17, 10, 4, 6, 14, 13, 3, 9, 15, 7, 8, 1, 5} R2 = {9, 15, 6, 1, 3, 12, 8, 2, 14, 7, 13, 11, 5, 17, 10, 16, 4, 0}	R1* = {12, 11, 16, 2, 0, 17, 10, 4, 13} R2* = {9, 15, 6, 1, 3, 8, 14, 7, 5} (47, 54) = 50.5 + 7 = 57.5	R1* = {10, 17, 12, 11, 4, 13, 14, 16, 0} R2* = {2, 9, 1, 3, 15, 6, 8, 7, 5} (**53, 53**) = **53 + 0 = 53.0**
		R1* = {12, 4, 16, 17, 0, 13, 14, 10, 11} R2* = {2, 8, 3, 1, 9, 15, 6, 5, 7} (**53, 53**) = **53 + 0 = 53.0**

For instance, in the first row in Table 1, given the rankings R1 = {3, 1, 2, 0} and R2 = {0, 3, 1, 2}, the teams obtained if the decision-makers follow the order established in the rankings are R1* = {3, 1} and R2* = {0, 2}, similarity of these teams respect to the ranking (according to expression 11) are 3 and 5 respectively, the value of the first term of expression (10) is 4 and the value of the second term is 2, the overall value according to expression (10) of this solution is 6. You can see that the solution shown in the third column obtained by the algorithms AS-TS and MMAS-TS is fairer because the second term of the expression (10) is 0; both decision-makers obtain similar solutions. In this case the algorithms AS-TS and MMAS-TS achieve equal solutions.

From the analysis in the tables, it can be concluded that the performance of the proposed methods allow obtaining teams that are fairer to both decision-makers, yet having the same overall quality compared to the method where the decision-makers selected alternatively based on their rankings. An important element is that, the larger the number of candidates, the proposed methods are more effective. The results obtained by the two proposed methods are similar, although in some cases MMAS-TS achieves better solutions.

Table 9 20 candidates

Rankings	Results according to the order established in the rankings	Results applying the algorithm AS-TS
		Results applying the algorithm MMAS-TS
R1 = {18, 17, 12, 19, 1, 15, 16, 13, 9, 7, 8, 2, 6, 14, 3, 5, 4, 10, 11, 0} R2 = {0, 12, 18, 17, 19, 15, 16, 10, 1, 6, 8, 2, 14, 3, 5, 4, 9, 7, 13, 11}	R1* = {18, 17, 19, 1, 13, 9, 7, 2, 3, 4} R2* = {0, 12, 15, 16, 10, 6, 8, 14, 5, 11} (83, 93) = 88 + 10 = 98.0	R1* = {5, 1, 13, 18, 17, 7, 19, 9, 2, 11} R2* = {3, 16, 10, 15, 0, 12, 6, 14, 8, 4} (86, 88) = 87 + 2 = 89.0
		R1* = {4, 18, 19, 17, 13, 2, 1, 9, 7, 11} R2* = {3, 0, 12, 16, 15, 8, 5, 10, 14, 6} **(87, 87) = 87 + 0 = 87.0**
R1 = {17, 19, 18, 12, 1, 15, 16, 13, 9, 7, 8, 2, 6, 14, 3, 5, 4, 10, 11, 0} R2 = {0, 12, 18, 17, 19, 15, 16, 10, 1, 6, 8, 2, 14, 3, 5, 4, 9, 7, 13, 11}	R1* = {17, 19, 18, 1, 13, 9, 7, 2, 3, 4} R2* = {0, 12, 15, 16, 10, 6, 8, 14, 5, 11} (82, 93) = 87.5 + 11 = 98.5	R1* = {1, 7, 19, 18, 13, 16, 14, 11, 8, 9} R2* = {6, 12, 0, 15, 17, 10, 2, 4, 3, 5} (88, 88) = 88 + 0 = 88.0
		R1* = {4, 19, 17, 1, 18, 13, 9, 7, 11, 2} R2* = {15, 12, 10, 0, 6, 8, 16, 3, 14, 5} **(86, 87) = 86.5 + 1 = 87.5**
R1 = {14, 4, 0, 5, 10, 1, 7, 19, 3, 16, 18, 15, 9, 12, 2, 8, 13, 6, 11, 17} R2 = {10, 5, 19, 14, 0, 2, 11, 1, 4, 7, 3, 16, 17, 18, 15, 9, 6, 12, 8, 13}	R1* = {14, 4, 0, 1, 7, 3, 18, 15, 12, 8} R2* = {10, 5, 19, 2, 11, 16, 17, 9, 6, 13} (81, 97) = 89 + 16 = 105.0	R1* = {12, 14, 4, 0, 8, 1, 13, 7, 16, 9} R2* = {19, 5, 2, 10, 11, 15, 3, 18, 6, 17} **(89, 89) = 89 + 0 = 89.0**
		R1* = {1, 14, 12, 4, 7, 3, 16, 18, 8, 9} R2* = {6, 10, 0, 19, 2, 5, 17, 15, 11, 13} **(89, 89) = 89 + 0 = 89.0**

7   Conclusions

The problem of personnel selection is of great importance for institutions to achieve a better performance. One case of this problem is called team selection, in which the aim is to select members of a team. Usually both, the personnel selection and the team selection are developed using the ranking of candidates.

In this paper we propose a new problem related to the team selection, where a set of candidates has to be divided in two teams taking into account the preferences of two decision-makers who will be responsible for a team. Each decision-maker defines a ranking expressing his preference on the candidates. Each decision-maker wants to form the team that is closest to his ranking. But as the preferences of decision-makers may be similar, the rankings established by them may be similar.

A common approach for developing the selection is to allow the decision-makers alternatively select a candidate according to their ranking. While this seems a fair approach, it does not necessarily in a fair set of teams as shown in the experimental session. To obtain high quality teams as well as a fair set of teams, two heuristic methods of solution are proposed based on the 2-type Ant Colony Optimization. The conducted experimental study shows that the proposed methods allow forming teams that are equally close to the preferences of both employers, yet are fair and that efficiency is more noticeable the larger the number of candidates.

References

1. Huselid MA (1995) The impact of human resource management practices on turnover, productivity, and corporate financial performance. Acad Manag J 38:635–672
2. Kangas A et al (2001) Outranking methods as tools in strategic natural resources planning. Silva Fennica 35:215–227
3. Robertson IT, Smith M (2001) Personnel selection. J occup Organ Psychol 74:441–472
4. Golec A, Kahya E (2007) A fuzzy model for competency-based employee evaluation and selection. Comput Ind Eng 52:143–161
5. Chien C-F, Chen L-F (2008) Data mining to improve personnel selection and enhance human capital: a case study in high-technology industry. Expert Syst Appl 34:280–290
6. Liao S-K, Chang K-L (2009) Selecting public relations personnel of hospitals by analytic network process. J Hosp Mark Public Relat 19:52–63
7. Kulik CT et al (2007) The multiple-category problem: category activation and inhibition in the hiring process. Acad Manag Rev 32:529–548
8. Canós L, Liern V (2008) Soft computing-based aggregation methods for human resource management. Eur J Oper Res 189:669–681
9. Canós L et al (2011) Personnel selection based on fuzzy methods. Revista de Matemática Teoría y Aplicaciones 18:177–192
10. Canós L et al (2014) Soft computing methods for personnel selection based on the valuation of competences. Int J Intell Syst 29:1079–1099
11. Hooper RS et al (1998) Use of an expert system in a personnel selection process. Expert Syst Appl 14:425–432
12. Lai Y-J (1995) IMOST: interactive multiple objective system technique. J Oper Res Soc 46:958–976
13. Mohamed F, Ahmed A (2013) Personnel training selection problem based on SDV-MOORA. Life Sci J 10
14. Chen C-T et al (2009) Applying multiple linguistic PROMETHEE method for personnel evaluation and selection. In: 2009 IEEE International Conference on Industrial Engineering and Engineering Management, pp. 1312–1316
15. Afshari AR et al (2010) Personnel selection using ELECTRE. J Appl Sci 10:3068–3075
16. Dağdeviren M (2010) A hybrid multi-criteria decision-making model for personnel selection in manufacturing systems. J Intell Manuf 21:451–460
17. El-Santawy MF, El-Dean RAZ (2012) On using VIKOR for ranking personnel problem. Life Sci J 9:1534–1536
18. Amit R, Belcourt M (1999) Human resources management processes: a value-creating source of competitive advantage. Eur Manag J 17:174–181
19. Huemann M et al (2007) Human resource management in the project-oriented company: a review. Int J Proj Manag 25:315–323
20. Zhang S-F, Liu S-Y (2011) A GRA-based intuitionistic fuzzy multi-criteria group decision making method for personnel selection. Expert Syst Appl 38:11401–11405
21. Baležentis A et al (2012) Personnel selection based on computing with words and fuzzy MULTIMOORA. Expert Syst Appl 39:7961–7967
22. Kabak M et al (2012) A fuzzy hybrid MCDM approach for professional selection. Expert Syst Appl 39:3516–3525

23. Kalugina E, Shvydun S (2014) An effective personnel selection model. Procedia Comput Sci 31:1102–1106
24. Kelemenis A, Askounis D (2010) A new TOPSIS-based multi-criteria approach to personnel selection. Expert Syst Appl 37:4999–5008
25. Mareschal B et al (1984) PROMETHEE: a new family of outranking methods in multicriteria analysis
26. Güngör Z et al (2009) A fuzzy AHP approach to personnel selection problem. Appl Soft Comput 9:641–646
27. Özdemir A (2013) A two-phase multi criteria dynamic programing approach for personnel selection process. Probl Perspect Manag 11:98–108
28. Dursun M, Karsak EE (2010) A fuzzy MCDM approach for personnel selection. Expert Syst Appl 37:4324–4330
29. Opricovic S (1998) Multicriteria optimization of civil engineering systems. Fac Civ Eng Belgrade 2:5–21
30. Boran FE et al (2011) Personnel selection based on intuitionistic fuzzy sets. Hum Factors Ergon Manuf Serv Ind 21:493–503
31. Wang D (2009) Extension of TOPSIS method for R&D personnel selection problem with interval grey number. Manag Serv Sci 1–4
32. Kelemenis AM, Askounis D (2009) An extension of fuzzy TOPSIS for personnel selection. In: SMC, pp 4704–4709
33. Tavana M et al (2013) A fuzzy inference system with application to player selection and team formation in multi-player sports. Sport Manag Rev 16:97–110
34. Wang J, Zhang J (2015) A win–win team formation problem based on the negotiation. Eng Appl Artif Intell 44:137–152
35. Ahmed F et al (2013) Multi-objective optimization and decision making approaches to cricket team selection. Appl Soft Comput 13:402–414
36. Hayano M et al (2014) Role and member selection in team formation using resource estimation for large-scale multi-agent systems. Neurocomputing 146:164–172
37. Dadelo S et al (2014) Multi-criteria assessment and ranking system of sport team formation based on objective-measured values of criteria set. Expert Syst Appl 41:6106–6113
38. Wi H et al (2009) A team formation model based on knowledge and collaboration. Expert Syst Appl 36:9121–9134
39. Strnad D, Guid N (2010) A fuzzy-genetic decision support system for project team formation. Appl Soft Comput 10:1178–1187
40. Agustín-Blas LE et al (2011) Team formation based on group technology: a hybrid grouping genetic algorithm approach. Comput Oper Res 38:484–495
41. Dorigo M et al (1996) Ant system: optimization by a colony of cooperating agents. IEEE Trans Syst Man Cybern Part B Cybern 26:29–41
42. Dorigo M et al (1999) Ant algorithms for discrete optimization. Artif Life 5:137–172
43. Dorigo M et al (2000) Ant algorithms and stigmergy. Futur Gener Comput Syst 16:851–871
44. Dorigo M, Stützle T (2004) Ant colony optimization. MIT Press, Cambridge, MA
45. Stützle T, Hoos HH (2000) MAX–MIN ant system. Futur Gener Comput Syst 16:889–914
46. Nowé A et al (2004) Multi-type ant colony: the edge disjoint paths problem. In: Ant colony optimization and swarm intelligence. Springer, pp 202–213

Probabilistic Linguistic Distance Measures and Their Applications in Multi-criteria Group Decision Making

Mingwei Lin and Zeshui Xu

Abstract The probabilistic linguistic term sets can express not only the decision makers' several possible linguistic assessment values, but also the weight of each linguistic assessment value, so they can preserve the original decision information and then have become an efficient tool for solving multi-criteria group decision making problems. To promote the wide applicability of probabilistic linguistic term sets in various fields, this chapter focuses on the distance measures for probabilistic linguistic term sets and their applications in multi-criteria group decision making. This chapter first defines the distance between two probabilistic linguistic term elements. Based on this, a variety of distance measures are proposed to calculate the distance between two probabilistic linguistic term sets. Then, these distance measures are further extended to compute the distance between two collections of probabilistic linguistic term sets by considering the weight information of each criterion. After that, the concept of the satisfaction degree of an alternative is given and utilized to rank the alternatives in multi-criteria group decision making. Finally, a real example is given to show the use of these distance measures and then compare the probabilistic linguistic term sets with hesitant fuzzy linguistic term sets.

Keywords Probabilistic linguistic term set · Distance measure · Multi-criteria group decision making

M. Lin
Faculty of Software, Fujian Normal University, Fuzhou 350108, Fujian, China
e-mail: linmwcs@163.com

Z. Xu (✉)
Business School, Sichuan University, Chengdu 610064, China
e-mail: xuzeshui@263.net

© Springer International Publishing AG 2018
M. Collan and J. Kacprzyk (eds.), *Soft Computing Applications for Group Decision-making and Consensus Modeling*, Studies in Fuzziness and Soft Computing 357, DOI 10.1007/978-3-319-60207-3_24

1 Introduction

Because of the fuzziness of human beings' thinking, the decision makers may prefer to assess the objects by using the linguistic terms rather than use the numerical values [1, 2]. For example, the decision makers may make use of the linguistic terms such as "bad", "medium", and "good" when the comfortable degree of a car is evaluated. In this case, Zadeh put forward the fuzzy linguistic approach to model the qualitative linguistic assessment information [3]. However, the fuzzy linguistic approach is built based on the fuzzy set theory [4] and then it shows some limitations on the linguistic information modeling and computing processes. To extend and improve the fuzzy linguistic approach, many linguistic models have been proposed, such as the membership function-based model [5], the type-2 fuzzy sets-based model [6], the ordinal scales-based model [7], the 2-tuple linguistic model [8], and the virtual linguistic model [9]. These models still have some limitations, especially on the linguistic information modeling. That is because they can only provide a single linguistic term to express the decision makers' linguistic assessment information regarding a linguistic variable. This case is not in conformity with human beings' thinking [10]. In some situations, the decision makers may hesitate among several possible linguistic terms to express their assessment information and the use of only a single linguistic term is insufficient to express their real assessment information accurately. Considering this fact, Rodriguez et al. put forward the concept of hesitant fuzzy linguistic term set (HFLTS) [11], which was inspired by the concept of hesitant fuzzy sets [12] and fuzzy linguistic approach. The HFLTS is a very powerful tool to represent the decision makers' linguistic assessment information in a more flexible way. It allows the decision makers to utilize some possible linguistic terms to evaluate a linguistic variable at the same time [13].

However, the HFLTS does not contain the weight information of each possible linguistic term and then all the possible linguistic terms are processed under the assumption that their weights are equal. Obviously, this is not the case. Although the decision makers hesitate among several possible linguistic terms, they may prefer some of them in some situations so that these linguistic terms have different weights. Considering this fact, Pang et al. put forward the concept of probabilistic linguistic term set (PLTS), which consists of several possible linguistic terms associated with their corresponding probabilities [14]. Since the PLTS can capture the weight of each possible linguistic term, it does not lose the linguistic assessment information and then it can achieve more reasonable decision results.

The distance measures are vital and they are widely used for decision making [15–17], pattern recognition [18–20], and clustering analysis [21–23]. However, there are very few research results on the distance measures of the PLTSs. Although there have been many studies about the distance measures of the fuzzy sets [4], interval-valued fuzzy sets [24], intuitionistic fuzzy sets [25], interval-valued intuitionistic fuzzy sets [26], hesitant fuzzy sets [12], interval-valued hesitant fuzzy sets [27], and hesitant fuzzy linguistic term sets [11], these methods cannot be employed to measure the distance between two PLTSs directly. Thus, the chapter investigates the distance measures for PLTSs and then applies them to multi-criteria group decision making.

The rest of this chapter is organized as follows: Sect. 2 gives the knowledge of the linguistic term sets and probabilistic linguistic term sets. In Sect. 3, the definition of distance measure is given for probabilistic linguistic term sets and then some distance measures are proposed to measure the distance between two PLTSs. In Sect. 4, the proposed distance measures are extended to compute the distance between two collections of PLTSs. In Sect. 5, the proposed distance measures are applied to multi-criteria group decision making and then a practical example is given to illustrate the applicability of the proposed distance measures and compare the probabilistic linguistic term sets with hesitant fuzzy linguistic term sets. Finally, the conclusions are drawn in Sect. 6.

2 Preliminaries

In this section, the concepts and operational laws about linguistic term sets and probabilistic linguistic term sets are given.

2.1 Linguistic Term Sets

The linguistic term sets are an important part of linguistic information modeling in the linguistic decision making and the decision makers can use them to give their assessment values over the considered objects. They consist of totally ordered linguistic terms, the number of which is finite. They are often defined as $S = \{s_\alpha | \alpha = 0, 1, \ldots, \tau\}$ [28], where s_α is a linguistic term and $\tau + 1$ denotes the cardinality of S. s_0 and s_τ are the minimum and maximum linguistic terms that can be utilized by the decision makers. When $\tau = 4$, $S = \{s_0 = none, s_1 = low, s_2 = medium, s_3 = high, s_4 = perfect\}$. For any S, the following conditions should be satisfied [29]:

(1) If $0 \leq \alpha < \beta \leq \tau$, then $s_\alpha < s_\beta$;
(2) There exists a negation operator $neg(s_\alpha) = s_\beta$, where $\alpha + \beta = \tau$.

However, during the process of aggregating decision making information, the aggregated results usually do not match any one element in the linguistic term set and then may result in the loss of decision information. Thus, Xu [30] extended it to be a virtual linguistic model $\overline{S_1} = \{s_\alpha | \alpha \in [0, q]\}$, where q is a positive integer much larger than τ. Given $s_\alpha \in \overline{S_1}$, if $s_\alpha \in S$, then it is an original linguistic term; otherwise, it is referred to as a virtual linguistic term. Although the virtual linguistic model can avoid the loss of decision information, the output virtual linguistic terms are not interpretable. In this case, Xu et al. [31] defined the syntax and semantics of virtual linguistic terms and reconstructed the computational model of virtual linguistic terms based on their semantics.

2.2 Probabilistic Linguistic Term Set

The concept of hesitant fuzzy linguistic term set is a very flexible tool for decision making, which can express a single decision maker's complex preference information. However, they do not contain the weight information of each linguistic term. In this case, the concept of probabilistic linguistic term set was proposed to express both a decision maker's several possible linguistic terms and the probabilistic information of each possible linguistic term [14]. In this subsection, the concept and operational laws of probabilistic linguistic term sets are reviewed briefly.

Definition 1 [14] Let $S = \{s_0, s_1, \ldots, s_\tau\}$ be a linguistic term set, a probabilistic linguistic term set (PLTS) is defined as:

$$L(p) = \left\{ L^{(k)}\left(p^{(k)}\right) | L^{(k)} \in S; p^{(k)} \geq 0; k = 1, 2, \ldots, \#L(p); \sum_{k=1}^{\#L(p)} p^{(k)} \leq 1 \right\}$$

where $L^{(k)}\left(p^{(k)}\right)$ denotes the linguistic term $L^{(k)}$ associated with its probability $p^{(k)}$ and $\#L(p)$ is the number of elements in $L(p)$. For convenience, $L^{(k)}\left(p^{(k)}\right)$ is called a probabilistic linguistic term element (PLTE). If $k = 1$ and $p^{(k)} = 1$, then a PLTS reduces to a linguistic term.

If $\sum_{k=1}^{\#L(p)} p^{(k)} = 0$, then the decision maker does not provide any assessment information. If $\sum_{k=1}^{\#L(p)} p^{(k)} = 1$, then the decision maker provides complete assessment information. If $0 < \sum_{k=1}^{\#L(p)} p^{(k)} < 1$, then it implies that the decision maker offers only partial assessment information and it is a very common case in the practical decision making problems. Therefore, Pang et al. normalized the PLTS by means of allocating the unknown $1 - \sum_{k=1}^{\#L(p)} p^{(k)}$ to all the linguistic terms in $L(p)$ averagely and then repeating the least linguistic term in the PLTS with the smaller cardinality until two PLTSs have the same number of elements.

Definition 2 [14] If $L(p)$ is a PLTS, where $0 < \sum_{k=1}^{\#L(p)} p^{(k)} < 1$, then the normalized

PLTS $L_n(p)$ can be defined as $L_n(p) = \left\{ L^{(k)}\left(p_n^{(k)}\right) | k = 1, 2, \ldots, \#L(p) \right\}$, where

$p_n^{(k)} = p^{(k)} / \sum_{k=1}^{\#L(p)} p^{(k)}$ and $k = 1, 2, \ldots, \#L(p)$.

Definition 3 [14] Given any two PLTSs, $L_1(p) = \left\{ L_1^{(k)}\left(p_1^{(k)}\right) | k = 1,$

$2, \ldots, \#L_1(p). \right\}$ and $L_2(p) = \left\{ L_2^{(k)}\left(p_2^{(k)}\right) | k = 1, 2, \ldots, \#L_2(p) \right\}$, where $\#L_1(p)$ and

$\#L_2(p)$ are the numbers of linguistic terms in $L_1(p)$ and $L_2(p)$ respectively. If
$\#L_1(p) > \#L_2(p)$, then $\#L_1(p) - \#L_2(p)$ linguistic terms are added to $L_2(p)$ until the
numbers of linguistic terms in $L_1(p)$ and $L_2(p)$ are equal. The added linguistic terms
are the least ones in $L_2(p)$ and their corresponding probabilities are zero.

However, the normalization of PLTSs changes the linguistic assessment information and then influences the decision making results.

Example 1 Given two PLTSs $L_1(p) = \{s_5(0.8)\}$ and $L_2(p) = \{s_5(0.9)\}$, then their
normalized PLTSs are $L_{n1}(p) = L_{n2}(p) = \{s_5(1)\}$ and then the distance between
$L_{n1}(p)$ and $L_{n2}(p)$ is zero. It is quite obvious that $L_1(p)$ and $L_2(p)$ are different and
their distance cannot be zero.

Thus, we only utilize Definition 3 to normalize the PLTSs in order to make two
PLTSs have the same number of PLTEs and then redefine the operational laws for
PLTSs as follows:

Definition 4 Given any two PLTSs, $L_1(p) = \left\{ L_1^{(k)}\left(p_1^{(k)}\right) | k_1 = 1, 2, \ldots, \#L_1(p) \right\}$
and $L_2(p) = \left\{ L_2^{(k)}\left(p_2^{(k)}\right) | k_2 = 1, 2, \ldots, \#L_2(p) \right\}$, then

$$L_1(p) \oplus L_2(p) = \cup_{L_1^{(k_1)}\left(p_1^{(k_1)}\right) \in L_1(p), L_2^{(k_2)}\left(p_2^{(k_2)}\right) \in L_2(p)} \left\{ L_1^{(k_1)} \oplus L_2^{(k_2)}\left(p_1^{(k_1)} p_2^{(k_2)}\right) \right\}$$

$$L_1(p) \otimes L_2(p) = \cup_{L_1^{(k_1)}\left(p_1^{(k_1)}\right) \in L_1(p), L_2^{(k_2)}\left(p_2^{(k_2)}\right) \in L_2(p)} \left\{ L_1^{(k_1)} \otimes L_2^{(k_2)}\left(p_1^{(k_1)} p_2^{(k_2)}\right) \right\}$$

where $L_1^{(k_1)}$ and $L_2^{(k_2)}$ are the k_1 th and k_2 th linguistic terms in $L_1(p)$ and $L_2(p)$
respectively, $p_1^{(k_1)}$ and $p_2^{(k_2)}$ are their corresponding probabilities.

Some other operational laws can be defined as follows:

$$\lambda L_1(p) = \cup_{L_1^{(k_1)}\left(p_1^{(k_1)}\right) \in L_1(p)} \left\{ \lambda L_1^{(k_1)}\left(p_1^{(k_1)}\right) | k_1 = 1, 2, \ldots, \#L_1(p) \right\}$$

$$(L_1(p))^\lambda = \cup_{L_1^{(k_1)}\left(p_1^{(k_1)}\right) \in L_1(p)} \left\{ \left(L_1^{(k_1)}\right)^\lambda \left(p_1^{(k_1)}\right) \Big| k_1 = 1, 2, \ldots, \#L_1(p) \right\}$$

3 Distance Measures Between PLTSs

Distance measures are very important tools that have been widely utilized for decision making, pattern recognition, and clustering analysis. Until now, many research results have been achieved, for example, Liu [32] proposed the axiom definitions of entropy and distance measure for fuzzy sets and discussed their relations. Chaudhuri et al. [33] modified the Hausdorff-like metric distance for fuzzy sets to contain a single term representing the geometric distance only. Balopoulos et al. [34] studied some normalized distance measures for fuzzy sets based on the binary operators and matrix norms. Heidarzade et al. [35] gave a new definition of centroid for an interval type-2 fuzzy set and then introduced a formulation for computing the distance between two interval type-2 fuzzy sets. Zhang et al. [36] introduced a new family of normalized distances to measure the distance between two interval-valued fuzzy sets. Zeng et al. [37] gave a normalized Hamming distance and a normalized Euclidean distance for interval-valued fuzzy sets. Xu et al. [38] provided an overview of distance measures for intuitionistic fuzzy sets and also defined several continuous distance measures. Hatzimichailidis et al. [39] utilized matrix norms and fuzzy implications to define a novel distance metric for intuitionistic fuzzy sets. Hung et al. [40] introduced the Hausdorff metric to calculate the distance between two intuitionistic fuzzy sets. Wang et al. [41] provided the axiom definition of distance measure for intuitionistic fuzzy sets and then gave several distance measures. Szmidt et al. [42] defined the distance between two intuitionistic fuzzy sets considering all three parameters. Duenci et al. [43] introduced the L_p norm and the level of uncertainty to develop a distance measure for interval-valued intuitionistic fuzzy sets. Li et al. [44] gave the definitions of the Hamming distance and Euclidean distance for interval intuitionistic fuzzy sets. Xu et al. [45] proposed a variety of distance measures to compute the distance between two hesitant fuzzy sets. Li et al. [46] gave the concept of hesitance degree of a hesitant fuzzy element and defined the distance measures for hesitant fuzzy sets considering the hesitance degree. Bai [47] gave the Hamming distance, Euclidean distance, Hausdorff distance, and generalized distance for interval-valued hesitant fuzzy sets. Liao et al. [48] proposed a variety of distance measures for hesitant fuzzy linguistic term sets and then used them to rank the alternatives in the multi-criteria decision making. Meng et al. [49] presented two generalized hesitant fuzzy linguistic weighted distance measures for hesitant fuzzy linguistic term sets.

To the best of our knowledge, currently, there is very little study on distance measures for probabilistic linguistic term sets. Moreover, since probabilistic linguistic term sets express the assessment information in a different way from other sets, the existing distance measures cannot be utilized for probabilistic linguistic term sets. Hence, this chapter focuses on the distance measures for probabilistic linguistic term sets.

Inspired by the above analysis, the axiom definition of distance measure is first given for PLTSs in this section and then a variety of distance measures are studied for PLTSs.

Motivated by the axiom definition of distance measure for HFLTSs given by Liao et al. [48], we give the axiom definition of distance measure for PLTSs as follows:

Definition 5 Let $L_1(p)$ and $L_2(p)$ be two PLTSs, then the distance measure between them is defined as $d(L_1(p), L_2(p))$, which satisfies the following three conditions:

(1) $0 \leq d(L_1(p), L_2(p)) \leq 1$;
(2) $d(L_1(p), L_2(p)) = 0$ if and only if $L_1(p) = L_2(p)$;
(3) $d(L_1(p), L_2(p)) = d(L_2(p), L_1(p))$.

Before giving the distance measures for PLTSs, this chapter gives the definition of the distance measure between two PLTEs as follows:

Definition 6 Let $L_1^{(k_1)}\left(p_1^{(k_1)}\right) \in L_1(p)$ and $L_2^{(k_2)}\left(p_2^{(k_2)}\right) \in L_2(p)$ be two PLTEs, then the distance measure between them is defined as:

$$d\left(L_1^{(k_1)}\left(p_1^{(k_1)}\right), L_2^{(k_2)}\left(p_2^{(k_2)}\right)\right) = \left| p_1^{(k_1)} \times \frac{I(L_1^{(k_1)})}{\tau} - p_2^{(k_2)} \times \frac{I(L_2^{(k_2)})}{\tau} \right|$$

where $I\left(L_1^{(k_1)}\right)$ and $I\left(L_2^{(k_2)}\right)$ are the subscripts of the linguistic terms $L_1^{(k_1)}$ and $L_2^{(k_2)}$, respectively.

Since $0 \leq I\left(L_1^{(k_1)}\right) \leq \tau$ and $0 \leq I\left(L_2^{(k_2)}\right) \leq \tau$, then $0 \leq \frac{I(L_1^{(k_1)})}{\tau} \leq 1$ and $0 \leq \frac{I(L_2^{(k_2)})}{\tau} \leq 1$. Also since $0 \leq p_1^{(k_1)} \leq 1$ and $0 \leq p_2^{(k_2)} \leq 1$, then $0 \leq d\left(L_1^{(k_1)}\left(p_1^{(k_1)}\right), L_2^{(k_2)}\left(p_2^{(k_2)}\right)\right) \leq 1$.

Given two PLTSs $L_1(p)$ and $L_2(p)$, $L_1(p) = \left\{ L_1^{(k_1)}\left(p_1^{(k_1)}\right) | k_1 = 1, 2, \ldots, \#L_1(p) \right\}$ and $L_2(p) = \left\{ L_2^{(k_2)}\left(p_2^{(k_2)}\right) | k_2 = 1, 2, \ldots, \#L_2(p) \right\}$ with $\#L_1(p) = \#L_2(p)$, then we can give the definitions of a normalized Hamming distance and a normalized Euclidean distance for PLTSs based on Definition 6.

The normalized Hamming distance between $L_1(p)$ and $L_2(p)$ can be given as:

$$d_{nhd}(L_1(p), L_2(p)) = \frac{1}{\#L_1(p)} \sum_{k=1}^{\#L_1(p)} d\left(L_1^{(k)}\left(p_1^{(k)}\right), L_2^{(k)}\left(p_2^{(k)}\right)\right) \qquad (2)$$

and the normalized Euclidean distance between $L_1(p)$ and $L_2(p)$ can be defined as:

$$d_{ned}(L_1(p), L_2(p)) = \left[\frac{1}{\#L_1(p)} \sum_{k=1}^{\#L_1(p)} \left(d\left(L_1^{(k)}\left(p_1^{(k)}\right), L_2^{(k)}\left(p_2^{(k)}\right)\right)\right)^2\right]^{1/2} \qquad (3)$$

Inspired by the generalized idea proposed by Yager [50], this chapter gives the generalized normalized distance as:

$$d_{gnd}(L_1(p), L_2(p)) = \left[\frac{1}{\#L_1(p)} \sum_{k=1}^{\#L_1(p)} \left(d\left(L_1^{(k)}\left(p_1^{(k)}\right), L_2^{(k)}\left(p_2^{(k)}\right)\right)\right)^{\lambda}\right]^{1/\lambda} \qquad (4)$$

where $\lambda > 0$.

In particular, if $\lambda = 1$, then the generalized normalized distance reduces to the normalized Hamming distance. If $\lambda = 2$, then it reduces to the normalized Euclidean distance.

Example 2 Given a linguistic term set $S = \{s_\alpha | \alpha = 0, 1, \ldots, 6\}$, as well as two PLTSs $L_1(p) = \{s_2(0.6), s_3(0.3)\}$ and $L_2(p) = \{s_4(0.8), s_5(0.2)\}$, then the generalized normalized distance between them is calculated as

$$d_{gnd}(L_1(p), L_2(p)) = \left[\frac{1}{2}\left(\left|0.6 \times \frac{2}{6} - 0.8 \times \frac{4}{6}\right|^\lambda + \left|0.3 \times \frac{3}{6} - 0.2 \times \frac{5}{6}\right|^\lambda\right)\right]^{1/\lambda}$$

If $\lambda = 1$, then the normalized Hamming distance is $d_{nhd}(L_1(p), L_2(p)) = 0.175$. If $\lambda = 2$, then the normalized Euclidean distance is $d_{ned}(L_1(p), L_2(p)) = 0.2360$.

The Hausdorff distance can also be given for PLTSs. Let $L_1(p)$ and $L_2(p)$ be any two PLTSs, then the generalized normalized Hausdorff distance between them can be defined as:

$$d_{gnhaud}(L_1(p), L_2(p)) = \left[\max_k \left(d\left(L_1^{(k)}\left(p_1^{(k)}\right), L_2^{(k)}\left(p_2^{(k)}\right)\right)\right)^\lambda\right]^{1/\lambda} \qquad (5)$$

where $\lambda > 0$.

Two special cases of the generalized normalized Hausdorff distance are given as follows:

(1) If $\lambda = 1$, then it reduces to the normalized Hamming-Hausdorff distance as:

$$d_{nhhaud}(L_1(p), L_2(p)) = \max_k \left(d\left(L_1^{(k)}\left(p_1^{(k)} \right), L_2^{(k)}\left(p_2^{(k)} \right) \right) \right) \qquad (6)$$

(2) If $\lambda = 2$, then it reduces to the normalized Euclidean-Hausdorff distance as:

$$d_{nehaud}(L_1(p), L_2(p)) = \left[\max_k \left(d\left(L_1^{(k)}\left(p_1^{(k)} \right), L_2^{(k)}\left(p_2^{(k)} \right) \right) \right)^2 \right]^{1/2} \qquad (7)$$

Example 3 If we adopt the data of Example 2 here, then their generalized normalized Hausdorff distance is

$$d_{gnhaud}(L_1(p), L_2(p)) = = \left[\max \left(\left| 0.6 \times \frac{2}{6} - 0.8 \times \frac{4}{6} \right|^{\lambda}, \left| 0.3 \times \frac{3}{6} - 0.2 \times \frac{5}{6} \right|^{\lambda} \right) \right]^{1/\lambda}$$

If $\lambda = 1$, then the normalized Hamming-Hausdorff distance measure between them is $d_{nhhaud}(L_1(p), L_2(p)) = 0.3333$.

If $\lambda = 2$, then the normalized Euclidean-Hausdorff distance measure between them is $d_{nehaud}(L_1(p), L_2(p)) = 0.3333$.

Additionally, some hybrid distance measures can be given for PLTSs by combining the above two types of distance measures as follows:

(1) The hybrid normalized Hamming distance between $L_1(p)$ and $L_2(p)$ is

$$d_{hnhd}(L_1(p), L_2(p)) = \frac{1}{2} \left[\frac{1}{\#L_1(p)} \sum_{k=1}^{\#L_1(p)} d\left(L_1^{(k)}\left(p_1^{(k)} \right), L_2^{(k)}\left(p_2^{(k)} \right) \right) + \max_k d\left(L_1^{(k)}\left(p_1^{(k)} \right), L_2^{(k)}\left(p_2^{(k)} \right) \right) \right] \qquad (8)$$

(2) The hybrid normalized Euclidean distance between $L_1(p)$ and $L_2(p)$ is

$$d_{hned}(L_1(p), L_2(p)) = \left\{ \frac{1}{2} \left[\frac{1}{\#L_1(p)} \sum_{k=1}^{\#L_1(p)} \left(d\left(L_1^{(k)}\left(p_1^{(k)} \right), L_2^{(k)}\left(p_2^{(k)} \right) \right) \right)^2 + \max_k \left(d\left(L_1^{(k)}\left(p_1^{(k)} \right), L_2^{(k)}\left(p_2^{(k)} \right) \right) \right)^2 \right] \right\}^{1/2} \qquad (9)$$

(3) The generalized hybrid normalized distance between $L_1(p)$ and $L_2(p)$ is

$$d_{ghnd}(L_1(p), L_2(p)) = \left\{ \frac{1}{2} \left[\frac{1}{\#L_1(p)} \sum_{k=1}^{\#L_1(p)} \left(d\left(L_1^{(k)}\left(p_1^{(k)} \right), L_2^{(k)}\left(p_2^{(k)} \right) \right) \right)^\lambda + \right. \right.$$

$$\left. \left. \max_k \left(d\left(L_1^{(k)}\left(p_1^{(k)} \right), L_2^{(k)}\left(p_2^{(k)} \right) \right) \right)^\lambda \right] \right\}^{1/\lambda}$$

$$(10)$$

Example 4 If we adopt the data of Example 2 here, then the generalized hybrid normalized distance is

$$d_{ghnd}(L_1(p), L_2(p)) = \left\{ \frac{1}{2} \left[\frac{1}{2} \left(\left(\left| 0.6 \times \frac{2}{6} - 0.8 \times \frac{4}{6} \right| \right)^\lambda + \left(\left| 0.3 \times \frac{3}{6} - 0.2 \times \frac{5}{6} \right| \right)^\lambda \right) + \right. \right.$$

$$\left. \left. \max \left(\left(\left| 0.6 \times \frac{2}{6} - 0.8 \times \frac{4}{6} \right| \right)^\lambda, \left(\left| 0.3 \times \frac{3}{6} - 0.2 \times \frac{5}{6} \right| \right)^\lambda \right) \right] \right\}^{1/\lambda}$$

If $\lambda = 1$, the hybrid normalized Hamming distance measure between them is $d_{hnhd}(L_1(p), L_2(p)) = 0.2542$.

If $\lambda = 2$, the hybrid normalized Euclidean distance measure between them is $d_{hned}(L_1(p), L_2(p)) = 0.2888$.

4   Distance Measures Between Two Collections of PLTSs

The distance measures between two PLTSs can only be used to measure the distance between two alternatives with respect to one criterion. However, in most cases, all the alternatives are usually assessed with respect to some different criteria, which have different weights. Then the distance between two alternatives with respect to a fixed number of criteria should be measured by using the distance measures between two collections of PLTSs by taking into account the weight information of criteria. Hence, in this section, we investigate the weighted distance measures between two collections of PLTSs in discrete and continuous cases.

4.1   Distance Measures Between Two Collections of PLTSs in Discrete Case

Given a LTS $S = \{s_\alpha | \alpha = 0, 1, \ldots, \tau\}$, two alternatives denoted as two collections of PLTSs $L_1 = \{L_{11}(p), L_{12}(p), \ldots, L_{1m}(p)\}$ and $L_2 = \{L_{21}(p), L_{22}(p), \ldots, L_{2m}(p)\}$ with the weight vector of criteria $w = (w_1, w_2, \ldots, w_m)^T$, where $0 \leq w_j \leq 1$ and

$\sum_{j=1}^{m} w_j = 1$, then a generalized weighted distance between two alternatives, namely, L_1 and L_2 is defined as:

$$d_{gwd}(L_1, L_2) = \left[\sum_{j=1}^{m} \frac{w_j}{\#L_{1j}(p)} \sum_{k_j=1}^{\#L_{1j}(p)} \left(d\left(L_{1j}^{(k_j)}\left(p_{1j}^{(k_j)} \right), L_{2j}^{(k_j)}\left(p_{2j}^{(k_j)} \right) \right) \right)^{\lambda} \right]^{1/\lambda} \quad (11)$$

and a generalized weighted Hausdorff distance between L_1 and L_2 is defined as:

$$d_{gwhaud}(L_1, L_2) = \left[\sum_{j=1}^{m} w_j \max_{k_j} \left(d\left(L_{1j}^{(k_j)}\left(p_{1j}^{(k_j)} \right), L_{2j}^{(k_j)}\left(p_{2j}^{(k_j)} \right) \right) \right)^{\lambda} \right]^{1/\lambda} \quad (12)$$

where $\lambda > 0$.

Particularly, if $\lambda = 1$, then the weighted Hamming distance between L_1 and L_2 is

$$d_{whd}(L_1, L_2) = \sum_{j=1}^{m} \frac{w_j}{\#L_{1j}(p)} \sum_{k_j=1}^{\#L_{1j}(p)} d\left(L_{1j}^{(k_j)}\left(p_{1j}^{(k_j)} \right), L_{2j}^{(k_j)}\left(p_{2j}^{(k_j)} \right) \right) \quad (13)$$

and the weighted Hamming-Hausdorff distance between L_1 and L_2 is

$$d_{whhaud}(L_1, L_2) = \sum_{j=1}^{m} w_j \max_{k_j} d\left(L_{1j}^{(k_j)}\left(p_{1j}^{(k_j)} \right), L_{2j}^{(k_j)}\left(p_{2j}^{(k_j)} \right) \right) \quad (14)$$

If $\lambda = 2$, then the weighted Euclidean distance between L_1 and L_2 is

$$d_{wed}(L_1, L_2) = \left\{ \sum_{j=1}^{m} \frac{w_j}{\#L_{1j}(p)} \sum_{k_j=1}^{\#L_{1j}(p)} \left(d\left(L_{1j}^{(k_j)}\left(p_{1j}^{(k_j)} \right), L_{2j}^{(k_j)}\left(p_{2j}^{(k_j)} \right) \right) \right)^{2} \right\}^{1/2} \quad (15)$$

and the weighted Euclidean-Hausdorff distance between L_1 and L_2 is

$$d_{wehaud}(L_1, L_2) = \left[\sum_{j=1}^{m} w_j \max_{k_j} \left(d\left(L_{1j}^{(k_j)}\left(p_{1j}^{(k_j)} \right), L_{2j}^{(k_j)}\left(p_{2j}^{(k_j)} \right) \right) \right)^{2} \right]^{1/2} \quad (16)$$

Based on the above weighted distances, several hybrid weighted distances are derived as follows:

(1) The hybrid weighted Hamming distance between L_1 and L_2 is defined as:

$$d_{hwhd}(L_1, L_2) = \sum_{j=1}^{m} \frac{w_j}{2} \left[\frac{1}{\#L_{1j}(p)} \sum_{k_j=1}^{\#L_{1j}(p)} d\left(L_{1j}^{(k_j)}\left(p_{1j}^{(k_j)}\right), L_{2j}^{(k_j)}\left(p_{2j}^{(k_j)}\right)\right) + \right.$$
$$\left. \max_{k_j} d\left(L_{1j}^{(k_j)}\left(p_{1j}^{(k_j)}\right), L_{2j}^{(k_j)}\left(p_{2j}^{(k_j)}\right)\right)\right]$$

$$(17)$$

(2) The hybrid weighted Euclidean distance between L_1 and L_2 is defined as:

$$d_{hwed}(L_1, L_2) = \left\{ \sum_{j=1}^{m} \frac{w_j}{2} \left[\frac{1}{\#L_{1j}(p)} \sum_{k_j=1}^{\#L_{1j}(p)} \left(d\left(L_{1j}^{(k_j)}\left(p_{1j}^{(k_j)}\right), L_{2j}^{(k_j)}\left(p_{2j}^{(k_j)}\right)\right)\right)^2 + \right. \right.$$
$$\left. \left. \max_{k_j} \left(d\left(L_{1j}^{(k_j)}\left(p_{1j}^{(k_j)}\right), L_{2j}^{(k_j)}\left(p_{2j}^{(k_j)}\right)\right)\right)^2 \right] \right\}^{1/2}$$

$$(18)$$

(3) The generalized hybrid weighted distance between L_1 and L_2 is defined as:

$$d_{ghwd}(L_1, L_2) = \left\{ \sum_{j=1}^{m} \frac{w_j}{2} \left(\frac{1}{\#L_{1j}(p)} \sum_{k_j=1}^{\#L_{1j}(p)} \left(d\left(L_{1j}^{(k_j)}\left(p_{1j}^{(k_j)}\right), L_{2j}^{(k_j)}\left(p_{2j}^{(k_j)}\right)\right)\right)^{\lambda} + \right. \right.$$
$$\left. \left. \max_{k_j} \left(d\left(L_{1j}^{(k_j)}\left(p_{1j}^{(k_j)}\right), L_{2j}^{(k_j)}\left(p_{2j}^{(k_j)}\right)\right)\right)^{\lambda} \right) \right\}^{1/\lambda}$$

$$(19)$$

where $\lambda > 0$.

4.2 Distance Measures Between Two Collections of PLTSs in Continuous Case

The above mentioned distance measures are specially proposed for the discrete case and then they cannot be applicable for the continuous case where the universe of discourse and the weights of PLTSs are continuous. Thus, this subsection focuses on the distance measures between two collections of PLTSs in continuous case.

Let $x \in X = [a, b]$ and the weights of PLTSs with respect to x be $w(x)$, where $w(x) \in [0, 1]$ and $\int_a^b w(x)\,dx = 1$, then we define the generalized continuous weighted distance between two collections of PLTSs L_1 and L_2 as:

$$d_{gcwd}(L_1, L_2) = \left\{ \int_a^b w(x) \frac{1}{l_x} \sum_{k=1}^{l_x} \left[d\left(L_{1x}^{(k)}\left(p_{1x}^{(k)}\right), L_{2x}^{(k)}\left(p_{2x}^{(k)}\right) \right) \right]^\lambda dx \right\}^{1/\lambda} \quad (20)$$

with

$$d\left(L_{1x}^{(k)}\left(p_{1x}^{(k)}\right), L_{2x}^{(k)}\left(p_{2x}^{(k)}\right) \right) = \left| p_1^{(k_1)}(x) \times \frac{I_1^{(k_1)}(x)}{\tau} - p_2^{(k_2)}(x) \times \frac{I_x^{(k_2)}(x)}{\tau} \right|$$

where $\lambda > 0$.

If $\lambda = 1$, then the generalized continuous weighted distance between L_1 and L_2 reduces to a continuous weighted Hamming distance:

$$d_{cwhd}(L_1, L_2) = \int_a^b w(x) \frac{1}{l_x} \sum_{k=1}^{l_x} d\left(L_{1x}^{(k)}\left(p_{1x}^{(k)}\right), L_{2x}^{(k)}\left(p_{2x}^{(k)}\right) \right) dx \quad (21)$$

If $\lambda = 2$, then the generalized continuous weighted distance between L_1 and L_2 becomes a continuous weighted Euclidean distance:

$$d_{cwed}(L_1, L_2) = \left\{ \int_a^b w(x) \frac{1}{l_x} \sum_{k=1}^{l_x} \left[d\left(L_{1x}^{(k)}\left(p_{1x}^{(k)}\right), L_{2x}^{(k)}\left(p_{2x}^{(k)}\right) \right) \right]^2 dx \right\}^{1/2} \quad (22)$$

If $w(x) = 1/(b-a)$ for all $x \in [a, b]$, then the generalized continuous weighted distance between two collections of PLTSs L_1 and L_2 reduces to the generalized continuous normalized distance:

$$d_{gcnd}(L_1, L_2) = \left\{ \frac{1}{b-a} \int_a^b \frac{1}{l_x} \sum_{k=1}^{l_x} \left[d\left(L_{1x}^{(k)}\left(p_{1x}^{(k)}\right), L_{2x}^{(k)}\left(p_{2x}^{(k)}\right) \right) \right]^\lambda dx \right\}^{1/\lambda} \quad (23)$$

Especially, if $\lambda = 1$, then the generalized continuous normalized distance becomes a continuous normalized Hamming distance:

$$d_{cnhd}(L_1, L_2) = \frac{1}{b-a} \int_a^b \frac{1}{l_x} \sum_{k=1}^{l_x} d\left(L_{1x}^{(k)}\left(p_{1x}^{(k)}\right), L_{2x}^{(k)}\left(p_{2x}^{(k)}\right) \right) dx \quad (24)$$

If $\lambda = 2$, then it reduces to a continuous normalized Euclidean distance:

$$d_{cned}(L_1, L_2) = \left[\frac{1}{b-a} \int_a^b \frac{1}{l_x} \sum_{k=1}^{l_x} \left(d\left(L_{1x}^{(k)}\left(p_{1x}^{(k)}\right), L_{2x}^{(k)}\left(p_{2x}^{(k)}\right) \right) \right)^2 dx \right]^{1/2} \quad (25)$$

where $\lambda > 0$.

By using the Hausdorff metric, we define a generalized continuous weighted Hausdorff distance:

$$d_{gcwhaud}(L_1(p), L_2(p)) = \left[\int_a^b w(x) \max_k \left(d\left(L_{1x}^{(k)}\left(p_{1x}^{(k)}\right), L_{2x}^{(k)}\left(p_{2x}^{(k)}\right) \right) \right)^\lambda dx \right]^{1/\lambda} \quad (26)$$

Especially, if $\lambda = 1$, then the generalized continuous weighted Hausdorff distance becomes a continuous weighted Hamming-Hausdorff distance:

$$d_{cwhhaud}(L_1(p), L_2(p)) = \int_a^b w(x) \max_k d\left(L_{1x}^{(k)}\left(p_{1x}^{(k)}\right), L_{2x}^{(k)}\left(p_{2x}^{(k)}\right) \right) dx \quad (27)$$

If $\lambda = 2$, then it reduces to a continuous weighted Euclidean-Hausdorff distance:

$$d_{cwehaud}(L_1(p), L_2(p)) = \left[\int_a^b w(x) \max_k \left(d\left(L_{1x}^{(k)}\left(p_{1x}^{(k)}\right), L_{2x}^{(k)}\left(p_{2x}^{(k)}\right) \right) \right)^2 dx \right]^{1/2} \quad (28)$$

If $w(x) = 1/(b-a)$ for all $x \in [a, b]$, then the generalized continuous weighted Hausdorff distance between two collections of PLTSs L_1 and L_2 reduces to the generalized continuous normalized Hausdorff distance:

$$d_{gcnhaud}(L_1(p), L_2(p)) = \left[\frac{1}{b-a} \int_a^b \max_k \left(d\left(L_{1x}^{(k)}\left(p_{1x}^{(k)}\right), L_{2x}^{(k)}\left(p_{2x}^{(k)}\right) \right) \right)^\lambda dx \right]^{1/\lambda} \quad (29)$$

If $\lambda = 1$, then the generalized continuous normalized Hausdorff distance reduces to the continuous normalized Hamming-Hausdorff distance:

$$d_{cnhhaud}(L_1(p), L_2(p)) = \frac{1}{b-a} \int_a^b \max_k d\left(L_{1x}^{(k)}\left(p_{1x}^{(k)}\right), L_{2x}^{(k)}\left(p_{2x}^{(k)}\right) \right) dx \quad (30)$$

If $\lambda = 2$, then it reduces to the continuous normalized Euclidean-Hausdorff distance:

$$d_{cnehaud}(L_1(p), L_2(p)) = \left[\frac{1}{b-a} \int_a^b \max_k \left(d\left(L_{1x}^{(k)}\left(p_{1x}^{(k)}\right), L_{2x}^{(k)}\left(p_{2x}^{(k)}\right) \right) \right)^2 dx \right]^{1/2} \quad (31)$$

Similarly, some hybrid continuous weighted distance measures can be defined by combing the above mentioned continuous distances. In what follows, we define the generalized hybrid continuous weighted distance:

$$d_{ghcwd}(L_1, L_2) = \left\{ \int_a^b \frac{w(x)}{2} \left[\frac{1}{l_x} \sum_{k=1}^{l_x} \left(d\left(L_{1x}^{(k)}\left(p_{1x}^{(k)}\right), L_{2x}^{(k)}\left(p_{2x}^{(k)}\right) \right) \right)^\lambda + \right. \right.$$
$$\left. \left. \max_k \left(d\left(L_{1x}^{(k)}\left(p_{1x}^{(k)}\right), L_{2x}^{(k)}\left(p_{2x}^{(k)}\right) \right) \right)^\lambda \right] dx \right\}^{1/\lambda} \tag{32}$$

If $\lambda = 1$, then the generalized hybrid continuous weighted distance reduces to the hybrid continuous weighted Hamming distance:

$$d_{hcwhd}(L_1, L_2) = \int_a^b \frac{w(x)}{2} \left[\frac{1}{l_x} \sum_{k=1}^{l_x} d\left(L_{1x}^{(k)}\left(p_{1x}^{(k)}\right), L_{2x}^{(k)}\left(p_{2x}^{(k)}\right) \right) + \right.$$
$$\left. \max_k d\left(L_{1x}^{(k)}\left(p_{1x}^{(k)}\right), L_{2x}^{(k)}\left(p_{2x}^{(k)}\right) \right) \right] dx \tag{33}$$

If $\lambda = 2$, then it reduces to the hybrid continuous weighted Euclidean distance:

$$d_{hcwed}(L_1, L_2) = \left\{ \int_a^b \frac{w(x)}{2} \left[\frac{1}{l_x} \sum_{k=1}^{l_x} \left(d\left(L_{1x}^{(k)}\left(p_{1x}^{(k)}\right), L_{2x}^{(k)}\left(p_{2x}^{(k)}\right) \right) \right)^2 + \right. \right.$$
$$\left. \left. \max_k \left(d\left(L_{1x}^{(k)}\left(p_{1x}^{(k)}\right), L_{2x}^{(k)}\left(p_{2x}^{(k)}\right) \right) \right)^2 \right] dx \right\}^{1/2} \tag{34}$$

where $\lambda > 0$.

If $w(x) = 1/(b-a)$ for all $x \in [a, b]$, then the generalized hybrid continuous weighted distance reduces to the generalized hybrid continuous normalized distance:

$$d_{ghcnd}(L_1, L_2) = \left\{ \frac{1}{2(b-a)} \int_a^b \left[\frac{1}{l_x} \sum_{k=1}^{l_x} \left(d\left(L_{1x}^{(k)}\left(p_{1x}^{(k)}\right), L_{2x}^{(k)}\left(p_{2x}^{(k)}\right) \right) \right)^\lambda + \right. \right.$$
$$\left. \left. \max_k \left(d\left(L_{1x}^{(k)}\left(p_{1x}^{(k)}\right), L_{2x}^{(k)}\left(p_{2x}^{(k)}\right) \right) \right)^\lambda \right] dx \right\}^{1/\lambda} \tag{35}$$

If $\lambda = 1$, then the generalized hybrid continuous normalized distance reduces to the hybrid continuous normalized Hamming distance:

$$d_{hcnhd}(L_1, L_2) = \frac{1}{2(b-a)} \int_a^b \left[\frac{1}{l_x} \sum_{k=1}^{l_x} d\left(L_{1x}^{(k)}\left(p_{1x}^{(k)}\right), L_{2x}^{(k)}\left(p_{2x}^{(k)}\right) \right) + \right.$$
$$\left. \max_k d\left(L_{1x}^{(k)}\left(p_{1x}^{(k)}\right), L_{2x}^{(k)}\left(p_{2x}^{(k)}\right) \right) \right] dx \tag{36}$$

If $\lambda = 2$, then it reduces to the hybrid continuous normalized Euclidean distance:

$$d_{hcned}(L_1, L_2) = \left\{ \frac{1}{2(b-a)} \int_a^b \left[\frac{1}{l_x} \sum_{k=1}^{l_x} \left(d\left(L_{1x}^{(k)}\left(p_{1x}^{(k)} \right), L_{2x}^{(k)}\left(p_{2x}^{(k)} \right) \right) \right)^2 + \max_k \left(d\left(L_{1x}^{(k)}\left(p_{1x}^{(k)} \right), L_{2x}^{(k)}\left(p_{2x}^{(k)} \right) \right) \right)^2 \right] dx \right\}^{1/2} \tag{37}$$

4.3 Ordered Weighted Distance Measures Between Two Collections of PLTSs

Ordered weighted distance measures were first proposed by Xu et al. to develop a group decision making approach [51]. The idea behind the ordered weighted distance measures aims to achieve a good trade-off between the influences of excessively large and small deviations on the aggregation results by assigning low weights to excessively large deviations and high weights to excessively small ones. Because of this excellent feature, many scholars have paid attention to the ordered weighted distance measures and then applied them to decision making and pattern recognition [52]. Until now, to our knowledge, there are no studies about the ordered weighted distance measures for PLTSs. Hence, in this subsection, we focus on the ordered weighted distance measures for two collections of PLTSs.

Inspired by the previous work on the ordered weighted distance measures, in the following, we define a generalized ordered weighted distance between two collections of PLTSs L_1 and L_2:

$$d_{gowd}(L_1, L_2) = \left\{ \sum_{j=1}^m \frac{w_j}{\#L_{1\delta(j)}(p)} \sum_{k_{\delta(j)}=1}^{\#L_{1\delta(j)}(p)} \left(d\left(L_{1\delta(j)}^{\left(k_{\delta(j)} \right)}\left(p_{1\delta(j)}^{\left(k_{\delta(j)} \right)} \right), L_{2\delta(j)}^{\left(k_{\delta(j)} \right)}\left(p_{2\delta(j)}^{\left(k_{\delta(j)} \right)} \right) \right) \right)^{\lambda} \right\}^{1/\lambda} \tag{38}$$

where $\lambda > 0$ and $\delta(j): (1, 2, \ldots, m) \to (1, 2, \ldots, m)$ is a permutation which satisfies

$$\frac{1}{\#L_{1\delta(j)}(p)} \sum_{k_{\delta(j)}=1}^{\#L_{1\delta(j)}(p)} \left(d\left(L_{1\delta(j)}^{\left(k_{\delta(j)} \right)}\left(p_{1\delta(j)}^{\left(k_{\delta(j)} \right)} \right), L_{2\delta(j)}^{\left(k_{\delta(j)} \right)}\left(p_{2\delta(j)}^{\left(k_{\delta(j)} \right)} \right) \right) \right)^{\lambda}$$

$$\geq \frac{1}{\#L_{1\delta(j+1)}(p)} \sum_{k_{\delta(j+1)}=1}^{\#L_{1\delta(j+1)}(p)} \left(d\left(L_{1\delta(j+1)}^{\left(k_{\delta(j+1)} \right)}\left(p_{1\delta(j+1)}^{\left(k_{\delta(j+1)} \right)} \right), L_{2\delta(j+1)}^{\left(k_{\delta(j+1)} \right)}\left(p_{2\delta(j+1)}^{\left(k_{\delta(j+1)} \right)} \right) \right) \right)^{\lambda} \tag{39}$$

where $j = 1, 2, \ldots, m$.

In particular, if $\lambda = 1$, then the generalized ordered weighted distance between two collections of PLTSs reduces to the ordered weighted Hamming distance:

$$d_{owhd}(L_1, L_2) = \sum_{j=1}^{m} \frac{w_j}{\#L_{1\delta(j)}(p)} \sum_{k_{\delta(j)}=1}^{\#L_{1\delta(j)}(p)} d\left(L_{1\delta(j)}^{(k_{\delta(j)})} \left(p_{1\delta(j)}^{(k_{\delta(j)})} \right), L_{2\delta(j)}^{(k_{\delta(j)})} \left(p_{2\delta(j)}^{(k_{\delta(j)})} \right) \right) \quad (40)$$

If $\lambda = 2$, then the generalized ordered weighted distance between two collections of PLTSs reduces to the ordered weighted Euclidean distance:

$$d_{owed}(L_1, L_2) = \left\{ \sum_{j=1}^{m} \frac{w_j}{\#L_{1\delta(j)}(p)} \sum_{k_{\delta(j)}=1}^{\#L_{1\delta(j)}(p)} \left(d\left(L_{1\delta(j)}^{(k_{\delta(j)})} \left(p_{1\delta(j)}^{(k_{\delta(j)})} \right), L_{2\delta(j)}^{(k_{\delta(j)})} \left(p_{2\delta(j)}^{(k_{\delta(j)})} \right) \right) \right)^2 \right\}^{1/2}$$

$$(41)$$

Considering the Hausdorff metric, here we define a generalized ordered weighted Hausdorff distance between two collections of PLTSs:

$$d_{gowhaud}(L_1, L_2) = \left[\sum_{j=1}^{m} w_j \max_{k_{\delta(j)}} \left(d\left(L_{1\delta(j)}^{(k_{\delta(j)})} \left(p_{1\delta(j)}^{(k_{\delta(j)})} \right), L_{2\delta(j)}^{(k_{\delta(j)})} \left(p_{2\delta(j)}^{(k_{\delta(j)})} \right) \right) \right)^\lambda \right]^{1/\lambda}$$

$$(42)$$

where $\lambda > 0$ and $\delta(j): (1, 2, \ldots, m) \rightarrow (1, 2, \ldots, m)$ is a permutation which satisfies

$$\max_{k_{\delta(j)}} \left(d\left(L_{1\delta(j)}^{(k_{\delta(j)})} \left(p_{1\delta(j)}^{(k_{\delta(j)})} \right), L_{2\delta(j)}^{(k_{\delta(j)})} \left(p_{2\delta(j)}^{(k_{\delta(j)})} \right) \right) \right)^\lambda \geq$$
$$\max_{k_{\delta(j+1)}} \left(d\left(L_{1\delta(j+1)}^{(k_{\delta(j+1)})} \left(p_{1\delta(j+1)}^{(k_{\delta(j+1)})} \right), L_{2\delta(j+1)}^{(k_{\delta(j+1)})} \left(p_{2\delta(j+1)}^{(k_{\delta(j+1)})} \right) \right) \right)^\lambda$$

$$(43)$$

where $j = 1, 2, \ldots, m$.

In particular, if $\lambda = 1$, then the generalized ordered weighted Hausdorff distance between two collections of PLTSs reduces to the ordered weighted Hamming-Hausdorff distance:

$$d_{owhhaud}(L_1, L_2) = \sum_{j=1}^{m} w_j \max_{k_{\delta(j)}} d\left(L_{1\delta(j)}^{(k_{\delta(j)})} \left(p_{1\delta(j)}^{(k_{\delta(j)})} \right), L_{2\delta(j)}^{(k_{\delta(j)})} \left(p_{2\delta(j)}^{(k_{\delta(j)})} \right) \right) \quad (44)$$

If $\lambda = 2$, then the generalized ordered weighted Hausdorff distance between two collections of PLTSs reduces to the ordered weighted Euclidean-Hausdorff distance:

$$d_{owehaud}(L_1, L_2) = \left[\sum_{j=1}^{m} w_j \max_{k_{\delta(j)}} \left(d\left(L_{1\delta(j)}^{(k_{\delta(j)})}\left(p_{1\delta(j)}^{(k_{\delta(j)})}\right), L_{2\delta(j)}^{(k_{\delta(j)})}\left(p_{2\delta(j)}^{(k_{\delta(j)})}\right)\right)\right)^2\right]^{1/2}$$

(45)

Combining the ordered weighted distances and the ordered weighted Hausdorff distances, some hybrid ordered weighted distances can be defined as follows:

(1) The hybrid ordered weighted Hamming distance between two collections of PLTSs L_1 and L_2 is

$$d_{howhd}(L_1, L_2) = \sum_{j=1}^{m} \frac{w_j}{2}\left[\frac{1}{\#L_{1\delta(j)}(p)}\sum_{k_{\delta(j)}=1}^{\#L_{1\delta(j)}(p)} d\left(L_{1\delta(j)}^{(k_{\delta(j)})}\left(p_{1\delta(j)}^{(k_{\delta(j)})}\right), L_{2\delta(j)}^{(k_{\delta(j)})}\left(p_{2\delta(j)}^{(k_{\delta(j)})}\right)\right) + \right.$$
$$\left. \max_{k_{\delta(j)}} d\left(L_{1\delta(j)}^{(k_{\delta(j)})}\left(p_{1\delta(j)}^{(k_{\delta(j)})}\right), L_{2\delta(j)}^{(k_{\delta(j)})}\left(p_{2\delta(j)}^{(k_{\delta(j)})}\right)\right)\right]$$

(46)

(2) The hybrid ordered weighted Euclidean distance between two collections of PLTSs L_1 and L_2 is

$$d_{howed}(L_1, L_2) = \left\{\sum_{j=1}^{m} \frac{w_j}{2}\left[\frac{1}{\#L_{1\delta(j)}(p)}\sum_{k_{\delta(j)}=1}^{\#L_{1\delta(j)}(p)} \left(d\left(L_{1\delta(j)}^{(k_{\delta(j)})}\left(p_{1\delta(j)}^{(k_{\delta(j)})}\right), L_{2\delta(j)}^{(k_{\delta(j)})}\left(p_{2\delta(j)}^{(k_{\delta(j)})}\right)\right)\right)^2 + \right.\right.$$
$$\left.\left. \max_{k_{\delta(j)}} \left(d\left(L_{1\delta(j)}^{(k_{\delta(j)})}\left(p_{1\delta(j)}^{(k_{\delta(j)})}\right), L_{2\delta(j)}^{(k_{\delta(j)})}\left(p_{2\delta(j)}^{(k_{\delta(j)})}\right)\right)\right)^2\right]\right\}^{1/2}$$

(47)

(3) The generalized hybrid ordered weighted distance between two collections of PLTSs L_1 and L_2 is

$$d_{ghowd}(L_1, L_2)$$
$$= \left\{\sum_{j=1}^{m} \frac{w_j}{2}\left[\frac{1}{\#L_{1\delta(j)}(p)}\sum_{k_{\delta(j)}=1}^{\#L_{1\delta(j)}(p)} \left(d\left(L_{1\delta(j)}^{(k_{\delta(j)})}\left(p_{1\delta(j)}^{(k_{\delta(j)})}\right), L_{2\delta(j)}^{(k_{\delta(j)})}\left(p_{2\delta(j)}^{(k_{\delta(j)})}\right)\right)\right)^{\lambda} + \right.\right.$$
$$\left.\left. \max_{k_{\delta(j)}} \left(d\left(L_{1\delta(j)}^{(k_{\delta(j)})}\left(p_{1\delta(j)}^{(k_{\delta(j)})}\right), L_{2\delta(j)}^{(k_{\delta(j)})}\left(p_{2\delta(j)}^{(k_{\delta(j)})}\right)\right)\right)^{\lambda}\right]\right\}^{1/\lambda}$$

(48)

where $\lambda > 0$ and $\delta(j): (1, 2, \ldots, m) \to (1, 2, \ldots, m)$ is a permutation which satisfies

$$\frac{1}{\#L_{1\delta(j)}(p)} \sum_{k_{\delta(j)}=1}^{\#L_{1\delta(j)}(p)} \left(d\left(L_{1\delta(j)}^{\left(k_{\delta(j)}\right)}\left(p_{1\delta(j)}^{\left(k_{\delta(j)}\right)}\right), L_{2\delta(j)}^{\left(k_{\delta(j)}\right)}\left(p_{2\delta(j)}^{\left(k_{\delta(j)}\right)}\right)\right)\right)^{\lambda} +$$

$$\max_{k_{\delta(j)}} \left(d\left(L_{1\delta(j)}^{\left(k_{\delta(j)}\right)}\left(p_{1\delta(j)}^{\left(k_{\delta(j)}\right)}\right), L_{2\delta(j)}^{\left(k_{\delta(j)}\right)}\left(p_{2\delta(j)}^{\left(k_{\delta(j)}\right)}\right)\right)\right)^{\lambda} \geq$$

$$\frac{1}{\#L_{1\delta(j+1)}(p)} \sum_{k_{\delta(j+1)}=1}^{\#L_{1\delta(j+1)}(p)} \left(d\left(L_{1\delta(j+1)}^{\left(k_{\delta(j+1)}\right)}\left(p_{1\delta(j+1)}^{\left(k_{\delta(j+1)}\right)}\right), L_{2\delta(j+1)}^{\left(k_{\delta(j+1)}\right)}\left(p_{2\delta(j+1)}^{\left(k_{\delta(j+1)}\right)}\right)\right)\right)^{\lambda} +$$

$$\max_{k_{\delta(j+1)}} \left(d\left(L_{1\delta(j+1)}^{\left(k_{\delta(j+1)}\right)}\left(p_{1\delta(j+1)}^{\left(k_{\delta(j+1)}\right)}\right), L_{2\delta(j+1)}^{\left(k_{\delta(j+1)}\right)}\left(p_{2\delta(j+1)}^{\left(k_{\delta(j+1)}\right)}\right)\right)\right)^{\lambda}$$

where $j = 1, 2, \ldots, m$.

5 The Application of Distance Measures in Multi-criteria Group Decision Making

In this section, multi-criteria group decision making problems under probabilistic linguistic environment are described and then the decision making procedure based on distance measures is designed.

5.1 Problem Description and Decision Making Procedure

Multi-criteria group decision making is a common activity in our daily life, which is a process of choosing a preferred alternative from a set of alternatives or ranking all the alternatives with respect to multiple criteria. A multi-criteria group decision making problem under probabilistic linguistic environment is described as follows:

There are a set of n alternatives, $A = \{a_1, a_2, \ldots, a_n\}$, and a set of m criteria, $C = \{c_1, c_2, \ldots, c_m\}$. The weight vector of criteria is $w = (w_1, w_2, \ldots, w_m)^T$, where $w_j \geq 0 \ (j = 1, 2, \ldots, m)$ and $\sum_{j=1}^{m} w_j = 1$. A group of decision makers are called to assess n alternatives with respect to m criteria by utilizing the linguistic term set to form a set of linguistic decision matrices and then these linguistic decision matrices are used to make up a probabilistic linguistic decision matrix as follows:

$$R = \left[L_{ij}(p) \right]_{n \times m} = \begin{bmatrix} L_{11}(p) & L_{12}(p) & \cdots & L_{1m}(p) \\ L_{21}(p) & L_{22}(p) & \cdots & L_{2m}(p) \\ \vdots & \vdots & \ddots & \vdots \\ L_{n1}(p) & L_{n2}(p) & \cdots & L_{nm}(p) \end{bmatrix}$$

where $L_{ij}(p) = \left\{ L_{ij}^{(k_{ij})} \left(p_{ij}^{(k_{ij})} \right) | k_{ij} = 1, 2, \ldots, \#L_{ij}(p) \right\}$ is a probabilistic linguistic

term set denoting the degree that the alternative a_i satisfies the criterion c_j.

Because all the PLTSs in the probabilistic linguistic decision matrix usually have different numbers of probabilistic linguistic elements, the PLTSs should be normalized as shown in Sect. 2.2.

Then we give the definitions of the probabilistic linguistic positive ideal solution x^+ and the probabilistic linguistic negative ideal solution x^- as follows:

Definition 7 Let $R = \left[L_{ij}(p) \right]_{n \times m}$ be a normalized probabilistic linguistic decision matrix with $L_{ij}(p) = \left\{ L_{ij}^{(k)} \left(p_{ij}^{(k)} \right) | k = 1, 2, \ldots, \#L_{ij}(p) \right\}$. Then the probabilistic linguistic positive ideal solution (PLPIS) of alternatives is

$$x^+ = \left(L_1(p)^+, L_2(p)^+, \ldots, L_m(p)^+ \right)$$

where $L_j(p)^+ = \left\{ \left(L_j^+ \right)(1) \right\}$ and $L_j^+ = \max_{i,k} \left(L_{ij}^{(k)} \right)$.

Definition 8 Let $R = \left[L_{ij}(p) \right]_{n \times m}$ be a normalized probabilistic linguistic decision matrix with $L_{ij}(p) = \left\{ L_{ij}^{(k)} \left(p_{ij}^{(k)} \right) | k = 1, 2, \ldots, \#L_{ij}(p) \right\}$. Then the probabilistic linguistic negative ideal solution (PLNIS) of alternatives is

$$x^- = \left(L_1(p)^-, L_2(p)^-, \ldots, L_m(p)^- \right)$$

where $L_j(p)^- = \left\{ \left(L_j^- \right)(1) \right\}$ and $L_j^- = \min_{i,k} \left(L_{ij}^{(k)} \right)$.

To choose a preferred alternative or rank all the alternatives, we should compute the distance between each alternative x_i and the probabilistic linguistic positive ideal solution x^+, and the distance between each alternative x_i and the probabilistic linguistic negative ideal solution x^-. Obviously, a better alternative should be closer to the probabilistic linguistic positive ideal solution and also farther from the probabilistic linguistic negative ideal solution. Inspired by the TOPSIS method [53, 54], we put forward a concept of the satisfaction degree, which takes the distances $d(x_i, x^+)$ and $d(x_i, x^-)$ of each alternative into consideration, as follows:

Definition 9 Given the distances $d(x_i, x^+)$ and $d(x_i, x^-)$ of an alternative x_i, then a satisfaction degree of this alternative with respect to multi-criteria is defined as:

$$s(x_i) = \frac{(1-\theta)d(x_i, x^-)}{\theta d(x_i, x^+) + (1-\theta)d(x_i, x^-)}$$

where the parameter $\theta \in [0, 1]$, which represents the risk preferences of decision makers. If $\theta < 0.5$, then it means that the decision makers are optimistic. If $\theta > 0.5$, then it means that they are pessimistic. The value of this parameter should be given by the decision makers in advance. Obviously, the higher the satisfaction degree, the better the alternative.

5.2 An Illustrative Example

In this section, a practical example adapted from [55] is employed to demonstrate the application of our proposed distance measures to multi-criteria group decision making:

Energy is a major input for overall socio-economic development of any society. Thus, the energy policy has a great impact on economic development and environment and then the selection of the most appropriate energy policy is very important. Suppose that there are five alternatives (energy projects) $\{a_1, a_2, a_3, a_4, a_5\}$ to be invested, and four criteria $\{c_1, c_2, c_3, c_4\}$ to be considered: c_1: technological; c_2 : environmental; c_3: socio-political; c_4: economic. The weight vector of four criteria is $w = (0.15, 0.3, 0.2, 0.35)$. Five decision makers $\{d_1, d_2, d_3, d_4, d_5\}$ are invited to assess all the alternatives by using the following linguistic term set:

$$S = \{s_0 = \text{none}, s_1 = \text{very low}, s_2 = \text{low}, s_3 = \text{medium},$$
$$s_4 = \text{high}, s_5 = \text{very high}, s_6 = \text{perfect}\}$$

For an alternative with respect to a criterion, all the decision makers provide their linguistic assessment information anonymously. Then all the linguistic assessment information and their weights are used to form a PLTS. For example, when the alternative a_1 with respect to c_1 is assessed, two decision makers provide the linguistic term s_3 and three decision makers provide the linguistic term s_4 as their evaluation values. Hence, the overall linguistic assessment information of the alternative a_1 over the criterion c_1 is denoted as a PLTS $L_{11}(p) = \{s_3(0.4), s_4(0.6)\}$. Then the probabilistic linguistic decision matrix given by these five decision makers are listed in Table 1.

To rank all the alternatives, Definitions 7 and 8 are used to obtain the probabilistic linguistic positive ideal solution $x^+ = \{\{s_5(1)\}, \{s_5(1)\}, \{s_5(1)\}, \{s_6(1)\}\}$ and the probabilistic linguistic negative ideal solution $x^- = \{\{s_2(1)\}, \{s_2(1)\}, \{s_1(1)\}, \{s_3(1)\}\}$. Then the distance $d(x_i, x^+)$ between each alternative x_i and the probabilistic linguistic positive ideal solution x^+ and the distance $d(x_i, x^-)$ between each alternative x_i and the probabilistic linguistic negative ideal solution x^- are computed. Finally, the satisfaction degree of each alternative x_i is calculated by using Definition 9. Without loss of generality, we set the value of the parameter θ as 0.5.

To consider the weight information of multiple criteria, the generalized weighted distance measure, the generalized weighted Hausdorff distance measure, and the generalized hybrid weighted distance measure are introduced to calculated the

Table 1 Probabilistic linguistic decision matrix of five decision makers

	c_1	c_2	c_3	c_4
a_1	$\{s_3(0.4), s_4(0.6)\}$	$\{s_2(0.2), s_4(0.8)\}$	$\{s_3(0.2), s_4(0.8)\}$	$\{s_3(0.4), s_5(0.6)\}$
a_2	$\{s_3(0.8), s_5(0.2)\}$	$\{s_2(0.2), s_3(0.4), s_4(0.2)\}$	$\{s_1(0.2), s_2(0.4), s_3(0.2)\}$	$\{s_3(0.8), s_4(0.2)\}$
a_3	$\{s_3(0.6), s_4(0.4)\}$	$\{s_3(0.6), s_4(0.2)\}$	$\{s_3(0.2), s_4(0.2), s_5(0.2)\}$	$\{s_4(0.8), s_6(0.2)\}$
a_4	$\{s_4(0.5), s_5(0.5)\}$	$\{s_3(0.4), s_4(0.4), s_5(0.2)\}$	$\{s_2(0.8), s_3(0.2)\}$	$\{s_3(0.4), s_4(0.6)\}$
a_5	$\{s_2(0.1), s_3(0.9)\}$	$\{s_3(0.4), s_5(0.6)\}$	$\{s_2(0.2), s_3(0.4), s_4(0.2)\}$	$\{s_4(0.5), s_5(0.5)\}$

Table 2 Satisfaction degrees and rankings of alternatives obtained by the generated weighted distance measures

	a_1	a_2	a_3	a_4	a_5	Rankings
$\lambda = 1$	0.5760	0.2376	0.1816	0.3600	0.4564	$a_1 \succ a_5 \succ a_4 \succ a_2 \succ a_3$
$\lambda = 2$	0.5649	0.2189	0.1972	0.3679	0.4532	$a_1 \succ a_5 \succ a_4 \succ a_2 \succ a_3$
$\lambda = 4$	0.5492	0.2131	0.2133	0.3684	0.4491	$a_1 \succ a_5 \succ a_4 \succ a_3 \succ a_2$
$\lambda = 6$	0.5400	0.2167	0.2219	0.3676	0.4434	$a_1 \succ a_5 \succ a_4 \succ a_3 \succ a_2$
$\lambda = 10$	0.5334	0.2238	0.2319	0.3671	0.4354	$a_1 \succ a_5 \succ a_4 \succ a_3 \succ a_2$

Table 3 Satisfaction degrees and rankings of alternatives obtained by the generated weighted Hausdorff distance measures

	a_1	a_2	a_3	a_4	a_5	Rankings
$\lambda = 1$	0.5626	0.1900	0.2056	0.3282	0.4457	$a_1 \succ a_5 \succ a_4 \succ a_3 \succ a_2$
$\lambda = 2$	0.5569	0.1972	0.2092	0.3412	0.4443	$a_1 \succ a_5 \succ a_4 \succ a_3 \succ a_2$
$\lambda = 4$	0.5468	0.2106	0.2163	0.3539	0.4389	$a_1 \succ a_5 \succ a_4 \succ a_3 \succ a_2$
$\lambda = 6$	0.5393	0.2199	0.2226	0.3594	0.4341	$a_1 \succ a_5 \succ a_4 \succ a_3 \succ a_2$
$\lambda = 10$	0.5305	0.2288	0.2318	0.3638	0.4280	$a_1 \succ a_5 \succ a_4 \succ a_3 \succ a_2$

Table 4 Satisfaction degrees and rankings of alternatives obtained by the generated hybrid weighted distance measures

	a_1	a_2	a_3	a_4	a_5	Rankings
$\lambda = 1$	0.5681	0.2086	0.1957	0.3413	0.4500	$a_1 \succ a_5 \succ a_4 \succ a_2 \succ a_3$
$\lambda = 2$	0.5599	0.2050	0.2048	0.3511	0.4475	$a_1 \succ a_5 \succ a_4 \succ a_2 \succ a_3$
$\lambda = 4$	0.5477	0.2114	0.2152	0.3587	0.4422	$a_1 \succ a_5 \succ a_4 \succ a_3 \succ a_2$
$\lambda = 6$	0.5395	0.2189	0.2223	0.3621	0.4370	$a_1 \succ a_5 \succ a_4 \succ a_3 \succ a_2$
$\lambda = 10$	0.5305	0.2273	0.2319	0.3648	0.4302	$a_1 \succ a_5 \succ a_4 \succ a_3 \succ a_2$

distances. Then the satisfaction degree of each alternative and the ranking of the alternatives are shown in Tables 2, 3 and 4.

As listed in Tables 2, 3 and 4, using different distance measures to obtain the same best alternative and using the generalized weighted distance measures and generalized hybrid weighted distance measures to get the same ranking of alternatives. As the parameter λ changes, the ranking of alternatives changes between a_2 and a_3 when the generalized weighted distance measures and the generalized hybrid weighted distance measures are employed, while the ranking of alternatives keeps unchanged when the generalized weighted Hausdorff distance measures are used.

Figures 1, 2 and 3 show that the satisfaction degree increases or decreases as the parameter λ increases. For example, when the generalized weighed distance measures are used, the satisfaction degrees of a_1, a_2, and a_5 decrease monotonically as the parameter λ increases, while the satisfaction degrees of a_3 and a_4 increase monotonically as the parameter λ increases. Therefore, from this point of view, the parameter λ can be considered as a decision maker's risk attitude and then the

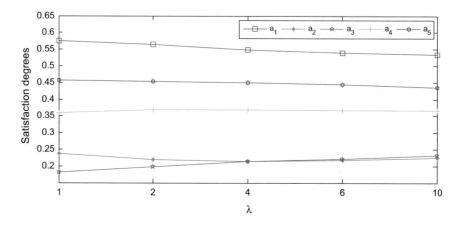

Fig. 1 Satisfaction degrees of alternatives obtained by the generated weighted distance measures

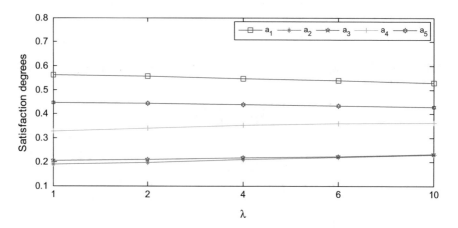

Fig. 2 Satisfaction degrees of alternatives obtained by the generated weighted Hausdorff distance measures

proposed distance measures provide the decision maker with more choices when the parameter value regarding to the decision maker's risk preference is given.

5.3 Discussions and Analysis

For the purpose of comparison, in this subsection, we use the distance measures to deal with the multi-criteria group decision making problems where the linguistic assessment information of an alternative over a criterion is represented by a hesitant fuzzy linguistic term set (HFLTS) not a PLTS.

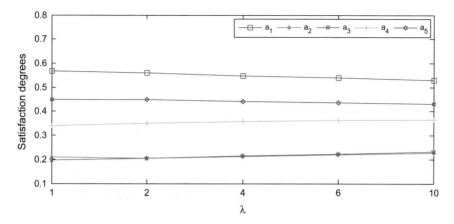

Fig. 3 Satisfaction degrees of alternatives obtained by the generated hybrid weighted distance measures

Table 5 Hesitant fuzzy linguistic decision matrix provided by five decision makers

	c_1	c_2	c_3	c_4
a_1	$\{s_3, s_4\}$	$\{s_2, s_4\}$	$\{s_3, s_4\}$	$\{s_3, s_5\}$
a_2	$\{s_3, s_5\}$	$\{s_2, s_3, s_4\}$	$\{s_1, s_2, s_3\}$	$\{s_3, s_4\}$
a_3	$\{s_3, s_4\}$	$\{s_3, s_4\}$	$\{s_3, s_4, s_5\}$	$\{s_4, s_6\}$
a_4	$\{s_4, s_5\}$	$\{s_3, s_4, s_5\}$	$\{s_2, s_3\}$	$\{s_3, s_4\}$
a_5	$\{s_2, s_3\}$	$\{s_3, s_5\}$	$\{s_2, s_3, s_4\}$	$\{s_4, s_5\}$

Then the hesitant fuzzy linguistic decision matrix provided by five decision makers is shown in Table 5.

Xu et al. [29] defined the distance between two linguistic terms as:

Definition 10 [29] Let $S = \{s_\alpha | \alpha = 0, 1, \ldots, \tau\}$ be a linguistic term set, s_a and s_b be two linguistic terms, then the distance between s_a and s_b is

$$d(s_a, s_b) = \frac{|a - b|}{\tau}$$

Based on Definition 10, Liao et al. [48] proposed a series of distance measures to compute the distance between two hesitant fuzzy linguistic term sets and the distance between two collections of hesitant fuzzy linguistic term sets. Without loss of generality, we utilized the generalized weighted distance measure, the generalized weighted Hausdorff distance measure, and the generalized hybrid weighted distance measure by Liao et al. to compute the distance between two collections of hesitant fuzzy linguistic term sets, which are defined as follows:

Given a linguistic term set $S = \{s_\alpha | \alpha = 0, 1, \ldots, \tau\}$ and two collections of HFLTSs $H_S^1 = \{H_{11}, H_{12}, \ldots, H_{1m}\}$ and $H_S^2 = \{H_{21}, H_{22}, \ldots, H_{2m}\}$, the weight vector of which is $w = (w_1, w_2, \ldots, w_m)^T$, where $0 \leq w_j \leq 1$ and $\sum_{j=1}^{m} w_j = 1$, then we define the generalized weighted distance between H_S^1 and H_S^2 as:

$$d_{gwd}\left(H_S^1, H_S^2\right) = \left(\sum_{j=1}^{m} \frac{w_j}{\#H_{1j}} \sum_{h=1}^{\#H_{1j}} \left(\frac{\left| I_{1j}^h - I_{2j}^h \right|}{\tau} \right)^\lambda \right)^{1/\lambda}$$

where $\#H_{1j}$ is the cardinality of the HFLTSs H_{1j} and H_{2j}, I_{1j}^h and I_{2j}^h are the subscripts of the j th linguistic terms in H_{1j} and H_{2j}, respectively, and $\lambda > 0$.

The generalized weighted Hausdorff distance between H_S^1 and H_S^2 is defined as:

$$d_{gwhaud}\left(H_S^1, H_S^2\right) = \left(\sum_{j=1}^{m} w_j \max_{h=1,2,\ldots,\#H_{1j}} \left(\frac{\left| I_{1j}^h - I_{2j}^h \right|}{\tau} \right)^\lambda \right)^{1/\lambda}$$

The generalized hybrid weighted distance between H_S^1 and H_S^2 is defined as:

$$d_{ghwd}\left(H_S^1, H_S^2\right) = \left(\sum_{j=1}^{m} \frac{w_j}{2} \left(\frac{1}{\#H_{1j}} \sum_{h=1}^{\#H_{1j}} \left(\frac{\left| I_{1j}^h - I_{2j}^h \right|}{\tau} \right)^\lambda + \max_{h=1,2,\ldots,\#H_{1j}} \left(\frac{\left| I_{1j}^h - I_{2j}^h \right|}{\tau} \right)^\lambda \right) \right)^{1/\lambda}$$

Tables 6, 7 and 8 show that by using different distance measures, we can achieve the same ranking of alternatives and the ranking of alternatives remains unchanged as the parameter λ increases when the HFLTSs are introduced to express the decision makers' linguistic assessment information, and the best alternative is a_3.

However, by using different distance measures, we can achieve the different rankings of alternatives and different best alternatives when the PLTSs are used to represent the decision makers' linguistic assessment information from that when the HFLTSs are employed to express the decision makers' linguistic assessment information. That is because the PLTSs can express not only the decision makers'

Table 6 Satisfaction degrees and rankings of alternatives obtained by the generated weighted distance measures

	a_1	a_2	a_3	a_4	a_5	Rankings
$\lambda = 1$	0.4297	0.3047	0.6172	0.4531	0.5000	$a_3 \succ a_5 \succ a_4 \succ a_1 \succ a_2$
$\lambda = 2$	0.4593	0.3530	0.5988	0.4616	0.5000	$a_3 \succ a_5 \succ a_4 \succ a_1 \succ a_2$
$\lambda = 4$	0.4622	0.3937	0.6035	0.4746	0.5082	$a_3 \succ a_5 \succ a_4 \succ a_1 \succ a_2$
$\lambda = 6$	0.4642	0.4125	0.6123	0.4820	0.5111	$a_3 \succ a_5 \succ a_4 \succ a_1 \succ a_2$
$\lambda = 10$	0.4724	0.4261	0.6265	0.4888	0.5097	$a_3 \succ a_5 \succ a_4 \succ a_1 \succ a_2$

Table 7 Satisfaction degrees and rankings of alternatives obtained by the generated weighted Hausdorff distance measures

	a_1	a_2	a_3	a_4	a_5	Rankings
$\lambda = 1$	0.4536	0.3711	0.5790	0.4667	0.5000	$a_3 \succ a_5 \succ a_4 \succ a_1 \succ a_2$
$\lambda = 2$	0.4537	0.3824	0.5881	0.4761	0.5063	$a_3 \succ a_5 \succ a_4 \succ a_1 \succ a_2$
$\lambda = 4$	0.4566	0.3986	0.6041	0.4866	0.5111	$a_3 \succ a_5 \succ a_4 \succ a_1 \succ a_2$
$\lambda = 6$	0.4615	0.4086	0.6165	0.4913	0.5111	$a_3 \succ a_5 \succ a_4 \succ a_1 \succ a_2$
$\lambda = 10$	0.4720	0.4188	0.6322	0.4950	0.5084	$a_3 \succ a_5 \succ a_4 \succ a_1 \succ a_2$

Table 8 Satisfaction degrees and rankings of alternatives obtained by the generated hybrid weighted distance measures

	a_1	a_2	a_3	a_4	a_5	Rankings
$\lambda = 1$	0.4441	0.3448	0.5944	0.4610	0.5000	$a_3 \succ a_5 \succ a_4 \succ a_1 \succ a_2$
$\lambda = 2$	0.4557	0.3719	0.5919	0.4705	0.5040	$a_3 \succ a_5 \succ a_4 \succ a_1 \succ a_2$
$\lambda = 4$	0.4585	0.3971	0.6039	0.4825	0.5101	$a_3 \succ a_5 \succ a_4 \succ a_1 \succ a_2$
$\lambda = 6$	0.4624	0.4098	0.6152	0.4883	0.5110	$a_3 \succ a_5 \succ a_4 \succ a_1 \succ a_2$
$\lambda = 10$	0.4721	0.4209	0.6305	0.4930	0.5088	$a_3 \succ a_5 \succ a_4 \succ a_1 \succ a_2$

some possible linguistic terms, but also their probabilistic information. Then, the decision makers' linguistic assessment information is not lost, and more reasonable decision results can be obtained.

6 Conclusions

In this chapter, we have put forward a variety of distance measures for PLTSs and applied them to multi-criteria group decision making. We have first put forward the axiom definition of distance measure for PLTSs and given the definition of distance between two PLTEs. Based on this, a new family of distance measures have been designed to compute the distance between two PLTSs. Because of the weighting information of multiple criteria, we have also proposed the weighted and ordered distance measures to calculate the distance between two collections of PLTSs. Motivated by the TOPSIS method, the distance measures have been used to define the concept of satisfaction degree of an alternative and then they have been utilized to solve the group decision making problems. Finally, a practical example concerning the selection of energy alternatives have been provided to illustrate the applicability of the proposed distance measures in multi-criteria group decision making and compare the PLTSs with the HFLTSs.

Acknowledgements This research work was partially supported by the National Natural Science Foundation of China (Nos. 61273209, 71571123).

References

1. Liu J, Li WJ, Chen SW, Xu Y (2014) An axiomatizable logical foundation for lattice-ordered qualitative linguistic approach for reasoning with words. Inf Sci 263:110–125
2. Wang JQ, Wang J, Chen QH, Zhang HY, Chen XH (2014) An outranking approach for multi-criteria decision-making with hesitant fuzzy linguistic term sets. Inf Sci 280:338–351
3. Zadeh LA (1975) Concept of a linguistic variable and its application to approximate reasoning —1. Inf Sci 8:199–249
4. Zadeh LA (1965) Fuzzy sets. Inf Control 8:338–353
5. Degani R, Bortolan G (1988) The problem of linguistic approximation in clinical decision making. Int J Approx Reason 2:143–162
6. Türkşen IB (2002) Type 2 representation and reasoning for CWW. Fuzzy Sets Syst 127:17–36
7. Yager RR (1981) A new methodology for ordinal multiobjective decision based on fuzzy sets. Decis Sci 12:589–600
8. Herrera F, Martinez L (2000) A 2-tuple fuzzy linguistic representation model for computing with words. IEEE Trans Fuzzy Syst 8:746–752
9. Xu ZS (2004) A method based on linguistic aggregation operators for group decision making with linguistic preference relations. Inf Sci 166:19–30
10. Wang H (2015) Extended hesitant fuzzy linguistic term sets and their aggregation in group decision making. Int J Comput Intell Syst 8:14–33
11. Rodriguez RM, Martinez L, Herrera F (2012) Hesitant fuzzy linguistic term sets for decision making. IEEE Trans Fuzzy Syst 20:109–119
12. Torra V (2010) Hesitant fuzzy sets. Int J Intell Syst 25:529–539
13. Rodriguez RM, Martinez L, Herrera F (2013) A group decision making model dealing with comparative linguistic expressions based on hesitant fuzzy linguistic term sets. Inf Sci 241:28–42
14. Pang Q, Wang H, Xu ZS (2016) Probabilistic linguistic term sets in multi-attribute group decision making. Inf Sci 369:128–143
15. Peng B, Ye CM, Zeng SZ (2014) Some intuitionist fuzzy weighted geometric distance measures and their application to group decision making. Int J Uncertain Fuzziness Knowl Based Syst 22:699–715
16. Liao HC, Xu ZS (2015) Approaches to manage hesitant fuzzy linguistic information based on the cosine distance and similarity measures for HFLTSs and their application in qualitative decision making. Expert Sys Appl 42:5328–5336
17. Xu ZS (2010) A method based on distance measure for interval-valued intuitionistic fuzzy group decision making. Inf Sci 180:181–190
18. Su Z, Xu ZS, Liu HF, Liu S (2015) Distance and similarity measures for dual hesitant fuzzy sets and their applications in pattern recognition. J Intell Fuzzy Syst 29:731–745
19. Papakostas GA, Hatzimichailidis AG, Kaburlasos VG (2013) Distance and similarity measures between intuitionistic fuzzy sets: a comparative analysis from a pattern recognition point of view. Pattern Recogn Lett 34:1609–1622
20. Cai M, Gong ZW, Wu DQ, Wu MJ (2014) A pattern recognition method based on linguistic ordered weighted distance measure. J Intell Fuzzy Syst 27:1897–1903
21. Zhang XL, Xu ZS (2015) Novel distance and similarity measures on hesitant fuzzy sets with applications to clustering analysis. J Intell Fuzzy Syst 28:2279–2296
22. Ye J (2014) Clustering methods using distance-based similarity measures of single-valued neutrosophic sets. J Intell Syst 23:379–389
23. Farhadinia B (2013) Information measures for hesitant fuzzy sets and interval-valued hesitant fuzzy sets. Inf Sci 240:129–144
24. Gorzalczany B (1983) Approximate inference with interval-valued fuzzy sets—an outline. In: Proceedings of the Polish symposium on interval and fuzzy mathematics, Poznan, Poland, pp 89–95

25. Atanassov K (1986) Intuitionistic fuzzy sets. Fuzzy Sets Syst 20:87–96
26. Atanassov K, Gargov G (1989) Interval-valued intuitionistic fuzzy sets. Fuzzy Sets Syst 31:343–349
27. Chen N, Xu ZS, Xia MM (2013) Interval-valued hesitant preference relations and their applications to group decision making. Knowl Based Syst 37:528–540
28. Herrera F, Herrera-Viedma E, Verdegay JL (1996) A model of consensus in group decision making under linguistic assessments. Fuzzy Sets Syst 78:73–87
29. Xu ZS (2005) Deviation measures of linguistic preference relations in group decision making. Omega 33:249–254
30. Xu ZS (2012) Linguistic decision making: theory and methods. Springer, Berlin
31. Xu ZS, Wang H (2016) On the syntax and semantics of virtual linguistic terms for information fusion in decision making. Inf Fus 34:43–48
32. Liu XC (1992) Entropy, distance measure and similarity measure of fuzzy sets and their relations. Fuzzy Sets Syst 52:305–318
33. Chaudhuri BB, Rosenfeld A (1999) Modified Hausdorff distance between fuzzy sets. Inf Sci 118:159–171
34. Balopoulos V, Hatzimichailidis AG, Papadopoulos BK (2007) Distance and similarity measures for fuzzy operators. Inf Sci 177:2336–2348
35. Heidarzade A, Mahdavi I, Mahdavi-Amiri N (2016) Supplier selection using a clustering method based on a new distance for interval type-2 fuzzy sets: a case study. Appl Soft Comput J 38:213–231
36. Zhang HY, Zhang WX, Mei CL (2009) Entropy of interval-valued fuzzy sets based on distance and its relationship with similarity measure. Knowl Based Syst 22:449–454
37. Zeng WY, Guo P (2008) Normalized distance, similarity measure, inclusion measure and entropy of interval-valued fuzzy sets and their relationship. Inf Sci 178:1334–1342
38. Xu ZS, Chen J (2008) An overview of distance and similarity measures of intuitionistic fuzzy sets. Int J Uncertain Fuzziness Knowl Based Syst 16, 529–555
39. Hatzimichailidis AG, Papakostas GA, Kaburlasos VG (2012) A novel distance measure of intuitionistic fuzzy sets and its application to pattern recognition problems. Int J Intell Syst 27:396–409
40. Hung WL, Yang MS (2004) Similarity measures of intuitionistic fuzzy sets based on Hausdorff distance. Pattern Recogn Lett 25:1603–1611
41. Wang WQ, Xin XH (2005) Distance measure between intuitionistic fuzzy sets. Pattern Recogn Lett 26:2063–2069
42. Szmidt E, Kacprzyk J (2000) Distances between intuitionistic fuzzy sets. Fuzzy Sets Syst 114:505–518
43. Duenci M (2016) A new distance measure for interval valued intuitionistic fuzzy sets and its application to group decision making problems with incomplete weights information. Appl Soft Comput J 41:120–134
44. Li YW, Shan YQ, Liu PD (2015) An extended TODIM method for group decision making with the interval intuitionistic fuzzy sets. Math Probl Eng 2015, Article ID 672140
45. Xu ZS, Xia MM (2011) Distance and similarity measures for hesitant fuzzy sets. Inf Sci 181:2128–2138
46. Li DQ, Zeng WY, Li JH (2015) New distance and similarity measures on hesitant fuzzy sets and their applications in multiple criteria decision making. Eng Appl Artif Intell 40:11–16
47. Bai ZY (2013) Distance similarity measures for interval-valued hesitant fuzzy sets and their application in multicriteria decision making. J Decis Syst 22:190–201
48. Liao HC, Xu ZS, Zeng XJ (2014) Distance and similarity measures for hesitant fuzzy linguistic term sets and their application in multi-criteria decision making. Inf Sci 271:125–142
49. Meng FY, Chen XH (2015) A hesitant fuzzy linguistic multi-granularity decision making model based on distance measures. J Intell Fuzzy Syst 28:1519–1531
50. Yager RR (2004) Generalized OWA aggregation operators. Fuzzy Optim Decis Mak 3:93–107

51. Xu ZS, Chen J (2008) Ordered weighed distance measure. J Syst Sci Eng 17:432–445
52. Xu ZS (2012) Fuzzy ordered distance measures. Fuzzy Optim Decis Mak 11:73–97
53. Hwang CL, Yoon K (1981) Multiple attribute decision making: methods and applications. Springer, Berlin
54. Chen SJ, Hwang CL (1992) Fuzzy multiple attribute decision making: methods and applications. Springer, Berlin
55. Kahraman C, Kaya I (2010) A fuzzy multicriteria methodology for selection among energy alternatives. Expert Syst Appl 37:6270–6281

Fuzzy Numbers and Consensus

Antonio Maturo and Aldo G.S. Ventre

Abstract Often in a group of decision makers there is a considerable variability in the scores that decision makers assign to the alternatives. In this paper we represent this variability with fuzzy numbers. Moreover we present an algorithm for the achievement of consensus based on suitable fuzzy numbers, on preorder and order relations in sets of fuzzy numbers, and on a procedure to decrease the spreads.

Keywords Fuzzy numbers and order/preorder relations · Multiperson decision making · Consensus

1 A Formalization of a Multiperson Decision Making Problem

Suppose that a set $D = \{d_1, \ldots, d_r, \ldots, d_p\}$ of decision makers, all having the same importance, should establish an order on a set $A = \{a_1, \ldots, a_i, \ldots, a_m\}$ of alternatives, evaluating the importance of the various alternatives with respect to a general objective indicated by OG.

If every decision maker establishes its own order, or, specifically a score for each alternative, then we must find a criterion for aggregating the various orders or the various scores in order to have a relationship of "social" order/pre-order among the alternatives.

Methods of aggregation for a "social order" can be found in [1, 4, 10, 11, 14, 43].

A. Maturo (✉)
Department of Architecture, University of Chieti-Pescara, Viale Pindaro 42, 65127 Pescara, Italy
e-mail: antomato75@gmail.com

A.G.S. Ventre
Accademia Pontaniana, 80100 Naples, Italy
e-mail: aldoventre@yahoo.it

© Springer International Publishing AG 2018   441
M. Collan and J. Kacprzyk (eds.), *Soft Computing Applications for Group Decision-making and Consensus Modeling*, Studies in Fuzziness and Soft Computing 357, DOI 10.1007/978-3-319-60207-3_25

The logic that each decision maker can follow to assign an order or a score in a rational way to the alternatives, for a social aggregation, is discussed and deepened, for example, in [3, 6, 13, 15, 21–24, 37, 38, 41]. An example of application in education is due to [7].

In some papers it is imposed the condition that the scores are measures or, at least, fuzzy measures (cfr [2, 12, 25–28, 42, 44, 46, 47]).

For decisions under uncertainty the scores of the alternatives are connected to the concepts of probability or coherent prevision [5, 6, 13] or to an extension of such concepts in Join Spaces (see [19]).

In this paper, following the AHP procedure by Saaty, we suppose that the overall objective OG is specified by a set of *sub-objectives* or *specific objectives*, and that they are specified by a set $C = \{c_1, \ldots, c_j, \ldots, c_n\}$ of *criteria*. It is supposed that decision makers agree on what are the specific objectives and criteria. Then it is defined the digraph associated with the problem of decision, having vertices located on 4 levels, where in the first level is the overall objective, in the second are specific objectives, in the third and fourth are criteria and alternatives, respectively. The arcs of the graph with the first extreme at a level $i \in \{1, 2, 3\}$ have the second extreme at the level $i + 1$.

Every decision maker, independently of the others, gives a score to each arc of the digraph with the following conditions:

(1) Scores are non-negative real numbers belonging to the interval $[0, 1]$;
(2) The sum of the scores attributed to the outgoing arcs from the general objective is equal to 1;
(3) The sum of the scores of the outgoing arcs from a sub-objective is equal to 1.

Regarding the scores of the arcs that connect a given criterion with the alternatives, it is not always appropriate to impose the condition that the sum of the scores is equal to 1. Advantages or disadvantages may occur and we must reason according to the context.

Once assigned the scores to the arcs, the score of a path of the digraph is defined as the *product* of the scores of the arcs in the path. Finally, the score of each vertex V different from the general objective is equal to the *sum* of the scores of the paths going from the general objective to the vertex V.

In conclusion, for each decision maker d_r, we obtain:

• a row vector $w_r = [w_{r1}, w_{r2}, \ldots, w_{rm}]$ of the criteria weights, which are positive real numbers with sum equal to 1;
• a matrix D_r with general term p_{rij} equal to the weight that the decision maker d_r assigns to the arc of the digraph with the first extreme the criterion c_j and the second extreme the alternative a_i.

The row vector $\pi_r = [\pi_{r1}, \pi_{r2}, \ldots, \pi_{rm}]$, of the scores of the alternatives assigned by decision maker d_r, is the matrix product of w_r by the transpose D_r' of the matrix D_r, i.e. $\pi_r = w_r D_r'$.

2 A Representation of the Problem of Consensus Among Decision Makers with Fuzzy Numbers

In various articles the consensus among decision makers has been addressed considering the decision makers as points of a metric space [4, 10, 11, 21–24]. Before any formula for the aggregation of the scores, dynamic algorithms are proposed to lead the points of the metric space, that represent the various decision makers, to reach consensus. Only when these points are fairly close then the diversity of views appears unbound by prejudice or very particular special interests, and thus the diversity of judgments seems to be related to objective assessments, although with some differences in points of view.

In this paper we consider an approach based on fuzzy numbers and the order and pre-order relationships between fuzzy numbers.

We recall the necessary definitions briefly (for further information see e.g. [12, 18, 20, 35, 45, 46, 47]).

Applications of fuzzy numbers for fuzzy regression and decision making problems are in [29, 31–34, 36].

Definition 2.1 (*Fuzzy number*) Let [a, b] be a compact interval of R. A *fuzzy number* with *base* [a, b] is a function u: R → [0, 1], having as domain the set of real numbers and with values in [0, 1], such that:

(BS) (*bounded support*) $u(x) = 0$ for $x \notin [a, b]$, and $u(x) > 0$ for x belonging to the open interval (a, b);

(CN) (*compactness and normality*) For every $r \in (0, 1]$ the set $[u]^r = \{x \in R: u(x) \geq r\}$ is a nonempty *compact* interval.

The numbers a, b, are called respectively, the *left* and the *right endpoint* of u, and the set $\{x \in [a, b]: u(x) > 0\}$ is said to be the *support* of u, denoted S(u). Moreover, u is said to be *degenerate* if $a = b$, that is S(u) is a singleton.

Definition 2.2 (r-*cuts*) For every r such that $0 \leq r \leq 1$ the set $[u]^r = \{x \in [a, b]: u(x) \geq r\}$ is said to be the r-*cut* of u. The left and right endpoints of $[u]^r$ are denoted, respectively, u_λ^r and u_ρ^r. The fuzzy number u is said to be *simple* if $c = d$, that is C(u) is a singleton.

In particular, for $r = 0$, $[u]^0 = [a, b]$, and, for $r = 1$, $[u]^1$ is a compact interval [c, d], called the *core* (or *central part*) of u, and denoted with C(u). The numbers c, d, are the *left* and the *right endpoint* of C(u).

Definition 2.3 (*Spreads*) The intervals [a, c) and (d, b] are, respectively, the *left part* and the *right part* of u. The real numbers $L(u) = c - a$, $M(u) = d - c$, and $R(u) = b - d$ are, the *left*, *middle*, and *right spreads* of u. Their sum $T(u) = b - a$ is the *total spread*.

Definition 2.4 (*Relation* ⊆) Let u: R → [0, 1] and v: R → [0, 1] two fuzzy numbers. We say that u is contained in v, we write $u \subseteq v$, if $u(x) \leq v(x)$, $\forall x \in R$.

Proposition 2.5 The relation $\subseteq$ is a partial order relation in the set of fuzzy numbers. A compact interval [a, b] is interpreted as a fuzzy set w: R $\rightarrow$ [0, 1] with base [a, b] and such that w(x) = 1, $\forall x \in$ [a, b]. Then it is the maximum fuzzy number with base [a, b] with respect to the order relation $\subseteq$.

Notations 2.6 We assume the following notations:

- (*endpoints notation*) u $\sim$ (a, c, d, b) stands for u is a fuzzy number with end-points a, b and core [c, d]; u $\sim$ (a, c, b) for simple u;
- (*spreads notation*) u $\sim$ [c, d, L, R] denotes that u is a fuzzy number with core [c, d] and left and right spreads L and R, respectively; u $\sim$ [c, L, R] denotes simple u;
- (*r-cut spreads notation*) the numbers $L^r(u) = (c - u_\lambda^r)$ and $R^r(u) = (u_\rho^r - d)$ are called the *r-cut left spread* and the *r-cut right spread* of u, we write $[u]^r = [c, d, L^r(u), R^r(u)]$, and, if u is simple, we write also $[u]^r = [c, L^r(u), R^r(u)]$.

Definition 2.7 We say that the fuzzy number u $\sim$ (a, c, d, b) is a *trapezoidal fuzzy number*, let us write u = (a, c, d, b), if:

$$\forall x \in [a, c), \ a < c \Rightarrow u(x) = (x - a)/(c - a), \tag{1}$$

$$\forall x \in (d, b], \ d < b \Rightarrow u(x) = (b - x)/(b - d), \tag{2}$$

A simple trapezoidal fuzzy number u = (a, c, c, b) is said to be a *triangular fuzzy number* and we write u = (a, c, b). A trapezoidal fuzzy number u = (c, c, d, d), with support equal to the core is said to be a *rectangular fuzzy number* and is identified with the compact interval [c, d] of R.

Proposition 2.8 In terms of r-cut left and right spreads u $\sim$ [c, d, L, R] is a trapezoidal fuzzy number, we write u = [c, c', L, R], iff:

$$L^r(u) = (1 - r)(c - a) = (1 - r)L, \qquad R^r(u) = (1 - r)(b - d) = (1 - r)R. \tag{3}$$

In particular, u $\sim$ [c, L, R] is a triangular fuzzy number, we write u = [c, L, R], iff:

$$L^r(u) = (1 - r)(c - a) = (1 - r)L, \qquad R^r(u) = (1 - r)(b - c) = (1 - r)R. \tag{4}$$

Several orderings can be defined in the set of fuzzy numbers [8, 9, 12, 16, 17, 18, 20]. We focus our attention on some fundamental orderings that play an important role when choices among social or economic actions are involved.

Let Φ be the set of all the fuzzy numbers.

Proposition 2.9 (*Main order*) The relation $\precsim_M$ on Φ such that:

$$\forall u, v \in \Phi, u \precsim_M v \Leftrightarrow \forall r \in [0, 1], [u]^r \leq [v]^r, \tag{5}$$

is a partial **order** relation on Φ, called the main order.

The main order is the basic ordering in the set of fuzzy numbers whatever the shape.

Proposition 2.10 (*Trapezoidal order*) The relation $\lesssim_T$ on Φ such that:

$$\forall u, v \in \Phi, u \lesssim_T v \Leftrightarrow [u]^0 \leq [v]^0, [u]^1 \leq [v]^1 \tag{6}$$

is a partial **preorder** relation on Φ, called the trapezoidal order.

Such a relation is mainly useful if the trapezoidal shape is preferred, because of the relative simplicity in handling these numbers. The restriction of $\lesssim_T$ to the set **T** of the trapezoidal fuzzy numbers is a partial **order** relation.

Proposition 2.11 (*Crisp order*) The relation $\lesssim_C$ such that:

$$\forall u, v \in \Phi, u \lesssim_C v \Leftrightarrow [u]^1 \leq [v]^1 \tag{7}$$

is a partial **preorder** relation on Φ, called the core order or crisp order.

The relation $\lesssim_C$ is useful when peripheral spreads are considered of marginal importance with respect to the central ones. Moreover the restriction of $\lesssim_C$ to the set of simple fuzzy numbers is a **total preorder** relation.

Proposition 2.12 (*Strict order*) The relation $\lesssim_S$ such that:

$$\forall u, v \in \Phi, u \lesssim_S v \Leftrightarrow (x \in S(u), y \in S(v)) \Rightarrow x \leq y \tag{8}$$

is a partial **preorder** relation on Φ, called the strict order.

Suppose that every decision maker d_r has attributed a row vector $\pi_r = [\pi_{r1}, \pi_{r2}, \ldots, \pi_{rm}]$ of scores to alternatives. We obtain the following matrix Π of scores of the alternatives, having rows for the decision makers and columns for the alternatives (Table 1).

The alternative a_i is associated with the column vector $[\pi_{1i}, \pi_{2i}, \ldots, \pi_{pi}]'$ of the scores of the different decision makers. Let $\alpha_i, \beta_i, \gamma_i$ be the minimum, the maximum and an average of these scores (such as the arithmetic mean, median, or generally the mean of order h > 0, with h fixed by decision makers), respectively. Then the "social" score of the alternative a_i assigned by the whole group D of decision makers can be represented by the triangular fuzzy number $\pi^*_i = (\alpha_i, \gamma_i, \beta_i)$ with left spread $\gamma_i - \alpha_i$ and right spread $\beta_i - \gamma_i$.

Table 1 Matrix of scores of decision makers

$\Pi =$		a_1	a_2		a_i		a_m
	d_1	π_{11}	π_{12}	.	π_{1i}	.	π_{1m}
	d_2	π_{21}	π_{22}	.	π_{2i}	.	π_{2m}
	.	.	.	.	.	.	.
	d_r	π_{r1}	π_{r2}	.	π_{ri}	.	π_{rm}
	.	.	.	.	.	.	.
	d_p	π_{p1}	π_{p2}	.	π_{pi}	.	π_{pm}
	D	π^*_1	π^*_2		π^*_i		π^*_m

There are many possibilities for constructing the triangular fuzzy number π^*_i depending on the chosen average. If we want to take into account of various averages, the minimum of which is γ_{i1} and the maximum γ_{i2}, the "social" score assigned to the alternative a_i by the whole group D of decision makers is the trapezoidal fuzzy number $\pi^{**}_i = (\alpha_i, \gamma_{i1}, \gamma_{i2}, \beta_i)$ with left spread $\gamma_{i1} - \alpha_i$, right spread $\beta_i - \gamma_{i2}$ and middle spread $\gamma_{i2} - \gamma_{i1}$.

As a particular case a triangular fuzzy number is obtained if $\gamma_{i2} = \gamma_{i1}$ and a rectangular fuzzy number, i.e. an interval, if $\alpha_i = \gamma_{i1}, \gamma_{i2} = \beta_i$.

3 Shared Ordering, and Consensus

Let us suppose, for instance, that social alternative scores are fuzzy triangular numbers $\pi^*_i = (\alpha_i, \gamma_i, \beta_i)$. Then, as a first approach to a ranking of alternatives the crisp order can be considered, in which the preferable alternative is the one with the highest core, and two alternatives with the same core are considered equally preferable.

This approach, however, does not take into account the views of decision makers "more peripheral" which attributed a score away from the central value γ_i to the alternative a_i. In particular, if for a couple of alternatives (a_i, a_h), it is $\pi^*_i \subset \pi^*_h$, the two alternatives are considered equivalent even if the spreads of the second alternative are greater than those of the first. The same reasoning can be extended to the case in which the social alternatives scores are trapezoidal fuzzy numbers or any shape fuzzy numbers.

Let us remark that, in the case the social alternatives scores are triangular fuzzy numbers or, in general, simple fuzzy numbers, we get a total pre-order relation between the alternatives is obtained.

We can adopt a more rigorous approach, which takes into account the left and right spreads, ordering the fuzzy triangular numbers $\pi^*_i = (\alpha_i, \gamma_i, \beta_i)$ (or, in general, the trapezoidal fuzzy numbers $\pi^{**}_i = (\alpha_i, \gamma_{i1}, \gamma_{i2}, \beta_i)$) with the main order. So we define $\pi^*_i \leq \pi^*_h$ if and only if $\alpha_i \leq \alpha_h, \gamma_i \leq \gamma_h, \beta_i \leq \beta_h$. The main order is an order relation, but in general it is not total. For example, two alternatives a_i, a_h such that $\pi^*_i \subset \pi^*_h$ are not comparable. Moreover, even with the main order, the opinion of the decision makers away from the core is scarcely taken into account.

A third approach, which takes into account the views of all decision makers, consists in ordering fuzzy numbers with the strict order.

Then, it seems reasonable to introduce the following definition:

Definition 3.1 We say that fuzzy numbers π^*_i (or in general π^{**}_i) identify a *shared ordering* among the decision makers if the strict order among the alternatives is a total ordering.

A second aspect to be considered is the consensus among decision makers.

We say "coalition" any non-empty subset A of D. The concept of social "score" can be extended to any coalition A of D.

Let only consider the scores assigned by a coalition A. We can associate each alternative to a triangular fuzzy number $\pi^*_{iA} = (\alpha_{iA}, \gamma_{iA}, \beta_{iA})$, called the *social score* of a_i assigned by A.

Following legal obligations or preliminary agreements among decision makers, we can fix an integer $k \in (p/2, p]$, called *level of majority* and a positive real number ε, called *level of consensus*. The coalition A is said to be [30, 39, 40]:

- *winning* if $|A| \geq k$;
- *loser* if $|A| \leq p - k$;
- *blocking* if $p - k < |A| < k$.

We denote by $\delta(A)$, called *diameter of the coalition*, the maximum of the spreads (right or left) of the fuzzy numbers π^*_{iA}. Given a real number $\varepsilon > 0$, we say that in a coalition A there is the consensus (to the level ε) if $\delta(A) \leq \varepsilon$.

The search for consensus among decision makers is to find, together with an impartial arbiter K, called the Demiurge, a procedure to identify a coalition A winning, in which there is consensus, and if possible maximal, i.e. such that $A \subset B \Rightarrow \delta(A) > \varepsilon$.

If K, with impartial procedure followed, cannot individuate a winning coalition in which there is consensus, then he declares the inability to reach consensus, and calls on all decision-makers to reformulate the decision problem, rethinking specific objectives, criteria, alternatives and scores.

If K is able to achieve a winning coalition in which there is consensus, then he determines whether it is or not maximal considering the diameters of coalitions $A \cup \{a_i\}$, with $a_i \in D - A$. If all these coalitions have a diameter greater than ε then A is maximal, otherwise K widens the coalition by bringing in the coalition A the decision maker a_i not belonging to A and such that $\delta(A \cup \{a_i\})$ is minimum.

A procedure for achieving a winning coalition in which there is consensus, or to decide that it is not possible to reach consensus, is as follows. First of all we identify a positive real number α, called the *minimum level of improvement*.

Empirical remark: the number α must be small enough to allow the procedure to move forward, but not too small, otherwise the process can become very long.

In step $s \geq 1$, let $D^{(s)}$ be the number of decision-makers remained after the previous steps. Evidently $D^{(1)} = D$. There are three possibilities:

(1) $D^{(s)} < k$, then the Demiurge K states that it is not possible to reach consensus, and the procedure ends;
(2) $D^{(s)} \geq k$ and $\delta(D^{(s)}) \leq \varepsilon$, then the Demiurge K states that consensus has been reached, and the procedure ends;
(3) $D^{(s)} \geq k$ and $\delta(D^{(s)}) > \varepsilon$.

In the third case, let $A = D^{(s)}$. The demiurge K evaluates all fuzzy triangular numbers $\pi^*_{iA} = (\alpha_{iA}, \gamma_{iA}, \beta_{iA})$. For each decision maker d_r belonging to A, and for each alternative a_i, K calculates the distance between the evaluation $\pi^{(s)}_{ri}$ given by the decision maker d_r at the beginning of step s, and the core γ_{iA}. The maximum of these numbers, at varying of i, is said the *variability* at step s of the decision maker d_r, indicated with $v^{(s)}(d_r)$.

K subsequently identifies decision makers with the highest variability, said *marginal decision makers*, in order to review, one by one, their evaluations in order to have new evaluations $\pi_{ri}^{(s+1)}$ with a new variability $v^{(s+1)}(d_r) \leq v^{(s)}(d_r) - \alpha$. If the decision maker adapts himself, then the Demiurge passes to the next step with the same set of decision makers. If he does not conform, the decision maker is excluded and the Demiurge considers the next step only with the remaining decision makers.

4 Conclusions

If the Demiurge can secure a maximal winning coalition A where there is consensus, then triangular fuzzy numbers $\pi^*_{iA} = (\alpha_{iA}, \gamma_{iA}, \beta_{iA})$ are the end result of the procedure for obtaining consensus. At this point the demiurge K can make its findings by ordering the alternatives according to the three order relations/preorder considered in Sect. 2.

These data, appropriately illustrated, will be a support for politicians and administrators for the choice of alternatives, or simply for their ordering by means of preference-indifference relations.

References

1. Arrow KJ (1951) Social choice and individual value. Yale University Press, New Haven
2. Banon G (1981) Distinction between several subsets of fuzzy measures. Int J Fuzzy Sets Syst 5:291–305
3. Boudon R (1979) La logique du social, Hachette, Paris
4. Carlsson C, Ehrenberg D, Eklund P, Fedrizzi M, Gustafsson P, Lindholm P, Merkurieva G, Riissanen T, Ventre AGS (1992) Consensus in distributed soft environments. Eur J Oper Res 61:165–185
5. Coletti G, Scozzafava R (2002) Probabilistic logic in a coherent setting. Kluwer Academic Publishers, Dordrecht
6. de Finetti B (1974) Theory of probability. J. Wiley, New York
7. Delli Rocili L, Maturo A, (2013) Teaching mathematics to children: social aspects, psychological problems and decision making models. In: Soitu, Gavriluta, Maturo (eds) Interdisciplinary approaches in social sciences, Editura Universitatii A.I. Cuza, Iasi, Romania
8. Dubois D, Prade H (1980) Fuzzy set and systems. Academic Press, New York
9. Dubois D, Prade H (1988) Fuzzy numbers: an overview. In: Bedzek JC (ed) Analysis of fuzzy information, vol 2 CRC-Press, Boca Raton, pp 3–39
10. Ehrenberg D, Eklund P, Fedrizzi M, Ventre AGS (1989) Consensus in distributed soft environments. Rep Comput Sci Math Åbo 88:1–24
11. Eklund P, Rusinowska A, De Swart H (2007) Consensus reaching in committees. Eur J Oper Res 178:185–193

12. Klir GJ, Yuan B (1995) Fuzzy sets and fuzzy logic. Prentice Hall, Jersey
13. Lindley DV (1985) Making decisions. Wiley, London
14. Luce RD, Raiffa H (1957) Games and decisions. Wiley, New York
15. March JG (1994) A primer on decision making: how decisions happen. The Free Press, New York
16. Mares M (1997) Weak arithmetic on fuzzy numbers. Fuzzy Sets Syst 91(2):143–154
17. Mares M (2001) Fuzzy cooperative games. Springer, New York
18. Maturo A (2009) Alternative fuzzy operations and applications to social sciences. Int J Intell Syst 24:1243–1264
19. Maturo A (2009) Coherent conditional previsions and geometric hypergroupoids. Fuzzy Sets, Rough Sets, Multivalued Oper Appl 1(1):51–62
20. Maturo A (2009) On some structures of fuzzy numbers. Iran J Fuzzy Syst 6(4):49–59
21. Maturo A, Ventre AGS (2008) Models for consensus in multiperson decision making. In: NAFIPS 2008 conference proceedings. Regular Papers 50014. IEEE Press, New York
22. Maturo A, Ventre AGS (2009) An application of the analytic hierarchy process to enhancing consensus in multiagent decision making. In: Proceeding of the international symposium on the analytic hierarchy process for multicriteria decision making, July 29–August 1, 2009, paper 48. University of Pittsburg, Pittsburgh, pp 1–12
23. Maturo A, Ventre AGS (2009) Aggregation and consensus in multi objective and multi person decision making. Int J Uncertain Fuzziness Knowl-Based Syst 17(4):491–499
24. Maturo A, Ventre A (2009) Multipersonal decision making, consensus and associated hyperstructures. In: Proceeding of AHA 2008. University of Defence, Brno, pp 241–250
25. Maturo A, Squillante M, Ventre AGS (2006) Consistency for assessments of uncertainty evaluations in non-additive settings. In: Amenta P, D'Ambra L, Squillante M, Ventre AGS (eds) Metodi, modelli e tecnologie dell'informazione a supporto delle decisioni. Franco Angeli, Milano, pp 75–88
26. Maturo A, Squillante M, Ventre AGS (2006) Consistency for nonadditive measures: analytical and algebraic methods. In: Reusch B (ed) Computational intelligence, theory and applications. Springer, Berlin, pp 29–40
27. Maturo A, Squillante M, Ventre AGS (2010) Coherence for fuzzy measures and applications to decision making. In: Greco S et al. (eds) Preferences and decisions, STUDFUZZ 257 Springer, Heidelberg, pp 291–304
28. Maturo A, Squillante M, Ventre AGS (2010) Decision making. Fuzzy Meas Hyperstruct Adv Appl Stat Sci 2(2):233–253
29. Maturo A, Maturo F (2013) Research in social sciences: fuzzy regression and causal complexity. Springer, Heidelberg, pp 237–249
30. Maturo A, Maturo F (2014) Finite geometric spaces, steiner systems and cooperative games. Analele Universitatii "Ovidius" Constanta. Seria Matematica 22(1): 189–205
31. Maturo A, Maturo F (2017) Fuzzy events, fuzzy probability and applications in economic and social sciences. Springer International Publishing, Cham, pp 223–233
32. Maturo A, Sciarra E, Tofan I (2008) A formalization of some aspects of the social organization by means of the fuzzy set theory. Ratio Sociologica 1(2008):5–20
33. Maturo F (2016) Dealing with randomness and vagueness in business and management sciences: the fuzzy probabilistic approach as a tool for the study of statistical relationships between imprecise variables. Ratio Mathematica 30:45–58
34. Maturo F (2016) La regressione fuzzy. Fuzziness: Teorie e Applicazioni. Aracne Editrice, Roma, Italy, pp 99–110
35. Maturo F, Fortuna F (2016) Bell-shaped fuzzy numbers associated with the normal curve. Springer, Cham, pp 131–144
36. Maturo F, Hošková-Mayerová S (2017) Fuzzy regression models and alternative operations for economic and social sciences. Springer, Cham, pp 235–247
37. Saaty TL (1980) The analytic hierarchy process. McGraw-Hill, New York

38. Saaty TL (2008) Relative measurement and its generalization in decision making, why pairwise comparisons are central in mathematics for the measurement of intangible factors, the analytic hierarchy/network process. Rev R Acad Cien Serie A Mat 102(2):251–318
39. Shapley LS (1953) A value for n-person games. Ann Math Stud (Princeton Univ) 28: 307–317
40. Shapley LS (1962) Simple games. An outline of the theory. Behav Sci 7:59–66
41. Simon HA (1982) Models of bounded rationality, Mit Press, Cambridge (Mass.)
42. Sugeno M (1974) Theory of fuzzy integral and its applications, PhD Thesis, Tokyo
43. Von Neumann J, Morgenstern O (1944) Theory of games and economic behavior. Princeton University Press, Princeton
44. Weber S (1984) Decomposable measures and integrals for Archimedean t-conorms. J Math Anal Appl 101(1):114–138
45. Yager R (1986) A characterization of the extension principle. Fuzzy Sets Syst 18:205—217
46. Zadeh L (1975) The concept of a linguistic variable and its application to approximate reasoning. Inf Sci, Part I 8:199–249, Part 2: 301–357
47. Zadeh L (1975) The concept of a linguistic variable and its applications to approximate reasoning, Part III. Inf Sci 9:43–80

Consensus in Multiperson Decision Making Using Fuzzy Coalitions

Fabrizio Maturo and Viviana Ventre

Abstract We consider the problem of group decisions, in which decision-makers have different opinions or interests. This study proposes various metric spaces for representing the movements of decision-makers for reaching consensus. We also introduce the concept of fuzzy coalition for developing an algorithm for building a feasible fuzzy coalition, which is defined as the union of winning maximum coalitions which solve the issue of consensus among decision-makers.

Keywords Multiperson decision making · Cooperative games · Mediation and consensus · Metric spaces · Cluster analysis

1 Introduction and Motivation

We consider a decision problem with different decision-makers, which is formalized as follows. A decision must be taken by a committee consisting of a set $D = \{d_1, \ldots, d_r, \ldots, d_p\}$ of decision-makers, that are supposed to have the same importance. The decision would establish a relationship of pre-order δ (total or, in general, partial) in a set $A = \{a_1, \ldots, a_i, \ldots, a_m\}$ of alternatives, with a general objective O, that is, for each pair of alternatives (a_i, a_j), the group of decision-makers poses $a_i \, \delta \, a_j$ if and only if the committee D evaluates the alternative a_i less desirable, or equally preferred, over alternative a_j.

We write $a_i \sim a_j$ if a_i and a_j are both considered equally preferable, i.e. both relations $a_i \, \delta \, a_j$ and $a_j \, \delta \, a_i$ apply. Finally, we write $a_i < a_j$ or $a_j > a_i$, if a_i is judged

F. Maturo (✉)
Department of Management and Business Administration,
University of Chieti-Pescara, Viale Pindaro 42, 65127 Pescara, Italy
e-mail: f.maturo@unich.it

V. Ventre
Department of Economics, Management and Quantitative Methods,
University of Sannio at Benevento, Via delle Puglie 82, 82100 Benevento, Italy
e-mail: ventre@unisannio.it

© Springer International Publishing AG 2018   451
M. Collan and J. Kacprzyk (eds.), *Soft Computing Applications for Group
Decision-making and Consensus Modeling*, Studies in Fuzziness and Soft
Computing 357, DOI 10.1007/978-3-319-60207-3_26

less preferable than a_j, i.e. the relation $a_i \, a_j$ holds but not $a_j \, \delta \, a_i$. Criteria to reach a relationship of social pre-order, starting from the relations of order of individual decision-makers are, for example, in [1]. The logic by which we obtain social decisions are further defined in [2].

Preferably, we want to assign a quantitative value $\mu(a_i)$ to each alternative a_i, which represents the extent to which, in the opinion of the group D, the alternative a_i satisfies the general objective O.

Primarily, in this study, we assume that, for each alternative a_i, $\mu(a_i)$ is a real number in [0, 1], so the relation of pre-order δ reduces to the usual relation of order in [0, 1].

In some parts of the paper, to take into account the uncertainty in attributing quantitative values, due both to the uncertainty of the individual decision-makers and the need to consider the variability of judgments of different decision-makers, we consider the general case in which a triangular fuzzy number $\mu^*(a_i)$ with support contained in [0, 1] is assigned to each alternative a_i. In this case, we indicate with $\mu(a_i)$ the core of $\mu^*(a_i)$ (see e.g. [17, 26, 28–31, 36–38]).

Afterwards, we denote with $u = (a_u, c_u, b_u)$ a triangular fuzzy number with core c_u and support in $[a_u, b_u]$. If T is the set of triangular fuzzy numbers with supports in [0, 1], we can consider various relationships of pre-order δ in T. The most important are the following [13, 17]:

(1) Crisp order, which is indicated with $\leq_c$: for each $u = (a_u, c_u, b_u)$, $v = (a_v, c_v, b_v)$,

$$u \leq_c v \text{ if and only if } c_u \leq c_v.$$

(2) Main order (or standard order), which is indicated with $\leq_s$: for each $u = (a_u, c_u, b_u)$, $v = (a_v, c_v, b_v)$,

$$u \leq_s v \text{ if and only if } a_u \leq a_v, \, c_u \leq c_v, \, b_u \leq b_v.$$

The crisp order is a relation of total pre-order, but it is not a relation of order. It is equivalent to consider that only the core of fuzzy numbers is relevant.

The main order is a relation of order in T, but not total. The couple (T, $\leq_s$) is a lattice. The minimum is the fuzzy set $\mathbf{0} = (0, 0, 0)$ whereas the maximum is $\mathbf{1} = (1, 1, 1)$. Furthermore, for each $u = (a_u, c_u, b_u)$, $v = (a_v, c_v, b_v)$:

$$\inf \{u, v\} = (\min\{a_u, a_v\}, \min\{c_u, c_v\}, \min\{b_u, b_v\}),$$
$$\sup\{u, v\} = (\max\{a_u, a_v\}, \max\{c_u, c_v\}, \max\{b_u, b_v\}).$$

The process for obtaining the values $m(a_i)$ and $m^*(a_i)$ should take into account the diversity of assessments of different decision-makers and the fact that the degree of satisfaction of the general objective O depends on the degree to which the criteria, that represent particular aspects of the objective O, are met. Therefore, we assume that the decision depends on a set $C = \{c_1, \ldots, c_j, \ldots, c_n\}$ of criteria, and that each decision-maker d_r, with its own procedure, assigns to each criterion c_j a weight w_{rj} which represents the degree of importance of c_j with respect to the objective O. Thus, a vector $w_r = (w_{r1}, \ldots, w_{rm})$, called *vector of criteria weights*, is associated with each decision-maker d_r.

In this paper, we assume that the weights w_{rj} are real numbers satisfying the following conditions:

(1) *Positivity:* for every decision-maker d_r and criterion c_j, $w_{rj} \geq 0$;
(2) *Normalization*: for every decision-maker d_r, $w_{r1} + \ldots + w_{rm} = 1$.

Every decision-maker d_r is called to attribute a score p_{rij} (or p^*_{rij}) to each couple (a_i, c_j) which is formed by an alternative and criterion. The score p_{rij} is a real number in [0, 1] indicating to what extent the alternative a_i satisfies the criterion c_j in the opinion of the decision-maker d_r. More generally, the decision-maker d_r assigns a score as a triangular fuzzy number p^*_{rij} with core p_{rij} and support in [0, 1].

In this paper, we analyze procedures and algorithms to get the scores p_{rij} (or p^*_{rij}) and weights w_{rj} for the summary evaluations $\mu(a_i)$ e $\mu^*(a_i)$, mediating between the opinions of different decision-makers.

2 Procedures for the Allocation of Weights and Scores

One of the most followed procedure, for the allocation of weights so that the conditions of positivity and normalization are met, is the AHP method illustrated by Saaty [32, 33]. Some applications are in [3, 6, 11, 12, 18–21].

Essentially, the general objective is stated (and also defined) by a set of specific objectives; each of the specific objectives is defined by sub-objectives, and each sub-objective by a set of criteria. Scores (of specific objectives with respect the overall objective, sub-objectives with respect to each of the specific objectives, criteria with respect to sub-objectives) are determined using mathematical calculations from the pairwise comparison matrices obtained from interviews to decision-makers [18, 19, 32, 33].

The next step is to assign, to each decision-maker, the scores of alternatives with respect to each criterion. Let p_{rij} be the score that the decision-maker d_r assigns to the alternative a_i with respect to the criterion c_j. Whatever the scale for allocating the scores, there are no difficulties to bring the scores to real numbers in [0, 1], with a minimum score of 0 and maximum score equal to 1. However, the normalization condition is not always appropriate:

$$\forall r, j, \ p_{r1j} + p_{r2j} + \cdots + p_{rmj} = 1. \tag{2.1}$$

Indeed, if the scores are utilities (or can be assumed as utilities), as required in various studies dealing with rational decisions [15, 16, 27, 34, 35], it may happen that all the alternatives have a high utility compared to a criterion and low utility with respect to another one. Furthermore, the normalization introduces a strict dependence of scores of some alternatives by scores of other alternatives, which is in general not possible for the utilities. The importance of having utilities is also useful in order to have a consistent evaluation of alternatives [4, 5, 14, 22–25]. Conditions for obtaining coherent subjective probabilities in decision problems under uncertainty are presented in [4].

In conclusion, the assignment of scores p_{rij} (interpretable as utilities) to alternative a_i, with respect to the criterion c_j, by the decision-maker d_r, leads us to obtain a number u_{rj}, which we call the *overall usefulness of the criterion* c_j, defined by the following formula:

$$u_{rj} = p_{r1j} + p_{r2j} + \cdots p_{rmj} \tag{2.2}$$

Thus, it follows that $u_{rj} > 0$, otherwise the criterion c_j would be totally useless and should be excluded. In general, it does not happen that $u_{rj} = 1$. However, the AHP procedure of the comparison in pairs among the alternatives with respect to each criterion, which would lead getting scores p'_{rij} such that for every criterion c_j

$$u'_{rj} = p'_{r1j} + p'_{r2j} + \cdots + p'_{rmj} = 1$$

is a very efficient procedure. Indeed, it seems to guarantee the obtaining of numbers proportional to the utility, even if it does not provide the real utility. If it were known the number u_{rj}, then the following numbers could be considered as *"coherent utility"* of alternative a_i with respect to the criterion c_j:

$$p''_{rij} = u_{rj} p'_{rij}. \tag{2.3}$$

As a compromise between the two requirements (to use an efficient procedure and get coherent utility), in this paper, we propose the following algorithm.

First phase: the decision maker d_r, as expert, or assisted by experts, attributes the p_{rij} scores and calculates u_{rj}, that is the overall usefulness of the criterion c_j.

Second phase: the decision maker d_r, using the AHP procedure by Saaty, calculates p'_{rij} numbers and then $p''_{rij} = u_{rj}\, p'_{rij}$.

Third step: setting a positive real number ε, we calculate the Euclidean distance $\delta^{(2)}(v_{rj}, v''_{rj})$ between the vectors $v_{rj} = (p_{rij}, ..., p_{rmj})$ and $v''_{rj} = (p''_{rij}, ..., p''_{rmj})$.

Fourth phase: if $\delta^{(2)}(v_{rj}, v''_{rj}) < \varepsilon$ we accept p_{rij} numbers as coherent utilities, otherwise we return to the first step and ask the decision-maker d_r to correct the scores p_{rij} so that $\delta^{(2)}(v_{rj}, v''_{rj})$ decreases.

At the end of the assignment procedure of the scores, to each decision-maker d_r is associated a matrix D_r, with general term p_{rij} and vector of criteria utility with respect to the decision maker d_r given by $u_r = (u_{r1}, ..., u_{rm})$ (as marginal row). In the marginal column of the matrix are inserted the final scores $\mu_r(a_i)$ of the alternatives according to the decision-maker d_r (Table 1).

Table 1 Matrix D_r of alternatives and criteria according to the decision-maker d_r

D_r	$c_1 (w_{r1})$	$c_2 (w_{r2})$	$c_j (w_{ri})$	$c_n (w_{rm})$	μ_r
a_1	p_{r11}	p_{r12}	p_{r1j}	p_{r1n}	$\mu_r(a_1)$
a_2	p_{r21}	p_{r22}	p_{r2j}	p_{r2n}	$\mu_r(a_2)$
a_i	p_{ri1}	p_{ri2}	p_{rij}	p_{rin}	$\mu_r(a_i)$
a_m	p_{rm1}	p_{rm2}	p_{rmj}	p_{rmn}	$\mu_r(a_m)$
u_r	u_{r1}	u_{r2}	u_{rj}	u_{rm}	

If such p_{rij} scores can be considered utilities, the scores of the alternatives $\mu_r(a_i)$ are obtained according to the linear formula:

$$m_r(a_i) = w_{r1}p_{ri1} + w_{r1}p_{ri1} + \cdots + w_{r1}p_{ri1} \tag{2.4}$$

3 Decision-Makers as Points of the M-Dimensional Space of the Alternatives

We set $\mu_{ri} = \mu_r(a_i)$. In various studies [3, 11, 12, 18], the decision-maker d_r is represented by the point $P_r = (\mu_{r1}, \ldots, \mu_{rm})$ of the m-dimensional space of the alternatives. In this space, the distance $\delta^{(2)}(P_r, P_s)$ between the decision-makers d_r and d_s has been defined with the Euclidean formula:

$$\delta_{rs}^{(2)} = \delta^{(2)}(P_r, P_s) = \{[(\mu_{r1} - \mu_{s1})^2 + \cdots + (\mu_{rm} - \mu_{sm})^2]/m\}^{1/2}. \tag{3.1}$$

More generally, for any real number $h \geq 1$ we can consider the distance of order h given by $\delta^{(h)}(P_r, P_s)$ between the decision-makers d_r and d_s with the formula:

$$\delta_{rs}^{(h)} = \delta^{(h)}(P_r, P_s) = \{[(\mu_{r1} - \mu_{s1})^h + \cdots + (\mu_{rm} - \mu_{sm})^h]/m\}^{1/h}. \tag{3.2}$$

As h tends to $+\infty$ we obtain, specifically

$$\delta_{rs}^{(\infty)} = \max\{|\mu_{r1} - \mu_{s1}|, \ldots, |\mu_{rm} - \mu_{sm}|\} \tag{3.3}$$

The choice of h depends on the importance that the commission wants to attribute to larger values $|\mu_{ri} - \mu_{si}|$ than to smaller ones. The larger h the more relevant are the greatest values. For $h = +\infty$ holds only the maximum value.

Decision-makers are points of the metric space $S^{(h)} = (R^m, \delta^{(h)})$ and, being $0 \leq \mu_{ri} \leq 1$, it results $\delta_{rs}^{(h)} \leq 1$; thus, given a decision-maker P_r, all the others are in the hypersphere of radius 1 and center P_r.

Fixed $h \geq 1$ and a value $\varepsilon > 0$, both defined by the commission D, we say that:

- There is consensus (at level ε) between the decision-makers d_r and d_s, and we write $P_r \, \gamma \, P_s$, if $\delta^{(h)}(P_r, P_s) \leq \varepsilon$;

- There is a strict consensus (at level ε) between the decision-makers d_r and d_s, and we write $P_r \, \sigma \, P_s$, if $\delta^{(h)}(P_r, P_s) \leq \varepsilon/2$.

Relations γ and σ in D are reflexive and symmetric, but not transitive. Furthermore, $\sigma \subseteq \gamma$. Transitivity is replaced by the weaker properties:

$$(P_r \, \sigma \, P_s, \; P_s \, \sigma \, P_t) \Rightarrow P_r \, \gamma \, P_t.$$

For each decision-maker P_r, we can finally define:

- The *degree of consensus* of P_r, indicated with $\gamma(P_r)$, as the number of decision-makers P_s in consensus with P_r;
- The *degree of the strict consensus* of P_r, indicated with $\sigma(P_r)$, as the number of decision-makers P_s in close consensus with P_r;
- The *variability* of P_r, indicated with $v(P_r)$, defined as the maximum distance of P_r from the other decision-makers.

We say that a coalition K has *internal consensus* (or *strict internal consensus*) if the distance of any two elements of the coalition does not exceed ε (or not exceed $\varepsilon/2$).

Evidently, set a decision-maker P_r, the decision-makers belonging to the hypersphere of center P_r and radius $\varepsilon/2$ (i.e. in close consensus with P_r) form a coalition K with internal consensus and $\sigma(P_r)$ elements.

An algorithm for reaching consensus is a procedure that allows the most distant decision-makers to modify their assessments of criteria weights and scores of the alternatives, so that the representative points of the decision-makers come close, gradually, to obtain a final coalition C with internal consensus, encompassing all decision-makers, or at least a majority of them.

In this process for reaching the final coalition, if there are f changes of assessments of decision-makers, the point P_r, which represents a decision-maker d_r belonging to C, moves as a finite sequence $(P_r^1, P_r^2, \ldots, P_r^f)$ of positions in the space R^m, from the initial situation $P_r^1 = (\mu_{r1}^1, \ldots, \mu_{rm}^1)$ to the final one $P_r^f = (\mu_{r1}^f, \ldots, \mu_{rm}^f)$.

Once the final coalition C and final score vectors $P_r^f = (\mu_{r1}^f, \ldots, \mu_{rm}^f)$ (that decision-makers belonging to the coalition C attribute to alternative) are obtained, we can get the group evaluation of alternatives such as an average of the assessments of individual decision-makers belonging to C.

The most common case is the calculation of the arithmetic mean, where the social score of alternative a_i is given by the formula

$$\mu(a_i) = (\delta_1 \mu_{1i}^f + \delta_2 \mu_{2i}^f + \cdots + \delta_p \mu_{pi}^f)/m \qquad (3.4)$$

where $\delta_r = 1$ if the decision-maker d_r belongs to C and $\delta_r = 0$ if d_r does not belong to C.

A more general procedure, introduced by Maturo and Ventre [20], can lead to obtain a set of $h > 1$ possible final coalitions. In this case, if n_r is the number of final coalitions to which d_r belongs, and m is the number of decision-makers that are part of at least one final coalition, we can define the power of d_r as the ratio:

$$\pi_r = n_r/h. \tag{3.5}$$

Placing $m = \pi_1 + \pi_2 + \ldots + \pi_q$, the formula (3.4) is replaced by the more general formula:

$$\mu(a_i) = (\pi_1 \mu_{1i}^f + \pi_2 \mu_{2i}^f + \cdots + \pi_p \mu_{pi}^f)/m \tag{3.6}$$

Evidently the (3.6) reduces to (3.4) in the case of only one possible final coalition.

A significant interpretation of the formula (3.6) is obtained in the context of fuzzy sets. In fact, we can think that the final set of decision-makers is a fuzzy set in which each decision-maker d_r has a degree of membership π_r.

In Fig. 1, we consider a set $D = \{a, b, c, d, e, f, g, h, i, l, m, n\}$ of 12 decision-makers and three final coalitions. According to this method, the decision-makers d and e have power 1, b has power 2/3, a, c, g, h, and i have power 1/3, whereas m, l, and n have zero power.

Fig. 1 Example of 12 decision-makers and three final coalitions

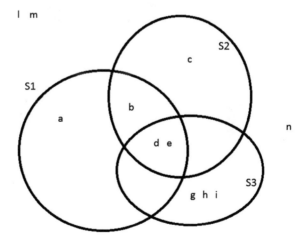

4   Decision-Makers as Points of the M × N Dimensional Space of Pairs (Alternatives, Criteria)

It may be useful to employ more analytical formulas for the distance (with respect to Eq. 3.2), in which it appears visible the evaluation of each decision-maker with respect to each pair (alternative, criterion), so that it is easier to show how the feedback of two decision-makers can get closer by setting new assessments of criteria and alternatives.

In this context, a decision-maker d_r is represented by the point P_r of the dimensional space m × n of the couples (alternative a_i, criterion c_j) of coordinates $(w_{rj} \, p_{rij}, \ i = 1, \ldots, m; \ j = 1, \ldots, n)$.

In this space, for any real number $h \geq 1$, we can consider the distance of order h $\lambda^{(h)}(P_r, P_s)$ between the decision-makers d_r and d_s, with the following formula:

$$\lambda^{(h)}(P_r, \ P_s) = \lambda^{(h)}(P_r, \ P_s) = \left\{ \left[\sum_{i=1}^{m} \left(\sum_{j=1}^{n} \frac{1}{n \times m} \left| w_{rj} p_{rij} - w_{sj} p_{sij} \right|^h \right) \right] \right\}^{1/h}. \qquad (4.1)$$

In particular, for $h = 2$ we have the Euclidean space whereas for $h = +\infty$ we have the distance:

$$\lambda^{(\infty)}(P_r, \ P_s) = \max \left\{ \left| w_{rj} p_{rij} - w_{sj} p_{sij} \right| \right\}, \ i = 1, \ \ldots, m; \ j = 1, \ \ldots, n. \qquad (4.2)$$

We can extend to these metric spaces all the considerations on procedures for achieving consensus that we presented in the previous paragraph.

It also seems interesting to consider the following formula:

$$v_{rs}^{(h)} = v^{(h)}(P_r, \ P_s) = \left\{ \left[\sum_{i=1}^{m} \left(\sum_{j=1}^{n} \frac{\varphi_{rsj}}{n \times m} \left| p_{rij} - p_{sij} \right|^h \right) \right] \right\}^{1/h}, \ h \geq 1 \qquad (4.3)$$

where $\varphi_{rsj} = \varphi(w_{rj}, w_{sj})$, and φ is a function defined in $[0, 1]^2$, with values in $[0, 1]$, symmetric, increasing with respect to each variable, strictly increasing in $(0, 1]^2$, and satisfying the boundary conditions:

- $$\varphi(0, \ 0) = 0;$$

- $$\sum_{j=1}^{n} \varphi_{rsj} = 1$$

Two special cases are:

$$\varphi_{rsj} = \left(w_{rj} + w_{sj}\right)/2 \,(\text{arithmetic mean});$$

$$\varphi_{rsj} = \left(w_{rj}w_{sj}\right)/\left(\sum_{k=1}^{n} w_{rk}\, w_{sk}\right).$$

5   An Algorithm for the Achievement of Consensus Among Decision-Makers or Coalitions

We present a dynamic algorithm with f steps, with f natural number to be determined. The set of decision-makers $D = \{d_1, ...,d_r, ..., d_p\}$, at the step s, is represented by the point cloud $N^s = \{P_1^s, ..., P_r^s, ..., P_p^s\}$ of a metric space (S, γ). In particular, if we consider the metric space of Sect. 3, it is $S = R^m$, $\gamma = \delta^{(h)}$ and, if we have the metric space of Sect. 4, then $S = R^{m \times n}$, $\gamma = \lambda^{(h)}$.

Following legal obligations or preliminary agreements among decision-makers, we can fix an integer $k \in (p/2, p]$, called *level of majority* and a positive real number ε, called *level of consensus*. We also assume to have an impartial referee A, called the "Demiurge", which, according to his knowledge and skills, looks for giving information and procedures for increasing consensus among decision-makers [3].

Let K^s be a coalition at the step s. We denote by $\omega(K^s)$, called *diameter of the coalition to the step s*, the maximum distance between two points representing K^s; let also be $|K^s|$ the number of elements of the coalition, and C^s, called the *center of the coalition*, the point of S such that

$$\sum_{P \in K^s} \gamma(P, C^s) = \min_{X \in S} \sum_{P \in K^s} \gamma(P, X).$$

The coalition K^s is said to be:

- *winning* if $|K^s| \geq k$;
- *loser* if $|K^s| \leq p - k$;
- *blocking* if $p - k < |K^s| < k$.

In addition, the coalition is said *feasible* if $\omega(K^s) \leq \varepsilon$.

We name *solution to the problem of consensus to the step s* every coalition that simultaneously meets three requirements: winning, feasible, and maximum.

The algorithm requires that, at each step, some decision-makers (who does not respect the rules set by the Demiurge (A) can be excluded from the decision-making power. We denote by $D^{(s)}$ the number of remaining decision-makers at the step s.

A positive real number α, called *level of improvement*, must be fixed. Thus, at the step s, the following cases may occur:

(C1) $D^{(s)}$ is a feasible winning coalition, or a union of feasible winning coalitions;
(C2) $D^{(s)}$ is not a feasible winning coalition, and there is only one solution to the problem of consensus;
(C3) $D^{(s)}$ is not a union of feasible winning coalitions, and there are several solutions;
(C4) there are no solutions.

In the case (C1), the algorithm ends and we set f = s. If $D^{(s)}$ is a feasible winning coalition, it is taken as final solution; instead, if $D^{(s)}$ is a union of feasible winning coalitions, we define, as *fuzzy feasible winning coalition,* the fuzzy set with domain $D^{(s)}$ in which each element P_r^s has a degree of membership equal to the ratio between the number of feasible winning coalitions to which it belongs, and the total number of feasible winning coalitions.

In the case (C2), let K^s be the solution; we calculate, for each decision-maker P_r^s which not belongs to the solution, the maximum distance β_r^s from an element of the solution, called *distance of P_r^s from the solution,* and each of these decision-makers is invited to review his assessments in order to move from the point P_r^s to a point P_r^{s+1} with distance from K^s equal to a number β_r^{s+1} (less than, or equal to, $\beta_r^s - \alpha$). The decision-makers who do not conform are excluded from the set $D^{(s+1)}$ of the remaining decision-makers to the step s + 1.

In the case (C3), for each decision-maker P_r^s not belonging to any solution, and for each solution K^s, we consider the maximum distance β_r^{sK} of P_r^s by an element of K^s, and β_r^s is set equal to the minimum of the values β_r^{sK} to vary of K^s in the set of solutions. The number β_r^s is said the *distance of P_r^s from all the solutions.* Then, each of these decision-makers is invited to review his assessments in order to move from the point P_r^s to a point P_r^{s+1} with distance from all of the solutions equal to a number β_r^{s+1} (less than, or equal to, $\beta_r^s - \alpha$). Also in this case, the decision-makers who do not conform are excluded from the set $D^{(s+1)}$ of the remaining decision-makers at the step s + 1.

In the case (C4), we consider, for each decision-maker P_r^s, the variability $v(P_r^s)$ and a degree of consensus $\gamma(P_r^s)$. Evidently, the maximum variability is equal to the diameter $\omega(D^s)$ of the coalition $D^{(s)}$. One at a time, the decision-makers with maximum variability and minimum consensus are invited to change their opinions in order to move from the point P_r^s to a point P_r^{s+1} with variability less than, or equal to, $\omega(D^s) - \alpha$, or with equal variability, but a greater degree of consensus. If there are different decision-makers with the same variability and degree of consensus, then we may choose the decision-maker that needs to change his position by lot, or by other criteria determined at the beginning by the Demiurge. If $|D^{(s)}| > k$, the decision-maker who does not conform will be excluded from the set $D^{(s+1)}$ of the remaining decision-makers at the step s + 1. If $|D^{(s)}| = k$ and a decision-maker who is called to modify his assessment do not adapt, then the Demiurge declares that it has not been possible to reach consensus, and invites all decision-makers to revise the decision problem from the beginning, in particular the set of criteria and alternatives.

6 The Possible Role of Cluster Analysis as an Aid to the Achievement of the Consensus

A possible help for an effective procedure for reaching consensus may come from cluster analysis. While cluster analysis algorithms are automatic, algorithms for and achieving consensus are not automatic because, at every step, may arise the willingness to change positions of decision-makers appropriately identified by the Demiurge; however, cluster analysis can help the Demiurge in the selection of individuals which should be invited to change their positions. One could also analyze the role of coalitions that, although no final or majority, appear to be "natural" or "spontaneous" because the groups should be obtained in few steps of cluster analysis.

Let us consider for example the figure, which represents 7 decision-makers in a two-dimensional space with two alternatives:

An algorithm for reaching consensus almost certainly would require a change of positions of decision-makers c and g because they are the most distant from each other, and it should try to move decision-makers close to d that is in the central position.

How can we move c (or g) without taking into account the natural coalition to which it belongs? How do we justify the power of d, which should attract the others, despite being an isolated decision-maker?

A cluster analysis algorithm should lead to three "natural" coalitions {a, b, c}, {d}, {e, f, g} (Fig. 2). In this case, this solution seems to take into account also "natural" or "spontaneous" coalitions. But how can we organize an effective and rational algorithm for reaching consensus after that we have individuated these three starting groups?

A possible solution could be asking smaller coalitions to move towards a larger one: e.g. the decision-making d moves towards {e, f, g} or toward {a, b, c}. In this case we can get a small majority. Another possibility is to ask the larger coalitions to agree and get closer to each other, such as approaching the centroids. In this way we can have a large majority.

In general, if we want to respect the coalitions, we could do the following. With a cluster analysis algorithm, we identify feasible "natural" or "spontaneous" coalitions. Then, for every coalition K we consider:

- the *variability* of K, indicated with v(K), defined as the maximum distance of K from the other coalitions;

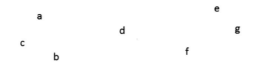

Fig. 2 Example of 7 decision-makers in a two-dimensional space with two alternatives

- the *power* of K, indicated with $\pi(K)$, as the number of decision-makers belonging to K;
- the *degree of the strict consensus* of K, indicated with $\sigma(K)$, as the maximum number of decision-makers P_s of K in close consensus with an element P_r of K.

Moreover, the Demiurge identifies a formula to measure the distance between two coalitions, e.g. the distance of order h between the centroids or, in general, between the centers (h is fixed at the beginning by the Demiurge). Then, the Demiurge identifies "spontaneous" coalitions with maximum variability and asks them, one at a time, to reduce their variability; first, the Demiurge asks the one with the least power to reduce its variability; in case there are more coalitions that have maximum variability and minimum power, the Demiurge asks the one with the least degree of strict consensus to reduce its variability.

More generally, the Demiurge may act as a mediator for negotiations between the coalitions, can explain the benefits of their approach, can illustrate the need to review the various positions.

7 Conclusions and Research Perspectives

In this paper, we have considered various representations of all the decision-makers and various possible algorithms for reaching consensus. The choice of the algorithms to be used can be the subject of prior agreements between the Demiurge and the Commission, or it can derive by special regulations.

A key element is to understand whether the decision-makers are independent of each other or if they are linked in such a way as to form spontaneous coalitions. In the first case, the algorithm of Sect. 5 seems the most suitable. In the second case, it seems that the help of cluster analysis and the theory of cooperative games can be an effective aid for reaching consensus.

A further problem is the respect for the diversity of the various stakeholders and the need to try to protect all users involved in the decisions. In this regard, it might be interesting to extend the social field the models used in the environmental field for the study of biodiversity [7–10]. The have results of such studies could be used in decision-making problems for the determination of the criteria, for the presentation of alternatives and for evaluation of the usefulness of each pair (alternative criteria). In addition, the consensus-algorithm may be modified taking into account, not only of consensus but also of respect for the diversity of views of decision makers linked to the need for social and economic.

References

1. Arrow KJ (1951) Social choice and individual value. Yale University Press, New Haven
2. Boudon R (1979) La logique du social. Hachette, Paris
3. Carlsson C, Ehrenberg D, Eklund P, Fedrizzi M, Gustafsson P, Lindholm P, Merkurieva G, Riissanen T, Ventre AGS (1992) Consensus in distributed soft environments. Eur J Oper Res 61:165–185
4. Coletti G, Scozzafava R (2002) Probabilistic logic in a coherent setting. Kluwer Academic Publishers, Dordrecht
5. de Finetti B (1974) Theory of probability. Wiley, New York
6. Delli Rocili L, Maturo A (2013) Teaching mathematics to children: social aspects, psychological problems and decision making models. In: Soitu, Gavriluta, Maturo (ed) Interdisciplinary approaches in social sciences, Editura Universitatii A.I. Cuza, Iasi, Romania.
7. Di Battista T, Fortuna F, Maturo F (2014) Parametric functional analysis of variance for fish biodiversity. In: Paper presented at the international conference on marine and freshwater environments, iMFE. www.scopus.com
8. Di Battista T, Fortuna F, Maturo F (2016) Environmental monitoring through functional biodiversity tools. Ecol Indic 60:237–247. doi:10.1016/j.ecolind.2015.05.056
9. Di Battista T, Fortuna F, Maturo F (2016) Parametric functional analysis of variance for fish biodiversity assessment. J Environ Inform. doi:10.3808/jei.201600348
10. Di Battista T, Fortuna F, Maturo F (2017) BioFTF: an R package for biodiversity assessment with the functional data analysis approach. Ecol Indic 73:726–732. doi:10.1016/j.ecolind. 2016.10.032
11. Ehrenberg D, Eklund P, Fedrizzi M, Ventre AGS (1989) Consensus in distributed soft environments. Rep Comput Sci Math A 88:1–24
12. Eklund P, Rusinowska A, De Swart H. (2007) Consensus reaching in committees. Eur J Oper Res 178:185–193
13. Klir GJ, Yuan B (1995) Fuzzy sets and fuzzy logic. Prentice Hall, Jersey
14. Lindley DV (1985) Making decisions. Wiley, London
15. Luce RD, Raiffa H (1957) Games and decisions. Wiley, New York
16. March JG (1994) A primer on decision making: how decisions happen. The Free Press, New York
17. Maturo A (2009) Alternative fuzzy operations and applications to social sciences. Int J Intel Syst 24:1243–1264
18. Maturo A, Ventre AGS (2008) Models for consensus in multiperson decision making. In: NAFIPS 2008 conference proceedings. Regular Papers 50014. IEEE Press, New York
19. Maturo A, Ventre AGS (2009) An application of the analytic hierarchy process to enhancing consensus in multiagent decision making. In: Proceeding of the international symposium on the analytic hierarchy process for multicriteria decision making, July 29–August 1, 2009, paper 48. University of Pittsburg, Pittsburgh, pp 1–12
20. Maturo A, Ventre AGS (2009) Aggregation and consensus in multi objective and multi person decision making. Int J Uncertain Fuzziness Knowl-Based Syst 17(4):491–499
21. Maturo A, Ventre AGS (2009) Multipersonal decision making, concensus and associated hyperstructures. In: Proceeding of AHA 2008. University of Defence, Brno, pp 241–250
22. Maturo A, Squillante M, Ventre AGS (2006) Consistency for assessments of uncertainty evaluations in non-additive settings. In: Amenta P, D'Ambra L, Squillante M, Ventre AGS (eds) Metodi, modelli e tecnologie dell'informazione a supporto delle decisioni. Franco Angeli, Milano, pp 75–88
23. Maturo A, Squillante M, Ventre AGS (2006) Consistency for nonadditive measures: analytical and algebraic methods. In: Reusch B (ed) Computational intelligence, theory and applications. Springer, Berlin, pp 29–40

24. Maturo A, Squillante M, Ventre AGS (2010) Coherence for fuzzy measures and applications to decision making. In: Greco S et al (eds) Preferences and decisions, STUDFUZZ 257 Springer, Heidelberg, pp 291–304
25. Maturo A, Squillante M, Ventre AGS (2010) Decision making, fuzzy measures, and hyperstructures. Adv Appl Stat Sci 2(2):233–253.
26. Maturo A, Maturo F (2013) Research in social sciences: fuzzy regression and causal complexity. In: Studies in fuzziness and soft computing. pp 237–249. doi:10.1007/978-3-642-35635-3_18
27. Maturo A, Maturo F (2014) finite geometric spaces, steiner systems and cooperative games. Analele Universitatii "Ovidius" Constanta—Seria Matematica 22(1). doi:10.2478/auom-2014-0015
28. Maturo A, Maturo F (2016) Fuzzy events, fuzzy probability and applications in economic and social sciences. In: Studies in systems, decision and control. pp 223–233. doi:10.1007/978-3-319-40585-8_20
29. Maturo F (2016) Dealing with randomness and vagueness in business and management sciences: the fuzzy probabilistic approach as a tool for the study of statistical relationships between imprecise variables. Ratio Mathematica 30:45–58
30. Maturo F, Fortuna F (2016) Bell-shaped fuzzy numbers associated with the normal curve. In: Topics on methodological and applied statistical inference, pp 131–144. doi:10.1007/978-3-319-44093-4_13
31. Maturo F, Hošková-Mayerová Š (2016) Fuzzy regression models and alternative operations for economic and social sciences. In: Studies in systems, decision and control, pp 235–247. doi:10.1007/978-3-319-40585-8_21
32. Saaty TL (1980) The analytic hierarchy process. McGraw-Hill, New York
33. Saaty TL (2008) Relative measurement and its generalization in decision making, why pairwise comparisons are central in mathematics for the measurement of intangible factors, the analytic hierarchy/network process. Rev R Acad Cien Serie A Mat 102(2):251–318
34. Simon HA (1982) Models of bounded rationality. Mit Press, Cambridge (Mass.)
35. von Neumann J, Morgenstern O (1944) Theory of games and economic behavior. Princeton University Press, Princeton, Jersey
36. Yager R (1986) A characterization of the extension principle. Fuzzy Sets Syst 18:205–217
37. Zadeh L (1975) The concept of a linguistic variable and its application to approximate reasoning. Inf Sci, Part I 8:199–249, Part 2: 301–357
38. Zadeh L (1975) The concept of a linguistic variable and its applications to approximate reasoning, Part III. Inf Sci 9:43–80

Reaching Consensus in a Group of Agents: Supporting a Moderator Run Process via Linguistic Summaries

Janusz Kacprzyk and Sławomir Zadrożny

Abstract We present some account of our works on consensus reaching processes in a set of agents (individuals, decision makers, etc.), notably those which are driven by a moderator, a "super-agent" who is in charge of running the process in an effective and efficient way. We assume the classic approach to the evaluation of a degree of consensus due to Kacprzyk and Fedrizzi [19–21] in which a soft degree of consensus has been introduced as a degree to which, for instance, "most of the important individuals agree as to almost all of the relevant options", using the fuzzy majority introduced into group decision making by Kacprzyk [17, 18] equated with a fuzzy linguistic quantifier (most, almost all, . . .) and handled via Zadeh's [53] classic calculus of linguistically quantified propositions or some other method, notably Yager's [50] OWA (ordered weighted average) operators. The consensus reaching process is run in a group of agents, which is assumed to be relatively small (e.g. human experts), by a moderator for whom some support, i.e. additional information may be useful. In our case, it is provided by a novel combination of, first, the use of the a soft degree of consensus alone within a decision support system setting along the lines of Fedrizzi et al. [5], Fedrizzi et al. [4], Kacprzyk and Zadrożny [28, 31]. Second, the linguistic data summaries in the sense of Yager [49], Kacprzyk and Yager [24], Kacprzyk et al. [25], in particular in its protoform based version proposed by Kacprzyk and Zadrożny [30, 32], are employed to indicate in a natural language some interesting relations between individuals and options to help the moderator identify crucial (pairs of) individuals and/options which pose some threats to the reaching of consensus. Third, we mention the use of some results obtained in our recent paper (Kacprzyk et al. [40]) on the use of a novel data min-

To Mario, Professor Mario Fedrizzi, a great and visionary scholar and researcher who has always been able to combine theory and practice to solve difficult real world problems, and a long time loyal friend.

J. Kacprzyk (✉) · S. Zadrożny
Systems Research Institute, Polish Academy of Sciences, ul. Newelska 6,
01–447 Warsaw, Poland
e-mail: kacprzyk@ibspan.waw.pl; Janusz.Kacprzyk@ibspan.waw.pl

S. Zadrożny
e-mail: zadrozny@ibspan.waw.pl

© Springer International Publishing AG 2018
M. Collan and J. Kacprzyk (eds.), *Soft Computing Applications for Group Decision-making and Consensus Modeling*, Studies in Fuzziness and Soft Computing 357, DOI 10.1007/978-3-319-60207-3_27

ing tool, a so-called action rule proposed by Raś and Wieczorkowska [46], which are meant in our context to find best concessions to be offered to the individuals for changing their preferences to increase the degree of consensus. Fourth, our new results of the use of the concepts of a consensory and dissensory agent (cf. Kacprzyk and Zadrożny [37]) are summarized.

Keywords Consensus · Consensus reaching support · Fuzzy preference · Fuzzy majority · Fuzzy logic · Linguistic quantifier · Linguistic summary · OWA (ordered weighted averaging) operator · Action rule · Consensory agent · Dissensory agent

1   Introduction

In view of the complexity of the present world plagued by economic, environmental, social, etc. problems, the process of reaching decisions that would be good, or even acceptable, to as many as possible members of a group, not to speak about the whole society, becomes extremely important. It is obvious that consensus, as a reflection of such a wide acceptance, has become a widely discussed and considered issue, both in the academic and scholarly community, and among economists, social scientists, political scientists, politicians, etc.

In most cases, decision making processes, maybe even serious discussions, start with a point of departure which exhibits a possibly high initial agreement of the agents, parties or institutions involved, though maybe yet far from a consensus. This can facilitate the process of reaching a proper decision (consensus) that will be acceptable to all the participants involved, providing some just and fair solutions. This clearly suggests that the initial *consensus* should then be further made deeper—through a consensus reaching process.

We will mainly focus our attention on the process of consensus reaching among agents whose testimonies may initially differ to a considerable extent but we assume the agents' rational commitment to consensus, i.e. a readiness to change testimonies to attain a higher extent of consensus. This will be done via suggestions of the moderator who will have some mechanisms and tools to propose arguments for changes. This is meant here as a *consensus reaching process* [2, 12] which is dynamic, iterative and interactive, meant to last over some time span.

In all nontrivial cases the consensus reaching process would need a computer based support, and a moderator to run it. The moderator is a special individual (agent), either human or software who is responsible for running the discussion within the group of agents, convincing agents to change their testimonies using analyses and argumentation, etc. until a proper state of agreement ("consensus") is reached that may facilitate a fair, just and acceptable collective decision to be made.

We assume the following basic setting. First, the agents present *individual* preference relations, which can be aggregated to a *social fuzzy preference relations*—cf., e.g., Nurmi [43]. Then, a fuzzy majority is assumed as proposed by Kacprzyk [17, 18]; for a comprehensive review, cf. Kacprzyk et al. [38, 39]. The fuzzy majority may be represented in a natural way via the so-called *linguistic quantifiers*: *most, almost all, much more than a half,* ... which can be handled by fuzzy logic using notably Zadeh's [53] calculus of linguistically quantified proposition which is simple and intuitively appealing; moreover, for instance, Yager's [50] OWA (ordered weighted average) operators can also be employed.

We assume a human-consistent approach to the definition of a *soft degree of consensus* introduced by Kacprzyk and Fedrizzi [19–21] as a degree to which, e.g., "*most* of the *relevant* (knowledgeable, expert, ...) individuals agree as to *almost all* of the *important* options". Then, we assume a moderator-run support of consensus reaching proposed by Fedrizzi et al. [5], and then further developed by Kacprzyk and Zadrożny [28, 31] that is based on the soft degrees of consensus mentioned above and some additional information, hints, clues, etc. derived.

To be more specific, in our former papers (cf. Kacprzyk and Zadrożny [29, 35]) we proposed the use of linguistic data summaries in the sense of Yager [49], but in their extended and implementable version of Kacprzyk and Yager [24] or Kacprzyk et al. [25] as a tool to get insight and information as to the group's testimonies, difficulties, dynamics, etc. exemplified by which individuals provide testimonies which pose problems, which pairs of options may pose some problems, etc.

Moreover, in our recent paper, cf. Kacprzyk and Zadrożny [37] we proposed a novel approach based on new concepts of a consensory and dissensory agent for whom the above mentioned additional information in the form of linguistic summaries is separately derived, and then used by the moderator as a step towards a fairness type attitude of the moderator.

Fuzzy majority is in common use by the humans, and—in our context—Loewer and Laddaga [41] statement is often cited:

> ...It can correctly be said that there is a *consensus* among biologists that Darwinian natural selection is an important cause of evolution though there is currently *no consensus* concerning Gould's hypothesis of speciation. This means that there is a *widespread agreement* among biologists concerning the first matter but *disagreement* concerning the second ...

and it is clear that a strict majority as, e.g., more than 75% would not reflect the intention.

The use of fuzzy linguistic quantifiers has been proposed by the authors to introduce a fuzzy majority for measuring (a degree of) consensus and deriving new solution concepts in group decision making (cf. Kacprzyk [17, 18], Kacprzyk and Fedrizzi [19–21]), as well as on voting (choice functions) by Kacprzyk and Zadrożny [27, 28, 32].

This soft degree of consensus is basically meant to overcome some "rigidity" of the conventional concept of consensus in which (full) consensus occurs only when "*all* the agents agree as to *all* the issues" which is unrealistic is practice. The new degree of consensus can be therefore equal to 1, which stands for full consensus, when, for instance, "*most* of the (important) agents agree as to *almost all* (of the relevant) options".

The agents provide their testimonies concerning (pairs of) options in question as *fuzzy preference relations* expressing preferences given in pairwise comparisons of options. For an effective and efficient group decision making, we need to first reach *consensus* among the agents. A discussion is run—supported by a moderator—to clarify points of view, exchange information, advocate opinions, etc., to imply some changes in the preferences, possibly getting them closer to consensus.

The moderator usually needs some support with some hints, clues, suggestions, etc. as to the most promising directions for a further discussion, troublesome options and/or individuals, etc.

Kacprzyk and Zadrożny [29] have proposed to use for this, first, linguistic data summaries to subsume what proceeds in the group with respect to the uniformity of preferences, in the sense of agents and options, which was then extended in Kacprzyk and Zadrożny [35]. To quote, for illustration some examples, linguistic summaries like: "Most individuals definitely prefer option o_{i1} to option o_{i2}, moderately prefer o_{i3} to o_{i4}, …", "Most individuals definitely preferring o_{i1} to o_{i2} also definitely prefer o_{i3} to o_{i4}", "Most individuals choose options $o_{i1}, o_{i2}, …$", "Most individuals reject options $o_{i1}, o_{i2}, …$", "Most options are dominated by option o_i in opinion of individual e_k", "Most options are dominated by option o_i in opinion of individual $e_{k1}, e_{k2}, …$", "Most options dominating alternative o_i in opinion of individual e_{k1} also dominate option o_i in opinion of individual e_{k2}", etc. can be given. Obviously, such summaries may help the moderator and the group of agents gain a deeper understanding and insight of relations within the agents and their testimonies (preferences). The derivation of such valuable summaries can be facilitated by Kacprzyk and Zadrożny's [27, 28] works on general choice functions in group decision making as well as, e.g. Herrera-Viedma et al. [14, 15] approach. The linguistic summaries will also be employed here though in a slightly different setting.

A novel and interesting approach in this context has been recently proposed by Kacprzyk et al. [40] based on the use of so-called *action rules*. The concept of an *action rule* was proposed by Raś and Wieczorkowska [46], in the context of Pawlak's [45] *information systems*, and then has been extensively further studied and developed by Raś and his collaborators. Roughly speaking, the purpose of an action rule is to show how a subset of flexible attributes should be changed to obtain an expected change of the decision attribute for a subset of objects characterized by some values of the subset of stable attributes. For example, in a bank context, an action rule may, e.g., indicate that offering a 20% reduction in a monthly bank account fee instead of a 10% reduction to a middle-aged customer is expected to increase his or her spendings from medium to high, and action rules are sought to attain as "cheaply as possible" a desired change of an attribute value. In our context, the essence is to find some cheapest concessions to eventually be offered to some agents to make them change preferences in a desired direction.

In general, the use of linguistic summaries, action rules, and other elements which can provide insight into the very essence of the structure of the group, testimonies, etc. may be described as tools and techniques for a better, more effective and efficient moderation in the consensus reaching process. This approach is very powerful but will not be used in this paper, for simplicity and clarity.

Recently, Kacprzyk and Zadrożny [37] have proposed to additionally use another approach. Namely, as already mentioned, the (computer based) support of the consensus reaching process as meant here (cf. Fedrizzi et al. [5], Fedrizzi et al. [4], Kacprzyk and Zadrożny [28, 31], and Zadrożny and Furlani [55]) is extended using the following reasoning: (1) at each consensus reaching stage, the agents present their current testimonies (individual fuzzy preference relations), a fuzzy majority based degree of consensus is calculated—if it is high enough, or a time limit is over, then the process is terminated and otherwise the subsequent steps are executed, (2) the current individual fuzzy preference relations are aggregated using, for instance, some averaging (cf. Nurmi [43], Kacprzyk et al. [38], etc.) to a current consensual social fuzzy preference relation, (3) normalized degrees of similarity of the particular individual fuzzy preference relations to the current consensual social fuzzy preference relation are calculated using some similarity measure, e.g. 1—the value of the Hamming or Euclidean distance, (4) normalized degrees of dissimilarity of the particular individual fuzzy preference relations to the current consensual social fuzzy preference relation are calculated using some dissimilarity measure (e.g. Hamming or Euclidean distance), (5) a fuzzy set of consensory agents is determined in which the degree of membership of a particular agent is the normalized degree of similarity of his/her individual fuzzy preference relation to the consensual social fuzzy preference relation, (6) a fuzzy set of dissensory agents is determined in which the degree of membership of a particular agent is the normalized degree of dissimilarity of his/her individual fuzzy preference relation to the consensual social fuzzy preference relation, (7) the moderator, supported by additional information provided by linguistic summaries, derived separately for the above two types of agents, consensory and dissensory, specifies sets of "troublesome" agents and options for which changes in the preferences should hopefully make the degree of consensus higher, and suggests these changes of preferences to those agents, and with respect to those options mentioned, and then the above steps are repeated until the end of the process.

Notice that the introduction of the concepts of a *consensory* and *dissensory* agent, then used separately in the moderated consensus reaching process, can clearly be viewed to follow the very idea of a *fairness*, or *equity* driven approach because we take into account both the "good" and "bad", "promising" and "nonpromising", "flexible and stubborn", etc. agents—cf. Gołuńska and Hołda [6], or Gołuńska and Kacprzyk [7]. This may provide a new way of a *equitable, fairness focused moderation* of the consensus reaching process. We will not consider this problem in more detail and refer the reader to Kacprzyk and Zadrożny [36].

Now, first, we will briefly present the calculus of linguistically quantified propositions, and the concept of a soft degree of consensus, followed by the essence of a consensus reaching process. We will then briefly show how linguistic data summaries can be used to help gain insight into the very essence of consensus reaching, and then we will briefly show the essence of a equitable, fairness focused moderation of the consensus reaching process.

2   A Soft Degree of Consensus Under Fuzzy Preferences and a Fuzzy Majority

First, to set the stage, we will outline Zadeh's calculus of linguistically quantified propositions.

2.1   Linguistic Quantifiers and a Fuzzy Logic Based Calculus of Linguistically Quantified Propositions

A *linguistically quantified proposition*, exemplified by "most individuals are convinced", may be generally written as

$$Qy\text{'s are } S \tag{1}$$

or, with importance R added, exemplified by "most of the important individuals are convinced", as

$$QRy\text{'s are } S \tag{2}$$

or more conveniently written as

$$Qy\text{'s are } (R, S) \tag{3}$$

The fuzzy predicates S and R are interpreted as fuzzy sets defined in the universe $Y = \{y_i\}$ under consideration.

Usually, a (proportional, nondecreasing) linguistic quantifier Q is used and is assumed to be a fuzzy set in $[0, 1]$, for instance

$$\mu_{\text{"most"}}(x) = \begin{cases} 1 & \text{for } x \geq 0.8 \\ 2x - 0.6 & \text{for } 0.3 < x < 0.8 \\ 0 & \text{for } x \leq 0.3 \end{cases} \tag{4}$$

The truth values (from $[0, 1]$) of (1) and (2) are calculated by using Zadeh's [53] calculus of linguistically quantified statements, respectively, as:

$$\text{truth}(Qy\text{'s are } S) = \mu_Q\left[\frac{1}{n}\sum_{i=1}^{n}\mu_S(y_i)\right] \tag{5}$$

$$\text{truth}(QRy\text{'s are } S) = \mu_Q\left[\frac{\sum_{i=1}^{n}(\mu_R(y_i) \wedge \mu_S(y_i))}{\sum_{i=1}^{n}\mu_R(y_i)}\right] \tag{6}$$

where "$\wedge$" (minimum) can be replaced by, e.g., a t-norm; n denotes the cardinality of the universe Y.

Some other methods to interpret linguistically quantified propositions can be used, exemplified by the OWA operators (cf. Yager [50]) which will not be, however, discussed here.

2.2 A Degree of Consensus Under Fuzzy Preferences and a Fuzzy majority

We have a set of n options, $O = \{o_1, \ldots, o_n\}$, and a set of m agents, $E = \{e_1, \ldots, e_m\}$. Each agent e_k, $k = 1, 2, \ldots, m$, provides his or her *individual fuzzy preference relation*, P_k, given by its membership function $\mu_{P_k} : O \times O \to [0, 1]$. If the cardinality of the set S is small enough (as assumed here), an individual fuzzy preference relation of individual e_k, P_k, may conveniently be represented by an $n \times n$ matrix $P_k = [r_{ij}^k]$, such that $r_{ij}^k = \mu_{P_k}(o_i, o_j)$; $i, j = 1, 2, \ldots, n$; $k = 1, \ldots, m$. P_k is commonly assumed (also here) to be reciprocal in that $r_{ij}^k + r_{ji}^k = 1$ for $i \neq j$; moreover, it is also normally assumed that $r_{ii}^k = 0$, for all i, j, k.

The P_k's are usually meant as

$$\mu_{P_k}(o_i, o_j) = \begin{cases} 1 & \text{if } o_i \text{ is definitely preferred to } o_j \\ c \in (0.5, 1) & \text{if } o_i \text{ is slightly preferred to } o_j \\ 0.5 & \text{in the case of indifference} \\ d \in (0, 0.5) & \text{if } o_j \text{ is slightly preferred to } o_i \\ 0 & \text{if } o_j \text{ is definitely preferred to } o_i \end{cases} \qquad (7)$$

The *relevance of options*, B, is assumed to be given as a fuzzy set B defined in the set of options O such that $\mu_B(o_i) \in [0, 1]$ is a *degree of relevance* of option o_i, from 0 for fully irrelevant to 1 for fully relevant, through all intermediate values.

The *relevance of a pair of options*, $(o_i, o_j) \in O \times O$, may be defined, for instance, as

$$b_{ij}^B = \frac{1}{2}[\mu_B(o_i) + \mu_B(o_j)] \qquad (8)$$

which is clearly the most straightforward option; evidently, $b_{ij}^B = b_{ji}^B$, and b_{ii}^B do not matter; for each i, j.

And analogously, the *importance of agents*, I, is defined as a fuzzy set in the set of agents E such that $\mu_I(e_k) \in [0, 1]$ is a *degree of importance* of agent e_k, from 0 for fully unimportant to 1 for fully important, through all intermediate values.

Then, the *importance of a pair of agents*, (e_{k_1}, e_{k_2}), b_{k_1,k_2}^I, may be defined in various ways, e.g., analogously as (8), i.e.

$$b^I_{k_1,k_2} = \frac{1}{2}[\mu_I(e_{k_1}) + \mu_I(e_{k_2})] \tag{9}$$

Using Zadeh's [53] calculus of linguistically quantified propositions, we derive the soft degree of consensus in the following steps, by consecutively calculating:

- the degree of strict agreement between agents e_{k_1} and e_{k_2} as to their preferences between options o_i and o_j

$$v_{ij}(e_{k_1}, e_{k_2}) = \begin{cases} 1 \text{ if } r^{k_1}_{ij} = r^{k_2}_{ij} \\ 0 \text{ otherwise} \end{cases} \tag{10}$$

where here and later on in this section, if not otherwise specified, $k_1 = 1, \dots, m - 1$; $k_2 = k_1 + 1, \dots, m$; $i = 1, \dots, n - 1$; $j = i + 1, \dots, n$.
- the degree of agreement between agents e_{k_1} and e_{k_2} as to their preferences between *all relevant* pairs of options is

$$v_B(e_{k_1}, e_{k_2}) = \frac{\sum_{i=1}^{n-1} \sum_{j=i+1}^{n} [v_{ij}(e_{k_1}, e_{k_2}) \wedge b^B_{ij}]}{\sum_{i=1}^{n-1} \sum_{j=i+1}^{n} b^B_{ij}} \tag{11}$$

- the degree of agreement between agents e_{k_1} and e_{k_2} as to their preferences between $Q1$ relevant pairs of options is

$$v^B_{Q1}(e_{k_1}, e_{k_2}) = \mu_{Q1}[v_B(e_{k_1}, e_{k_2})] \tag{12}$$

- the degree of agreement of *all important* pairs of agents as to their preferences between $Q1$ pairs of relevant options

$$v^{I,B}_{Q1} = \frac{2}{m(m-1)} \frac{\sum_{k_1=1}^{m-1} \sum_{k_2=k_1+1}^{m} [v^B_{Q1}(e_{k_1}, e_{k_2}) \wedge b^I_{k_1,k_2}]}{\sum_{k_1=1}^{m-1} \sum_{k_2=k_1+1}^{m} b^I_{k_1,k_2}} \tag{13}$$

and, finally,
- the degree of agreement of $Q2$ important pairs of agents as to their preferences between $Q1$ relevant pairs of options, called the *degree of $Q1/Q2/I/B$-consensus*

$$con(Q1, Q2, I, B) = \mu_{Q2}(v^{I,B}_{Q1}) \tag{14}$$

Since the strict agreement (10) may be viewed too rigid, we can use the degree of sufficient agreement (at least to degree $\alpha \in (0, 1]$ of agents e_1 and e_2 as to their preferences between options o_i and o_j, as well as to explicitly introduce the strength of agreement into (10), and analogously define the degree of strong agreement of agents e_1 and e_2 as to their preferences between options o_i and o_j. This will not be considered in this paper, for simplicity.

Obviously, instead of the classic minimum, "∧", we can use any t-norm (cf. Kacprzyk et al. [38]).

2.3 Linguistic Data Summaries

A linguistic data summary is meant as a natural language like (short) sentence that subsumes the very essence (from a certain point of view) of a (numeric) set of data, too large to be comprehensible by humans. The original Yagers approach to the linguistic summaries (cf. Yager [49], Kacprzyk and Yager [24], Kacprzyk et al. [25] and Kacprzyk and Zadrożny [30]) may be briefly presented as follows: $Y = \{y_1, \ldots, y_n\}$ is a set of objects, $A = \{A_1, \ldots, A_m\}$ is a set of attributes characterizing objects from Y, and $A_j(y_i)$ denotes a value of attribute A_j for object y_i.

A linguistic summary of set Y consists of:

- a summarizer S, i.e. an attribute together with a linguistic value (label) defined on the domain of attribute A_j;
- a quantity in agreement Q, i.e. a linguistic quantifier (e.g. most);
- truth (validity) T of the summary, i.e. a number from the interval $[0, 1]$ assessing the truth (validity) of the summary (e.g., 0.7),

and, optionally, a qualifier R may occur, i.e. another attribute together with a linguistic value (label) defined on the domain of attribute A_k.

In our context we may identify objects with agents and their attributes with their preferences over various pairs of options. Then, the linguistic summary may be exemplified by

$$T(Most \text{ agents prefer option } o_1 \text{ to } o_2) = 0.7 \tag{15}$$

A richer form of the summary, which will be of importance for our work, may include a qualifier as in, e.g.,

$$T(Most \text{ of } important \text{ agents prefer option } o_1 \text{ to } o_2) = 0.7 \tag{16}$$

Thus, the core of a linguistic summary is a *linguistically quantified proposition* in the sense of Zadeh [53] which were presented in the previous section.

The linguistic summaries (15) and (16) may be written in a more general form as:

$$Qy\text{'s are } S \tag{17}$$

$$QRy\text{'s are } S \tag{18}$$

or more conveniently as

$$Qy\text{'s are } (R, S) \tag{19}$$

Then, T, i.e. its truth (validity), directly corresponds to the truth value of (17) or (18).

Using Zadeh's [53] fuzzy logic based calculus of linguistically quantified propositions, a (proportional, nondecreasing) linguistic quantifier Q is assumed to be a fuzzy set in $[0, 1]$. Then, the truth values (from $[0, 1]$) of (17) and (18) are calculated, respectively, as:

$$\text{truth}(Qy\text{'s are } S) = \mu_Q[\frac{1}{n}\sum_{i=1}^{n} \mu_S(y_i)] \tag{20}$$

$$\text{truth}(QRy\text{'s are } S) = \mu_Q[\frac{\sum_{i=1}^{n}(\mu_R(y_i) \wedge \mu_S(y_i))}{\sum_{i=1}^{n} \mu_R(y_i)}] \tag{21}$$

where "$\wedge$" (minimum) can be replaced by, e.g., a t-norm.

The fuzzy predicates S and R are assumed here to be of a simplified, atomic form referring to just one attribute as, e.g., *importance*, or to the preferences with respect to *one* pair of options, in the context considered here. They can be extended to cover more sophisticated summaries involving some confluence of various attribute values as, e.g., *young* and *well paid* for *age* and *salary*. Clearly, the most interesting are non-trivial, human-consistent summarizers (concepts) as, e.g.: productive workers, difficult orders, etc. Their definition may require a complicated combination of attributes, possibly with a hierarchy of the attributes imposed (not all attributes are of the same importance for a concept in question)—cf. Kacprzyk and Zadrożny [30, 34], etc.

Notice that the very concept of a linguistic summary is obviously closely related to the definitions of degrees of consensus discussed in previous sections. However, the specific setting of linguistic data summaries will be more convenient for our discussion of how some additional information (or knowledge) can be used for helping the moderator run a consensus reaching session.

3 A Consensus Reaching Process

The basic setting of the consensus reaching process follows Fedrizzi et al. [5], Fedrizzi et al. [4], Kacprzyk and Zadrożny [28, 31], and Zadrożny and Furlani [55] approach though a different perspective, mainly represented by the Spanish researchers, notably, Chiclana, Herrera, Herrera-Viedma et al., etc. [15] can also be used.

Basically, there is a group of participating agents (human, software agents, small groups, organizations, institutions, etc.) and a moderator who is meant to effectively and efficiently run the consensus reaching session. The individuals and the moderator exchange information and opinions, provide argumentation, operating in a network as shown in Fig. 1.

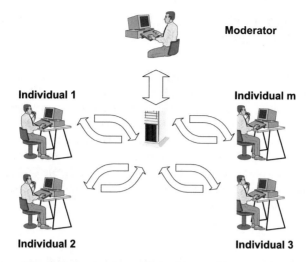

Fig. 1 Agents and the moderator in a consensus reaching session

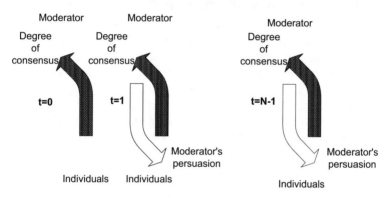

Fig. 2 Dynamics of the consensus reaching process

The consensus reaching proceeds dynamically, step by step. In the beginning, at $t = 0$, the agents present their initial fuzzy preference relations which may differ from each other to a large extent. The moderator tries to persuade them to change their preference relations using some argumentation, concessions, etc. The agents are, obviously, assumed to be rationally committed to reaching consensus, that is, are willing to change their preferences, if necessary and to some extent (Fig. 2).

The moderator should be supported, e.g., via an effective and efficient human-computer interface, enhanced communication capabilities, advanced presentation tools for the visualization or verbalization of results obtained, etc. We will use here some natural language based tools.

4 Supporting the Moderator to Run a Consensus Reaching Session Using Linguistic Data Summaries

In this work we propose a new approach whose essence is, in its first part, a natural language based support—suggested in Kacprzyk and Zadrożny [34]—that is based on the verbalization of results obtained by using linguistic summaries of data in the sense of Yager [49], but in their implementable and extended version proposed by Kacprzyk and Yager [24] and Kacprzyk et al. [25], or even more so in the sense of a protoform based analysis proposed by Kacprzyk and Zadrożny [30], and an NLG (natural language generation) based perspective suggested by Kacprzyk and Zadrożny [34], in which an extremely powerful and far reaching relation to natural language generation (NLG) is shown.

4.1 Using Linguistic Data Summaries to Help the Moderator Run a Consensus Reaching Session

We use a set of indicators assessing how far the group is from consensus, what are the obstacles in reaching consensus, which preference matrix may be a candidate for a consensual one, etc. These indicators are presented as some linguistic data summaries.

The original definition of a soft degree of consensus, i.e. the degree to which "*Most* of the *important* individuals agree in their preferences as to *almost all* of the *important options*", may be more formally expressed as follows:

$$Qh\text{'s are } (B', Qq\text{'s are } (I', \text{sim}(p_q^{h_1}, p_q^{h_2})) \tag{22}$$

where: $h = (h_1, h_2) \in E \times E$ is a pair of individuals, B' represents importance of a pair of individuals (related to B, an importance of particular individuals), $q \in O \times O$ is a pair of options, I' represents importance of a pair of options (related to I, an importance of particular options), $p_q^{h_i}$ is a preference degree of individual i of pair h for pair of options q, and $\text{sim}(\cdot, \cdot)$ is a measure of similarity between two preference degrees.

Thus, the very concept of the consensus degree may be seen as corresponding to a specific linguistic summary truth value. Other linguistic summaries may be instrumental in supporting consensus reaching process, too. Their general structure is such that summarizer S and qualifier R are composed of features of either agents or options (depending on the perspective adopted, to be discussed later) and fuzzy values (labels) expressing the degree of preferences or importance weights of individuals/options.

4.1.1   The First Case: Agents as Objects

The objects of a linguistic summary may be identified with agents and their attributes are preference degrees for the particular pairs of options and importance degrees of the agents. Formally, referring to Sect. 2.3, we have:

$$Y = E \tag{23}$$

and

$$A = \{\mathcal{P}_{ij}\} \cup \{\mathcal{B}\} \tag{24}$$

where attributes $\mathcal{P}_{ij}$ correspond to preference degrees over pairs of options (o_i, o_j), and $\mathcal{B}$ represents the importance.

Then, the following types of summaries may be useful for the moderator to run the consensus reaching session.

4.1.2   Consensus Indicating/Building Summaries

They correspond to a flexible definition of consensus (cf. (22)) that states that *most of the individuals express similar preferences*, for instance "Most individuals definitely prefer o_{i1} to o_{i2}, moderately prefer o_{i3} to o_{i4}, ...", etc. formally written as

$$Qe_k \, (p_{i1,i2}^k = definite) \wedge (p_{i3,i4}^k = moderate) \wedge \ldots \tag{25}$$

If the list of conjuncts is long enough, then a high value of truth of (25) means that there is a consensus among the individuals as to their preferences.

Clearly, this type of a linguistic summary may be used as another definition of consensus. Similarly to (22), importance weights of individuals and/or options may be added.

If the list of conjuncts is short, such a summary may be treated as an indication that no sufficient consensus has been reached but can provide an indication of opinions that are shared by the group of individuals. Thus, they may be either further discussed to reach more agreement in the group or assumed to be ready to proceed to other issues.

4.1.3   Discussion Targeting Summaries

They may be used to direct a further discussion in the group, and may disclose some patterns of understanding. For instance:

Most individuals definitely preferring o_{i1} to o_{i2} also definitely prefer o_{i3} to o_{i4}

to be formally expressed as

$$Qe_k \, (p^k_{i1,i2} = definite, p^k_{i3,i4} = moderate) \tag{26}$$

The discovery of association expressed with such a summary may trigger a further discussion enabling a better understanding of the decision problem.

4.1.4   Option Choice Oriented Summaries

Though we have assumed so far that the goal is to agree upon the content of the preference matrices, an ultimate goal is often to select either an option or a set of options preferred by the group. This can also be supported by proper summaries. To generate summaries taking that into account we have to assume a constructive definition of an option preferred by an agent as implied by his or her fuzzy preference relation, and we can apply here some *choice functions* considered by Kacprzyk and Zadrożny [27, 28, 32]. They are based on the concept of the classic choice function, C, that may be defined in a slightly simplified general form as:

$$C(S,P) = O_0, \quad O_0 \subseteq S \tag{27}$$

that may be exemplified by

$$C(S,P) = \{o_i \in O : \forall_{i \neq j} P(o_i, o_j)\} \tag{28}$$

where P denotes a classical crisp preference relation.

In the case of a fuzzy preference relation, P, we assume C to be a fuzzy set of options defined as:

$$\mu_C(o_i) = \min_j \mu_P(o_i, o_j) \tag{29}$$

which may lead to a more flexible formula by replacing the strict min operator with a linguistic quantifier Q (e.g., "most") yielding:

$$\mu_C(o_i) = T(Qo_j \, P(o_i, o_j)) \tag{30}$$

For our further discussion the specific form of the choice function is not important, and we assume (30). In fact, it is possible that each individual adopts a different choice function, and hence a choice function assigned to each individual is denoted as C_k.

Now, we can define a linguistic summary selecting a set of collectively preferred options, for instance as:

Most individuals choose options $o_{i1}, o_{i2}, \ldots$

to be formally expressed as, e.g.,

$$Qe_k \, (\mu_{C_k}(o_{i1}) = high) \wedge (\mu_{C_k}(o_{i2}) = very \ high) \wedge \ldots \tag{31}$$

where the membership degrees to a choice set are discretized and expressed using linguistic labels.

The options referred to in such a summary qualify as a consensus solution if the goal of the group is to arrive at a subset of collectively preferred options. Therefore, such a summary may be viewed an alternative indicator of consensus.

On the other hand, a summary exemplified by

Most individuals reject options $o_{i1}, o_{i2}, \ldots$

to be formally expressed as, e.g.,

$$Qe_k (\mu_{C_k}(o_{i1}) = low) \wedge (\mu_{C_k}(o_{i2}) = very\ low) \wedge \ldots \tag{32}$$

make it possible to exclude the options concerned from a further consideration.

Therefore, by using the concept of a choice function we can get a constructive and practical definition of a consensus degree. Namely, both (22) and (25) refer to the preferences of the individuals over all pairs of options, possibly with importance weights. Since these importance weights are set independently of the current "standing" of the options implied by individual preference relations, a more rational definition should put more emphasis on preferences related to the options preferred by individuals and less on those rejected by them. Thus, the importance weights of pairs of options in (22) may be assumed as:

$$\mu_{I'_{kl}}(o_i, o_j) = f(\mu_{C_k}(o_i), \mu_{C_l}(o_i), \mu_{C_k}(o_j), \mu_{C_l}(o_j)) \tag{33}$$

that is, importance weights of pairs of options are specific for each pair of individuals. Function f may be exemplified by a simple arithmetic average.

4.1.5 Second Case: Options as Objects

Objects of linguistic summaries may also be equated with options and, then, their attributes are preference degrees over other options as expressed by particular agents adding, possibly, importance degrees of the options. Formally, we have:

$$Y = O \tag{34}$$

and

$$A = \{\mathcal{P}^k_{ij}\} \cup \{\mathcal{I}\} \tag{35}$$

where attributes $\mathcal{P}^k_{ij}$ correspond to preference degrees over other options and $\mathcal{I}$ represents importance.

This perspective may give an additional insight into the structure of preferences of both the entire group and particular individuals. For example, a summary:

Most options are dominated by option o_i in opinion of individual k

formally expressed as, e.g.,

$$Qo_j \, p_{ij}^k = definite \tag{36}$$

directly corresponds to the choice function mentioned earlier. Namely, if such a summary is valid, then it means that option o_i belongs to the choice set of individual k.

On the other hand, a summary like:

Most options are dominated by option o_i in opinion of individual $k1, k2, \ldots$

formally expressed as, e.g.,

$$Qo_j \, (p_{ij}^{k1} = definite) \wedge (p_{ij}^{k2} = definite) \wedge \ldots \tag{37}$$

indicates option o_i as a candidate for a consensual solution.

Interesting patterns in the group may be grasped via linguistic summaries exemplified by:

Most options dominating option o_i in opinion of individual $k1$ also dominate option o_i in opinion of individual $k2$

to be formally expressed as, e.g.,

$$Qo_j \, (p_{ji}^{k1} = definite, p_{ji}^{k2} = definite) \tag{38}$$

Such a summary indicates a similarity of preferences of individuals $k1$ and $k2$. This similarity is here limited to just a pair of options but may be much more convincing in case of:

$$Qo_j \, (p_{ji_1}^{k1} = definite \wedge p_{ji_2}^{k1} = definite \wedge \ldots \, ,$$
$$p_{ji_1}^{k2} = definite \wedge p_{ji_2}^{k2} = definite \wedge \ldots)$$

Another perspective may be obtained assuming a different set of attributes for options. Namely, we can again employ the concept of a choice set and characterize each option o_i by a vector:

$$[\mu_{C_1}(o_i), \mu_{C_2}(o_i), \ldots, \mu_{C_m}(o_i)] \tag{39}$$

Then, a summary like:

Most options are preferred by individual e_k

formally expressed as, e.g.,

$$Qo_i \, \mu_{C_k}(o_i) = high \tag{40}$$

indicates individual e_k as being rather indifferent in his/her preferences, while a summary like:

Most options are rejected by individual e_l

formally expressed as, e.g.,

$$Qo_i \ \mu_{C_l}(o_i) = low \tag{41}$$

suggests that individual e_l exposes a clear preference towards a limited subset of options.

The second representation of options as objects may be seen as a kind of a compression of the first. Namely, for a given option o_i all p_{ij}^k's related to individual e_k which represent o_i in (35) are compressed into one number $\mu_{C_k}(o_i)$ in (39), i.e.,

$$[p_{i1}^k, p_{i2}^k, \ldots, p_{in}^k] \longrightarrow \mu_{C_k}(o_i) \tag{42}$$

Another compression is possible by aggregating, for a given option o_i, all p_{ij}^k's related to option o_j which represent o_i in (35) into one number, i.e.,

$$[p_{ij}^1, p_{ij}^2, \ldots, p_{ij}^m] \longrightarrow \text{aggregation}(p_{ij}^k)_{k=1,m} \tag{43}$$

The aggregation operator may take various forms, including a linguistic quantifier guided aggregation. The representation of options as objects obtained thus far may be used to generate summaries with interpretations similar to (39), but with slightly different semantics. The difference is related to the *direct* and *indirect* approaches to group decision making as proposed and discussed in Kacprzyk [17, 18].

This subsumes some basic possible verbalized types of an additional information, which is based on linguistic summaries, that can be of a great help in supporting the moderator to effectively and efficiently run a consensus reaching session. Among other approaches in a related spirit one should also cite Herrera-Viedma et al. [14, 15].

4.1.6 Remarks on Using the Concepts of a Consensory and Dissensory Agents

As we have already indicated, the recent new approach to the support of consensus reaching processes has been proposed by Kacprzyk and Zadrożny [37]. It is based on a new concept of a consensory and dissensory agent. Basically, the moderator tries to induce changes at preferences of agents and pairs of options that are the most "promising" to possibly faster arrival at a higher degree of consensus. This is often equivalent to the concentration of agents whose testimonies are already close to consensus, and the disregard of agents whose testimonies are far from consensus. A much more promising policy (cf. Gołuńska and Kacprzyk [7], Gołuńska et al. [8], Gołuńska et al. [9–11]) should most probably be the one which would be *fair* in the sense that it would guarantee that both more promising and less promising agents, that is, in our context both the consensory and dissensory agents, would be taken into account.

Therefore, we derive the linguistic summaries shown in this section separately for the consensory and dissensory agents, that is, just to give some examples:

- for the consensory agents:

  - for the linguistic summary *most of the important consensory agents express similar preferences*, for instance: "Most of the important consensory agents definitely prefer o_{i1} to o_{i2}, moderately prefer o_{i3} to o_{i4}, ...", "most of the important consensory agents definitely preferring o_{i1} to o_{i2} also definitely prefer o_{i3} to o_{i4}, ...",

- for the dissensory agents:

  - for the linguistic summary *most of the important dissensory agents express similar preferences*, for instance "Most of the important dissensory agents definitely prefer o_{i1} to o_{i2}, moderately prefer o_{i3} to o_{i4}, ...", etc., "most of the important dissensory agents definitely preferring o_{i1} to o_{i2} also definitely prefer o_{i3} to o_{i4}, ...",

The moderator can now see more in detail how the preferences of the consensory and dissensory agents look like and are distributed, and therefore has much more information and clues on which changes of preferences to suggest to which agents. By taking into account testimonies of both the consensory and dissensory agents, i.e. taking into account all agents and not neglecting any agent, we obtain an important step towards a fair treatment of all participating agents.

5   Concluding Remarks

The purpose of this paper was to extend the setting of a moderator run consensus reaching process under individual fuzzy preference relations, fuzzy majority and a soft degree of consensus as proposed by the authors in previous papers by some new elements. First, to support the moderator, the system provides a possibility to generate linguistic summaries of how the preferences and their required and current changes proceed, in the context of agents and options, extending former approaches by the authors. Then, using a new concept of a set of consensory and dissensory agents recently introduced by the authors, a new approach was proposed in which the linguistic summaries for the support of the moderator while running a consensus reaching session were generated separately for the consensory and dissensory agents so that the moderator was provided with a deeper view that makes it possible to treat the agents more fairly. This had proved to be very effective and efficient in the running of real world consensus reaching sessions.

Acknowledgements Authors of this publication acknowledge the contribution of the Project 691249, *RUC-APS: Enhancing and implementing Knowledge based ICT solutions within high Risk and Uncertain Conditions for Agriculture Production Systems* (https://www.ruc-aps.eu), funded by the European Union under their funding scheme H2020-MSCA-RISE-2015.

References

1. Cabrerizo FJ, Alonso S, Herrera-Viedma E (2009) A consensus model for group decision making problems with unbalanced fuzzy linguistic information. Int J Inf Technol Decis Mak 8(1):109–131
2. Fedrizzi M, Kacprzyk J, Zadrożny S (1988) An interactive multi-user decision support system for consensus reaching processes using fuzzy logic with linguistic quantifiers. Decis Support Syst 4(3):313–327
3. Fedrizzi M, Kacprzyk J, Nurmi H (1993) Consensus degrees under fuzzy majorities and fuzzy preferences using OWA (ordered weighted average) operators. Control Cybern 22:71–80
4. Fedrizzi M, Kacprzyk J, Owsiński JW, Zadrożny S (1994) Consensus reaching via a GDSS with fuzzy majority and clustering of preference profiles. Ann Oper Res 51:127–139
5. Fedrizzi M, Kacprzyk J, Zadrożny S (1988) An interactive multi-user decision support system for consensus reaching processes using fuzzy logic with linguistic quantifiers. Decis Support Syst 4:313–327
6. Gołuńska D, Hołda M (2013) The need of fairness in the group consensus reaching process in a fuzzy environment, Technical Transactions, Automatic Control, vol 1-AC/2013, pp 29–38
7. Gołuńska D, Kacprzyk J (2013) The conceptual framework of fairness in consensus reaching process under fuzziness. In: Proceedings of the 2013 joint IFSA world congress and NAFIPS annual meeting. Edmonton, Canada, pp 1285–1290
8. Gołuńska D, Kacprzyk J, Herrera-Viedma E (2013) Modeling different advising attitudes in a consensus focused process of group decision making. In: Angelov P et al (eds) Intelligent Systems'2014—proceedings of the 7th IEEE international conference intelligent systems IS2014, September 24–26, 2014, Warsaw, Poland, Vol 1; Mathematical foundations, theory, analyses. AISC 322. Springer, Heidelberg, pp 279–288
9. Gołuńska D, Kacprzyk J, Zadrożny S (2014) A consensus reaching support system based on concepts of ideal and anti-ideal point. In: Proceedings of the 2014 North American fuzzy information processing society conference (NAFIPS'2014), pp 1–6
10. Gołuńska D, Kacprzyk J, Zadrożny S (2014) A model of efficiency-oriented group decision and consensus reaching support system in a fuzzy environment. In: Proceedings of the 15th international conference on information processing and management of uncertainty in knowledge-based systems, IPMU-2014. France, Montpellier, pp 424–433
11. Gołuńska D, Kacprzyk J, Zadrożny S (2014) On efficiency-oriented support of consensus reaching in a group of agents in a fuzzy environment with a cost based preference updating approach. In: Proceedings of SSCI-2014: 2014 IEEE symposium series on computational intelligence. USA, IEEE Press, Orlando, pp 15–21
12. Herrera F, Herrera-Viedma E, Verdegay JL (1996) A model of consensus in group decision making under linguistic assessments. Fuzzy Sets Syst 78:73–88
13. Herrera F, Herrera-Viedma E (2000) Choice functions and mechanisms for linguistic preference relations. Eur J Oper Res 120:144–161
14. Herrera-Viedma E, Mata F, Martinez L, Pérez LG (2005) An adaptive module for the consensus reaching process in group decision making problems. Proc MDAI 2005:89–98
15. Herrera-Viedma E, Martinez L, Mata F, Chiclana F (2005) A consensus support system model for group decision-making problems with multi-granular linguistic preference relations. IEEE Trans Fuzzy Syst 13(5):644–658
16. Herrera-Viedma E, Cabrerizo F, Kacprzyk J, Pedrycz W (2014) A review of soft consensus models in a fuzzy environment. Inf Fusion 17:4–13
17. Kacprzyk J (1985) Group decision—making with a fuzzy majority via linguistic quantifiers. Part I: a consensory—like pooling; Part II: a competitive—like pooling. Cybern Syst Int J 16:119–129 (Part I), 131– 144 (Part II)
18. Kacprzyk J (1986) Group decision making with a fuzzy linguistic majority. Fuzzy Sets Syst 18:105–118
19. Kacprzyk J, Fedrizzi M (1986) 'Soft' consensus measures for monitoring real consensus reaching processes under fuzzy preferences. Control Cybern 15:309–323

20. Kacprzyk J, Fedrizzi M (1988) A 'soft' measure of consensus in the setting of partial (fuzzy) preferences. Eur J Oper Res 34:315–325
21. Kacprzyk J, Fedrizzi M (1989) A 'human-consistent' degree of consensus based on fuzzy logic with linguistic quantifiers. Math Social Sci 18:275–290
22. Kacprzyk J, Fedrizzi M, Nurmi H (1992) Group decision making and consensus under fuzzy preferences and fuzzy majority. Fuzzy Sets Syst 49:21–31
23. Kacprzyk J, Nurmi H (1991) On fuzzy tournaments and their solution concepts in group decision making. Eur J Oper Res 51:223–232
24. Kacprzyk J, Yager RR (2001) Linguistic summaries of data using fuzzy logic. Int J Gen Syst 30:33–154
25. Kacprzyk J, Yager RR, Zadrożny S (2000) A fuzzy logic based approach to linguistic summaries of databases. Int J Appl Math Comput Sci 10(2000):813–834
26. Kacprzyk J, Zadrożny S (2001) Computing with words in decision making through individual and collective linguistic choice rules. Int J Uncertain Fuzziness Knowl Based Syst 9:89–102
27. Kacprzyk J, Zadrożny S (2002) Collective choice rules in group decision making under fuzzy preferences and fuzzy majority: a unified OWA operator based approach. Control Cybern 31:937–948
28. Kacprzyk J, Zadrożny S (2003) An internet-based group decision support system. Management VII(28):2–10
29. Kacprzyk J, Zadrożny S (2004) Linguistically quantified propositions for consensus reaching support. In: Proceedings of the IEEE international conference on fuzzy systems, Budapest, Hungary, pp 1135–1140
30. Kacprzyk J, Zadrożny S (2005) Computing with words in intelligent database querying: stand-alone and Internet-based applications. Inf Sci 134:71–109
31. Kacprzyk J, Zadrożny S (2008) On a concept of a consensus reaching process support system based on the use of soft computing and Web techniques. In: Ruan D, Montero J, Lu J, Martnez L, D'hondt P, Kerre EE (eds) Computational intelligence in decision and control. World Scientific, Singapore, pp 859–864
32. Kacprzyk J, Zadrożny S (2009) Towards a general and unified characterization of individual and collective choice functions under fuzzy and nonfuzzy preferences and majority via the ordered weighted average operators. Int J Intell Syst 24(1):4–26
33. Kacprzyk J, Zadrożny S (2009) Protoforms of linguistic database summaries as a human consistent tool for using natural language in data mining. Int J Softw Sci Comput Intell 1(1):100–111
34. Kacprzyk J, Zadrożny S (2010) Computing with words is an implementable paradigm: fuzzy queries, linguistic data summaries and natural language generation. IEEE Trans Fuzzy Syst 18(3):461–472
35. Kacprzyk J, Zadrożny S (2010) Soft computing and Web intelligence for supporting consensus reaching. Soft Comput 14:833–846
36. Kacprzyk J, Zadrożny S (2016) On a fairness type approach to consensus reaching support under fuzziness via linguistic summaries. In: Proceedings of 2016 IEEE international conference on fuzzy systems, FUZZ-IEEE'2016, Vancouver, BC, Canada, IEEE Press, pp 1999–2006
37. Kacprzyk J, Zadrożny S (2016) Towards a fairness-oriented approach to consensus reaching support under fuzzy preferences and a fuzzy majority via linguistic summaries. Transactions on Computational Collective Intelligence XXIII, Springer, pp 189–211
38. Kacprzyk J, Zadrożny S, Fedrizzi M, Nurmi H (2008) On group decision making, consensus reaching, voting and voting paradoxes under fuzzy preferences and a fuzzy majority: a survey and a granulation perspective. In: Pedrycz W, Skowron A, Kreinovich V (eds) Handbook of granular computing. Wiley, Chichester, pp 906–929
39. Kacprzyk J, Zadrożny S, Fedrizzi M, Nurmi H (2008) On group decision making, consensus reaching, voting and voting paradoxes under fuzzy preferences and a fuzzy majority: a survey and some perspectives. In: Bustince H, Herrera F, Montero J (eds) Fuzzy sets and their extensions: representation, aggregation and models. Springer, Heidelberg, pp 263–295

40. Kacprzyk J, Zadrożny S, Raś ZW (2010) How to support consensus reaching using action rules: a novel approach. Int J Uncertain Fuzziness Knowl-Based Syst 18(4):451–470
41. Loewer B, Laddaga R (1985) Destroying the consensus. In: Loewer B, Guest (ed.) Special issue on consensus. Synthese, vol 62(1), pp 79–96
42. Mata F, Martínez L, Herrera-Viedma E (2009) An adaptive consensus support model for group decision-making problems in a multigranular fuzzy linguistic context. IEEE Trans Fuzzy Syst 17(2):279–290
43. Nurmi H (1981) Approaches to collective decision making with fuzzy preference relations. Fuzzy Sets Syst 6:187–198
44. Nurmi H, Kacprzyk J, Fedrizzi M (1996) Probabilistic, fuzzy and rough concepts in social choice. Eur J Oper Res 95:264–277
45. Pawlak Z (1981) Information systems theoretical foundations. Inf Syst J 6(3):205–218
46. Raś ZW, Wieczorkowska A (2000) Action rules: how to increase profit of a company. LNAI 1910:587–592
47. Szmidt E, Kacprzyk J (2002) Using intuitionistic fuzzy sets in group decision making. Control Cybern 31:1055–1057
48. Szmidt E, Kacprzyk J (2003) A consensus-reaching process under intuitionistic fuzzy preference relations. Int J Intell Syst 18(7):837–852
49. Yager RR (1982) A new approach to the summarization of data. Inf Sci 28:69–86
50. Yager RR (1988) On ordered weighted averaging operators in multicriteria decision making. IEEE Trans Syst Man Cybern SMC-18:183–190
51. Yager RR, Kacprzyk J (eds) (1997) The ordered weighted averaging operators: theory and applications. Kluwer, Boston
52. Yager RR, Kacprzyk J, Beliakov G (eds) (2011) Recent developments in the ordered weighted averaging operators: theory and practice. Springer, Heidelberg
53. Zadeh LA (1983) A computational approach to fuzzy quantifiers in natural languages. Comput Math Appl 9:149–184
54. Zadrożny S (1997) An approach to the consensus reaching support in fuzzy environment. In: Kacprzyk J, Nurmi H, Fedrizzi M (eds) Consensus under fuzziness. Kluwer, Boston, pp 83–109
55. Zadrożny S, Furlani P (1991) Modelling and supporting of the consensus reaching process using fuzzy preference relations. Control Cybern 20:135–154
56. Zadrożny S, Kacprzyk J (2003) An Internet-based group decision and consensus reaching support system. In: Yu X, Kacprzyk J (eds) Applied decision support with soft computing. Springer, Heidelberg, pp 263–275

Curriculum Vitae

© Springer International Publishing AG 2018

M. Collan and J. Kacprzyk (eds.), *Soft Computing Applications for Group Decision-making and Consensus Modeling*, Studies in Fuzziness and Soft Computing 357, DOI 10.1007/978-3-319-60207-3

Professor Mario Fedrizzi

This volume is dedicated to Professor Mario Fedrizzi, Full Professor of the Department of Economics and Management at the University of Fedrizzi received the M.Sc. Degree in Mathematics in 1973 from the University of Padua, Italy, and the Ph.D. in Operations Research from the University of Venice (Cà Foscari). He served as a Head of the Institute of Informatics and as a Dean of the Faculty of Economics and Business Administration of the University of Trento from 1985 to 1995, and as a Deputy Rector of the University of Trento from 2004 to 2008. His research has been focused on utility and risk theory, group decision making, fuzzy decision analysis, consensus modeling and decision support systems in fuzzy environments. He has authored and co-authored books and a number of papers, appeared in peer reviewed international journals (among others, European Journal of Operational Research, Fuzzy Sets and Systems, Decision Support Systems, Information Fusion, Expert Systems with Applications, Mathematical Social Sciences). He is member of the editorial board of International Journal of Uncertainty Fuzziness and Knowledge-Based Systems, Group Decision and Negotiation, International Journal of General Systems, Control and Cybernetics, and Applied Computational Intelligence and Soft Computing. He is member of the international advisory board of KEDRI, Auckland University of Technology (NZ) and of the Systems Research Institute, Polish Academy of Sciences, Warsaw (PL). He was also involved in consulting activities in the areas of information systems and DSSs, quality control, project management, and e-learning. From 1995 to 2006 he was appointed as chairman of a regional bank, and of a real estate company, and as a member of the board of directors of one of the largest European banking groups.

Printed in the United States
By Bookmasters